PERIODIC TABLE OF THE ELEMENTS

H 1	He 2					

I	II	III	IV	V	VI	VII	VIII
Li 3	Be 4	B 5	C 6	N 7	O 8	F 9	Ne 10
Na 11	Mg 12	Al 13	Si 14	P 15	S 16	Cl 17	A 18

K 19	Ca 20	Sc 21	Ti 22	V 23	Cr 24	Mn 25	Fe 26	Co 27	Ni 28	Cu 29	Zn 30	Ga 31	Ge 32	As 33	Se 34	Br 35	Kr 36
Rb 37	Sr 38	Y 39	Zr 40	Nb 41	Mo 42	Tc 43	Ru 44	Rh 45	Pd 46	Ag 47	Cd 48	In 49	Sn 50	Sb 51	Te 52	I 53	Xe 54
Cs 55	Ba 56	La † 57	Hf 72	Ta 73	W 74	Re 75	Os 76	Ir 77	Pt 78	Au 79	Hg 80	Tl 81	Pb 82	Bi 83	Po 84	At 85	Rn 86
Fr 87	Ra 88	Ac 89	Th 90	Pa 91	U 92	Np 93	Pu 94	Am 95	Cm 96	Bk 97	Cf 98						

Metals

| | II | III | IV | V | VI | VII | VIII |

† Rare earth metals

Ce 58	Pr 59	Nd 60	Pm 61	Sm 62	Eu 63	Gd 64	Tb 65	Dy 66	Ho 67	Er 68	Tm 69	Yb 70	Lu 71

The elemental semiconductors are in Group IV. Most dopant impurities (for the elemental semiconductors) are in Groups III and V.

DEVICE ELECTRONICS FOR INTEGRATED CIRCUITS

A 32-bit microprocessor integrated circuit. The chip occupies an area of only 1.5 cm², yet has as much processing power as some minicomputers. (*Courtesy AT&T Bell Laboratories*)

DEVICE ELECTRONICS
FOR
INTEGRATED CIRCUITS

Second Edition

Richard S. Muller

University of California
Berkeley, California

Theodore I. Kamins

Hewlett-Packard Laboratories
Palo Alto, California

JOHN WILEY & SONS

New York   Chichester   Brisbane   Toronto   Singapore

Library of Congress Cataloging-in-Publication Data:
Muller, Richard S.
   Device electronics for integrated circuits.

   Includes index.
   1. Semiconductors.   2. Integrated circuits.

I. Kamins, Theodore I.   II. Title.
TK7871.85.M86 1986   621.38′73   85-22774
ISBN 0-471-88758-7

Printed and bound in the United States of America

20 19 18 17 16 15 14

To our TEACHERS. . .
who opened new vistas,
and to our STUDENTS. . .
who will see further than we.

About the Authors

Richard S. Muller, Professor
Department of Electrical Engineering and Computer Sciences
University of California, Berkeley

Richard S. Muller received the degrees of Mechanical Engineer from the Stevens Institute of Technology and the M.S./E.E. and Ph.D. from the California Institute of Technology. He has served as a Professor in the Department of Electrical Engineering and Computer Sciences at the University of California, Berkeley since 1962. Prior to this, Dr. Muller had been a Member of the Technical Staff of the Hughes Aircraft Company and an Instructor at the California Institute of Technology and the University of Southern California. In 1968 he was awarded a NATO fellowship, and in 1982 he was the recipient of a Fulbright Senior Research Professorship at the Technical University and Fraunhofer Institute in Munich, Germany. Dr. Muller has conducted research on solid-state devices for over twenty years and his current research concentrates on silicon sensor devices and systems. He has been a consultant for a number of industrial laboratories and has participated in and organized post-graduate seminars and short courses around the world. He is a member of the Sensors Advisory Board for the Electron-Devices Society of the Institute of Electrical and Electronic Engineers, has served as Chairman of the Detectors, Sensors and Displays Committee for the International Electron Devices Meeting, as the Technical Program Chairman for the annual Sensors Meeting, and as a frequent reviewer and member of the organizing committees of international technical meetings. Dr. Muller is a member of the editorial board for *SENSORS and ACTUATORS* as well as of Tau Beta Pi, Sigma Xi, Pi Delta Epsilon and Eta Kappa Nu. He and his wife, the former Joyce E. Regal, have two sons, Paul and Thomas.

Theodore I. Kamins, Project Leader
Hewlett-Packard Laboratories
Palo Alto, California

Theodore I. Kamins received the Bachelor of Science degree in Electrical Engineering from the University of California, Berkeley (1963), and both the M.S. and Ph.D. degrees in Solid-State Electronics (1964 and 1968) from the University of California, Berkeley. Currently a Project Leader at the Hewlett-Packard Laboratories in Palo Alto, California, he conducts research on devices and materials for integrated circuits. From 1969 to 1974 he served in the Research and Development

Laboratory of Fairchild Semiconductor. Dr. Kamins has also been a consultant at the Stanford Electronics Laboratory, a Visiting Lecturer at Stanford University, an Acting Assistant Professor at the University of California, Berkeley, and has presented short courses for the University of California, the Institute of Electrical and Electronics Engineers (IEEE), and the American Vacuum Society. He has written or coauthored numerous technical papers and holds a number of patents. Dr. Kamins is a Senior Member of the Institute of Electrical and Electronics Engineers and a member of the Electrochemical Society.

PREFACE
TO THE FIRST EDITION

If, as we believe, the past is guide to the future, then major advances in integrated circuits will be made by individuals who have a firm grasp of device electronics. For this reason, a course emphasizing the concepts that underlie the operation and use of integrated-circuit devices has been offered to students at the University of California, Berkeley for several years. This book is the result of teaching students whose diverse interests range from basic device electronics to circuit design. About ten years' experience in teaching the course has given us the opportunity to experiment with a number of approaches to making manageable and, more important, making *digestible* the many facts of solid-state physics that are fundamental to device electronics.

After trying several alternatives, the course (and this book) evolved in the following way. First, we narrowed our focus to the most important devices used in silicon-integrated circuits. This selection left us with material that could be adequately introduced in roughly two quarters of the university calendar. Furthermore, this topical coverage has strong student appeal, a fact consistent with a precept that was well-phrased by Anatole France: "The whole art of teaching is only the art of awakening the natural curiosity of young minds."

Our next step in organizing the material was to prepare a list of the devices of interest and to order the list in a sequence determined by the number of physical concepts necessary for the reader to understand the underlying basic ideas. The book was then put together by alternating a discussion of physical concepts with a description of the application of these concepts to specific devices. Device applications are dealt with in the concluding sections of each chapter. This approach has the following major advantages:

1. It exposes the reader to material that has more permanence than does detailed description of a particular device.
2. It emphasizes the association between the facts of science and the techniques of engineering.
3. It presents the material in a *fact-application* sequence that stimulates understanding and retention.

However, in using this technique, it is difficult to maintain an even level of discussion. The problem arises because the operation of some devices can be explained in terms of only a few concepts, even though from many aspects the device is still quite complex. For example, only a few physical principles are needed to explain the Schottky-barrier diodes covered in Chapter 2, but the complexity of analysis required for rigor in the discussion can make the topic seem unduly confusing. To avoid this problem and to provide a more uniform level of discussion, we have marked certain sections of the text with daggers (†). These sections contain material of generally higher sophistication than that in the unmarked sections and can be omitted by the reader who is interested mainly in first principles. Whenever results from the more sophisticated topics are necessary for the continuity of the presentation, we have included a more easily grasped explanation outside of the sections marked with a dagger.

A summary is given at the end of each chapter. It is intended to help the student review the material and to focus his attention on the essential concepts developed in the chapter.

Most of the problems at the end of each chapter give the reader an opportunity to apply the concepts that have been developed. There are only a few straightforward "plug-in" problems that follow directly from formulas developed in the text. These have usually been included to give the student a feeling for the magnitudes of various quantities. Some problems ask him to carry out portions of derivations only partially developed in the text. These can be used gainfully to emphasize certain treatments that may hold special interest. Challenging problems are occasionally included to extend the reader's comprehension and to deduce new principles. We have marked with a dagger both these problems and those that emphasize material from a section marked with a dagger. An instructor's manual containing solutions to all of the problems is available from the publisher. The manual contains suggestions on how to use some of the problems for lecture discussions.

The students using this book at the University of California, Berkeley, are mostly seniors, but some juniors and a few graduate students are also enrolled each quarter. All students are required to have a standard sophomore background in the sciences that encompasses basic chemistry and physics including introductory quantum mechanics, courses in differential and integral calculus, a first course in electric circuit principles, and a course in the properties of materials that contains a discussion of basic crystal structure. We have found that the most important background needed for the course is an understanding of electromagnetic theory and of the elements of modern physics, particularly the notion of allowed energy values in systems of electrons.

The development of this book was greatly aided by the supportive comments of many colleagues at the University of California. Earlier versions of the manuscript were used in teaching by Professors T. Van Duzer, R. W. Brodersen, T. E. Everhart, and C. Hu. Each made useful suggestions for improvement. Professors A. R. Neureuther, R. G. Meyer, and D. A. Hodges provided helpful criticism of the later chapters in the book. The manuscript was also used at Stanford University by one of the authors (T. I. Kamins) and by Professors J. B. Angell, R. W. Dutton, and J. G. Linvill. Several helpful suggestions came from colleagues in industrial laboratories. These include Dr. J. Kerr at Signetics Corporation, Dr. J. Graul at Siemens Research Laboratory, Munich, Germany, Mr. E. H. Nicollian at the Bell Telephone Laboratories, Murray Hill, New Jersey, Mr. H. Jakobsen at the Central Institute for Industrial Research, Oslo, Norway, and Mr. T. Masuhara at the Hitachi Research Laboratory, Tokyo, Japan. Many comments from students were influential in improving the manuscript, and special appreciation is expressed to several graduate students at the University of California, Berkeley, who provided useful, detailed criticism. These include R. Amantea, S. H. Kwan, W. Loesch, K. W. Yeh, S. P. Fan, and especially R. Coen who very thoroughly checked several of the chapters and who diligently read and solved many of the problems. The problems were checked further and the solutions manual was written with meticulous care by K. C. Hsieh. F. Kashkooli of Signetics Corporation and J. Solomon of National Semiconductor responded generously to requests for illustrations.

As any author knows, textbooks are only harvested after months of nurture. It seems appropriate to acknowledge here the contribution of the University sabbatical system which afforded the opportunity to carry a rather large effort to completion. The staff at Wiley, our editor G. Davenport and his assistant Mrs. E. King, our copy editor E. Patti, and our production manager P. Klein have all been extremely helpful in guiding this project through its publication phase. Our appreciation is also extended to Bettye Fuller who ably typed several sections. Mrs. Joyce R. Muller gave devoted effort to shaping the manuscript and typed most of it several times. Her careful reading, copy editing, typing, and suggestions for revision and rewording have been responsible for improvements throughout the entire book.

RICHARD S. MULLER
THEODORE I. KAMINS

PREFACE
TO THE SECOND EDITION

The astonishingly rapid pace of integrated-circuit development that characterized the 1970s has continued unabated into this decade. In 1977, when the first edition was published, designers were working on integrated circuits composed of tens-of-thousands of elements. In the mid 1980s, million-element circuits have been demonstrated with performance complexities typified by the 32-bit microprocessor shown in the frontispiece.

The decade just passed has seen MOS technologies and especially CMOS become the major focus for *IC* design and development. Intense research on silicon processing has continued, and ever-increasing automation in design and manufacture has led to lowered costs and higher fabrication yields. Despite the appearance of relatively small-scale integrated circuits made on gallium arsenide, the predominance of silicon as the material for *IC* production seems assured for the forseeable future. The basic device structures and operating principles that characterized the *IC*s of 1977 are still fundamental in today's products, but continued reductions in device sizes have required the consideration of new effects in these devices.

This capsule view of developments has influenced our decisions in this revision and expansion of the book. We have also responded to requests from our readers that specific examples be given (throughout the text) as well as answers to some* of the problems (at the end of the book). As in the first edition, we conclude each chapter with a detailed device discussion illustrating applications of basic physical principles.

There are ten chapters, two more than in the first edition. The added material expands the discussion of *IC* technology, and provides a more detailed discussion of *VLSI* MOSFET electronics. Most instructors will find that too much material is provided for a typical one-semester course. In an introductory course, much of Chapters 2, 7, and 10 may be omitted. Alternatively, students having a strong solid-state physics background may find Chapter 1 unnecessary. Throughout the book, more advanced topics and problems, indicated by a dagger superscript (†), can be omitted without loss of continuity.

The topics covered in each of the chapters are described in the following.

Chapter 1 SEMICONDUCTOR ELECTRONICS. This chapter gives an overview of the physical electronics of semiconductors: energy-band theory, doping principles, free-carrier statistics, and drift and diffusion. Only topics that are needed for later device discussions are presented and physical pictures rather than

---

* those marked by an asterisk

mathematical formulations are emphasized. Integrated Hall-effect magnetic sensing circuits are the subject of a concluding section on specific *IC* applications.

Chapter 2 SILICON TECHNOLOGY. Silicon crystal growth, wafer preparation, oxidation, wafer doping by diffusion and ion implantation, pattern definition, lead attachment and packaging; (in short, the techniques of the modern planar process) are described in this chapter. There is considerable detail in some sections and these are marked with a dagger. By omitting these sections, the chapter can be covered in shorter time while still affording the student a broad overview of silicon technology. The device section discusses diffused resistors in *IC*s.

Chapter 3 METAL-SEMICONDUCTOR CONTACTS. In this chapter the student is introduced to the concept of thermal equilibrium between dissimilar solids. The electronics of Schottky-barrier contacts are described in detail. Optional sections present a rigorous derivation of the current-voltage relationship for the junction, the nature of Schottky ohmic contacts, and surface effects at metal-semiconductor contacts. Practical Schottky diodes are considered in the device section.

Chapter 4 *pn* JUNCTIONS. The topic of this chapter is the electronics associated with inhomogeneously doped semiconductors. The behavior of *pn* junction under reverse bias is described in detail. A section on junction breakdown is included, but may be omitted without loss of continuity. In the concluding section the junction field-effect transistor is discussed.

Chapter 5 CURRENTS IN *pn* JUNCTIONS. The continuity equations for free carriers are derived. The concepts of generation and recombination are introduced and then applied to describe the electronics of *pn* junctions under forward (injecting) bias. A discussion of minority-carrier storage and circuit modeling of diodes provides a foundation for the later development of transistor equivalent circuits. Optional sections include a full description of Shockley-Hall-Read recombination theory and space-charge-region generation and recombination. The device discussion focuses on the *IC* uses of junction diodes, particularly for the junction isolation of devices.

Chapter 6 BIPOLAR TRANSISTORS I: Basic Operation. The framework for the discussion of junction transistors in this chapter is the integrated-charge model. The familiar homogeneously doped transistor is treated as a special case of the general theory. This approach to bipolar transistors creates a smooth transition into topics such as bias ranges, current gain, and the Ebers-Moll model. The device discussion focuses on properties of planar *IC* transistors for amplifying and switching purposes. Only one section of this chapter, that dealing with reciprocity in the transistor, is properly optional to further discussion.

Chapter 7 BIPOLAR TRANSISTORS II: Limitations and Models. The topical coverage in this chapter is oriented toward computer modeling of the transistor. For example, the Early effect is introduced in terms of the Early voltage, and effects at low and high emitter-base bias are considered with the use of the integrated-charge model. The same approach is taken to describe the electronics of the base transit time. This topic leads naturally to the charge-control model for the transistor and to examples of its use. The small-signal transistor model (hybrid-pi) is deduced from the charge-control model, and equivalences between the various representations for the transistor are discussed. The final model discussed is for computer simulation. The device discussion focuses on *pnp* transistors for integrated circuits, and introduces circuit applications such as integrated-injection logic.

Chapter 8 PROPERTIES OF THE OXIDE-SILICON SYSTEM. This chapter provides the necessary additional background in physical electronics that is needed for a discussion of MOS devices. The control exercised by the gate to accumulate, deplete, or invert the semiconductor surface is described, and the concepts underlying the flat-band and threshold voltages in an MOS system are presen,ed. An optional section describes surface effects on *pn* junctions. The chapter concludes with a discussion of MOS capacitors for integrated circuits and charge-coupl d devices (CCDs).

Chapter 9 MOS FIELD-EFFECT TRANSISTORS I: Basic Theories and Models. The properties of MOSFETs are deduced initially from an approximate charge-control model, and then more rigorously from a distributed model. The parameters of the device and a model for circuit design are described with a detailed discussion given to the threshold voltage. The technological evolution of MOSFETs and aspects of MOS memory provide a practical perspective in this rapidly developing field. The design concepts and device aspects of CMOS are discussed in the final device section of the chapter.

Chapter 10 MOS FIELD-EFFECT TRANSISTORS II: Limitations and Perspectives. This chapter takes up several effects that are usually not considered in basic descriptions of MOSFET operation: subthreshold current, channel velocity limitations, geometric effects on the threshold voltage, hot-carrier effects, MOSFET breakdown, and scaling MOSFETS to smaller sizes. An example is presented to show techniques for the numerical simulation of MOSFETs. A final section describes the design and properties of ion-implanted, *n*-channel MOSFETs and an introduction to a model for depletion-mode MOSFETs.

Many comments from readers have helped us to shape this revision. Along with the individuals named in the "Preface to the First Edition," we thank P. K. Ko, N. C. Cheung, and W. G. Oldham for useful suggestions and material for problems and examples. We appreciate the careful criticism of F. C. Hsu, S. K. Tam, and T. Masuhara. Finally, we once again express grateful acknowledgement to Joyce R. Muller whose editing and typing have helped immensely to bring a formidable job to completion.

Berkeley, California   RICHARD S. MULLER
Palo Alto, California   THEODORE I. KAMINS

CONTENTS

3
METAL-SEMICONDUCTOR CONTACTS

4
pn JUNCTIONS

5
CURRENTS IN *pn* JUNCTIONS

1

SEMICONDUCTOR
ELECTRONICS

From everyday experience we know that the electrical properties of materials vary widely. If we measure the current I flowing through a bar of homogeneous material with uniform cross section when a voltage V is applied across it, we can find its resistance $R = V/I$. The resistivity ρ—a basic electrical property of the material comprising the bar—is related to the resistance of the bar by a geometric ratio

$$\rho = R\frac{A}{L} \tag{1.1}$$

where L and A are the length and cross-sectional area of the sample.

The resistivities of common materials used in solid-state devices cover a wide range of values. An example is the range of resistivities encountered at room temperature for the materials used to fabricate typical silicon integrated circuits. Deposited metal strips, made from very low-resistivity materials, connect elements of the integrated circuit; aluminum, which is most frequently used, has a resistivity at room temperature of about 10^{-6} Ω-cm. On the other end of the resistivity scale are insulating materials such as silicon dioxide, which serve to isolate portions of the integrated circuit. Silicon dioxide has a resistivity 22 orders of magnitude higher than does aluminum—about 10^{16} Ω-cm. The resistivity of the plastics often used to encapsulate integrated circuits may be as high as 10^{18} Ω-cm. Thus, a typical integrated circuit may contain materials with resistivities varying over 24 orders of magnitude—an extraordinarily wide range for a common physical property.

For electrical applications, materials are generally classified according to their resistivities. Those with resistivities less than about 10^{-2} Ω-cm are called _conductors_, while those having resistivities greater than about 10^5 Ω-cm are called _insulators_. In the intermediate region is a class of materials called semiconductors that have profound electrical importance because their resistivities can be varied by design under precise control. Equally important, they can be made to conduct by one of two types of current carriers. The utilization of these unique properties of semiconductors is the primary topic of this book.

In the first portion of this chapter we consider what gives rise to the enormous range of resistivities in solids. We then briefly review some important concepts from semiconductor physics. To focus on devices, however, our discussion of materials is brief, and we assume that the reader has had some exposure to basic semiconductor and solid-state physics such as that contained in reference 1. The treatment here is merely a review of some of the more important concepts in a form useful for later application.

1.1 Physics of Semiconductor Materials

An understanding of the physics of electrons in solids can be attained by first considering electrons in an isolated atom. By logically perturbing the electrons in this system, we can learn about some important electronic properties of solids.

We therefore begin by considering the allowed energies of an electron influenced only by an isolated atomic core. Then we look at the effect of bringing other atoms near the first atom so that the core of one atom influences the electrons associated with the other atoms, as is true in a solid material. In this manner we can study the effect of the atomic cores in a crystal on the behavior of the associated electrons. We then investigate the effect on the electrons of applying an electric field to the solid material.

Two complementary models are used in the discussion of semiconductor physics: the _energy-band model_ and the _crystal-bonding model_. Let us first consider the energy-band description and bring in the bonding model later to introduce other concepts.

Band Model of Solids

We know that an electron acted on by the Coulomb potential of an atomic nucleus may have only certain allowed energies (Figure 1.1). In particular, the electron can occupy one of a series of energy levels

$$E_n = \frac{-Z^2 m_0 q^4}{8\epsilon_0^2 h^2 n^2} \tag{1.1.1}$$

below a reference energy taken as 0.

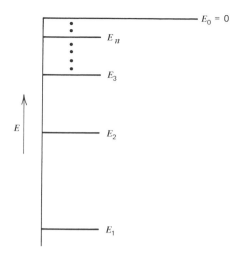

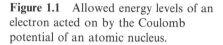

Figure 1.1 Allowed energy levels of an electron acted on by the Coulomb potential of an atomic nucleus.

The electron carries charge $-q$ so that q in Equation 1.1.1 is a positive number. This is our convention throughout the book. The other symbols in Equation 1.1.1 are Z, the number of protons in the nucleus; m_0, the free electron mass; ϵ_0, the permittivity of free space; h, Planck's constant; and n, which stands for a positive integer. For the hydrogen atom with $Z = 1$, the allowed energies are $-2.19 \times 10^{-18}/n^2$ J or $-13.6/n^2$ electron volts (eV) below the zero reference level.* At low temperatures, when more than one electron are associated with the atom, the electrons fill the allowed levels starting with the lowest energies. By the Pauli exclusion principle, at most two electrons (of opposite spins) may occupy any energy level.

Let us now consider the electron in the highest occupied energy level of an atom and neglect the lower filled levels. When two isolated atoms are separated by a large distance, the electron associated with each atom has the energy E_n given by Equation 1.1.1. If two atoms approach one another, however, the atomic core of the first atom exerts a force on the second electron. The potential that determines the energy levels of the electron is therefore changed. All allowed energy levels for the electron are consequently modified because of this change in potential.

We must also consider the Pauli exclusion principle when describing a two-atom system. An energy level E_n, which can contain at most two electrons of opposite spin, is associated with each isolated atom so that the total system can contain at most four electrons. When the two atoms are brought together to form one system, however, only two electrons can be associated with the allowed energy level E_n. Therefore, the allowed energy level E_n of the isolated atoms must split into two levels with slightly different energies in order to retain space for a total of four electrons. Therefore, bringing two atoms close together not only slightly perturbs each energy level of the isolated atom but also splits each of the energy levels of

---

* The use of the unit electron volt (1 eV = 1.6×10^{-19} J; see the conversion factors on the inside front cover) is convenient in semiconductor physics since it often avoids cumbersome exponents. Although not an MKS unit, the electron volt is widely employed and is used extensively in our discussion.

the isolated atoms into two slightly separated energy levels. As the two atoms are brought closer together, stronger interactions are expected and the splitting becomes greater.

As more atoms are added to form a crystalline structure, the forces encountered by each electron are altered further, and additional changes in the energy levels occur. Again, the Pauli exclusion principle demands that each allowed electron energy level have a slightly different energy so that many distinct, closely spaced energy levels characterize the crystal. Each of the original quantized levels of the isolated atom is split many times, and each contains one level for each atom in the system. When N atoms are included in the system, the original *energy level E_n* splits into N different allowed levels, forming an *energy band*, which may contain at most $2N$ electrons (because of spin degeneracy). Since the number of atoms in a crystal is generally large—of the order of 10^{22} cm^{-3}—and the total extent of the energy band is of the order of a few electron volts, the separation between the N different energy levels within each band is much smaller than the thermal energy possessed by an electron at room temperature, and the electron may easily move between levels. Thus, we may speak of a continuous band of allowed energies containing space for $2N$ electrons. This allowed band is bounded by maximum and minimum energies, and it may be separated from adjacent allowed bands by *forbidden-energy gaps*, as shown in Figure 1.2a, or it may overlap other bands. The detailed behavior of the bands (whether they overlap or form gaps and, if so, the size of the gaps) fundamentally determines the electronic properties of a given material. It is the essential feature differentiating conductors, insulators, and semiconductors.

Although each energy level of the original isolated atom splits into a band composed of $2N$ levels, the range of allowed energies of each band may be different. The higher energy bands generally span a wider energy range than do those at lower energies. The cause for this difference can be seen by considering the Bohr radius

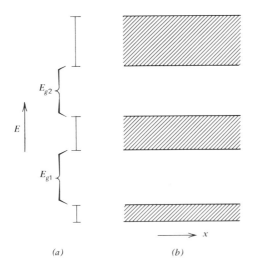

(a)   (b)

Figure 1.2 Broadening of allowed energy levels into allowed energy bands separated by forbidden-energy gaps as more atoms influence each electron in a solid: (*a*) one-dimensional representation; (*b*) two-dimensional diagram in which energy is plotted versus distance.

r_n associated with the nth energy level:

$$r_n = \frac{n^2 \epsilon_0 h^2}{Z \pi m_0 q^2} = \frac{n^2}{Z} \times 0.0529 \ nm \qquad (1.1.2)$$

For higher energy levels (larger n) the electron is less tightly bound and may wander farther from the atomic core. If the electron is less tightly confined, it comes closer to the adjacent atoms and is more strongly influenced by them. This greater interaction, of course, causes a larger change in the energy levels so that the wider energy bands correspond to the higher energy electrons of the isolated atoms.

We represent the energy bands at the equilibrium spacing of the atoms by the one-dimensional picture in Figure 1.2a. The highest allowed level in each band is separated by an energy gap E_g from the lowest allowed level in the next band. It is convenient, however, to extend the band diagram into a two-dimensional picture (Figure 1.2b) where the vertical axis still represents the electron energy while the horizontal axis now represents position in the semiconductor crystal. This version of the diagram emphasizes that electrons in the bands are not associated with any of the individual nuclei, but are confined only by the crystal boundaries. This type of diagram is especially useful when we consider combining materials with different energy-band structures into semiconductor devices. In this heuristic derivation of the basis for energy bands in elemental solids, we have implicitly assumed that each atom is like its neighbor in all respects including orientation; that is, we are considering *perfect crystals*. In practice, a very high level of crystalline order, in which defects are measured in parts per billion or less, is normal for device-quality semiconductor materials.

The formation of energy bands from discrete levels occurs whenever the atoms of any element are brought together to form a solid. However, differing numbers of electrons within the energy bands of different solids strongly influence their electrical properties. Consider first, for example, an alkali metal composed of N atoms each with one valence electron in the outer shell. When the atoms are brought close together, an energy band forms from this energy level. In the simplest case this band has space for $2N$ electrons. The N available electrons then fill the lower half of the energy band (Figure 1.3a), and there are empty states just above the filled states. The electrons near the top of the filled portion of the band can easily gain small amounts of energy from an applied electric field and move into these empty states. They behave, therefore, almost as free electrons and can be transported through the crystal by an externally applied electric field. In general, metals are characterized by partially filled energy bands similar to the case just described. They are, therefore, highly conductive.

Markedly different electrical behavior occurs in materials in which the valence (or outermost shell) electrons completely fill an allowed energy band and there is an energy gap to the next higher band. In this case, characteristic of insulators, the closest allowed band above the filled band is completely empty at low temperatures as shown in Figure 1.3b. Thus, the lowest-energy empty states are separated from the highest filled states by the energy gap E_g. In insulating material E_g is generally

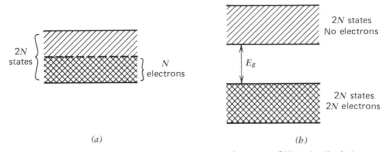

(a)   (b)

Figure 1.3   Energy-band diagrams: (a) N electrons filling half of the $2N$ allowed states, as might occur in a metal. (b) A completely empty band separated by an energy gap E_g from a band whose $2N$ states are completely filled by $2N$ electrons, representative of an insulator.

greater than 5 eV ($\sim$9 eV for SiO_2),* much larger than typical thermal or field-imparted energies (tenths of an eV or less). In the idealization that we are considering there are no electrons close to empty allowed states and, therefore, none available to gain small energies from an externally applied field. Consequently, no electrons may carry an electric current and the material is an insulator.

An analogy may be helpful. Consider a horizontal glass tube with sealed ends representing the allowed energy states and fluid in the tube representing the electrons in a solid. In the case analogous to a metal, the tube is partially filled (Figure 1.4). When a force (gravity in this case) is applied by tipping the tube, the fluid can easily move along the tube. In the situation analogous to an insulator, the tube is completely filled with fluid (Figure 1.5). When the filled tube is tipped, the fluid cannot flow since there is no empty volume into which it can move; that is, there are no empty allowed states.

Semiconductors have band structures similar to insulators. The difference between these two classifications arises from the size of the forbidden energy gap and the ability to populate a nearly empty band by adding conductivity-enhancing impurities to a semiconductor. In a semiconductor the energy gap separating the highest band that is filled at absolute zero temperature from the lowest empty

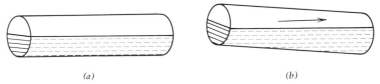

(a)   (b)

Figure 1.4   Electron motion in an allowed band is analogous to fluid motion in a glass tube with sealed ends; the fluid can move in a half-filled tube just as electrons can move in a metal.

---

* Specifying band gaps in amorphous materials may cause anxiety to theorists, but short-range order in such materials leads to effects that can be interpreted with the aid of an energy-band description.

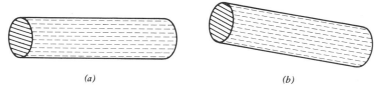

(a)   *(b)*

Figure 1.5 No fluid motion can occur in a completely filled tube with sealed ends.

band is typically of the order of 1 eV (silicon: 1.1 eV; germanium: 0.7 eV). In an impurity-free semiconductor the uppermost filled band is populated by electrons that were the valence electrons of the isolated atoms; this band is known as the *valence band*.

The band structure of a semiconductor is shown in Figure 1.6. At any temperature above absolute zero, the valence band is not entirely filled because a small number of electrons possess enough thermal energy to be excited across the forbidden gap into the next allowed band. The smaller the energy gap and the higher the temperature, the greater the number of electrons that can jump between bands. The electrons in the upper band can easily gain small amounts of energy and can respond to an applied electric field to produce a current. This band is called the *conduction band* because the electrons that populate it are conductors of electricity. The current density J (current per unit area) flowing in the conduction band can be found by summing the charge $(-q)$ times the net velocity (v_i) of each electron populating the band. The summation is then taken over all electrons per unit volume in the conduction band.

$$J_{cb} = \frac{I}{A} = \sum_{cb}(-q)v_i \tag{1.1.3}$$

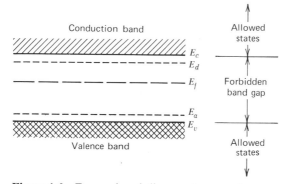

Figure 1.6 Energy-band diagram for a semiconductor showing the lower edge of the conduction band (E_c), a donor level (E_d) within the forbidden gap, the Fermi level (E_f), an acceptor level (E_a), and the top edge of the valence band (E_v).

Since only a small number of electrons exist in this band, however, the current for a given field is considerably less than that in a metal.

Holes

When electrons are excited into the conduction band, empty states are left in the valence band. If an electric field is then applied, nearby electrons can respond to it by moving into these empty states to produce a current. This current can be formulated by summing the motion of all electrons per unit volume in the valence band:

$$J_{vb} = \sum_{vb} (-q)v_i \tag{1.1.4}$$

Since there are many electrons in the valence band but only a few empty states, it is easier to describe the conduction resulting from electrons interacting with these empty states than to describe the motion of all the electrons in the valence band. Mathematically, we may describe the current in the valence band as the current that would flow if the band were completely filled minus that associated with the missing electrons. Again, summing over the populations per unit volume, we find

$$J_{vb} = \sum_{vb} (-q)v_i = \sum_{Filled\ band} (-q)v_i - \sum_{Empty\ states} (-q)v_i \tag{1.1.5}$$

Since no current can flow in a completely filled band (because no net energy can be imparted to the electrons populating it), the current in the valence band can be written as

$$J_{vb} = 0 - \sum_{Empty\ states} (-q)v_i = \sum_{Empty\ states} qv_i \tag{1.1.6}$$

where now the summation is over the empty states per unit of volume. We have found in Equation 1.1.6 that we can express the motion of charge in the valence band in terms of the vacant states by treating the states as if they were particles with positive charge. These "particles" are called *holes*; they can only be discussed in connection with the energy bands of a solid and cannot exist in free space. Note that energy-band diagrams, such as those shown in Figures 1.3 and 1.6, are drawn for electrons so that the energy of an electron increases as we move toward the top of the diagram. However, because of its opposite charge, the energy of a hole increases as we move downward on this same diagram.*

The concept of holes can be illustrated by continuing our analogy of fluids in glass tubes. We start with two sealed tubes—one completely filled, and the other completely empty (Figure 1.7a). When we apply a force by tipping the tubes, no motion can occur (Figure 1.7b). If we transfer a small amount of fluid from the lower tube to the upper tube (Figure 1.7c), however, the fluid in the upper tube can move when the tube is tilted (Figure 1.7d). This flow corresponds to electron

---

* As in Figure 1.6 we often simplify the energy-band diagram by showing only the upper bound of the valence band (denoted by E_v) and the lower bound of the conduction band (denoted by E_c) since we are primarily interested in states near these two levels and in the energy gap E_g separating them.

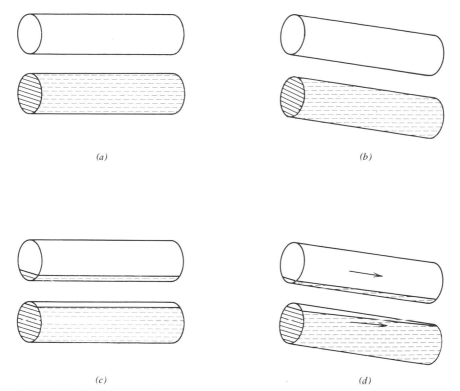

Figure 1.7 Fluid analogy for a semiconductor. (*a*) and (*b*) No flow can occur in either the completely filled or completely empty tube. (*c*) and (*d*) Fluid can move in both tubes if some of it is transferred from the filled tube to the empty one, leaving unfilled volume in the lower tube.

conduction in the conduction band of the solid. In the lower tube, a bubble is left because of the fluid we removed. This bubble is analogous to holes in the valence band. It cannot exist outside of the nearly filled tube, just as it is only useful to discuss holes in connection with a nearly filled valence band. When the tube is tilted, the fluid in the tube moves downward, but the bubble moves in the opposite direction, as if it has a mass of opposite sign to that of the fluid. Similarly the holes in the valence band move in a direction opposite to that of the electrons, as if they had a charge of the opposite sign. Just as it is easier to describe the motion of the small bubble than that of the large amount of fluid, it is easier to discuss the motion of the few holes rather than the motion of the electrons, which nearly fill the valence band.

Bond Model

The discussion of free holes and electrons in semiconductors can also be phrased in terms of the behavior of completed and broken electronic bonds in a semiconductor crystal. This viewpoint, which is often called the *bond model*, fails to account for important quantum-mechanical constraints on the behavior of electrons in crystals, but it does illustrate several useful qualitative concepts.

To discuss the bond model, we consider the diamond-type crystal structure that is common to silicon and germanium (Figure 1.8). In the diamond structure, each atom has covalent bonds with its four nearest neighbors. There are two tightly bound electrons associated with each bond—one from each atom. At absolute zero temperature, all electrons are held in these bonds, and therefore none are free to move about the crystal in response to an applied electric field. At higher temperatures, thermal energy breaks some of the bonds and creates nearly free electrons, which can then contribute to the current under the influence of an applied electric field. This current corresponds to the current associated with the conduction band in the energy-band model.

After a bond is broken by thermal energy and the freed electron moves away, an empty bond is left behind. An electron from an adjacent bond can then jump into the vacant bond, leaving a vacant bond behind. The vacant bond, therefore,

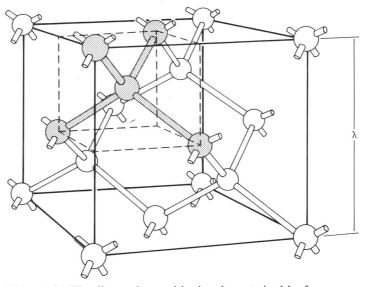

Figure 1.8 The diamond-crystal lattice characterized by four covalently bonded atoms. The lattice constant, denoted by λ, is 0.356, 0.543 and 0.565 nm for diamond, silicon, and germanium, respectively. Nearest neighbors are spaced $(\sqrt{3}\,\lambda/4)$ units apart. Of the 18 atoms shown in the figure, only 8 belong to the volume λ^3. Since the 8 corner atoms are each shared by 8 cubes, they contribute 1 atom; the 6 face atoms are each shared by 2 cubes and thus contribute 3 atoms, and there are 4 atoms inside the cube. The atomic density is therefore $8/\lambda^3$, which corresponds to 17.7, 5.00, and 4.43×10^{22} cm^{-3}, respectively. After W. Shockley: *Electrons and Holes in Semiconductors*, Van Nostrand, Princeton, N.J., 1950.

moves in the opposite direction to the electrons. If a net motion is imparted to the electrons by an applied field, the vacant bond can continue moving in the direction opposite to the electrons as if it had a positive charge.* This vacant bond corresponds to the hole associated with the valence band in the energy-band picture.

A crystal lattice with many similarities to the diamond structure is the *zincblende* lattice which characterizes several important compound semiconductors composed of atoms in the third and fifth columns of the periodic table (called III-V semiconductors). Some III-V semiconductors, particularly gallium arsenide (GaAs) and gallium phosphide (GaP), have important device applications. Many properties of the elemental and compound semiconductors are given in Table 1.3, which appears at the end of this chapter. Table 1.3 also includes data for some insulating materials used in the manufacture of integrated circuits. A second table at the end of the chapter (Table 1.4) contains additional properties of the most important semiconductor, silicon.

Donors and Acceptors

Thus far we have discussed a pure semiconductor material in which each electron excited into the conduction band leaves a vacant state in the valence band. Consequently, the number of electrons n in the conduction band equals the number of holes p in the valence band. Such a material is called an *intrinsic* semiconductor, and the densities of electrons and holes in it (carriers cm^{-3}) are usually subscripted i, as n_i and p_i. However, the most important uses of semiconductors arise from the interaction of adjacent semiconductor materials having differing densities of the two types of charge carriers. We can achieve such a structure either by physically joining two materials with different band gaps or by varying the number of carriers in one semiconductor material.

The most useful means for controlling the number of carriers in a semiconductor is by incorporating *substitutional impurities*; that is, impurities that occupy lattice sites in place of the atoms of the pure semiconductor. For example, if we replace one silicon atom (four valence electrons) with an impurity atom from group V in the periodic table, such as phosphorus (five valence electrons), then four of the valence electrons from the impurity atom fill bonds between the impurity atom and the adjacent silicon atoms. The fifth electron, however, is not covalently bonded to near neighbors; it is only weakly bound to the impurity atom by the excess positive charge on the nucleus. Only a small amount of energy is required to break this weak bond so that the fifth electron can wander about the crystal and contribute to electrical conduction. Since the substitutional group V impurities donate electrons to the silicon, they are known as *donors*.

In order to estimate the amount of energy needed to break the bond to a donor atom, we consider the net Coulomb potential that the electron experiences because of the core of its parent atom. We assume that the electron is attracted by the

---

* A positive charge is associated with the vacant bond because there are insufficient electrons in its vicinity to balance the proton charges on the atomic nucleus.

single net positive charge of the impurity atom core weakened by the polarization effects of the background of silicon atoms. The energy binding the electron to the core is then

$$E = \frac{m_n^* q^4}{8h^2 \epsilon_0^2 \epsilon_{rs}^2} = \frac{13.6 \, m_n^*}{\epsilon_{rs}^2 \, m_0} \text{ eV} \qquad (1.1.7)$$

where ϵ_{rs} is the relative permittivity of the semiconductor and m_n^* is the effective mass of the electron in the semiconductor conduction band. The use of an effective mass accounts for the influence of the crystal lattice on the motion of an electron. For silicon with $\epsilon_{rs} = 11.7$ and $m_n^* = 0.26 \, m_0$, $E \approx 0.03$ eV, which is only about 3% of the silicon band-gap energy (1.1 eV). (More detailed calculations and measurements indicate that the binding energy for typical donors is somewhat more: 0.044 eV for phosphorus, 0.049 eV for arsenic, and 0.039 eV for antimony.) The small binding energies make it much more likely that the weak bond connecting the fifth electron to the donor will be broken than will the silicon-silicon bonds.

n-type Semiconductors. According to the energy-band model, it requires only a small amount of energy to excite the electron from the donor atom into the conduction band, while a much greater amount of energy is required to excite an electron from the valence band to the conduction band. Therefore, we may represent the state corresponding to the electron when bound to the donor atom by a level E_d about 0.05 eV below the bottom of the conduction band E_c (Figure 1.6). The density of donors (atoms cm^{-3}) is generally designated by N_d. Thermal energy at temperatures greater than about 150 K is generally sufficient to excite electrons from the donor atoms into the conduction band. Once the electron is excited into the conduction band, a fixed, positively charged atom core is left behind in the crystal lattice. The allowed energy state provided by a donor (*donor level*) is, therefore, neutral when occupied by an electron and positively charged when empty.

If most impurities are of the donor type, the number of electrons in the conduction band is much greater than the number of holes in the valence band. Electrons are then called the *majority carriers*, and holes are called the *minority carriers*. The material is said to be an *n-type* semiconductor since most of the current is carried by the *negatively charged* electrons. A graph showing the conduction electron concentration versus temperature for silicon and germanium is sketched in Figure 1.9. Since the hole density is at most equal to n_i, this figure shows clearly that electrons are far more numerous than holes when the temperature ranges from a value sufficient to ionize the donor atoms (about 150 K) up to values that free many electrons from silicon-silicon bonds (about 600 K).

p-type Semiconductors. In an analogous manner, an impurity atom with three valence electrons, such as boron, can replace a silicon atom in the lattice. The three electrons fill three of the four covalent silicon bonds, leaving one bond vacant. If another electron moves to fill this vacant bond from a nearby bond, the vacant bond is moved, carrying with it positive charge and contributing to hole conduction. Just as a small amount of energy was necessary to initiate the conduction process in the case of a donor atom, only a small amount of energy is needed to excite an electron

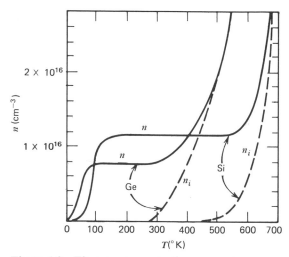

Figure 1.9 Electron concentration versus temperature for two doped semiconductors: (a) Silicon doped with 1.15×10^{16} arsenic atoms cm^{-3},[1] (b) Germanium doped with 7.5×10^{15} arsenic atoms cm^{-3}.[2]

from the valence band into the vacant bond caused by the trivalent impurity. This energy is represented by an energy level E_a slightly above the top of the valence band E_v (Figure 1.6). An impurity that contributes to hole conduction is called an *acceptor impurity* since it leads to vacant bonds, which easily accept electrons. The acceptor concentration (atoms cm^{-3}) is denoted as N_a. If most of the impurities in the solid are acceptors, the material is called a *p-type* semiconductor since most of the conduction is carried by *positively charged* holes. An acceptor level is neutral when empty and negatively charged when occupied by an electron.

Semiconductors in which condition results primarily from carriers contributed by impurity atoms are said to be *extrinsic*. The donor and acceptor impurity atoms, which are intentionally introduced to change the charge-carrier concentration, are called *dopant* atoms.

In compound semiconductors, such as gallium arsenide, certain group IV impurities can be substituted for either element. Thus, silicon incorporated as an impurity in gallium arsenide contributes holes when it substitutes for arsenic and electrons when it substitutes for gallium. This undesirable *amphoteric* doping behavior is not found for all group IV impurities; for example, the group IV element tin is incorporated almost exclusively in place of gallium in gallium arsenide and is therefore a useful *n*-type dopant. Impurities from group VI that substitute for arsenic, such as tellurium, selenium, or sulfur, are also used to obtain *n*-type gallium arsenide, whereas group II elements like zinc or cadmium have been used extensively to obtain *p*-type material.

Other impurity atoms or crystalline defects may provide charge-neutral energy states in which electrons are tightly bound; this means that it takes appreciable energy to excite an electron from a bound state to the conduction band. Such *deep donors* may be represented by energy levels well below the conduction-band edge in contrast to the *shallow donors* previously discussed, which had energy levels only a few times the thermal energy kT below the conduction-band edge. Similarly, *deep acceptors* are located well above the valence-band edge. Since deep levels are not always related to impurity atoms in the same straightforward manner as shallow donors and acceptors, the distinction between the terms donor and acceptor is made on the basis of the possible charge states the level may take. A deep level is called a donor if it is neutral when occupied by an electron and positively charged when empty, while a deep acceptor is neutral when empty and negative when occupied by an electron.

Compensation. The intentional doping of silicon with shallow donor impurities, to make it *n*-type, or with shallow acceptor impurities, to make it *p*-type, is the most important processing step in the fabrication of silicon devices. An especially useful feature of the doping process is that one may *compensate* a doped silicon crystal (for example an *n*-type sample) by subsequently adding the opposite type of dopant impurity (a *p*-type dopant in this example). Reference to Figure. 1.6 may help to clarify the process. Notice that in Figure 1.6 donor atoms add allowed energy states to the energy-band diagram at E_d, close to the conduction-band energy E_c, whereas acceptor atoms add allowed energy states at E_a, close to the valence-band energy E_v. At typically useful temperatures for silicon devices, each donor atom has lost an electron and each acceptor atom has gained an electron. Since the acceptor atoms provide states at lower energies than those either in the conduction band or at the donor levels, the electrons from the donor levels transfer (or "fall") to the lower-energy acceptor sites as long as any of these remain unfilled. Hence, in a doped semiconductor, the effective dopant concentration is equal to the magnitude of the difference between the donor and acceptor concentrations $|(N_d - N_a)|$; the semiconductor is *n*-type if N_d exceeds N_a and *p*-type if N_a exceeds N_d. Although in theory one may achieve a zero effective dopant density through compensation (with $N_d = N_a$), such exact control of the dopant concentrations is technically impractical. As we shall see in Chapter 2 (where technology is discussed) compensation doping usually involves adding a dopant density that is about an order of magnitude higher than the density of dopant that is initially present.

EXAMPLE   Donors and Acceptors

A silicon crystal is known to contain 10^{-4} atomic percent of arsenic (As) as an impurity. It then receives a uniform doping of 3×10^{16} cm^{-3} phosphorus (P) atoms and a subsequent uniform doping of 10^{18} cm^{-3} boron (B) atoms. A thermal annealing treatment then completely activates all impurities.

   (a) What is the conductivity type of this silicon sample?
   (b) What is the density of the majority carriers?

Solution

Arsenic is a group V impurity, and acts therefore as a donor. Since silicon has 5×10^{22} atoms cm^{-3} (Table 1.3), 10^{-4} atomic percent implies that the silicon is doped to a concentration of

$$5 \times 10^{22} \times 10^{-6} = 5 \times 10^{16} \text{ As atoms } cm^{-3}$$

The added doping of 3×10^{16} P atoms cm^{-3} increases the donor doping of the crystal to $8 \times 10^{16} cm^{-3}$.

Additional doping by B (a group III impurity) converts the silicon from *n*-type to *p*-type because the density of acceptors now exceeds the density of donors. The net acceptor density is, however, less than the density of B atoms owing to the *donor compensation*.

(a) Hence, the silicon is *p*-type.

(b) The density of holes is equal to the net dopant density:

$$p = N_a(\text{B}) - [N_d(\text{As}) + N_d(\text{P})]$$
$$= 10^{18} - [5 \times 10^{16} + 3 \times 10^{16}]$$
$$= 9.2 \times 10^{17} \text{ } cm^{-3}$$

Thermal-Equilibrium Statistics

Before proceeding to a more detailed discussion of electrical conduction in a semi-conductor, let us consider three additional concepts: first, the concept of thermal equilibrium; second, the relationship at thermal equilibrium between the majority- and minority-carrier concentrations in a semiconductor; and third, the use of Fermi statistics and the Fermi level to specify the carrier concentrations.

Thermal Equilibrium. We have seen that free-carrier densities in semiconductors are related to the populations of allowed states in the conduction and valence bands. The densities are therefore dependent upon the net energy in the semiconductor. This energy is stored in crystal-lattice vibrations (phonons) as well as by the electrons. Although a semiconductor crystal may be excited by external sources of energy such as incident photoelectric radiation, many situations exist where the total energy is a function only of the crystal temperature. In this case the semiconductor will spontaneously (but not instantaneously) reach a state known as *thermal equilibrium*. Thermal equilibrium is a dynamic situation in which every process is balanced by its inverse process. For example, at thermal equilibrium if electrons are being excited from a lower energy E_1 to a higher energy E_2, then there must be equal traffic of electrons from the states at E_2 to those at E_1. Likewise, if energy is being transferred into the electron population from the crystal vibrations (phonons), then at thermal equilibrium an equal flow of energy is occurring in the opposite direction. A useful thought picture for thermal equilibrium is that a moving picture taken of any event can be run either backward or forward without the viewer being able to detect any difference; thermal equilibrium means that time can run toward the past as well as into the future! In the following we consider some properties of hole and electron populations in semiconductors at thermal equilibrium.

Mass-Action Law. At most temperatures of interest to us, there is sufficient thermal energy to excite some electrons from the valence band to the conduction band. A dynamic equilibrium exists in which some electrons are constantly being excited into the conduction band while others are losing energy and falling back across the energy gap to the valence band. The excitation of an electron from the valence band to the conduction band corresponds to the generation of a hole and an electron, while an electron falling back across the gap corresponds to electron-hole recombination, since it annihilates both carriers. The generation rate of electron-hole pairs G depends on the temperature T but is, to first order, independent of the number of carriers already present. We therefore write

$$G = f_1(T) \qquad (1.1.8)$$

where $f_1(T)$ is a function determined by crystal physics and temperature. The rate of recombination R, on the other hand, depends on the concentration of electrons n in the conduction band and also on the concentration of holes p in the valence band, since both species must interact for recombination to occur. We therefore represent the recombination rate as a product of these concentrations as well as other factors that are included in $f_2(T)$:

$$R = npf_2(T) \qquad (1.1.9)$$

At equilibrium the generation rate must equal the recombination rate. Equating G and R in Equations 1.1.8 and 1.1.9, we have

$$npf_2(T) = f_1(T)$$

or

$$np = \frac{f_1(T)}{f_2(T)} = f_3(T) \qquad (1.1.10)$$

Equation 1.1.10 expresses the important result that at thermal equilibrium the product of the hole and electron densities in a given semiconductor is a function only of temperature.

In an intrinsic (i.e., undoped) semiconductor all carriers result from excitation across the forbidden gap. Consequently, $n = p = n_i$, where the subscript i reminds us that we are dealing with intrinsic material. Applying Equation 1.1.10 to intrinsic material, we have

$$n_i p_i = n_i^2 = f_3(T) \qquad (1.1.11)$$

The intrinsic carrier concentration depends on temperature because thermal energy is the source of carrier excitation across the forbidden energy gap. The intrinsic concentration is also a function of the size of the energy gap since fewer electrons can be excited across a larger gap. We shall soon be able to show that under most conditions n_i^2 is given by the expression

$$n_i^2 = N_c N_v \exp\left(\frac{-E_g}{kT}\right) \tag{1.1.12}$$

where N_c and N_v are related to the density of allowed states near the edges of the conduction band and valence band, respectively. Although N_c and N_v vary some-what with temperature, n_i is much more temperature dependent because of the exponential term in Equation 1.1.12. For silicon with $E_g = 1.1$ eV, n_i doubles for every 8°C increase in temperature near room temperature. Since the intrinsic carrier concentration n_i is constant for a given semiconductor at a fixed temperature, it is useful to replace $f_3(T)$ by n_i^2 in Equation 1.1.10. Therefore the relation

$$np = n_i^2 \tag{1.1.13}$$

holds for both extrinsic and intrinsic semiconductors; it shows that increasing the number of electrons in a sample by adding donors causes the hole concentration to decrease so that the product np remains constant. This result, often called the *mass-action law*, has its counterpart in the behavior of interacting chemical species, such as the concentrations of hydrogen and hydroxyl ions (H^+ and OH^-) in acidic or basic solutions. As we see from our derivation, the law of mass action is a straight-forward consequence of equating generation and recombination, that is, of thermal equilibrium.

In the neutral regions of a semiconductor (i.e., regions free of field gradients), the number of positive charges must be exactly balanced by the number of negative charges. Positive charges exist on ionized donor atoms and on holes, while negative charges are associated with ionized acceptors and electrons.* If there is charge neutrality in a region where all dopant atoms are ionized,

$$N_d + p = N_a + n \tag{1.1.14}$$

Rewriting Equation 1.1.14 and using the mass-action law (Equation 1.1.13), we obtain the expression

$$n - \frac{n_i^2}{n} = N_d - N_a \tag{1.1.15}$$

which may be solved for the electron concentration n:

$$n = \frac{N_d - N_a}{2} + \left[\left(\frac{N_d - N_a}{2}\right)^2 + n_i^2\right]^{1/2} \tag{1.1.16}$$

In an n-type semiconductor $N_d > N_a$. From Equation 1.1.16 we see that the electron density depends on the net excess of ionized donors over acceptors. Thus, a piece of p-type material containing N_a acceptors may be converted into n-type material

---

* When we speak of *electrons* in our discussion of devices in subsequent chapters, we refer generally to electrons in the conduction band; exceptions are explicitly noted. The term *holes* always denotes vacancies in the valence band.

by adding an excess of donors so that $N_d > N_a$. In Chapter 2 we shall see how this conversion is carried out in fabricating silicon integrated circuits.

For silicon at room temperature, n_i is 1.45×10^{10} cm^{-3} while the net donor density in n-type silicon is typically about 10^{15} cm^{-3} or greater: Hence $(N_d - N_a) \gg n_i$ and Equation 1.1.16 reduces to $n \approx (N_d - N_a)$. Consequently, from Equation 1.1.13,

$$p = \frac{n_i^2}{n} \approx \frac{n_i^2}{N_d - N_a} \tag{1.1.17}$$

Thus, for $N_d - N_a = 10^{15}$ cm^{-3} we have $p = 2 \times 10^5$ cm^{-3}, nearly 10 orders of magnitude below the majority-carrier population. In general, the concentration of one type of carrier is many orders of magnitude greater than that of the other in extrinsic semiconductors.

Fermi Level. The numbers of free carriers (electrons and holes) in any macroscopic piece of semiconductor are relatively large—usually large enough to make use of the laws of statistical mechanics to determine physical properties.* One important property of electrons in crystals is their distribution at thermal equilibrium among the allowed energy states. Basic considerations of ways to populate allowed energy states with particles subject to the Pauli exclusion principle leads to a distribution function for electrons in energy that is called the Fermi-Dirac distribution function. It is denoted by $f_D(E)$ and has the form

$$f_D(E) = \frac{1}{1 + \exp[(E - E_f)/kT]} \tag{1.1.18}$$

where E_f is a reference energy called the *Fermi energy* or *Fermi level*. From Equation 1.1.18, we see that $f_D(E_f)$ always equals $\frac{1}{2}$. The Fermi-Dirac distribution function, often called simply the *Fermi function*, describes the probability that a state at energy E is filled by an electron. As shown in Figure 1.10a, the Fermi function approaches unity at energies much lower than E_f, indicating that the lower energy states are mostly filled. It is very small at higher energies, indicating that few electrons are found in high-energy states at thermal equilibrium—in agreement with physical intuition. At absolute zero temperature all allowed states below E_f are filled and all states above it are empty. At finite temperatures, the Fermi function does not change so abruptly; there is a small probability that some states above the Fermi level are occupied and some states below it are empty.

The Fermi function represents only a probability of occupancy. It does not contain any information about the states available for occupancy and, therefore, cannot by itself specify the electron population at a given energy. An application of quantum physics to a given system will reveal information about the density of available states as a function of energy. We denote this function by $g(E)$. A sketch of $g(E)$ for an intrinsic semiconductor is shown in Figure 1.10b. Note that it is zero

---

* However, in some devices made with submicrometer dimensions, the number of dopant atoms in the active regions is so small that statistical fluctuations in the number of dopant atoms can affect device characteristics.

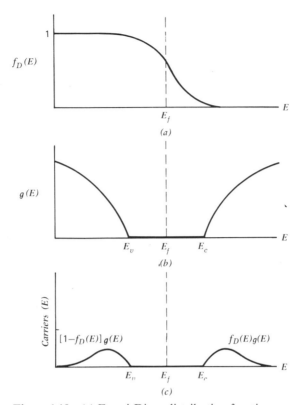

Figure 1.10 (*a*) Fermi-Dirac distribution function
describing the probability that an allowed state at
energy E is occupied by an electron. (*b*) The density
of allowed states for a semiconductor as a function of
energy; note that $g(E)$ is zero in the forbidden gap
between E_v and E_c. (*c*) The product of the distribution
function and the density of states function.

in the forbidden gap $(E_c > E > E_v)$, but it rises sharply within both the valence
band $(E < E_v)$ and the conduction band $(E > E_c)$. The actual distribution of electrons as a function of energy can then be found from the product of the density
of allowed states $g(E)$ within a small energy interval dE and the probability that
these states are filled $f_D(E)$. The total density of electrons in the conduction band
may be obtained by multiplying the density-of-states function in the conduction
band $g(E)$ by the Fermi function and integrating over the conduction band:

$$n = \int_{cb} f_D(E)g(E)\,dE \qquad (1.1.19)$$

Similarly, the density of holes in the valence band is found by multiplying the
density-of-states function in the valence band by the probability that these states
are empty $[1 - f_D(E)]$ and integrating over the valence band.

In n-type material that is not too highly doped, few of the allowed states in the conduction band are filled. The Fermi function in the conduction band is very small, and the Fermi level is well below the bottom of the conduction band. Then $(E_c - E_f) \gg kT$, and the Fermi function given by Equation 1.1.18 reduces to the mathematically simpler Maxwell-Boltzmann distribution function:

$$f_M(E) = \exp\left[\frac{-(E - E_f)}{kT}\right] \tag{1.1.20}$$

This thermal-equilibrium distribution function can also be derived independently by omitting the limitations imposed by the Pauli exclusion principle; that is, the Boltzmann function applies to the case that any number of electrons may exist in an allowed state. At energies well above the Fermi level the fraction of available states that are occupied is so small that the exclusion-principle limitation has no practical effect and Maxwell-Boltzmann statistics are applicable.

Using Equation 1.1.20 in the integration described in Equation 1.1.19 and making several approximations, the carrier concentration in the conduction band can be expressed in terms of the Fermi level by

$$n = N_c \exp\left[-\frac{(E_c - E_f)}{kT}\right] \tag{1.1.21}$$

Similarly, in moderately doped p-type material, the Fermi level is significantly above the top of the valence band, and

$$p = N_v \exp\left[-\frac{(E_f - E_v)}{kT}\right] \tag{1.1.22}$$

where $(E_c - E_f)$ is the energy between the bottom edge of the conduction band and the Fermi level, and $(E_f - E_v)$ is the energy separation from the Fermi level to the top of the valence band. The quantities N_c and N_v, called the effective densities of states at the conduction- and valence-band edges, respectively, are given by the expressions

$$N_c = 2\left(\frac{2\pi m_n^* kT}{h^2}\right)^{3/2} \tag{1.1.23}$$

and

$$N_v = 2\left(\frac{2\pi m_p^* kT}{h^2}\right)^{3/2} \tag{1.1.24}$$

where m_n^* and m_p^* denote the effective masses of electrons and holes. These effective masses are related to m^* as introduced in Equation 1.1.7 but differ slightly from it because of factors we shall not consider here. As can be seen from Equations 1.1.21 and 1.1.22, the quantities N_c and N_v effectively concentrate all of the distributed conduction- and valence-band states at E_c and E_v. They may be used to calculate thermal-equilibrium densities whenever the Fermi level is a few kT or more removed from a band edge.

Except for slight differences in the m^* values, all terms in Equations 1.1.23 and 1.1.24 are equal so that $N_c \simeq N_v$. Hence, in an n-doped material for which $n \gg p$,

$(E_c - E_f) \ll (E_f - E_v)$; this means that the Fermi level is much closer to the conduction band than it is to the valence band. Similarly, the Fermi level is nearer to the valence band than to the conduction band in a p-type semiconductor.

In an intrinsic semiconductor $n = p$. Therefore, $(E_c - E_f) \approx (E_f - E_v)$ and the Fermi level is nearly at the middle of the forbidden gap $[E_f = (E_c + E_v)/2]$. We denote this *intrinsic Fermi level* by the symbol E_i. Just as the quantity n_i is useful in relating the carrier concentrations even in an extrinsic semiconductor (Equation 1.1.13), E_i is frequently used as a reference level when discussing extrinsic semiconductors. In particular, since

$$n_i = N_c \exp\left[\frac{-(E_c - E_i)}{kT}\right] = N_v \exp\left[\frac{-(E_i - E_v)}{kT}\right] \tag{1.1.25}$$

the expressions for the carrier concentrations n and p in an extrinsic semiconductor Equations 1.1.21 and 1.1.22 may be rewritten in terms of the intrinsic carrier concentration and the intrinsic Fermi level:

$$n = n_i \exp\left[\frac{(E_f - E_i)}{kT}\right] \tag{1.1.26}$$

and

$$p = n_i \exp\left[\frac{(E_i - E_f)}{kT}\right] \tag{1.1.27}$$

Thus, the energy separation from the Fermi level to the intrinsic Fermi level is a measure of the departure of the semiconductor from intrinsic material. Since E_f is above E_i in an n-type semiconductor, $n > n_i > p$, as we found before.

When the semiconductor contains a large dopant concentration $[N_d \rightarrow N_c$ or $N_a \rightarrow N_v (\sim 10^{19}$ cm^{-3} for Si$)]$, we may no longer ignore the limitations imposed by the Pauli exclusion principle. That is, the Fermi-Dirac distribution may not be approximated by the Maxwell-Boltzmann distribution function. Equations 1.1.21–22 and 1.1.26–27 are then no longer valid. More exact expressions must be used or the limited validity of the simplified expressions must be realized. Very highly doped semiconductors $(N_d \gtrsim N_c$ or $N_a \gtrsim N_v)$ are called *degenerate* semiconductors because the Fermi level is within the conduction or valence band. Therefore, allowed states for electrons exist very near the Fermi level, just as is the case in metals. Consequently, many of the electronic properties of semiconductors *degenerate* under heavy doping into those of metals.

EXAMPLE   Thermal Equilibrium Statistics

Find the equilibrium electron and hole concentrations and the location of the Fermi level (with respect to the intrinsic Fermi level E_i) in silicon at 300 K if the silicon contains 8×10^{16} cm^{-3} arsenic (As) atoms and 2×10^{16} cm^{-3} boron (B) atoms.

Solution

Since the donor (As) density exceeds the acceptor (B) density, the crystal is *n*-type. The net doping concentration is the difference between the donor dopant density (8×10^{16}) and the acceptor dopant density (2×10^{16}) and is 6×10^{16} cm^{-3}.
   The electron density equals the net dopant concentration.

$$n = 6 \times 10^{16} \text{ cm}^{-3}$$

The hole density is (from Equation 1.1.13)

$$p = \frac{n_i^2}{n} = 3.5 \times 10^3 \text{ cm}^{-3}$$

From Equation 1.1.26,

$$E_f - E_i = kT \ln(n/n_i)$$
$$= 0.0258 \ln(6 \times 10^{16}/1.45 \times 10^{10})$$
$$= 0.393 \text{ eV}$$

   Note that the Fermi level could be located with respect to the conduction band by using Equation 1.1.21

$$E_c - E_f = kT \ln(N_c/n)$$
$$= 0.0258 \ln (2.8 \times 10^{19}/6 \times 10^{16})$$
$$= 0.159 \text{ eV}$$

The sum of these two energies is 0.55 eV, half the bandgap energy of Si

Quasi-Fermi Levels.[†] We have already found the Fermi level to be a useful concept to explain the behavior of semiconducting materials; we will see many further applications as we extend our discussion to devices. The Fermi level arises from the statistics of an ensemble of electrons at thermal equilibrium, and in fact only for thermal equilibrium is there a fundamental physical definition for this energy. Often, however, thermal equilibrium is disturbed by causes such as incident radiation or the application of bias to *pn* junctions. To analyze these nonequilibrium cases, it is useful to introduce two related parameters called *quasi-Fermi levels*.*

   * Some authors have also used the term *Imref* which can be taken to mean "imaginary reference" as well as being *Fermi* spelled backwards.

We define the quasi-Fermi levels in a manner that preserves the relationship between the intrinsic-carrier density and the electron and hole densities as expressed for thermal equilibrium in Equations 1.1.26 and 1.1.27. Under nonequilibrium conditions similar equations can only be written if two different quasi-Fermi levels are defined, one level for electrons and one for holes.

These conditions are met if we define the quasi-Fermi level for electrons E_{fn} (and its corresponding quasi-Fermi potential $\varphi_{fn} = -E_{fn}/q$), and the quasi-Fermi level for holes E_{fp} (and corresponding potential $\varphi_{fp} = -E_{fp}/q$), by

$$E_{fn} = E_i + kT \ln(n/n_i) \quad \text{or} \quad \varphi_{fn} = \varphi_{fi} - \frac{kT}{q} \ln(n/n_i) \quad (1.1.28)$$

and

$$E_{fp} = E_i - kT \ln(p/n_i) \quad \text{or} \quad \varphi_{fp} = \varphi_{fi} + \frac{kT}{q} \ln(p/n_i) \quad (1.1.29)$$

where φ_{fi} is the potential associated with E_i, $\varphi_{fi} = -E_i/q$. Under nonequilibrium conditions, the np product is not equal to the thermal equilibrium value n_i^2 but is a function of the separation of the two quasi-Fermi levels. From Equations 1.1.28 and 1.1.29, we can derive

$$np = n_i^2 \exp[(E_{fn} - E_{fp})/kT] \quad (1.1.30)$$

The separation between the two quasi-Fermi levels is, therefore, a measure of the deviation from thermal equilibrium of the semiconductor free-carrier populations, and is identically zero at thermal equilibrium.

The concept of quasi-Fermi levels is especially useful in the consideration of photoconduction in which excess electrons and holes are generated by light. In general, it is helpful to use quasi-Fermi levels to discuss generation and recombination, as we shall see in more detail in Chapter 5.

Photoconduction[†] The covalent bonds holding electrons at atomic sites in the lattice can be broken by incident radiant energy (photons) if the photon energy is sufficient. When the bonds are broken, both the freed electrons and the vacancies left behind are able to move through the semiconductor crystal and act as current carriers. In terms of the energy-band picture this process of free-carrier production, called *photogeneration*, is equivalent to exciting electrons from the valence band into the conduction band leaving free holes behind. The required photon energy for photogeneration is thus at least equal to the bandgap energy, and the number of holes created equals the number of generated electrons. The band gap in silicon (1.124 eV) is energetically equivalent to photons in the far infrared portion of the electromagnetic spectrum (1.10 μm wavelength).

The radiation incident on the semiconductor surface is absorbed as it penetrates into the crystal lattice. The amount of energy ΔI absorbed in each small increment of length Δx along the flux path of the radiation is described by an *absorption coefficient* α

$$\Delta I = I(x + \Delta x) - I(x) = I(x) \times \alpha \Delta x \quad (1.1.31)$$

where $I(x)$ is the energy reaching the position x. Treating Δx as an infinitesimal, Equation 1.1.31 can be rewritten as a differential equation whose solution is

$$I(x) = I_o \exp(-\alpha x) \qquad (1.1.32)$$

where I_o is the energy that enters the solid at $x = 0$.

The absorption coefficient is typically a strong function of photon energy, as can be seen in the plot of α vs wavelength (and photon energy)* for silicon shown in Figure 1.11. High-energy ultraviolet (UV) light is absorbed with a characteristic length (equal to α^{-1}) that is less than 10 nm, while light of 1 μm wavelength (in

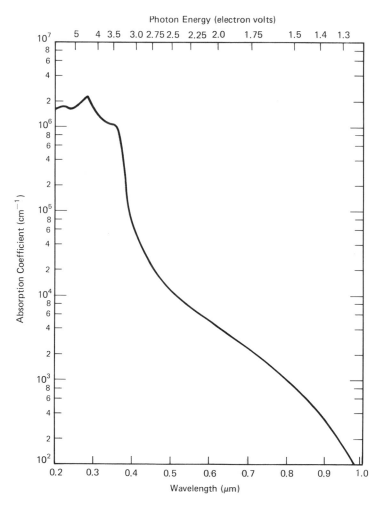

Figure 1.11   Absorption coefficient of light in silicon.

* Electromagnetic wavelength λ is related to photon energy E by the equation $\lambda = hc/E$ where hc is the product of Planck's constant and the speed of light. For λ in μm, and E in eV, the conversion equation is $\lambda = 1.24/E$.

free space) is not efficiently absorbed and penetrates about $100 \ \mu m$ into silicon before decaying appreciably. Absorption of photons having energies greater than the bandgap is almost entirely due to the generation of holes and electrons. The specific shape of the light-absorption curve can be understood in terms of details of the energy-band picture for silicon, but a full discussion of this important topic is better reserved for a fundamental course in solid-state physics.

When photogeneration occurs in silicon, the incident radiation supplies energy which is additional to the thermal energy of the crystal. Hence, the silicon is not at thermal equilibrium, and quasi-Fermi levels are appropriate measures for the free-carrier densities.

EXAMPLE   Photogeneration and Quasi-Fermi Levels

A silicon wafer is doped with $10^{15} \ cm^{-3}$ donor atoms.

(a) Find the electron and hole concentrations and the location of the Fermi level with respect to the intrinsic Fermi level.

(b) Light irradiating the wafer leads to a steady-state photogenerated density of electrons and holes equal to $10^{12} \ cm^{-3}$. We assume that the wafer is thin when compared to the absorption depth for the light so that the free carriers are uniformly distributed throughout its volume. Find the overall electron and hole concentrations in the wafer and calculate the positions of the quasi-Fermi levels for the two carrier types.

(c) Repeat the calculations of (b) under the condition that the light intensity is increased so that the photogeneration produces $10^{18} \ cm^{-3}$ electron-hole pairs.

Solution

(a)
$$n = N_d = 10^{15} \ cm^{-3}$$

$$p = \frac{n_i^2}{n} = 2.1 \times 10^5 \ cm^{-3}$$

$$E_f - E_i = kT \ln(n/n_i) = 0.29 \ eV$$

(b)
$$n = 10^{15} + 10^{12} \approx 10^{15} \ cm^{-3}$$

$$p = 2.1 \times 10^5 + 10^{12} \approx 10^{12} \ cm^{-3}$$

$$E_{fn} - E_i = kT \ln(n/n_i) = 0.29 \ eV$$

$$E_i - E_{fp} = kT \ln(p/n_i) = 0.11 \ eV$$

(c)
$$n = 10^{15} + 10^{18} \approx 10^{18} \ cm^{-3}$$

$$p = 2.1 \times 10^5 + 10^{18} \approx 10^{18} \ cm^{-3}$$

$$E_{fn} - E_i = kT \ln(n/n_i) = 0.47 \ eV$$

$$E_i - E_{fp} = kT \ln(p/n_i) = 0.47 \ eV$$

In part (b), photogeneration is shown to change the minority-carrier concentration by seven orders of magnitude without causing any appreciable variation in the majority-carrier density. Consequently, the electron quasi-Fermi level is close to the thermal-equilibrium Fermi level, but the hole quasi-Fermi level is displaced by 0.40 eV. When the light intensity increases as in part (c), both the hole and electron densities are affected, and both quasi-Fermi levels are strongly displaced from the thermal-equilibrium position. The two carrier densities in this case are nearly equal, as would be the case in an intrinsic semiconductor at high temperatures. Most instances of photogeneration in doped semiconductors are similar to case (b) in the example; that is, the minority-carrier densities are greatly changed by the incident radiation while the majority-carrier concentrations are essentially unaffected.

Heavy Doping.[†] The statistical expressions developed for the densities of holes and electrons in semiconductors were simplified by assuming that only a small fraction of the available electron states in the conduction band are full and only a small fraction of valence-band states are empty. Under these assumptions, for example, we were able to make approximations for the integral in Equation 1.1.19 in order to define the "effective density of conduction-band states" N_c and to approximate the Fermi-Dirac statistics for electron density by the simpler Maxwell-Boltzmann statistics (Equation 1.1.21). These approximations become invalid when a crystal is doped with impurities at densities that approach N_c. In addition to expressing the free-carrier statistics correctly, other, more basic, effects need to be considered when a semiconductor is heavily doped.

If moderate concentrations of dopant impurities are present (for example, the bulk doping in a silicon wafer), the individual impurity atoms do not interact with one another, and they do not perturb the band structure of the host crystal. For example, a dopant density of 5×10^{15} cm^{-3} represents only about one atom of dopant in 10^7 atoms of silicon. Each of the dopant atoms then adds a discrete allowed donor energy level in the silicon bandgap. If the dopant density is increased sufficiently to become a significant fraction of the silicon-atom density, however, the band structure itself begins to be perturbed.

The most significant perturbation is a reduction in the size of the silicon bandgap. The reduced bandgap energy causes the product of the free-carrier densities p and n to increase. This effect is usually expressed in terms of a value for the pn product in the form

$$pn = n_i^2 \exp(\Delta E_g/kT) = n_{ie}^2 \qquad (1.1.33)$$

where ΔE_g expresses the effective bandgap narrowing caused by heavy doping, and n_{ie} is an effective value of the intrinsic-carrier density. Measurements of bandgap narrowing indicate that this effect is negligible for dopant densities less than 10^{18} cm^{-3}, but at higher dopant densities it can become sizable. Some experimental data showing ΔE_g as a function of the free-electron density n in silicon are plotted in Figure 1.12. Heavy doping effects begin to be noticeable between 10^{18} and 10^{19}

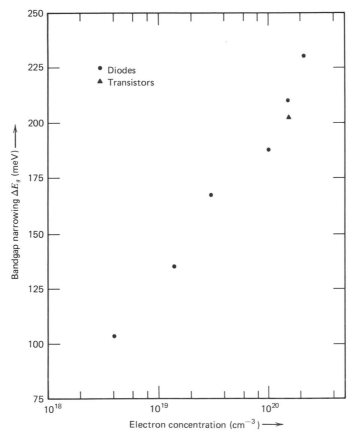

Figure 1.12 Energy-gap narrowing ΔE_g as a function of electron concentration. [A. Neugroschel, S. C. Pao, and F. A. Lindholm, IEEE Trans. Electr. Devices, ED-29, 894 (May 1982)].

electrons cm^{-3}, and already at an electron concentration of 10^{19}, ΔE_g is more than 10% of the bandgap energy.

A detailed study of the effect of heavy doping on the semiconductor band structure shows that as the dopant densities increase, the energy levels they introduce are no longer distinct, but instead broaden into bands. These *impurity bands* can overlap the adjacent conduction or valence bands so that no energy is required to ionize the dopant atoms and provide free carriers. Therefore, under heavy doping conditions, the formulas derived earlier in this chapter for silicon doping need modification.

The influence of heavy doping on basic silicon properties is presently the subject of intensive research; its most important device effect is to limit the achievable current gain of bipolar transistors.

1.2 Free Carriers in Semiconductors

Our first reference to the electronic properties of solids earlier in this chapter was to the familiar linear relationship that is often found between the current flowing through a sample and the voltage applied across it. This relationship is known as Ohm's law: $V = IR$. Although a thorough derivation of the physics of ohmic-current flow can be quite complex, an approximate representation of the process will provide adequate background for our purposes. To accomplish this we first develop a picture of the kinetic properties of free electrons without any external fields. We then consider the imposition first of low to moderate fields, characteristic of many device applications, and finally we discuss the high-field case.

We begin by recalling that electrons (and holes) in semiconductors are almost "free particles" in the sense that they are not associated with any particular lattice site. The influences of crystal forces are incorporated in an *effective mass* that differs somewhat from the free-electron mass. Using the laws of statistical mechanics, we can assert that electrons and holes have the thermal energy associated with classical free particles: $\frac{1}{2}kT$ units of energy per degree of freedom where k is Boltzmann's constant and T is the absolute temperature. This means that electrons in a crystal are not stationary, but rather move about with random velocities. Furthermore, the mean-square thermal velocity of the electrons is approximately* related to the temperature by the equation

$$\frac{1}{2} m_n^* v_{th}^2 = \frac{3}{2} kT \tag{1.2.1}$$

where m_n^* is the effective mass of conduction-band electrons. For silicon $m_n^* = 0.26\, m_0$ (where m_0 is the free electron rest mass), and v_{th} as calculated from Equation 1.2.1 is therefore 2.3×10^7 cm s^{-1} at $T = 300$ K. The electrons may be pictured as moving in random directions through the lattice, colliding among themselves and with the lattice. At thermal equilibrium the motion of the system of electrons is completely random so that the net current in any direction is zero. Collisions with the lattice result in energy transfer between the electrons and the atomic cores that form the lattice. The time interval between collisions averaged over the entire electron population is τ_{cn}, the mean scattering time for electrons. These considerations all apply to the field-free, thermal-equilibrium crystal.

Drift Velocity

Let us now consider a small electric field applied to the lattice. The electrons are accelerated along the field direction during the time between the collisions. Figure 1.13a is a sketch of the motion typical of a crystal electron in response to a small applied field $\mathscr{E}$. Note in the figure that the field-directed motion is a small

---

* The formulation of Equation 1.2.1 is slightly in error because of improper averaging. Only the order of magnitude of the result is of consequence to us, however, and it is not worthwhile to make an exact formulation. At 300 K, v_{th} is typically taken to be 10^7 cm s^{-1} for electrons or holes in silicon.

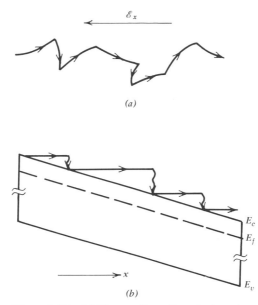

Figure 1.13 (a) The motion of an electron in a solid under the influence of an applied field. (b) Energy-band representation of the motion, indicating the loss of energy when the electron undergoes a collision.

perturbation on the random thermal velocity. Therefore τ_{cn}, the mean scattering time, is not altered appreciably by the applied field.

In Figure 1.13b electron motion in a small applied field is schematically represented on a band diagram. A constant applied field results in a linear variation in the energy levels in the crystal. Electrons (which move downward on energy-level diagrams) tend to move to the right on the diagram, as is appropriate for a field directed in the negative x direction. The electrons exchange energy when colliding with the lattice and drop toward their thermal-equilibrium positions. If the field is small, the energy exchanged is also small, and the lattice is not appreciably heated by the passage of the current. The slope of the energy bands and the energy losses associated with lattice collisions are exaggerated in Figure 1.13b to show the process schematically. In fact, the incremental energy lost in each collision is much less than the mean thermal energy of the electrons. An electron at rest in the conduction band would be at the band edge E_c. The kinetic energy of electrons is therefore measured by $(E - E_c)$, and the mean value at thermal equilibrium for all the electrons is just $(E - E_c) = \frac{3}{2}kT$ or about 0.04 eV at 300 K. This is less than 4% of the bandgap energy.

The net carrier velocity in an applied field is called the *drift velocity*, v_d. It can be found by equating the impulse (force × time) applied to an electron during its free flight between collisions with the momentum gained by the electron in the same period. This equality is valid because steady state is reached when all momentum gained between collisions is lost to the lattice in the collisions. The force on an electron is $-q\mathscr{E}$ and the momentum gained is $m_n{}^*v_d$. Thus (cf footnote referring to Equation 1.2.1)

$$-q\mathscr{E}\tau_{cn} = m_n{}^*v_d \tag{1.2.2}$$

or

$$v_d = -\frac{q\mathscr{E}\tau_{cn}}{m_n{}^*} \tag{1.2.3}$$

Equation 1.2.3 states that the electron drift velocity v_d is proportional to the field with a proportionality factor that depends on the mean scattering time and the effective mass of the quasi-free electron. The proportionality factor is an important property of the electron called the *mobility*. It is designated by the symbol μ_n.

$$\mu_n = \frac{q\tau_{cn}}{m_n{}^*} \tag{1.2.4}$$

Since $v_d = -\mu_n\mathscr{E}$, the mobility describes how strongly the motion of an electron is influenced by an applied field.

From Equation 1.1.3 the current density flowing in the direction of the applied field can be found by summing the product of the charge on each electron times its velocity over all electrons per unit volume n.

$$J_n = \sum_{i=1}^{n} -qv_i = -nqv_d = nq\mu_n\mathscr{E} \tag{1.2.5}$$

Entirely analogous arguments apply to holes. A hole with zero kinetic energy resides at the valence-band edge E_v. Its kinetic energy is therefore measured by $(E_v - E)$. If a band edge is tilted, the hole moves upward on an electron energy-band diagram. The hole mobility μ_p is defined as $\mu_p = q\tau_{cp}/m_p{}^*$. The total current can be written as the sum of the electron and hole components:

$$J = J_n + J_p = (nq\mu_n + pq\mu_p)\mathscr{E} \tag{1.2.6}$$

The term in parentheses in Equation 1.2.6 is defined as the conductivity σ of the semiconductor:

$$\sigma = q\mu_n n + q\mu_p p \tag{1.2.7}$$

In extrinsic semiconductors, only one of the components in Equation 1.2.7 is generally significant because of the large ratio between the two carrier densities. The resistivity, which is the reciprocal of the conductivity, is shown as a function of the dopant concentration in Figure 1.14 for phosphorus-doped, n-type silicon and for boron-doped, p-type silicon. There are slight variations in the resistivities obtained for different dopant species, particularly in the heavy-doping range. In most practical cases, however, Figure 1.14 can be used for any dopant species.

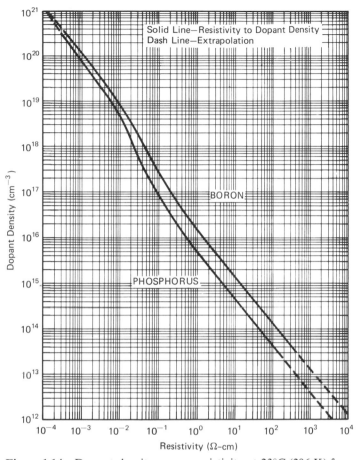

Figure 1.14 Dopant density versus resistivity at 23°C (296 K) for silicon doped with phosphorus and with boron. The curves can be used with little error to represent conditions at 300 K. [W. R. Thurber, R. L. Mattis, and Y. M. Liu, National Bureau of Standards Special Publication 400–64, 42 (May 1981)].

A property of a solid that is closely related to its conductivity is its *dielectric relaxation time*. The dielectric relaxation time is a measure of the time it takes for charge in a semiconductor to become neutralized by conduction processes. It is small in metals and can be large in semiconductors and insulators. One of its uses is to obtain qualitative insight. Several device concepts may be quite readily interpreted according to the relative sizes of the charge transit time through a material and the dielectric relaxation time in the same material. Problem 1.12 provides further introduction to the dielectric relaxation time.

Mobility and Scattering

Quantum-mechanical calculations indicate that a perfectly periodic lattice would not scatter free carriers; that is, the carriers would not interchange energy with a stationary, perfect lattice. However, at any temperature above absolute zero the atoms that form the lattice vibrate. These vibrations disturb periodicity and allow energy to be transferred between the carriers and lattice.* The interactions with lattice vibrations can be viewed as collisions with energetic "particles" called *phonons*. Phonons, like photons, have energies quantized in units of hv, where v is the lattice-vibrational frequency and h is Planck's constant. The theories of thermal and electrical conduction can often be simplified when formulated in terms of phonon interactions. For silicon at room temperature and moderate electric fields the lowest vibrational mode for the lattice corresponds to a phonon energy of 0.063 eV, and the energy of an electron is changed by this amount when it interacts with these lowest-energy phonons. Theoretical analysis indicates that the mobility should decrease with temperature in proportion to T^{-n} with n between 1.5 and 2.5 when lattice scattering is dominant. Experimentally, values of n range from 1.66 to 3, with $n = 2.5$ being common.

In addition to lattice vibrations, dopant impurities also cause local distortions in the lattice and scatter free carriers. However, unlike scattering from lattice vibrations, scattering from ionized impurities becomes less significant at higher temperatures. Since the carriers are moving faster at higher temperatures, they remain near the impurity atom for a shorter time and are therefore less effectively scattered. Consequently, when impurity scattering is dominant, the mobility increases with increasing temperature. Scattering may also be caused by collisions with unintentional impurities and by crystal defects. These defects can arise from poor control of the quality of the semiconductor, or they may be related to boundaries between grains of a polycrystalline material. An example of the latter would be the thin films of polycrystalline silicon used to form portions of many MOS integrated circuits (Chapter 10). The grain boundaries and defects in polycrystalline material can reduce the mobility to a small fraction of its value in single-crystal material with the same dopant concentration.

Two or more of the scattering processes discussed above may be important at the same time, and their combined effect on mobility must be assessed. To do this we consider the number of particles that are scattered in a time dt. The probability that a carrier is scattered in a time interval dt by process i is dt/τ_i where τ_i is the average time between scattering events resulting from process i. The total probability dt/τ_c that a carrier is scattered in the time interval dt is then the sum of the probabilities of being scattered by each mechanism:

$$\frac{dt}{\tau_c} = \sum_i \frac{dt}{\tau_i} \tag{1.2.8}$$

---

* Since this energy is supplied to the carriers by the applied field, scattering processes lead to heating of the semiconductor. The dissipation of this heat is often a limiting factor in the size of semiconductor devices. A device must be large enough to avoid heating to temperatures at which it no longer functions.

This is a reasonable way to combine the scattering probabilities since the average scattering time resulting from all of the processes acting simultaneously is less than that resulting from any one process, and is dominated by the shortest scattering time. Since the mobility μ equals $q\tau_c/m^*$, we may write

$$\frac{1}{\mu} = \sum_i \frac{1}{\mu_i} \tag{1.2.9}$$

The mobility of a carrier subjected to several different scattering mechanisms can thus be found by combining the reciprocal mobilities as determined by each scattering type; the resultant mobility is therefore smaller than that determined by any of the individual scattering mechanisms. Because of the reciprocal relation for combining mobility components (Equation 1.2.9), the overall mobility is dominated by the process for which τ_i is smallest.

We can apply these considerations to the mobilities of electrons and holes in silicon at room temperature which are plotted in Figure 1.15. The figure represents

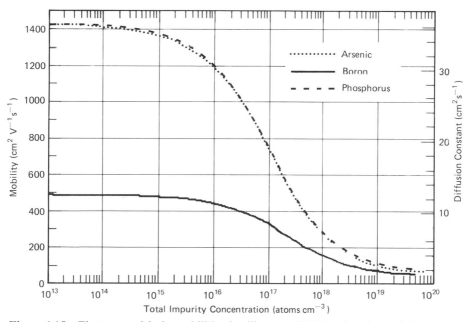

Figure 1.15 Electron and hole mobilities in silicon at 300 K as functions of the total dopant concentration. The values plotted are the results of curve fitting measurements from several sources. The mobility curves can be generated using Equation 1.2.10 with the following parameter values:[3]

Parameter	Arsenic	Phosphorus	Boron
μ_{min}	52.2	68.5	44.9
μ_{max}	1417	1414	470.5
N_{ref}	9.68×10^{16}	9.20×10^{16}	2.23×10^{17}
α	0.680	0.711	0.719

a "best fit" to measured data reported in a number of different sources. In lightly-doped material, the mobility resulting from ionized impurity scattering is higher than that due to lattice scattering. Hence, for silicon having impurity concentrations less than about 10^{15} cm^{-3}, the mobilities both for holes and electrons remain nearly constant as the dopant concentration is varied. At higher dopant concentrations, however, scattering by ionized impurities becomes comparable to that resulting from lattice vibrations, and the total mobility decreases.

An important practical consequence of the dependence of mobility on total impurity concentration is observed if a semiconductor is converted from one type to the other (p to n or n to p) by compensating the dopant impurity atoms already present. While the carrier densities depend on the difference between the concentrations of the two types of dopant impurities, $(N_d - N_a)$ (Equations 1.1.16 and 1.1.17), the scattering depends on the sum of the ionized impurity concentrations $(N_d + N_a)$. Thus, the mobilities in a compensated semiconductor may be markedly lower than those in an uncompensated material with the same net carrier density

Figure 1.15 represents an empirical formulation to fit measured data for electron and hole mobilities in silicon[3]. The equation is

$$\mu = \mu_{\text{min}} + \frac{\mu_{\text{max}} - \mu_{\text{min}}}{1 + (N/N_{ref})^{\alpha}} \qquad (1.2.10)$$

where N is the total dopant concentration in the silicon and the four parameters μ_{max}, μ_{min}, N_{ref}, and α have different values for each dopant species. Values of these parameters for the most common dopants in silicon are included in the caption for Figure 1.15. Table 1.1 gives numerical values of the mobility (as calculated from Equation 1.2.10) at decade values of N.

Table 1.1   **Mobilities in Silicon (cm^{2}v^{-1}s^{-1})**

N	Arsenic	Phosphorus	Boron
10^{13}	1423	1424	486
10^{14}	1413	1416	485
10^{15}	1367	1374	478
10^{16}	1184	1194	444
10^{17}	731	727	328
10^{18}	285	279	157
10^{19}	108	115	72

The dependence of mobility on dopant type is seen, from Figure 1.15, to be slight for total impurity concentrations less than 10^{19} cm^{-3}. In the high doping range (for $N > 10^{19}$ cm^{-3}), the mobility of phosphorus-doped silicon is 10 to 20% greater than that of silicon doped with arsenic. At very high dopant densities (above about 10^{20} cm^{-3}), measured mobilities drop below the minimum values shown in Figure 1.15.

Temperature Dependence. The different scattering mechanisms that affect free-carrier mobilities have varying dependences on temperature. For example, scattering by ionized impurities becomes less effective as the temperature increases because the faster moving carriers interact less effectively with stationary impurities. However, scattering by lattice vibrations (phonon collisions) becomes more effective at higher temperatures. Because of this, at lower temperatures the mobility characteristically increases with rising temperatures (since impurity-scattering predominates), while at higher temperatures the mobility decreases (because phonon collisions predominate). These competing temperature variations lead to a characteristic maximum in the mobility versus temperature relation as seen in the experimental data shown in Figure 1.16. At the peak in mobility, the two temperature variations are balanced, and the mobility has its minimum temperature sensitivity.

For design and analysis, equations for the dependence of mobility on temperature and dopant concentration are useful. Such expressions have been derived empirically for silicon and can be written[4]

$$\mu_n = 88 \ T_n^{-0.57} + \frac{7.4 \times 10^8 \ T^{-2.33}}{1 + [N/(1.26 \times 10^{17} \ T_n^{2.4})] \, 0.88 \ T_n^{-0.146}}$$

$$\mu_p = 54.3 \ T_n^{-0.57} + \frac{7.4 \times 10^8 \ T^{-2.23}}{1 + [N/(2.35 \times 10^{17} \ T_n^{2.4})] \, 0.88 \ T_n^{-0.146}} \qquad (1.2.11)$$

where $T_n = T/300$ with T measured in K (Kelvin scale), and N is the total dopant density in the silicon. Equation 1.2.11 is useful up to dopant densities of 10^{20} cm^{-3} and for temperatures between 250 and 500 K.

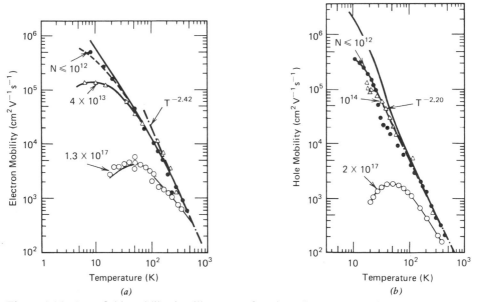

Figure 1.16 Low-field mobility in silicon as a function of temperature for electrons (*a*), and for holes (*b*). The continuous lines represent the theoretical predictions for pure lattice scattering.[5]

Velocity Limitations. In the simplified treatment presented thus far, we have assumed (by taking τ_c to be insensitive to $\mathscr{E}$) that the velocity imparted to the free carriers by the applied field is much less than the random thermal velocity, which we found from Equation 1.2.1 to be approximately 10^7 cm s^{-1} at room temperature. For electrons in silicon with $\mu_n \approx 1400$ cm^2 V^{-1} s^{-1} the drift velocity at a typical field of 100 V cm^{-1} is roughly 1.5% of the thermal velocity, and the applied field does not appreciably change the total velocity or energy of the carrier. At high fields, however, the drift velocity becomes comparable to the random thermal velocity, and can no longer be considered as a small increment to the thermal motion. The total energy of the carrier then increases significantly as the fields are increased. When carriers attain energies above the ambient thermal energy, they are often characterized as *hot carriers* with an effective temperature T_e that rises with increasing fields and corresponds to the increased kinetic energy of the carriers.[6]

If T_e increases sufficiently to cause substantial transfer of energy from the electrons (and thereby from the field) to the lattice, the mobility decreases from its low-field value. When the energy of the *hot electrons* reaches a critical value, a new scattering process (collisions with high energy or "optical" phonons) becomes important. This new scattering process is very effective in transferring energy from the hot carriers to the lattice, and it is a major reason that the drift velocity approaches a limiting value v_l at high fields. Figure 1.17 shows measured values of drift velocity for electrons (at 77 K and 300 K) and holes (at 300 K) in silicon as functions of the applied field. At low voltages, the curves are linear, indicating

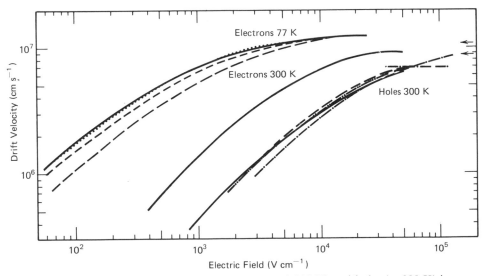

Figure 1.17   Drift velocities of electrons (at 77 K and 300 K) and holes (at 300 K) in silicon as functions of the applied field showing velocity saturation at high fields. The presence of several curves indicates the variation in reported data. An empirical "best fit" to these curves is given in Equation 1.2.12 and Table 1.2.[5]

constant mobility. At fields above a few thousand volts per centimeter, however, there are noticeable deviations from constant mobility. Because fields of this magnitude are frequently present in integrated-circuit devices (equivalent to a few hundred millivolts across a micrometer), velocity saturation must be considered in analyzing many practical devices.

As a useful approximation, the data shown in Figure 1.17 can be modeled empirically by the expression[5]

$$|v_d| = v_l \frac{\mathscr{E}}{\mathscr{E}_c} \left[\frac{1}{1 + (\mathscr{E}/\mathscr{E}_c)^\beta} \right]^{1/\beta} \tag{1.2.12}$$

The parameters v_l, $\mathscr{E}_c$, and β in Equation 1.2.12 are given (as a function of the absolute temperature T) in Table 1.2.

Table 1.2 **Parameters for Field Dependence of Drift Velocity**

Parameter	Electrons		Holes	
	Expression	at 300 K	Expression	at 300 K
v_l cm s^{-1}	$1.53 \times 10^9 \, T^{-0.87}$	1.07×10^7	$1.62 \times 10^8 \, T^{-0.52}$	8.34×10^6
$\mathscr{E}_c$ V cm^{-1}	$1.01 \, T^{1.55}$	6.91×10^3	$1.24 \, T^{1.68}$	1.45×10^4
β	$2.57 \times 10^{-2} \, T^{0.66}$	1.11	$0.46 \, T^{0.17}$	2.637

EXAMPLE   Velocity Limitations

Use Equation 1.2.12 to find the field (at 300 K) at which the effective electron mobility (defined as the ratio of the drift velocity to the field) is reduced to half its low-field value.

Solution

At low fields, the drift velocity v_d is proportional to the field (Equation 1.2.3) and $|v_d| = \mu_n \mathscr{E}$. Applying Equation 1.2.12 at low fields ($\mathscr{E}/\mathscr{E}_c \ll 1$), we find that the low-field mobility can be expressed in terms of $\mathscr{E}_c$ and v_l; $\mu_n = |v_d/\mathscr{E}| = |v_l/\mathscr{E}_c|$. If we use the values from Table 1.2 in this expression, we calculate $\mu_n = 1548$ cm^2 V^{-1} s^{-1} for the low-field mobility which is about 10% higher than the value shown in Figure 1.15. This lack of correspondence is not unusual when parameters are obtained by curve-fitting.

To find the field at which the effective mobility $\mu_n(\mathscr{E})$ is reduced by 50% from the low-field value, we note that $\mu_n(\mathscr{E}) = |v_d/\mathscr{E}| = (1/2) \times |v_l/\mathscr{E}_c|$. Hence, from Equation 1.2.12

$$\frac{1}{2} = \left[\frac{1}{1 + (\mathscr{E}/\mathscr{E}_c)^\beta} \right]^{1/\beta}$$

where $\beta = 1.11$ and $\mathscr{E}_c = 6.91 \times 10^3$ V cm^{-1}. Solving this equation, we find $\mathscr{E}/\mathscr{E}_c = 1.142$ or $\mathscr{E} = 7.89 \times 10^3$ V cm^{-1}.

Before leaving the topic of carrier transport at high fields, we should make a final point about hot electrons. The hot-electron temperature T_e, introduced earlier in our discussion, describes an ensemble of electrons that is interchanging energy by colliding with the lattice. Some electrons may, by chance, avoid collisions for relatively long times, and thus achieve velocities exceeding v_l and kinetic energies corresponding to temperatures greater than T_e. Although these electrons are few in number, they can have important physical effects because of their very high energies. We shall discuss some of these effects in Chapter 10.

Diffusion Current

In the previous section we discussed drift current, which flows when an electric field is applied and which gives rise to Ohm's law behavior. Ohmic behavior is observed in metals and semiconductors and is probably familiar from direct experience. In semiconductors, however, another important component of current can exist if there is a spatial variation of carrier energies or densities within the material. This component of current is called *diffusion current*. Diffusion current is generally not an important consideration in metals because of their high conductivities. The lower conductivity and the possibility of nonuniform densities of carriers and of carrier energies, however, often makes diffusion a very important process affecting current flow in semiconductors.

To understand the origin of diffusion current, we consider the hypothetical case of an *n*-type semiconductor with an electron density that varies only in one dimension (Figure 1.18). We assume that the semiconductor is at a uniform temperature so that the average energy of electrons does not vary with *x*; only the

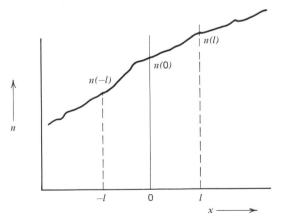

Figure 1.18 Electron concentration *n* versus distance *x* in a hypothetical one-dimensional solid. Boundaries are demarcated *l* units on either side of the origin where *l* is a collision mean-free path.

density $n(x)$ is variable. We consider the number of electrons crossing the plane at $x = 0$ per unit time per unit area. Because they are at finite temperature, the electrons have random thermal motion along the single dimension, but we assume that no electric fields are applied. On the average, the electrons crossing the plane $x = 0$ from the left in Figure 1.18 start at approximately $x = -l$ after a collision where l is the *mean-free path* of an electron, given by $l = v_{th}\tau_{cn}$. The average rate (per unit area) of electrons crossing the plane $x = 0$ from the left, therefore, depends upon the density of electrons that started at $x = -l$ and is

$$\tfrac{1}{2}n(-l)v_{th} \tag{1.2.13}$$

The factor $(\tfrac{1}{2})$ appears since half of the electrons travel to the left and half travel to the right after a collision at $x = -l$. Similarly, the rate at which electrons cross the plane $x = 0$ from the right is given by

$$\tfrac{1}{2}n(l)v_{th} \tag{1.2.14}$$

so that the net rate of particle flow from the left (denoted F) is

$$F = \tfrac{1}{2}v_{th}[n(-l) - n(l)] \tag{1.2.15}$$

Approximating the densities at $x = \pm l$ by the first two terms of a Taylor series expansion, we find

$$F = \tfrac{1}{2}v_{th}\left\{\left[n(0) - \frac{dn}{dx}l\right] - \left[n(0) + \frac{dn}{dx}l\right]\right\} = -v_{th}l\frac{dn}{dx} \tag{1.2.16}$$

Since each electron carries a charge, $-q$, the particle flow corresponds to a current

$$J_n = -qF = qlv_{th}\frac{dn}{dx} \tag{1.2.17}$$

We thus see that diffusion current is proportional to the spatial derivative of the electron density and arises because of the random thermal motion of charged particles in a concentration gradient. For an electron density that increases with x the gradient is positive, as is the current. Because we expect electrons to flow from the higher density region at the right to the lower density region at the left, and current flows in the direction opposite to that of the electrons, the direction of current indicated by Equation 1.2.17 is physically reasonable.

We can write Equation 1.2.17 in a more useful form by applying the theorem for the equipartition of energy to this one-dimensional case. This allows us to write

$$\tfrac{1}{2}m_n^*v_{th}^2 = \tfrac{1}{2}kT \tag{1.2.18}$$

We now use the relationship $l = v_{th}\tau_{cn}$ together with Equation 1.2.4 to write Equation 1.2.17 in the form

$$J_n = q\left(\frac{kT}{q}\mu_n\right)\frac{dn}{dx} \tag{1.2.19}$$

The quantity in parentheses on the right-hand side of Equation (1.2.19) is defined as the *diffusion constant D_n* and our short derivation has established that

$$D_n = \left(\frac{kT}{q}\right)\mu_n \tag{1.2.20}$$

Equation (1.2.20) is known as the *Einstein relation*. It relates the two important constants that characterize free-carrier transport by drift and by diffusion in a solid. Its validity can be established rigorously by considering the statistical mechanics of solids in detail. Our derivation is aimed at intuition, not physical exactness.

If a field were present, then drift, as well as a diffusion, would occur. In that case, the total current would be the sum of both the drift and diffusion currents.

$$J_{nx} = q\mu_n n\mathscr{E}_x + qD_n\frac{dn}{dx} \tag{1.2.21}$$

where $\mathscr{E}_x$ is the component of the electric field in the x-direction.

A similar expression may be found for the hole diffusion current so that the total hole current is written as

$$J_{px} = q\mu_p p\mathscr{E}_x - qD_p\frac{dp}{dx} \tag{1.2.22}$$

where the negative sign arises because of the positive charge of a hole. The Einstein relation (Equation 1.2.20) also applies between D_p and μ_p.

Our discussion of diffusion current has been framed in terms of nonuniform carrier concentrations. Although this is the most frequent situation encountered in device analysis and Equations 1.2.21 and 1.2.22 are usually adequate, diffusion will also occur if carriers are equal in density in a semiconductor but are, at the same time, more energetic in one region than in another. In this case, other formulations must be used. A practical situation in which this is the case forms the basis for Problem 1.16.

EXAMPLE   Diffusion Current

An electric field has a non-zero value at plane x_1 (perpendicular to the x-axis) inside a silicon crystal. At x_1, the electron density is 10^6 cm^{-3} and the electron density is nonuniform in a direction perpendicular to the plane. It is observed that no electron current flows across the plane.

(a) Explain why no current is flowing.
(b) If the electric field is -10^3 V cm^{-1} (i.e., 10^3 V cm^{-1} in the negative x-direction), what is the electron gradient perpendicular to the plane?

Solution

(a) An electric field $\mathscr{E}$ would result in a drift current of magnitude $J_n = q\mu_n n\mathscr{E}$ according to Equation 1.2.5. Since no current flows, there must be a diffusion-current component that is equal in magnitude but opposite in direction. The balance of these two components leads to zero electron current.

(b) Using Equation 1.2.21, we have

$$J_n = 0 = q\mu_n \, n\mathscr{E}_x + qD_n \frac{dn}{dx}$$

$$\frac{dn}{dx} = -\frac{\mu_n}{D_n} \, n\mathscr{E}$$

$$= -\frac{q}{kT} \, n\mathscr{E}$$

$$= -\frac{10^6 \times (-10^3)}{0.0258}$$

$$= 3.88 \times 10^{10} \ \text{cm}^{-4}$$

Total currents and quasi-Fermi Levels.[†] Quasi-Fermi levels, defined in Equations 1.1.28 and 1.1.29, are useful in the analysis of semiconductors that are not at thermal equilibrium. A semiconductor in which a current is flowing is such a nonequilibrium case, and the quasi-Fermi levels can be used to define both drift and diffusion currents in a compact form.

To illustrate this, we first develop an expression for the electric field in a semiconductor in terms of the electron energy. The presence of an electric field causes charged particles to have position-dependent energies. Consequently, when a field is present, the energy bands for electrons are inclined away from the horizontal (constant energy), as seen in Figure 1.13. It is useful and convenient to express the field in terms of the intrinsic Fermi level E_i by using the fact that the electron energy is obtained by multiplying the potential by the charge $-q$. The field $\mathscr{E}$, which is the negative derivative of the potential, can therefore be written (in one dimension)

$$\mathscr{E} = \frac{1}{q}\frac{dE_i}{dx} = -\frac{d\varphi_{fi}}{dx} \qquad (1.2.23)$$

where $\varphi_{fi} = -E_i/q$ was introduced in Equation 1.1.28. The electron and hole concentrations are expressed in terms of the quasi-Fermi levels by

$$n = n_i \exp(E_{fn} - E_i)/kT = n_i \exp q(\varphi_{fi} - \varphi_{fn})/kT$$
$$p = n_i \exp(E_i - E_{fp})/kT = n_i \exp q(\varphi_{fp} - \varphi_{fi})/kT \qquad (1.2.24)$$

Using Equations 1.2.23 and 1.2.24 in Equation 1.2.21 together with the Einstein relation (Equation 1.2.20), we obtain

$$J_n = \mu_n n \frac{dE_{fn}}{dx} = -q\mu_n n \frac{d\varphi_{fn}}{dx} \qquad (1.2.25)$$

A similar derivation leads to

$$J_p = \mu_p p \frac{dE_{fp}}{dx} = -q\mu_p p \frac{d\varphi_{fp}}{dx} \qquad (1.2.26)$$

Equations 1.2.25 and 1.2.26 show that the total current (the sum of both drift and diffusion components) for each carrier is proportional to the gradient of the quasi-Fermi level of the respective carrier type. This compact representation can be very helpful in using energy-band diagrams to visualize the total current in a device. The equations are also useful for mathematical analysis, and their form simplifies a number of complex problems.

1.3 Device: Hall-Effect Magnetic Sensor

As will be our pattern throughout the book, we conclude this chapter with a discussion of an integrated-circuit device, in this case a Hall-effect sensor for magnetic fields. Hall-effect sensors are perhaps unconventional integrated-circuit devices, but they are important commercially. Hundreds of millions of integrated Hall circuits are in use, mainly as contactless switches (e.g. in computer-terminal keyboards) and as mechanical proximity detectors. Hall-effect, magnetic sensing, integrated circuits are, in fact, highly successful examples of *integrated sensors*, that is, integrated circuits having intentional sensitivity to nonelectrical signals. This sensitivity is achieved by incorporating sensing elements on a silicon chip together with bias, amplifying, and signal-processing circuitry. The field of integrated sensors is developing rapidly because it capitalizes on the extraordinary refinements already achieved by purely electrical integrated circuits and economically applies these circuits when the input signals are not electrical quantities. Microprocessors integrated with nonelectrical sensors are, for example, expected to be powerful new components in control systems. In addition to magnetic fields, other nonelectrical inputs for which integrated sensors show appreciable promise include visible and infrared radiation, temperature, pressure, force, acceleration, chemical vapors and humidity.

Physics of the Hall Effect

The Hall effect, named for American physicist E. H. Hall who discovered it in 1879, is a direct consequence of the force exerted on charged carriers moving in a magnetic field. The force on a particle having charge q and moving in a magnetic field $\vec{B}$ with velocity $\vec{v}$ (both variables being vector quantities) is written

$$\vec{F} = q\vec{v} \times \vec{B} \tag{1.3.1}$$

where the vector cross product ($\times$) signifies the product of the vector magnitudes times the sine of the angle between them.

The resultant Hall effect is illustrated schematically in Figure 1.19. The Hall effect is usually employed in an extrinsic semiconductor so that one carrier is dominant, and the other has a negligible density. To aid our discussion, however, both electrons and holes are shown in Figure 1.19.

As expressed in Equation 1.3.1, the current carriers (electrons or holes) in a conductor experience a force in a direction perpendicular to both the magnetic field and the carrier velocity. In the steady state, this force is balanced by an induced electric field which results from a slight charge redistribution. These forces must balance because there can be no net steady-state motion of the carriers in the transverse direction. The induced electric field is called the *Hall field* $\mathscr{E}_H$. Integrating the Hall field with respect to position across the width of the conductor produces the *Hall voltage* V_H, which can be detected by contacts placed at opposite sides of the conductor. For a uniform conductor and a uniform magnetic field, the magnitude of the Hall voltage is just the product of the Hall field and the conductor width W: $\mathbf{V}_H = \mathscr{E}_H W$.

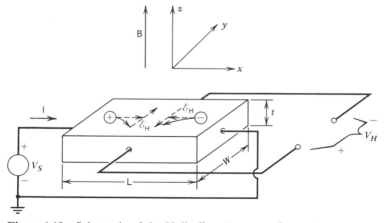

Figure 1.19 Schematic of the Hall effect. A current flows along the positive x-direction. A magnetic field B along the positive z-direction deflects holes and electrons along the negative y-direction. This leads to a Hall field $\mathscr{E}_H$ along the positive y-direction for holes and along the negative y-direction for electrons.

In Figure 1.19, current in the positive x-direction can be carried either by holes flowing toward positive x or else by electrons moving in the opposite direction. Since both the charge and the velocity are of opposite signs for each carrier type, the magnetic force on both holes and electrons has the same sign for a given current direction. The induced Hall voltage is, therefore, of opposite polarity for holes and electrons. The sign dependence of the Hall voltage can thus be used to determine whether a semiconductor is p- or n-type.

We can derive the basic theory for the Hall-effect using the quantities shown in Figure 1.19. In the figure, current flow is in the positive x-direction, the magnetic field is in the positive z-direction and the Hall field is therefore directed along the y-direction. The magnetic force deflects both holes and electrons in the negative y-direction and induces a Hall field toward positive y for holes, and in the opposite direction for electrons. Consider that the current carriers have a drift velocity v_d. Equating the magnitudes of the magnetic and Hall-field forces, we have $q\mathscr{E}_H = qv_dB$. The velocity v_d is related in magnitude to the current by $v_d = J_x/qp$, for holes or by $v_d = -J_x/qn$, for electrons. Thus, the Hall field can be written in terms of the current and the applied magnetic field as

$$\mathscr{E}_H = \frac{J_x B}{qp} \tag{1.3.2}$$

for holes, and,

$$\mathscr{E}_H = -\frac{J_x B}{qn} \tag{1.3.3}$$

for electrons. Both Equations 1.3.2 and 1.3.3 can be expressed as

$$\mathscr{E}_H = R_H J_x B \tag{1.3.4}$$

where R_H, the *Hall coefficient* is equal to $1/qp$ for holes and to $-1/qn$ for electrons in this simplified derivation. In practice, Equation 1.3.4 predicts the Hall field accurately if the Hall coefficient is modified to account properly for statistical variations in the velocities of free carriers. This modification introduces a new factor r into the expression for the Hall coefficient which now becomes

$$R_H = r/qp \tag{1.3.5}$$

for holes, and

$$R_H = -r/qn \tag{1.3.6}$$

for electrons. The factor r is typically between 1 and 2 (theoretically 1.18 for lattice scattering and 1.93 for ionized impurity scattering.)

The Hall voltage V_H is given by the product of $\mathscr{E}_H$ and W which can be written in terms of the total current I as

$$V_H = \frac{R_H I B}{10^8 t} \tag{1.3.7}$$

with R_H measured in cm^3 C^{-1}, I in amperes, B in Gauss, t in cm and V_H in volts.

(The factor 10^8 is needed to convert the *MKS* units meter and Tesla (or Webers m^{-2}) to more conventional semiconductor units centimeter and Gauss.)

From Equation 1.3.7 we see that in an unknown semiconductor, the Hall coefficient can be determined by measuring the Hall voltage for given magnetic field and current conditions. Equations 1.3.5 and 1.3.6 then permit the calculation of the unknown carrier types and densities. From the carrier densities and known currents, the material conductivity and *Hall mobility* ($\mu = \sigma|R_H|$) can then be found. The Hall effect thus provides a powerful experimental technique for the study of semiconductors, and it is very frequently used for this purpose.[7]

Integrated Hall-Effect Magnetic Sensor

To use the Hall effect in an integrated circuit, it is necessary to isolate a conducting pattern similar to the region sketched in Figure 1.19. This is typically accomplished by using epitaxy and junction isolation (procedures described in Chapter 2) as shown in Figure 1.20. For the simplest Hall-effect theory to apply, the width W should be much greater than the length L so that the current density J is uniform over the sample cross section. As a practical matter, however, shorting of the Hall voltage by the ohmic end contacts is reduced if $L \gg W$. In production integrated circuits, W is usually made comparable to L. For a rectangular geometry the theory that we have derived can be corrected by multiplying the expression for V_H (Equation 1.3.7) by a factor K that is typically nearly unity and is a function only if the ratio W/L.[8] This refinement is not included in our treatment of the Hall effect.

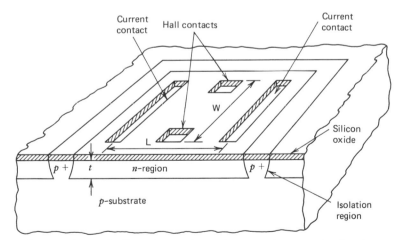

Figure 1.20 Hall-effect element for an integrated-sensor circuit. The element is fabricated in high-resistivity *n*-type silicon which is isolated by *p*-type regions (described in Chapters 2 and 4). The sketch shows the element prior to the application of contact metal.

One important consideration when designing a Hall element for an integrated circuit is the power dissipated in the device. To consider power consumption, it is useful to express the resistance of the Hall element in terms of the Hall coefficient. For a p-type element, for example, we can write an expression for the resistance R

$$R = \frac{\rho L}{A} = \frac{L}{q\mu_p p W t} = \frac{LR_H}{r\mu_p W t} \tag{1.3.8}$$

Hence, for a supply voltage V_S, we can write $I = V_S/R$ or

$$I = \frac{rV_S\mu_p W t}{R_H L} \tag{1.3.9}$$

which allows us to write Equation 1.3.7 as

$$V_H = r\mu_p V_S \frac{W}{L} B \times 10^{-8} \tag{1.3.10}$$

where, as in Equation 1.3.7, B is measured in Gauss and length is measured in cm.

EXAMPLE   Hall Element Figure of Merit

(a) Derive a figure of merit M_H for a Hall element that expresses the Hall voltage per unit of magnetic field per unit of power dissipation. Consider a p-type element having $R_H = 8 \times 10^3$ cm³ C^{-1}, $W/L = 1$, $r = 1.2$, and $t = 8$ μm.

(b) Calculate the resistance of the element and the value of M_H if $B = 500$ Gauss and the power dissipated in the element is 1.43 mW.

Solution

(a) From Equation 1.3.10, we have V_H in terms of the supply voltage V_S. The dissipated power P_H is equal to $V_S \times I$.
Therefore

$$P_H = \frac{rV_S^2\mu_p W t}{R_H L}$$

and

$$M_H = \frac{V_H}{P_H B} = \frac{R_H \times 10^{-8}}{V_S t} = \frac{r \times 10^{-8}}{qpV_S t}$$

in units of volts per Gauss watt. From this derived result, we see that M_H is improved by decreasing the supply voltage and reducing both the dopant density and the thickness of the Hall element.

(b) Calculating the parameters, we have

1. $p = r/qR_H = 9.38 \times 10^{14}$.
2. From Figure 1.15, $\mu_p = 475 \text{ cm}^2 \text{ V}^{-1} \text{ s}^{-1}$.
3. From Equation 1.3.8, $R = 17.5 \text{ k}\Omega$.
4. Since $P_H = V_S^2/R$, $V_S = \sqrt{P_H R}$, or $V_S = 5 \text{ V}$.
5. Hence, $M_H = \dfrac{8 \times 10^3 \times 10^{-8}}{5 \times 8 \times 10^{-4}}$ or $M_H = 0.02$.

As seen in this example, the performance of the Hall sensor is improved by decreasing the dopant density. As the dopant density is reduced, the mobilities of both holes and electrons increase (cf Figure 1.15) leading to an increased Hall voltage for a given bias (cf Equation 1.3.10). Most *IC* Hall sensors are fabricated in *n*-type Si doped with about 10^{15} donors cm^{-3}. From Figure 1.15, we see that at this concentration we are below the region in which the mobility varies strongly with the impurity density. Hence, for this dopant concentration, unintentional processing variations in the dopant density lead to only small variations in mobility, and thus in the Hall coefficient. Typical sensitivities obtained for Hall elements are of the order of 30 μV per Gauss with excellent linearity up to tens of kilo Gauss.

The major use of integrated Hall circuits is for sensing the position of a device or element. Most of the circuits produced are used in contactless keyboard switches. In a typical keyboard switching application, a permanent magnet on a plunger is activated by depressing a key. This action introduces a magnetic field of the order of 500 Gauss near the sensing Hall element, producing a Hall voltage of roughly 15 mV which is easily detected by an on-chip amplifier. Integrated Hall sensors are also produced for magnetometer applications in which a signal proportional to the magnetic field is desired. Figure 1.21 shows a commercial Hall-effect integrated sensor which has been designed for magnetometer use.

Summary

One of the cornerstones of solid-state electronics is the band structure of solids. This key concept is related to the quantized energy levels of isolated atoms. Huge differences in the electrical conductivities of metals, semiconductors, and insulators result from basic differences in the band structures of these three classes of materials.

By considering the band structure of semiconductors, we can deduce the existence of two types of current carriers, *holes* and *electrons*. In most cases of practical interest, holes and electrons can be considered as classical free particles inside the semiconductor crystal. The motions of the holes and electrons and the statistical distributions that characterize their energies can be calculated if their effective masses are modified from those of truly free particles. The populations of holes

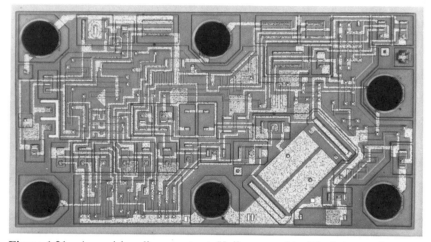

Figure 1.21   A precision, linear-output, Hall-sensor chip. The integrated circuit contains bias elements, temperature-compensation circuitry, and on-chip amplification. The chip area is 1.12 by 1.98 mm^2 and the Hall element (large pattern on the lower right-hand side) measures 230 by 335 μm^2. (*Courtesy: G. B. Hocker, Honeywell Corporation*)

and electrons can be controlled by adding impurities to otherwise pure semiconductor crystals. In this process, known as *doping*, extra holes can be added to single-crystal silicon, for example, by introducing ions with valence three, which are incorporated in the lattice in place of silicon, which has valence four. The resultant crystal is known as *p*-type silicon. To add electrons and thereby obtain *n*-type silicon by doping, impurities with valence five can be substituted for silicon atoms in the lattice.

An understanding of many electronic properties can be gained by applying the laws of statistical physics to the populations of electrons and holes in semiconductors. The *Fermi level* is an important parameter used to describe the concentrations of free carriers in a crystal at thermal equilibrium. Under usual conditions, the Fermi level in a semiconductor exists within a *forbidden gap* of energies that is ideally free of any allowed energy states for electrons. Semiconductors in this condition are described as *nondegenerate*. If the Fermi level lies within a band of allowed electronic states, the semiconductor is called *degenerate* because its conduction properties degenerate towards those of a metal. When thermal equilibrium does not apply, it is useful to define two *quasi-Fermi levels* which are different for holes and electrons. The separation of the two quasi-Fermi levels is a measure of the degree to which thermal equilibrium in the semiconductor is disturbed. When the semiconductor returns to thermal equilibrium, the hole and electron quasi-Fermi levels merge into a single Fermi level that characterizes both carrier densities. Quasi-Fermi levels are useful in considering photogeneration of electrons and holes in semiconductors, and in simplifying the representation of free-carrier diffusion and drift.

In the basic theory of electrons and holes in semiconductors, it is assumed that only very low concentrations of dopants are present and that they do not influence the band structure of the host crystal. At *heavy-doping* concentrations, the impurity atoms are no longer situated far apart from one another, and their presence causes the band structure to be modified. One of the most influential effects of heavy doping is that n_{ie} the effective value of the *intrinsic-carrier density* becomes an increasing function of the doping density.

When free carriers in a semiconductor are subjected to a low or moderate electric field, they move with a constant *drift* velocity. The velocity per unit field is known as the *mobility* and given the symbol μ. The magnitude of the mobility is determined by the nature of the scattering events experienced by free carriers as they interact with the lattice. The thermal energy of free carriers in solids causes them to move randomly and results in a net flux of carriers if either the carrier densities or their energies are not homogeneous in space. The net motion resulting from this process is called *diffusion*. When the crystal is at a uniform temperature, diffusion occurs from regions where the carriers are more dense to regions where their density is lower. In nondegenerate semiconductors, the *diffusion constant D* for a given type of carrier is related to the mobility μ by the *Einstein relation*, $D/\mu = kT/q$.

At high electric fields, scattering of electrons and holes is no longer determined mainly by the thermal velocity, and the drift motion is no longer proportional to the applied field. Mobility, instead of being constant, begins to decrease as the field increases. At fields in the range of 10^4 to 10^5 V cm^{-1}, additional modes of free-carrier scattering are introduced that lead to upper limits for hole and electron drift velocities.

The force on a moving charged particle in a magnetic field gives rise to the *Hall effect* when a semiconductor carries a current perpendicular to a magnetic field. Hall-effect experiments provide a powerful means of studying free-carrier properties in semiconductors. *Integrated-sensor* circuits in which Hall elements, together with bias circuits, temperature-compensating circuits, and amplifiers are all fabricated on a single silicon chip are important commercial products.

References

1. (a) F. Morin and J. P. Maita, *Phys. Rev.* **96**, 28 (1954).
   (b) A. S. Grove, *Physics and Technology of Semiconductor Devices.* Wiley, New York, (1967).
2. P. P. Debye and E. M. Conwell, *Phys. Rev.* **93**, 693 (1954).
3. (a) D. M. Caughey and R. F. Thomas, *Proc. IEEE* **55**, 2192 (December 1967).
   (b) G. Masetti, M. Severi, and S. Solmi, *IEEE Trans. Electr. Devices* **ED-30**, 764 (July 1983).
4. N. D. Arora, J. R. Hauser, and D. J. Roulston, *IEEE Trans. Electr. Devices* **ED-29**, 292 (February 1982).
5. C. Jacoboni, C. Canali, G. Ottaviani, and A. A. Quaranta, *Solid-State Electronics* **20**, 77 (February 1977).

6. J. L. Moll, *Physics of Semiconductors*. McGraw-Hill, New York (1964), p. 198.

7. E. H. Putley, *The Hall Effect and Related Phenomena*. Butterworths, London (1960).

8. J. T. Maupin and M. L. Geske, *The Hall Effect in Silicon Circuits*, Symposium on the Hall Effect, Johns Hopkins University, Baltimore, Md. (1981), p. 421.

9. S. Wang, *Solid-State Electronics*. McGraw-Hill, New York (1966), p. 263.

10. S. M. Sze and J. C. Irvin, *Solid-State Electronics* **11**, 599 (1968).

11. J. L. Moll, *Physics of Semiconductors*. McGraw-Hill, New York (1964), p. 99.

12. R. B. Adler, A. C. Smith, and R. L. Longini, *SEEC*, Vol. 1, *Introduction to Semiconductor Physics*. Wiley, New York (1964).

Textbooks

B. G. Streetman, *Solid-State Electronic Devices*, 2nd ed. Prentice-Hall, Englewood Cliffs, N.J. (1980).

R. F. Pierret, *Semiconductor Fundamentals*, Vol. I, *Modular Series on Solid-State Devices*, Addison-Wesley, Reading, Mass. (1983).

J. P. McKelvey, *Solid-State and Semiconductor Physics*. Original publisher Harper & Row, New York (1966), reprinted by Dover Publications, New York (1984).

W. E. Beadle, J. C. C. Tsai, and R. D. Plummer Editors, *Quick Reference Manual for Silicon Integrated-Circuit Technology*, Wiley-Interscience, New York (1985) (for silicon properties).

Problems

1.1 Phosphorus donor atoms with a concentration of 10^{16} cm^{-3} are added to a pure sample of silicon. Assume that the phosphorus atoms are distributed homogeneously throughout the silicon. The atomic weight of phosphorus is 31.

(a) What is the sample resistivity at 300 K?

(b) What proportion by weight does the donor impurity comprise?

(c) If 10^{17} atoms cm^{-3} of boron are included in addition to the phosphorus, and distributed uniformly, what is the resultant resistivity and conductivity type (i.e., p- or n-type material)?

(d) Sketch the energy-band diagram under the condition of (c) and show the position of the Fermi level.

1.2* Find the equilibrium electron and hole concentrations and the location of the Fermi level for silicon at 27°C if the silicon contains the following concentrations of shallow dopant atoms:

(a) 1×10^{16} cm^{-3} boron atoms

(b) 3×10^{16} cm^{-3} arsenic atoms and 2.9×10^{16} cm^{-3} boron atoms

1.3 An n-type sample of silicon has a uniform density $N_d = 10^{16}$ atoms cm^{-3} of arsenic, and a p-type silicon sample has $N_a = 10^{15}$ atoms cm^{-3} of boron. For each semiconductor material determine the following:

(a) The temperature at which half the impurity atoms are ionized. Assume that all mobile electrons and holes come from dopant impurities. [*Hint.* Use Equations 1.1.21 and 1.1.22 together with the fact that $f_D(E_f) = \frac{1}{2}$]

---

* Answers to problems marked by an asterisk appear at the end of the book.

(b) The temperature at which the intrinsic concentration n_i exceeds the impurity density by a factor of 10. See Table 1.4 for $n_i(T)$.

(c) The equilibrium minority-carrier concentrations at 300 K. Assume full ionization of impurities.

(d) The Fermi level referred to the valence-band edge E_v in each material at 300 K. The Fermi level if both types of impurities are present in a single sample.

1.4* A piece of n-type silicon has a resistivity of 5 Ω-cm at 27°C. Find the thermal-equilibrium hole concentrations at 27, 100, and 500°C. (Refer to the tables and figures for needed quantities.)

1.5† Grain boundaries and other structural defects introduce allowed energy states deep within the forbidden gap of polycrystalline silicon. Assume that each defect introduces two discrete levels: an acceptor level 0.51 eV above the top of the valence band and a donor level 0.27 eV above the top of the valence band. (Note that $E_a > E_d$ and that these are *not* shallow levels.) The ratio of the number of defects with each charge state $(+, -, \text{ or neutral})$ at thermal equilibrium is given by the relation (see Ref. 11)

$$N_d{}^+ : N_o : N_a{}^- = \exp \frac{E_d - E_f}{kT} : 1 : \exp \frac{E_f - E_a}{kT}$$

(a) Sketch the densities of these three species as the Fermi level moves from E_v to E_c. Which species dominates in heavily doped p-type material? In n-type material?

(b) What is the effect of the defects on the majority-carrier concentrations?

(c) Using the above information, determine the charge state of the defect levels and the position of the Fermi level in a silicon crystal containing no shallow dopant atoms. Is the sample p- or n-type?

(d) What are the electron and hole concentrations and the location of the Fermi level in a sample with 2×10^{17} cm^{-3} phosphorus atoms and 5×10^{16} cm^{-3} defects?

1.6* Two scattering mechanisms are important in a piece of semiconductor. If only the first scattering process were present, the mobility would be 800 cm^2 (Vs)$^{-1}$. If only the second were present, the mobility would be 200 cm^2 (Vs)$^{-1}$. What is the mobility considering both scattering processes?

1.7 Find the mobility of electrons in aluminum with resistivity 2.8×10^{-6} Ω-cm and density 2.7 g cm^{-3}. The atomic weight of aluminum is 27. Of the three valence electrons in aluminum, on the average 0.9 electrons are free to participate in conduction at room temperature. If $m^* = m_0$, find the mean time between collisions and compare this value to the corresponding value in lightly doped silicon.

1.8* An electron is moving in a piece of lightly doped silicon under an applied field at 27°C so that its drift velocity is one-tenth of its thermal velocity. Calculate the average number of collisions it will experience in traversing by drift a region 1μm wide. What is the voltage applied across this region?

1.9 In an experiment the voltage across a uniform, 2 μm long region of 1 Ω-cm, n-type silicon is doubled, but the current only increases by 50%. Explain. (The silicon remains neutral at both current levels.)

1.10* The electron concentration in a piece of uniform, lightly-doped, n-type silicon at room temperature varies linearly from 10^{17} cm^{-3} at $x = 0$ to 6×10^{16} cm^{-3} at $x = 2$ μm. Electrons are supplied to keep this concentration constant with time. Calculate the electron current density in the silicon if no electric field is present. Assume $\mu_n = 1000$ cm^2 V^{-1} s^{-1} and T = 300 K.

1.11[†] Silicon atoms are added to a piece of gallium arsenide. The silicon can replace either trivalent gallium or pentavalent arsenic atoms. Assume that silicon atoms act as fully ionized dopant atoms and that 5% of the 10^{10} cm^{-3} silicon atoms added replace gallium atoms and 95% replace arsenic atoms. The sample temperature is 300 K.
   (a) Calculate the donor and acceptor concentrations.
   (b) Find the electron and hole concentrations and the location of the Fermi level.
   (c) Find the conductivity of the gallium arsenide assuming that lattice scattering is dominant.
   See Table 1.3 for properties of GaAs.

1.12[†] (Dielectric relaxation in solids.) Consider a homogeneous one-carrier conductor of conductivity σ and permittivity ϵ. Imagine a given distribution of the mobile charge density $\rho(x,y,z; t = 0)$ in space at $t = 0$. We know the following facts from electromagnetism, provided we neglect diffusion current:

$$\nabla \cdot D = \rho; \quad D = \epsilon\mathscr{E}; \quad J = \sigma\mathscr{E}; \quad \nabla \cdot J = \frac{-d\rho}{dt}$$

   (a) Show from these facts that $\rho(x,y,z; t) = \rho(x,y,z; t = 0)\, e^{-t/(\epsilon/\sigma)}$. This result shows that uncompensated charge cannot remain in a uniform conducting material, but must accumulate at discontinuous surfaces or other places of nonuniformity.
   (b) Compute the value of the *dielectric relaxation time* ϵ/σ for intrinsic silicon; for silicon doped with 10^{16} donors cm^{-3}; and for thermal SiO_2 with $\sigma = 10^{-16}$ $(\Omega\text{-cm})^{-1}$.[12]

1.13[†] Because of their thermal energies, free carriers are continually moving throughout a crystal lattice. While the net flow of all carriers across any plane is zero at thermal equilibrium, it is useful to consider the directed components that balance to a null. The component values are physically significant in that they measure the quantity of current that can be delivered by diffusion alone. This would be relevant if, for example, one were able to unbalance the thermal equilibrium condition by intercepting all carriers flowing in a given direction. By considering that $J_x = -qn_0v_x$, show that the current in a solid in any random direction resulting from thermal processes is

$$J = \frac{-qn_0v_{th}}{4}$$

   where v_{th} is the mean thermal velocity and n_0 is the free-electron density.
   (*Hint.* Consider the flux through a solid angle of 2π steradians.)[12]

1.14* Calculate the wavelengths of radiation needed to create hole-electron pairs in intrinsic germanium, silicon, gallium arsenide, and SiO_2. Identify the spectrum range (e.g., infrared, visible, UV, and X ray) for each case.

1.15[†] The relation between D and μ is given by

$$\frac{D}{\mu} = \frac{1}{q}\frac{dE_f}{d(\ln n)}$$

   for a material that may be *degenerate*. (That is, the Fermi-Dirac distribution function must be used since the Fermi level may enter an allowed energy band.) Show that this relation reduces to the simpler Einstein relation $D/\mu = kT/q$ if the material is *non-degenerate* so that Boltzmann statistics can be used.

1.16 A hot probe setup is a useful laboratory apparatus. It is used to determine the conductivity type of a semiconductor sample. It consists of two probes and an ammeter that indicates the direction of current flow. One of the probes is heated (most simply, a soldering iron tip is used), and the other is at room temperature. No voltage is applied, but a current flows when the probes touch the semiconductor. Considering the role of diffusion currents, explain the operation of the apparatus and draw diagrams to indicate the current directions for p- and n-type semiconductor samples, respectively.

1.17 Consider a simple model for diffusion in one space dimension. Assume that all particles must move only at discrete, regularly spaced times, one time unit apart. When they move, each particle may only jump one space unit either to the left or to the right, with *equal* probability. Start with 1024 particles at $x = 0$, $t = 0$, and build up, step by step, the pattern of particles in space after 10 time units. Plot the distribution in space after each jump and measure the corresponding width W between points of one-half maximum concentration. Then plot W^2 versus time. What does your result suggest about the rate of "spread" of particles by the diffusion process (in *one* space dimension at least)?[12]

1.18 The values in Tables 1.3 and 1.4 are not entirely consistent. To see that this is the case,
   (a) Calculate E_g (T = 300 K) using the equation in Table 1.4 and compare the result with E_g in Table 1.3.
   (b) Use both values of E_g to calculate n_i from Equation 1.1.25 with N_c and N_v as given in Table 1.3.
   (c) Calculate n_i at 300 K using the temperature-variation formula in Table 1.4.
   [The calculated inconsistencies reflect the fact that values in the tables are determined from measurements using a variety of experimental methods. Based on many experiments, a value of n_i equal to 1.45×10^{10} cm^{-3} for silicon at 300 K is widely accepted. The degree of precision of many other parameters is unknown.]

1.19[†] In a nearly intrinsic material, the motions of both electrons and holes are significant for the Hall effect. Show that the Hall coefficient in the case that both holes and electrons are present becomes

$$R_H = \frac{\mathscr{E}_H}{J_x B} = \frac{\mu_p^2 p - \mu_n^2 n}{q(\mu_p p + \mu_n n)^2}$$

Prove that this result is consistent with the simpler theory derived in Section 1.3.

Table 1.3 **Properties of Semiconductors and Insulators (at 300 K Unless Otherwise Noted)**

Property	Symbol	Units	Si	Ge	GaAs	GaP	SiO$_2$	Si$_3$N$_4$
Crystal structure			Diamond	Diamond	Zincblende	Zincblende	Amorphous for most IC applications	
Atoms per unit cell			8	8	8	8		
Atomic number	Z		14	32	31/33	31/15	14/8	14/7
Atomic or molecular weight	MW	g/g-mole	28.09	72.59	144.64	100.70	60.08	140.28
Lattice constant	a_0	nm	.54307	.56575	.56532	.54505		.775
Atomic or molecular density	N_0	cm^{-3}	5.00×10^{22}	4.42×10^{22}	2.21×10^{22}	2.47×10^{22}	2.20×10^{22}	1.48×10^{22}
Density		g cm^{-3}	2.328	5.323	5.316	4.13	2.19	3.44
Energy gap 300K	E_g	eV	1.124	0.67	1.42	2.24	~8 to 9	4.7
0K	E_g	eV	1.170	0.744	1.52	2.40		
Temperature dependence	$\Delta E_g / \Delta T$	eV K^{-1}	-2.7×10^{-4}	-3.7×10^{-4}	-5.0×10^{-4}	-5.4×10^{-4}		
Relative permittivity	ϵ_r		11.7	16.0	13.1	10.2	3.9	7.5
Index of refraction	n		3.44	3.97	3.3	3.3	1.46	2.0
Melting point	T_m	°C	1412	937	1237	1467	~1700	~1900
Vapor pressure		Torr (mm Hg) (at °C)	10^{-7} (1050) 10^{-5} (1250)	10^{-9} (750) 10^{-7} (880)	1 (1050) 100 (1220)	10^{-6} (770) 10^{-4} (920)		
Specific heat	C_p	J (g K)$^{-1}$	0.70	0.32	0.35		1.4	0.17
Thermal conductivity	κ	W(cm K)$^{-1}$	1.412	0.606	0.455	0.97	0.014	0.185(?)
Thermal diffusivity	D_{th}	cm^2 s^{-1}	0.87	0.36	0.44		0.004	0.32(?)

Property	Symbol	Units						
Coefficient of linear thermal expansion	α'	K^{-1}	2.5×10^{-6}	5.7×10^{-6}	5.9×10^{-6}	5.3×10^{-6}	5×10^{-7}	2.8×10^{-6}
Intrinsic carrier concentration	n_i	cm^{-3}	1.45×10^{10}	2.4×10^{13}	9.0×10^{6}			
Lattice mobility Electron	μ_n	$cm^2\,(V\,s)^{-1}$	1417	3900	8800	300	20	
Hole	μ_p	$cm^2\,(V\,s)^{-1}$	471	1900	400	100	$\sim 10^{-8}$	
Effective density of states Conduction band	N_c	cm^{-3}	2.8×10^{19}	1.04×10^{19}	4.7×10^{17}			
Valence band	N_v	cm^{-3}	1.04×10^{19}	6.0×10^{18}	7.0×10^{18}			
Electric field at breakdown	$\mathscr{E}_1$	$V\,cm^{-1}$	3×10^{5}	8×10^{4}	3.5×10^{5}		$6-9 \times 10^{6}$	
Effective mass Electron	m_n^*/m_0		1.08^a 0.26^b	0.55^a 0.12^b	0.068	0.5		
Hole	m_p^*/m_0		0.81^a 0.386^b	0.3	0.5	0.5		
Electron affinity	qX	eV	4.05	4.00	4.07	~ 4.3	1.0	
Average energy loss per phonon scattering		eV	0.063	0.037	0.035			
Optical phonon mean-free path Electron	l_{ph}	nm	6.2	6.5	3.5			
Hole	l_{ph}	nm	4.5	6.5	3.5			

Sources: A. S. Grove, Physics and Technology of Semiconductor Devices, Wiley, New York (1967); S. M. Sze, Physics of Semiconductor Devices, 2nd ed., Wiley, New York (1981); D. E. Hill, Some Properties of Semiconductors (table), Monsanto Co., St. Peters, Mo., (1971); H. Wolf, Semiconductors, Wiley, New York (1971)

[a] Used in density-of-state calculations.
[b] Used in conductivity calculations.[9]

Table 1.4 Additional Properties of Silicon (Lightly Doped, at 300 K Unless Otherwise Noted)

Property	Symbol	Units	Values
Tetrahedral radius		nm	.117
Pressure coefficient of energy gap	$\Delta E_g/\Delta p$	eV (atm)$^{-1}$	-1.5×10^{-6}
Intrinsic carrier density[a]	n_i	(T in K)	$3.87 \times 10^{16}\, T^{3/2} \exp\left[\dfrac{-7014}{T}\right]$
Bandgap	E_g	eV	$1.16 - \dfrac{7.02 \times 10^{-4}\, T^{3/2}}{T + 1108}$
Temperature coefficient of lattice mobility			
Electron	$\Delta\mu_n/\Delta T$	cm^2(V s K)$^{-1}$	-11.6
Hole	$\Delta\mu_p/\Delta T$	cm^2(V s K)$^{-1}$	-4.3
Diffusion constant			
Electron	D_n	cm^2 s^{-1}	34.6
Hole	D_p	cm^2 s^{-1}	12.3
Hardness	H	Mohs	7.0
Elastic constants	c_{11}	dyne cm^{-2}	1.656×10^{12}
	c_{12}		0.639×10^{12}
	c_{44}		0.796×10^{12}
Young's modulus ($\langle 111 \rangle$ direction)	Y	dyne cm^{-2}	1.9×10^{12}
Surface tension (at 1412°C)	σ_0	dyne cm^{-2}	720
Latent heat of fusion	H_F	eV	0.41
Expansion on freezing		%	9.0
Cut-off frequency of lattice vibrations	ν_0	Hz	1.39×10^{13}

Main Source: H. F. Wolf, *Semiconductors*, Wiley, New York (1971), p. 45.

Energy Levels of Elemental Impurities in Si[b]

Top labels: Li Sb P As Bi   Ni   S Mn   Ag Pt Hg

Li 0.033; Sb 0.039; P 0.044; As 0.049; Bi 0.069; S 0.18

0.35 A; S 0.37; Ag 0.33; Pt 0.37; Hg 0.33

Gap center (Si): 0.54; 0.55; 0.52 A; D 0.55; Mn 0.53; (dashed line — gap center)

0.39; 0.31; 0.37; 0.35 D; D 0.40; Ag 0.34; Hg 0.36

0.26; 0.24; 0.22

0.16

Sb 0.057; 0.065; B 0.045; Fe 0.03

Bottom labels: B Al Ga In Tl Co Zn Cu Au   Fe   O

[a] a plot of n_i versus T at higher temperatures is given in Figure 2.10.

[b] The levels below the gap center are measured from the top of the valence band and are acceptor levels unless indicated by D for donor level. The levels above the gap center are measured from the bottom of the conduction band and are donor levels unless indicated by A for acceptor level.[10]

2

SILICON TECHNOLOGY

Successful engineering rests on two foundations. One is a mastery of underlying physical concepts; a second foundation, at least of equal importance, is a perfected technology—a means to translate engineering concepts into useful structures. In Chapter 1 we reviewed the physical principles needed for integrated-circuit electronics. In this chapter we discuss the technology to produce devices in silicon, a technology now so powerful that the entire realm of engineering has felt its impact. In addition to describing *IC* technology as it is now practiced, we attempt to provide a perspective on continuing developments in this fast-moving area.

Technological evolution toward integrated circuits began in earnest with development of an understanding of diode action and the invention of the transistor in the late 1940s. At that time the semiconductor of greatest interest was germanium. Experiments with germanium produced important knowledge about the growth of large single crystals having a chemical purity and crystalline perfection that was previously unachievable.

Germanium is an element that crystallizes in a diamond-like lattice structure in which each atom forms covalent bonds with its four nearest neighbors. The crystal structure is shown in Figure 1.8. Germanium has a band gap of 0.67 eV and an intrinsic carrier density equal to 2.5×10^{13} cm^{-3} at 300 K. Because of the relatively small band gap in germanium, its intrinsic-carrier density increases rapidly with temperature, growing roughly to 10^{15} cm^{-3} at 400 K (*cf.* Figure 1.9).

Because most devices are no longer useful when the intrinsic-carrier concentration of a semiconductor becomes comparable to the dopant density, germanium devices are limited to operating temperatures below about 70°C (343 K). As early as 1950, the temperature limitations of germanium devices had spurred research efforts on several other semiconductors that crystallize with similar lattice structures but are suitable for higher temperatures. For only two semiconductors, the element silicon ($E_g = 1.12$ eV) and the compound gallium arsenide ($E_g = 1.42$ eV), has sufficient technological capability evolved for application to integrated circuits. The use of gallium arsenide in integrated circuits is, however, still limited; silicon is the semiconductor of overwhelming interest.

Properties of germanium, silicon, gallium arsenide, and several other useful electronic materials are summarized in Table 1.3 (page 54). Table 1.4 (page 56) contains some additional properties of silicon.

2.1 The Silicon Planar Process

In addition to the good semiconducting properties of silicon, the major reason for its widespread use is the ability to form on it a stable, controllable oxide film (silicon dioxide SiO_2) that has excellent insulating properties. This capability, which is not matched by any other semiconductor-insulator combination, makes it possible to introduce controlled amounts of dopant impurities into small, selected areas of a silicon sample. The ability to dope small regions of the silicon is the key to producing dense arrays of devices in integrated circuits.

Two chemical properties of the Si–SiO_2 system are of basic importance to silicon technology. First, *selective etching* is possible using etchants that attack only one of the two materials. For example, hydrofluoric acid dissolves silicon dioxide but not silicon. Second, silicon dioxide can be used to shield an underlying silicon crystal from dopant impurity atoms conveyed to the surface either by high-energy ion beams or from a gaseous diffusion source at elevated temperatures.

These features make possible the introduction of dopant atoms into areas on the silicon that have not been shielded by patterns of silicon dioxide. The shielded regions can be accurately defined by using photosensitive polymer films exposed with photographic masks. The polymer pattern protects selected oxide regions on the silicon surface when it is immersed in a hydrofluoric-acid bath, allowing formation of a surface consisting of bare silicon *windows* in a silicon-dioxide layer. This selective etching process, originally developed for lithographic printing applications, permits delineation of very small patterns. When the silicon sample is placed in an ambient that deposits dopant atoms on the surface, these atoms can enter the silicon only at the exposed silicon windows.

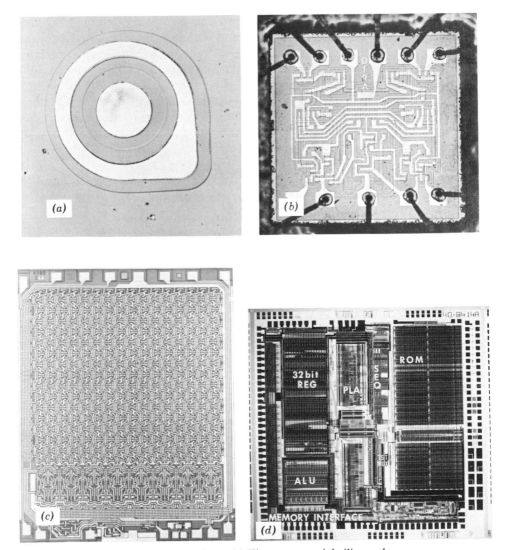

Figure 2.1 Evolution of *IC* technology. (*a*) First commercial silicon planar transistor (1959) (outer diameter 0.87 mm). (*b*) Diode-transistor logic (*DTL*) circuit (1964) (chip size 1.9 mm square). (*c*) 256-bit bipolar random access memory (*RAM*) circuit (1970) (chip size 2.8 × 3.6 mm). (*d*) VLSI central-processor computer chip containing 450,000 transistors (1981). The numerous functions carried out by the *IC* are labeled on the figure (chip size 6.3 mm square). [(*a*), (*b*), (*c*) *courtesy of B.E. Deal—Fairchild Semiconductor*, (*d*) *Hewlett-Packard Co.*]

Proper sequencing and repetition of the oxidation, patterning, and dopant-addition operations just described can be used to introduce *p*- and *n*-type dopant atoms selectively into regions on a surface having dimensions ranging down to the submicrometer range. These steps are the basic elements of the *silicon planar process*, so called because it is a process that produces device structures through a sequence of steps carried out near the surface plane of the silicon crystal. Refinement of the planar process and, along with it, the growth of silicon-based electronics has been amazingly rapid. Figure 2.1 is intended to give some perspective to its development over the past 25 years.

The silicon crystals used for the planar process consist of slices (called *wafers*) that are made from a large single crystal. Dopant atoms are typically added to the silicon by *depositing* them in selected regions on or near the wafer surface, and then *diffusing* them into the silicon. Since dopant atoms are introduced from the surface and typical diffusion dimensions are very small, the active regions of the planar-processed devices are themselves within a few micrometers* of the wafer surface. The remaining thickness (generally a few hundred micrometers) serves simply as a mechanical support for the important surface region.

A major advantage of the planar process is that each fabrication step (prior to packaging) is typically applied to the entire wafer. Hence, it is possible to make and interconnect many devices with high precision to build an *integrated circuit* (*IC*). At present, individual *IC*s typically measure about 5 mm on a side, so that a wafer (most often 10 cm in diameter or larger) can contain many *IC*s. There is clear economic advantage to increase the area of individual wafers and, at the same time, to reduce the area of each integrated circuit.

The most important steps in the planar process, shown in Figure 2.2, include (*a*) formation of a masking oxide layer, (*b*) its selective removal, (*c*) deposition of dopant atoms on or near the wafer surface, and (*d*) their diffusion into the exposed silicon regions. These processes determine the location of the dopants which, in turn, determines the electrical characteristics of the devices and *IC*s.

From the discussion to this point, we see that the planar process is the foundation for the production of silicon integrated circuits. For this reason, the past 20 years has seen a huge research and development investment in this technology, which has led to its present advanced state. With suitable modifications, silicon planar technology promises to dominate electronics through the remainder of this century. Making the best use of its many degrees of freedom, requires a fairly thorough understanding of the basic elements of silicon technology. The rest of this chapter is directed toward providing such an understanding.

2.2 Crystal Growth

Silicon used for the production of integrated circuits consists of large, high quality, single crystals. What is meant by "high quality" can be better understood from a brief consideration of the ultimate requirements placed on the silicon.

---

* 1 micrometer = 1 μm, sometimes called a *micron*, equals 10^{-4} cm = 10^3 nm = 10^4 Å (angstrom).

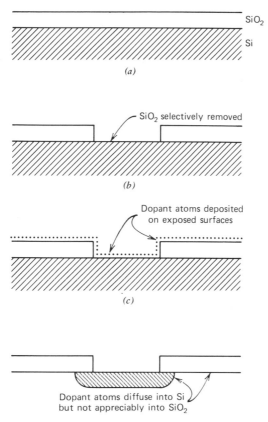

Figure 2.2 Basic fabrication steps in the silicon planar process: (a) oxide formation, (b) selective oxide removal, (c) deposition of dopant atoms on wafer, (d) diffusion of dopant atoms into exposed regions of silicon.

Typical *IC* applications require dopant concentrations ranging from roughly 10^{15} to 10^{20} atoms cm^{-3}. To control device properties, any unintentional or background concentrations of electrically active impurity atoms should be at least two orders of magnitude lower than the minimum intentional dopant concentration— that is, about 10^{13} cm^{-3} or lower. Since silicon contains roughly 5×10^{22} atoms cm^{-3}, the presence of only about one unintentional, electrically active, impurity atom per billion silicon atoms can be tolerated. This high purity is well beyond that required for the raw material in virtually any other industry; the routine production of material of this quality for the manufacture of integrated circuits is testimony to the extraordinary refinement of silicon technology.

High-purity silicon is obtained from two common materials: silicon dioxide (found in common sand) and elemental carbon. In a high-temperature ($\sim 2000°C$)

electric-arc furnace, the carbon reduces the silicon dioxide to elemental silicon which condenses as about 90% pure, metallurgical-grade silicon—still not pure enough for use in semiconductor devices. The metallurgical-grade silicon is purified by converting it into trichlorosilane ($SiHCl_3$), which can be purified. Selective distillation (fractionation) separates the trichlorosilane from any other chloride complexes. The purified trichlorosilane is then reduced by hydrogen to form high-purity, semiconductor-grade, solid silicon. At this point, the silicon is *polycrystalline*, composed of many small (micrometer-sized) crystals with generally random orientations. This elemental silicon, also called *polysilicon*, is usually deposited on a high-purity rod of semiconductor-grade silicon to avoid contamination.

The silicon must now be formed into a large (about 10 cm diameter), nearly perfect, single crystal since grain boundaries and other crystalline defects degrade device performance. Sophisticated techniques are needed to obtain single crystals of such high quality. These crystals can be formed by either the *Czochralski (CZ)* technique or the *float-zone (FZ)* method.

Czochralski Silicon. In the Czochralski technique, which is most widely used to form the starting material for integrated circuits, pieces of the polysilicon rod are first melted in a fused-silica crucible in an inert atmosphere (typically argon) and held at a temperature just above 1412°C, the melting point of silicon (Figure 2.3.).

A high-quality seed crystal with the desired crystalline orientation is then lowered into the melt while being rotated, as indicated in Figure 2.3a. The crucible is simultaneously turned in the opposite direction to induce mixing in the melt and

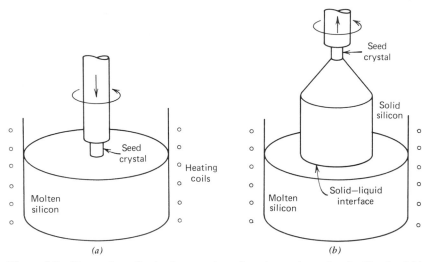

Figure 2.3 Formation of a single-crystal semiconductor ingot by the Czochralski process: (*a*) initiation of the crystal by a seed held at the melt surface, (*b*) withdrawal of the seed "pulls" a single crystal.

to minimize temperature nonuniformities. A portion of the seed crystal is dissolved in the molten silicon to remove the strained outer portions and to expose fresh crystal surfaces. The seed is then slowly raised (or *pulled*) from the melt. As it is raised, it cools, and material from the melt adheres to it, thereby forming a larger crystal, as shown in Figure 2.3*b*. Under the carefully controlled conditions maintained during growth, the new silicon atoms continue the crystal structure of the already solidified material. The desired crystal diameter is obtained by controlling the pull rate and temperature with automatic feedback mechanisms. In this manner cylindrical, single-crystal *ingots* of silicon can be fabricated. As crystal-growing

Figure 2.3 (*c*) banks of crystal pullers in a semiconductor factory (*Courtesy Motorola, Inc.*).

technology has developed, the diameter of the cylindrical crystals has increased progressively from a few mm to the 10 or 15 cm common today.

For many integrated-circuit processes, an initial dopant density of about 10^{15} cm^{-3} is desired in the silicon. This dopant concentration is obtained by incorporating a small, carefully controlled quantity of the desired dopant element, such as boron or phosphorus, into the melt. Typically, dopant impurities weighing about one-tenth milligram must be added to each kilogram of silicon. For accurate control, small quantities of heavily doped silicon are usually added to the undoped melt. The dopant concentration in the pulled crystal of silicon is always less than that in the melt because dopant is rejected from the crystal into the melt as the silicon solidifies. This *segregation* causes the dopant concentration in the melt to increase as the crystal grows. The *seed* end of the crystal is, therefore, less heavily doped than is the *tail* end. Slight dopant-concentration gradients also exist along the crystal radius in Czochralski silicon.

Czochralski silicon contains a substantial quantity of oxygen, resulting from the slow dissolution of the fused-silica (silicon dioxide) crucible which holds the molten silicon. Oxygen does not contribute significantly to the net dopant density in the moderately doped wafers used for silicon integrated circuits. However, the typical inclusion of about 10^{18} cm^{-3} oxygen atoms (the solid solubility of oxygen in silicon at the solidification temperature) in silicon crystals produced by the Czochralski process can be used to control the movement of unintentional impurities (typically metals) in *IC* wafers. At the temperatures used in integrated-circuit processing, oxygen can precipitate, forming sites on which other impurities tend to accumulate. If the oxygen precipitates are located within an active device, they can degrade its performance. On the other hand, if they are remote from active devices, precipitates can function as *gettering* sites for unwanted impurities, attracting them away from electrically active regions and thereby improving device properties. Control of the location and precipitation of the oxygen is important in determining the uniformity of device properties in high-density integrated circuits.

A small fraction of the included oxygen, about 0.01% of the total, may act as donors after certain heat treatments are carried out. This small concentration (about 10^{14} cm^{-3}) does not appreciably alter the resistivity of most integrated-circuit silicon. However, for the high-resistivity (20–100 Ω-cm) silicon used to produce power devices and for other specialized applications, it can become a problem. For this reason, high-resistivity silicon is usually formed by the *float-zone process*.

In this process, a rod of cast polycrystalline silicon is held in a vertical position and rotated while a melted zone (between 1 and 2 cm long) is slowly passed from the bottom of the rod to the top, as shown in Figure 2.4. The melted region is heated [usually by a radio-frequency (*RF*) induction heater] and moved through the rod starting from a *seed crystal* that initiates crystallization. The impurities tend to segregate in the molten portion and volatilize so that the silicon rod is purified. In contrast to Czochralski-processed silicon, oxygen is not introduced into float-zone silicon because the original material is not held in an oxygen-containing crucible. Although it is a more costly process, the silicon produced by

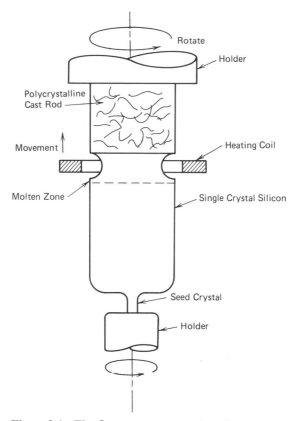

Figure 2.4 The float-zone process. A molten zone
passes through a cast polycrystalline-silicon rod,
and a single crystal grows from a seed at the
bottom end.

the float-zone process typically contains about 1% as much oxygen as that produced
by the Czochralski method, and other impurity concentrations are also reduced.
Multiple passes of the molten zone can be used to produce silicon with resistivities
above 10^4 Ω-cm.

Wafer Production. Once the single-crystal ingot is grown, it is sliced with a
diamond saw into thin, circular *wafers*. The wafers are chemically etched to remove
sawing damage and then polished with successively finer polishing grits and chemical etchants until a defect-free, mirror-like surface is obtained. The wafers are now
ready for device fabrication.

Before slicing, index marks are placed on the wafer to facilitate the orientation
of processed circuits along specific crystal directions. In particular, edges of an
area on the wafer called a *die* or a *chip* are typically aligned in directions along
which the wafer readily breaks so that the *dice* can be separated from one another

Table 2.1   **Location of Secondary Wafer Flat**

Crystal Orientation	Conductivity Type	Secondary Flat (Relative to Primary Flat)
(100)	n	180°
(100)	p	90°
(111)	n	45°
(111)	p	No secondary flat

after planar processing is completed. This separation is often accomplished by *scribing* between them with a stylus and *breaking* them apart, so the orientation of the easy cleavage planes is important. The crystal directions are indicated by grinding a *flat* on the wafer, usually perpendicular to an easy cleavage direction. In most cases, the *primary flat* is formed along a $\langle 110 \rangle$ direction. A smaller *secondary flat* is added to identify the orientation and conductivity type of the wafer, according to the specification shown in Table 2.1.

2.3 Thermal Oxidation

An oxide layer about 2 nm thick is quickly formed on the surface of a bare silicon wafer even in room-temperature air. The thicker (typically 20 nm to 1 μm) silicon dioxide layers used to shield the silicon surface during dopant incorporation can be formed either by *thermal oxidation* or by *deposition*. When silicon dioxide is formed by deposition, both silicon and oxygen are conveyed to the wafer surface and reacted there. In thermal oxidation, however, there is a direct reaction between atoms near the surface of the wafer and oxygen supplied in a high-temperature furnace. Thermally grown oxides are generally of a higher quality than deposited oxides. Although their structure is amorphous, they typically have an exact stoichiometric ratio (SiO_2), and they are strongly bonded to the silicon surface. The interface between silicon and thermally grown SiO_2 has stable and controllable electrical properties. As we shall see in Chapter 8, the quality of this excellent semiconductor-insulator interface is fundamental to the successful production of metal-oxide-semiconductor (MOS) transistors.

To form a thermal oxide, the wafer is placed inside a quartz tube that is set within the cylindrical opening of a resistance-heated furnace (Figure 2.5). Temperatures in the range of 850 to 1100°C are typical, the reaction proceeding more rapidly at higher temperatures. Silicon itself does not melt until the temperature reaches 1412°C, but oxidation temperatures are kept considerably lower to inhibit the generation of crystalline defects and the movement of previously introduced dopant atoms in the crystal. In addition, the quartz furnace tube and other fixtures start to soften and degrade above 1150°C.

The oxidizing ambient can be dry oxygen, or it can contain water vapor, which is generally produced by reaction of oxygen and hydrogen in the high-temperature furnace. Use of such *pyrogenic steam* requires careful safety procedures to handle

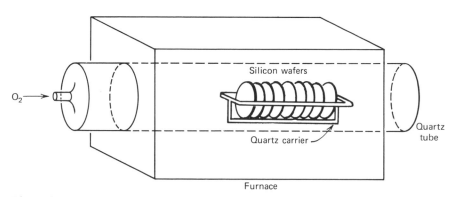

Figure 2.5 An insulating layer of silicon dioxide is grown on silicon wafers by exposing them to oxidizing gases in a high-temperature furnace.

explosive hydrogen gas, and a slight excess of oxygen is generally employed to avoid the presence of unreacted hydrogen in the furnace. A steam environment can also be formed by passing high-purity, dry oxygen or nitrogen through water heated almost to its boiling point. The overall oxidation reactions are

$$Si(s) + O_2(g) \rightarrow SiO_2(s) \qquad (2.3.1a)$$

and

$$Si(s) + 2H_2O(g) \rightarrow SiO_2(s) + 2H_2(g) \qquad (2.3.1b)$$

Oxidation proceeds much more rapidly in the steam ambient, which is consequently used for the formation of thicker protective layers of silicon dioxide. Growth of a thick oxide in the slower, dry-oxygen environment can lead to undesirable movement (*redistribution*) of impurities introduced into the wafer during previous processing.

Oxidation takes place at the Si–SiO$_2$ interface so that oxidizing species must diffuse through any previously formed oxide and then react with silicon at this interface (Figure 2.6). At lower temperatures and for thinner oxides, the surface reaction rate at the Si–SiO$_2$ interface limits the growth rate, and the thickness of the oxide layer is a linear function of the oxidation time.

At higher temperatures and for thicker oxides, the oxidation process is limited by diffusion of the oxidizing species through the previously formed oxide. In this case the grown oxide thickness is roughly proportional to the square root of the oxidation time. This square-root dependence is characteristic of diffusion processes and sets a practical upper limit on the thickness that can be conveniently obtained.

Oxidation Kinetics

For most of the oxidation process (after formation of a thin layer which obeys different oxidation kinetics), the relation between the grown oxide thickness x_{ox}, oxidation time t, and temperature T can be found by equating the rates at which

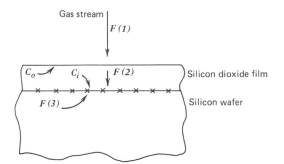

Figure 2.6   Three fluxes that characterize the
oxidation rate: $F(1)$ the flow from the gas stream
to the surface, $F(2)$ the diffusion of oxidizing
species through the already formed oxide, and
$F(3)$ the reaction at the Si–SiO$_2$ interface. The
concentration of the oxidizing species varies in
the film from C_o near the gas interface to C_i
near the silicon interface.

oxygen atoms (1) transfer from the gas phase to the growing oxide, (2) move through
the already formed oxide, and (3) react according to Equation 2.3.1 at the Si–SiO$_2$
interface (Figure 2.6).

First, we consider transfer of the oxidizing species (either oxygen or water vapor)
from the gas phase to the outer layer of the already formed oxide. This transfer
rate is proportional to the difference between the actual concentration C_o of the
oxidizing species in the solid at its surface and C^*, which is the concentration that
would be in equilibrium with the oxidant in the gas phase:

$$F(1) = h(C^* - C_o) \qquad (2.3.2)$$

where h is the gas-phase, mass-transfer coefficient, and the concentrations C^* and
C_o are related to corresponding gas-phase partial pressures p by the ideal gas law
$C = p/kT$.

Transport of the oxidizing species across the growing oxide to the Si–SiO$_2$
interface occurs by diffusion, a process analogous to the hole and electron diffusion
discussed in Section 1.2 (Equation 1.2.15). The flux of the diffusing species can be
written as the product of the concentration gradient across the oxide $(C_o - C_i)/x_{ox}$
(Figure 2.6) and the diffusivity D. (C_i is the concentration of the oxidizing species
in the oxide near the Si–SiO$_2$ interface.) The diffusing flux is therefore

$$F(2) = D \frac{(C_o - C_i)}{x_{ox}} \qquad (2.3.3)$$

Reaction of the oxidizing species at the Si–SiO$_2$ interface is characterized by a
rate constant k_s so that the reaction rate $F(3)$ of the oxidant is

$$F(3) = k_s C_i \qquad (2.3.4)$$

In steady state $F(1) = F(2) = F(3) = F$. The oxidation rate can be found from the flux and expressed in terms of N_{ox}, the density of oxidant molecules per unit volume of oxide. Eliminating C_o and C_i from Equations 2.3.2–2.3.4, we find the oxide growth rate R to be

$$R = \frac{dx_{ox}}{dt} = F/N_{ox} = \frac{k_s C^*/N_{ox}}{(1 + k_s/h + k_s x_{ox}/D)} \tag{2.3.5}$$

Equation 2.3.5 can be solved to find the oxide thickness grown in a time t (Problem 2.6).[8]

$$x_{ox} = \frac{A}{2}\left[\sqrt{1 + \frac{(t + \tau)}{A^2/4B}} - 1\right] \tag{2.3.6}$$

where

$$A = 2D\left[\frac{1}{k_s} + \frac{1}{h}\right] \tag{2.3.7}$$

and

$$B = \frac{2DC^*}{N_{ox}} \tag{2.3.8}$$

The parameter τ depends mainly on the thickness of the oxide that is initially present. For short oxidation times the surface reaction rate $F(3)$ limits oxide growth, and Equation 2.3.6 can be approximated by a linear relationship between x_{ox} and t

$$x_{ox} = \frac{B}{A}(t + \tau) \tag{2.3.9}$$

The proportionality factor B/A in Equation 2.3.9, called the *linear rate coefficient*, is related to breaking bonds at the Si–SiO$_2$ interface [$F(3)$], and therefore depends on crystal orientation. The most commonly used crystals for ICs are (100)- or (111)-oriented. The linear rate coefficient is larger for the (111) orientation which has fewer bonds between adjacent planes than does (100)-oriented silicon.

For long oxidation times a square-root relationship is obtained from Equation 2.3.6

$$x_{ox} = \sqrt{B(t + \tau)} \approx \sqrt{Bt} \tag{2.3.10}$$

The coefficient B in Equation 2.3.10, called the *parabolic rate coefficient*, depends on diffusion across the already formed oxide [flux $F(2)$], and is independent of the orientation of the silicon crystal. Values for the linear rate coefficient B/A and the parabolic rate coefficient B as obtained from experiments are shown in Figures 2.7a and 2.7b.

In practice, values of A, B, and τ are determined experimentally from measurements of oxide thickness versus time at various temperatures. Examples of these

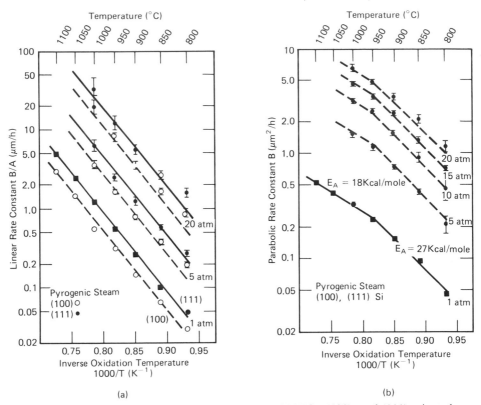

Figure 2.7 (*a*) Linear rate coefficient B/A versus $1000/T$ for (100)- and (111)-oriented silicon oxidized in pyrogenic steam at 1, 5, and 20 atm. (*b*) Parabolic rate coefficient B versus $1000/T$ for (100)- and (111)-oriented silicon oxidized in pyrogenic steam at 1, 5, 10, 15, and 20 atm.[7]

measurements are shown in Figures 2.8*a* (for dry oxidation) and *b* (for steam oxidation). The data in both figures apply to oxides grown on (111)-oriented Si. The oxide thicknesses would be somewhat less for (100)-oriented material, especially for thin oxides and lower oxidation temperatures at which growth is limited for a longer time by the surface oxidation rate [$F(3)$]. As an example of the use of Figures 2.8*a* and *b*, we can compare the time it takes to grow 300 nm (0.3 μm) of oxide in dry oxygen at 1100°C (4.5 h from Figure 2.8*a*) to the growth time in steam (0.18 h $\approx$ 11 min from Figure 2.8*b*). Oxide thicknesses of a few hundred nanometers are often used, with 1 to 2 μm being the upper practical limit when conventional oxidation techniques are employed.

In *IC* fabrication we must frequently determine the thickness added to an already formed oxide layer during additional oxidation. To do this, we can use the parameter τ in Equation 2.3.6 as the time needed to grow an oxide of the thickness already present *under the conditions of the additional oxidation*. The final thickness is then determined by adding the oxidation time t and calculating the oxide thickness corresponding to the total time $(t + \tau)$.

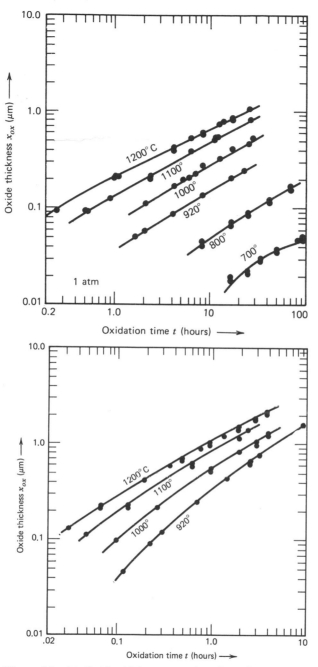

Figure 2.8 (*a*) Oxide thickness as a function of reaction time in dry oxygen for several commonly used oxidation temperatures for (111)-oriented silicon. (*b*) Oxide thickness x_{ox} as a function of time in a steam ambient for (111)-oriented silicon (typically obtained by reacting hydrogen and oxygen near the entrance of the oxidation furnace). *Adapted from reference 8.*

EXAMPLE   Calculation of Oxide Thickness

Find the final oxide thickness after an additional oxidation in dry oxygen for 2 hours at 1000°C of a region that is covered initially by 100 nm of SiO_2.

Solution

Figure 2.8*a*

$$x_i = 100 \text{ nm} \qquad \tau = 1.9 \text{ h}$$
$$(t + \tau) = 3.9 \text{ h} \qquad x_{ox} = 154 \text{ nm}$$

Note that the oxide thickness increases less than linearly with time. The initial 100 nm of SiO_2 would have grown in 1.9 h at 1000°C, whereas another 2 h of oxidation only forms an additional 54 nm of oxide.

During thermal oxidation, silicon is consumed from the wafer and incorporated in the growing oxide. Since SiO_2 contains 2.2×10^{22} molecules cm^{-3} (and an equal number of silicon atoms), while pure silicon contains 5.0×10^{22} atoms cm^{-3}, the thickness of silicon consumed is 0.44 times the thickness of SiO_2 formed. This relation holds for all orientations of silicon and also for polycrystalline silicon because it depends only on volume densities.

High-Pressure Oxidation. Growth of thick oxides is time consuming because oxide thickness increases only as the square root of time (*cf* Equation 2.3.10). Excessive movement of dopant atoms during extended oxidation often makes more rapid oxide formation desirable. This rapid formation can be accomplished by using high pressures of the oxidizing species.

Although atmospheric-pressure oxidation processes are described by Equation 2.3.6 with well-known values of A and B, the same equation is also valid at other pressures (with different values of A and B). The oxidation rate for thick oxides (Equation 2.3.10) is determined by the parameter B. As seen from Equation 2.3.8, B can be increased if C^*, the concentration of oxidizing species in equilibrium with the gas phase is increased. This, in turn, is accomplished by supplying the oxidant at a high pressure. Typical elevated pressures are 10 to 20 atmospheres, causing a 10- to 20-fold increase in the parameter B. Although the diffusivity D of the oxidant in oxide may depend somewhat on pressure, the major effect on oxidation rate is through the increased concentration C^*.

High-pressure oxidation allows formation of a desired oxide thickness at the same temperature in a shorter time than at one atmosphere; alternatively, a desired thickness can be formed in the same time at a lower temperature. A lower oxidation temperature can help reduce the number of crystal defects introduced during thermal oxidation.

Concentration-Enhanced Oxidation.[†] Heavily doped *n*-type silicon oxidizes more rapidly than does lightly doped silicon (Figure 2.9). The differential oxidation rate of heavily and lightly doped regions of an *IC* can be used for selective definition

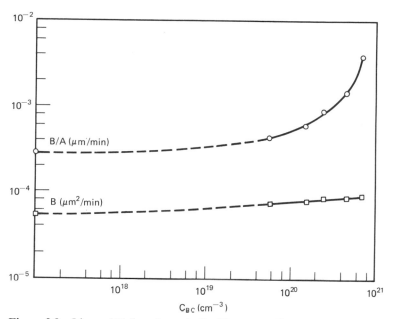

Figure 2.9 Linear (B/A) and parabolic (B) rate coefficients as functions of the initial phosphorus concentration in the substrate for oxidation at 900°C.[9]

of desired areas; conversely, it can make uniform etching of oxides grown on different parts of the circuit difficult.

High concentrations of some dopants cause isolated *point defects* in the silicon lattice consisting of voids (silicon *vacancies*) or *interstitial* silicon atoms. These point defects affect the surface reaction rate (and thereby the linear rate coefficient B/A) if their concentration becomes greater than the intrinsic carrier concentration n_i *at the oxidation temperature.* As shown in Figure 2.10, the value of n_i increases rapidly with temperature and is 10^{19} cm^{-3} at 1000°C.

The parabolic rate coefficient B depends on properties of the SiO$_2$, so it is not directly affected by point defects in the silicon. However, B does depend on the diffusivity of the oxidizing species in the oxide. Since some dopant from the silicon can enter the oxide during oxidation and weaken its structure, the diffusivity (and hence the parabolic rate coefficient B) can increase. This increase, however, is small compared to the increase in the linear rate coefficient B/A. Thus, for thicker oxides, oxidation-rate enhancement owing to high concentrations of dopant in the silicon causes only a small effect.

Chlorine Oxidation.[†] The oxidation rate is also increased (typically by 10–20%) when a chlorine-containing species is added to the oxidant. The chlorine, usually obtained from HCl, Cl$_2$, or organic compounds, is incorporated near the Si–SiO$_2$ interface where it can improve the electrical properties of devices, especially MOS devices to be discussed in Chapter 8. The addition of chlorine to the oxidant to

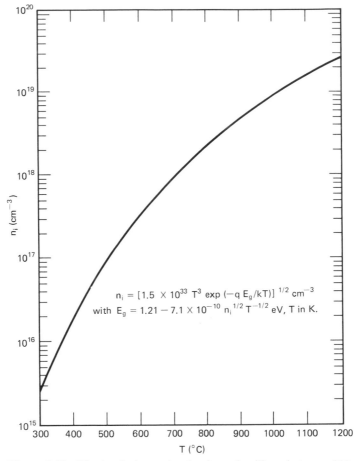

Figure 2.10 The intrinsic carrier density n_i in silicon between 300 and 1200°C.[10]

improve the properties of the Si–SiO$_2$ system must be done very carefully. Only a small improvement is possible when the concentration is too low, but it is easy to provide an overabundance that can form gas bubbles at the Si–SiO$_2$ interface and rupture the oxide.

2.4 Lithography and Pattern Transfer

Photolithography. Once the protective layer of SiO$_2$ has been formed on the silicon wafer, it must be selectively removed from those areas in which dopant atoms are to be introduced. Selective removal is usually accomplished by the use of a light-sensitive polymer material called a *resist*. The oxidized wafer is first coated with the liquid resist by placing a few drops on a rapidly spinning wafer. After

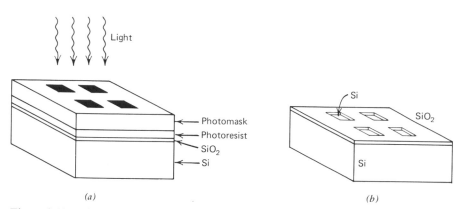

Figure 2.11 The areas from which the oxide is to be etched are defined by polymerizing a light-sensitive resist through a photographic negative or mask.

drying the resist, a partially transparent photographic negative (called a *mask* or *photomask*) is placed over the wafer, as shown in Figure 2.11*a*, and aligned using a microscope. The resist is then exposed to ultraviolet or near-ultraviolet light that changes its structure; the molecules of a *negative resist* become cross linked (polymerized) in areas that are exposed to light. For a *positive resist*, molecular bonds are broken where the resist is illuminated. The resist is not affected in regions that are shielded by the opaque portions of the mask. The unpolymerized areas of the resist are then selectively dissolved, using a solvent such as trichloroethylene, so that the acid-resistant, polymerized, protective coating replicates the mask pattern on the SiO_2 (Figure 2.11*b*). Similarly, resist patterns can be formed on top of other layers used in the *IC* process.

In the most straightforward implementation of planar processing, the patterns on all the dice on a wafer are exposed simultaneously. However, as device dimensions become smaller, not only does the minimum feature size that must be resolved decrease, but the registration of one layer of pattern to another must become more exact. During heat cycles, the wafer can become slightly distorted because of thermal stress or stress from incorporated dopant atoms or added layers of material. Thus, a subsequent photomask may not exactly match a pattern previously formed on the wafer. This distortion can limit the accuracy of registration from one masking level to another. One tool to improve the registration accuracy is the *optical wafer stepper*, which exposes one die on a wafer at a time. After one die is exposed, the wafer is moved or *stepped* to the next die which is then exposed. Although this technique is mechanically more complex and slower than full-wafer exposure, it allows improved registration. In addition, by demagnifying the pattern (about five or ten times) from a large mask or *reticle* containing the mask for a single die, smaller device features can be defined.

Advanced Lithography. If the minimum feature size approaches the wavelength of the light used in optical exposure systems (≈ 400 nm), diffraction can limit the attainable resolution. Alternatives to conventional photolithography are being developed to overcome this limitation. A simple approach is to use ultraviolet light of a shorter wavelength ("deep UV") that correspondingly reduces the diffraction limits for the light source. The advantages of this technique are limited, however, and methods using electron, x-ray, or ion beams to expose the resist are being developed. Of these advanced techniques, the greatest progress has been made with *electron-beam lithography.*

In electron-beam lithography a focused stream of electrons delivers energy to the resist and exposes it. Rather than exposing all features of a complex pattern simultaneously, as is done in optical lithography, the electron beam is deflected to expose the desired pattern sequentially. The information necessary to guide the electron beam is stored in a computer and no mask is needed. The most frequent use of electron-beam lithography thus far has been to fabricate fine geometry masks, which are then used with more conventional photolithographic exposure techniques. The electron beam can be finely focused to a size much smaller than the minimum feature size and moved across the die, or it can be formed into a rectangular shape and the pattern built up by repeated block-like exposures. In either case the sequential nature of electron-beam exposure limits the speed with which a wafer can be processed. The increased time needed to expose a wafer is, however, at least partially compensated by the higher circuit density which can be obtained.

A more recent development is the use of *x-ray lithography.* In this technique, a beam of x rays is passed through a mask to expose a resist layer. X-ray lithography, like photolithography, exposes many features simultaneously, but the short x-ray wavelengths allow the formation of finer features. X-ray lithography is less well developed than either photolithography or electron-beam lithography and is limited by the demanding requirements of x-ray sources and masks. Most common x-ray sources are point sources which cause the beam to diverge as it travels to the mask and wafer, thus limiting the possible mask-to-wafer spacing. At present, synchrotron sources, which produce partially collimated beams, are undergoing commercial development.

X-ray masks are also difficult to fabricate. Since the beam must readily penetrate the "transparent" sections of a mask, a thin membrane is generally used that makes handling difficult. A more fundamental limitation to the use of x rays is that the area exposed at one time may be limited by wafer distortion during processing. Any large-area exposure technique suffers from this limitation. As with optical exposure, single-die exposure may allow adequate registration of fine features between layers, even when a small amount of wafer distortion is present. Finally, possible damage to the active device regions in the silicon by high-energy x rays must be considered.

In any lithographic technique the minimum feature size can also be limited by the nonuniform surface of the partially fabricated *IC* chip. A line crossing a high step formed by previous processing will probably be poorly defined. Better line definition can be obtained with a resist structure consisting of two or three layers. The thick first layer fills in the recessed portions of the chip, producing a more uniform surface. Fine features can then be exposed in the thin resist layer on top. An example of the improvement that can be achieved with a *multilayer-resist technique* is shown in Figure 2.12*a*.

Pattern Transfer. After the pattern is formed in the resist, the unprotected regions of the SiO_2 or other material are removed to transfer the pattern to the chip. If SiO_2 is being defined, the exposed layer can be dissolved in hydrofluoric acid (HF) to expose the bare silicon surface, and the resist is then removed from the remaining areas. At this point, portions of the wafer are protected by SiO_2, and bare silicon is exposed at windows through which the dopant impurities are subsequently introduced (Figure 2.11). There are a large number of different chemical reagent solutions that etch materials selectively so that there is very little attack of underlying materials. This *selectivity* is an advantage of liquid-based or *wet etching*.

Dry Etching. As the dimensions to be defined on an *IC* die are reduced, several limitations of the wet-etching process become apparent. A major problem is that

Figure 2.12 Fine-line resist patterning. (*a*) Uniformly wide stripes of exposed resist formed by a tri-layer resist technique over a 1 μm oxide step on an *IC*.[11]

wet etching is usually isotropic, proceeding laterally under the masking layer as well as vertically toward the silicon surface. Thus, the etched features are generally larger than the dimensions on the mask. Dry (*plasma* or *reactive-ion*) etching techniques can overcome this difficulty.

In dry etching, the masked wafer is exposed to a plasma, which is an almost neutral mixture of energetic molecules, ions, and electrons that have been excited in a radio-frequency electric field. The excited species interact chemically with exposed regions of the material to be etched, while the ions in the plasma bombard the surface and knock away the exposed atoms.

A suitable reactive gas is chosen for the material to be etched. The dominant consideration is that the reaction products must be volatile. Silicon and its compounds are etched by gases containing fluorine, whereas aluminum is removed with chlorine-containing species. Organic resists are dry etched in oxygen plasmas to produce water vapor and carbon dioxide. By proper choice of the reactor conditions, the etching reaction can be made *anisotropic* so that nearly vertical walls are fabricated in the etched material as shown in Figure 2.12*b*. Low pressures and directional electric fields tend to enhance ion bombardment and anisotropic etching.

Figure 2.12   (*b*) Anisotropically etched lines 500 nm wide spaced 1.5 μm apart. Resist covers the double-layer structure consisting of 180 nm $TaSi_2$ over 260 nm of polycrystalline silicon. Note the uniformity of the vertical surface through the various layers (*Courtesy: G. Dorda, Siemens Corporation*).

The selectivity of dry-etching processes is not, however, as great as that of wet etching, so that the masking material can be significantly attacked during the etching process. The limited selectivity also causes attack of the material under the layer being etched. Therefore, the dry etching process must be terminated as soon as the desired layer has been removed. This can be done effectively by monitoring the emission of characteristic light from the reaction. The process designer using dry etching must also consider the reactive environment in which dry etching occurs. The excited ions and high-energy photons in the reactor bombard the silicon surface and can damage the devices being fabricated.

2.5 Dopant Addition and Diffusion

The distribution of the dopant atoms added to the silicon (*dopant profile*) is generally determined in two steps. First, the dopant atoms are placed on or near the surface of the wafer by *ion implantation*, *gaseous deposition*, or possibly by coating the wafer with a layer containing the desired dopant impurity. This step is followed by a *drive-in diffusion* that transports the dopant atoms into the wafer. The shape of the resulting dopant distribution is determined primarily by the manner in which the dopant is placed near the surface, whereas the diffusion depth depends chiefly on the temperature and time of the drive-in diffusion. Since the characteristics of semiconductor devices depend strongly on the dopant profiles, it is useful to discuss the processes used for dopant addition in some detail.

Ion Implantation

The technique of *ion implantation* provides a highly controlled means for the introduction of dopants into semiconductors. In ion implantation, the desired dopant atoms are first ionized and then accelerated by an electric field to a high energy (typically from 25 to 200 keV). A beam of these high-energy ions strikes the semiconductor surface (Figure 2.13) and penetrates into exposed regions of the wafer. The penetration is typically less than 1 μm below the surface, and considerable damage is done to the crystal during implantation. Consequently, an

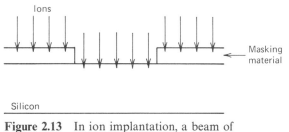

Figure 2.13   In ion implantation, a beam of high-energy ions strikes selected regions of the semiconductor surface, penetrating into these exposed regions.

annealing step is necessary to restore the lattice quality and to ensure that the implanted dopant atoms are located on substitutional sites, where they act as donors or acceptors. After the ions are implanted they may be redistributed by a subsequent diffusion if desired.

Ion implantation makes possible precise control of the area density (atoms cm^{-2}) of dopants that enter the wafer. Since the dopants are conveyed as electrically charged ions, they can be counted by a relatively simple charge-sensing apparatus placed adjacent to the beam path during the implantation. The beam can then be turned off when a predetermined number of ions has been introduced.

Dopant *doses* (number of dopant atoms cm^{-2}) ranging from mid-10^{11} to more than 10^{16} cm^{-2} can be introduced into a wafer with fine control by ion implantation. The lowest doses permit precise tailoring of device properties, while the upper limit is comparable to the quantity of dopant that can be introduced by a gaseous deposition. Typical ion-beam currents in an implanter are of the order of 1 mA, which corresponds to a flux of 6.25×10^{15} singly charged ions s^{-1}. For high doses, implanters capable of currents as high as 15 mA are used to reduce the time needed for an implantation.

Besides controlling very exactly the total dose, ion implantation also makes it possible to assure that the dopant species are extremely pure. The purity is achieved by the use of a mass spectrometer near the dopant source to sort ionic species and allow only the desired dopant species to reach the wafer.

Because the implanted ions penetrate beneath the exposed surface, ion implantation can be used to introduce dopant atoms through an overlying layer of another material such as SiO_2. This capability is useful because the implanted atoms can be added *after* the high temperature heat cycles needed to form a thermal oxide. The location of the dopant atoms is then determined by the energy of the implant, rather than by subsequent annealing. Only a mild heat cycle is needed to remove the implantation damage and to activate the dopant. The lateral spread of the dopant atoms is, therefore, also reduced.

The distribution of implanted ions in an amorphous layer is nearly Gaussian in shape [i.e., having the form $A \exp{-(x/\lambda)^2}$]. The Gaussian distribution reaches a maximum value beneath the surface at a mean penetration depth called the *projected range* R_p. The width of the distribution is described by its *standard deviation* ΔR_p. The total distribution as a function of the depth x below the surface is

$$C(x) = C_p \exp\left[-\frac{(x - R_p)^2}{2\,\Delta R_p{}^2} \right] \qquad (2.5.1)$$

where C_p the peak dopant concentration (atoms cm^{-3}) is related to the implanted dose N' (atoms cm^{-2}) by

$$C_p = \frac{N'}{\sqrt{\pi}\,(\sqrt{2}\,\Delta R_p)} \qquad (2.5.2)$$

As seen from Equation 2.5.1, $\sqrt{2}\,\Delta R_p$ is a characteristic length describing the spatial extent or *spread* of the implanted distribution. Extensive tables have been published giving values of R_p and ΔR_p.[1] Values of these parameters for the three most common dopants in silicon are shown in Figures 2.14 through 2.16. The penetration of an ion through several layers can be approximated by considering the energy lost in each successive layer.

As stated above, Equations 2.5.1 and 2.5.2 apply strictly only to implantation in amorphous material in which the ion scattering can be assumed to be isotropic. In crystalline material the ion penetration can be significantly larger than R_p if the ions enter the lattice in a direction that allows them to *channel* along planes containing few or no atoms. Channeling is undesirable for most *IC* processing because of its sensitivity to the angle of incidence of the ion beam on the wafer surface. It can be reduced if the wafer is tilted until the beam and the crystal axis make an angle of roughly 7 degrees during the implantation or if the wafer surface is coated with a thin amorphous layer such as SiO_2. When either of these measures is taken, the range parameters for amorphous implantation can be used. The many different orientations in a polycrystalline material make it impossible to avoid channeling in all crystallites by tilting the wafer.

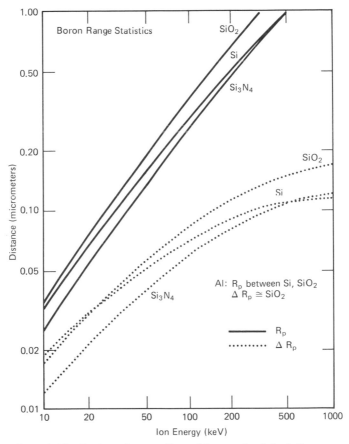

Figure 2.14 Projected range R_p and its standard deviation ΔR_p for implantation of boron into Si, SiO_2, Si_3N_4, and Al.[1]

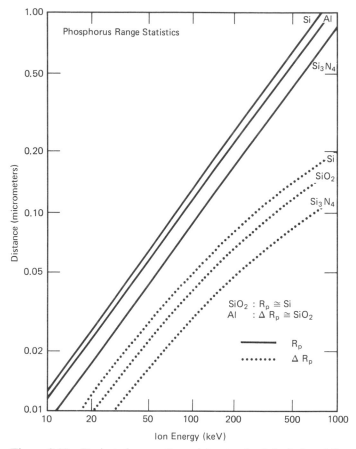

Figure 2.15 Projected range R_p and its standard deviation ΔR_p for implantation of phosphorus into Si, SiO$_2$, Si$_3$N$_4$ and Al.[1]

Although masks for ion implantation can be made of thick layers of SiO$_2$, it is also possible to use metals or organic resists because high temperatures are typically not experienced by the wafer during implantation. The use of resist materials is convenient, but care must be taken to insure that the implant itself does not heat the organic materials sufficiently to make them flow and lose definition. This is a special hazard for high-current implants. The energy in the implant beam may also change the resist structure so that it cannot subsequently be removed easily with wet chemicals. In this case it necessary to etch reactively in an oxygen plasma.

The excellent control over the quantity, purity, and position of implanted dopant atoms has made ion implantation an extremely important *IC* fabrication procedure for the introduction of dopant atoms. We discuss implantation further in Chapter 10, where it is seen to be an essential part of advanced MOS processing.

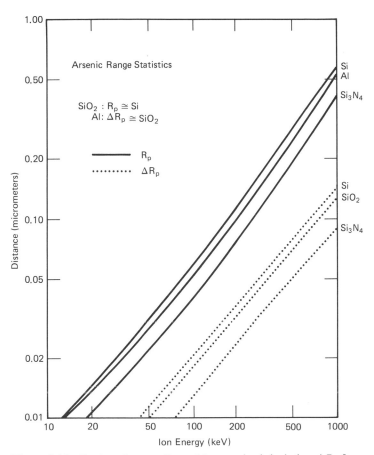

Figure 2.16 Projected range R_p and its standard deviation ΔR_p for implantation of arsenic into Si, SiO_2, Si_3N_4 and Al.[1]

Diffusion

Dopant atoms introduced into a silicon wafer migrate through the crystal if it is heated sufficiently, being transferred from the region of high concentration where they are first deposited toward regions of lower concentration that are typically deeper in the wafer. The diffusion of dopant atoms is similar to the diffusion flow of free carriers discussed in Section 1.2. The primary difference between the two cases is the temperature necessary to cause appreciable transfer to occur. Dopant atoms must typically make their way through the lattice by meeting point defects (generally silicon vacancies) whereas free carriers in the valence or conduction band move without interacting with point defects. Temperatures of the order of 1000°C are required before there is appreciable diffusion of typical dopant atoms.

The change in the concentration C of the dopant atoms with time in a narrow region of width dx at a depth x from the surface (Figure 2.17) can be written

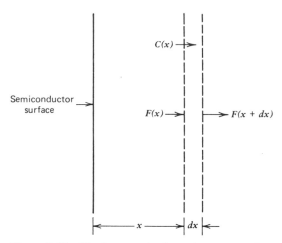

Figure 2.17   The increase in dopant concentration in a region dx is related to the net flux of atoms into the region: $F(x) - F(x + dx)$.

as the difference between the flux of dopant atoms entering the region from the left and that leaving at the right

$$\frac{\partial C}{\partial t} dx = F(x) - F(x + dx) \tag{2.5.3}$$

We can approximate the last term by the first two terms of a Taylor-series expansion.

$$F(x + dx) \approx F(x) + \left(\frac{\partial F}{\partial x}\right) dx \tag{2.5.4}$$

to obtain

$$\frac{\partial C(x)}{\partial t} = -\frac{\partial F}{\partial x} \tag{2.5.5}$$

As we saw in Section 1.2, the first-order expression for the flux is proportional to the concentration gradient

$$F = -D \frac{\partial C}{\partial x} \tag{2.5.6}$$

The parameter D, called the *diffusivity*, describes the ease with which the dopant atoms move in the lattice and is, therefore, a function of temperature. The diffusivities of several common dopant impurities are shown in Figure 2.18. The simple model considered here neglects many aspects of diffusion that become important at higher dopant concentrations. Consequently, the values given in Figure 2.18 are only valid at low and moderate dopant concentrations. They must be used with

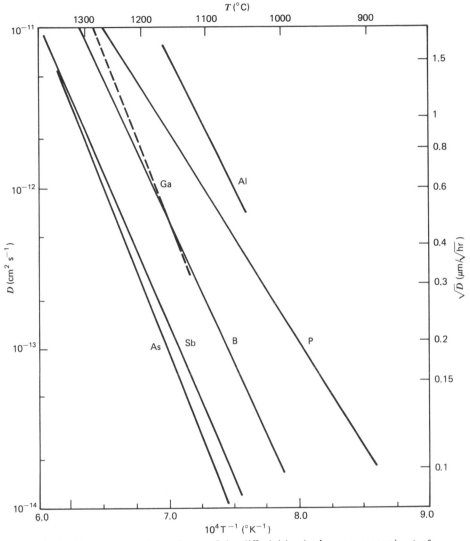

Figure 2.18 Temperature dependence of the diffusivities (at low concentrations) of commonly used dopant impurities in silicon.[2]

caution when the dopant densities become high because the diffusivity itself is then a function of the dopant concentration. Phosphorus is especially troublesome in this respect. More details of these second-order considerations are discussed later. Combining Equations 2.5.5 and 2.5.6, we find that

$$\frac{\partial C}{\partial t} = D \frac{\partial^2 C}{\partial x^2} \tag{2.5.7}$$

Equation 2.5.7 (sometimes called *Fick's second law*) can be solved explicitly for $C(x,t)$.

Because Equation 2.5.7 is a partial differential equation, it has an infinity of
solutions. A unique solution is obtained for a given problem by matching a known
boundary condition. For the diffusion conditions most common in semiconductor
processing, only two boundary conditions for Equation 2.5.7 are widely used, and
therefore, only two functions are encountered. A complementary error function
applies for diffusion with a fixed surface dopant concentration, and a Gaussian
distribution describes the redistribution of a constant total number of diffusing
atoms (Figure 2.19).

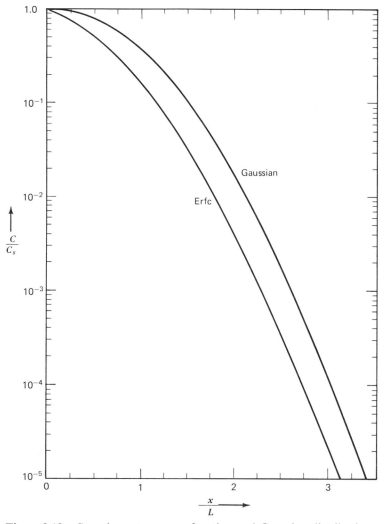

Figure 2.19 Complementary-error-function and Gaussian distributions;
the vertical axis is normalized to the peak concentration C_s, while the
horizontal axis is normalized to the characteristic length $L = 2\sqrt{Dt}$.

Gaseous Deposition. When a gaseous deposition source is used to introduce dopant atoms into a semiconductor, the patterned wafer is placed in a diffusion furnace similar to the furnace used for oxidation, and a gas containing the desired dopant impurity—typically phosphorus or boron—is passed over it. The quantity of dopant that enters the wafer is limited to values less than or equal to the solid solubility of the dopant in silicon at the furnace temperature.

The best way to control the number of dopant atoms entering the silicon is to adjust the gas flows so that the dopant concentration at the silicon surface is at its solid solubility. The solid solubilities in silicon for several common dopant species are shown in Figure 2.20.

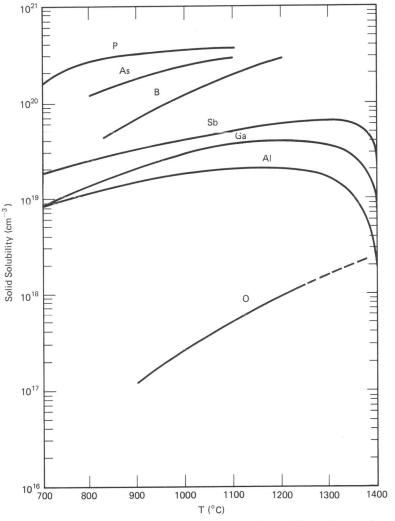

Figure 2.20 Temperature dependence of the solid solubilities of several elements in silicon.[12]

Because of the relatively low temperatures and short times typically used, the penetration of the dopant atoms during this deposition step is generally very small. A drive-in diffusion is subsequently used to distribute the deposited atoms over the desired depth.

The silicon surface is exposed to a constant concentration of dopant atoms during deposition. For this condition, the solution of Equation 2.5.7 indicates that the dopant atoms have a complementary-error-function distribution along x (the dimension measured away from the surface) after deposition.

$$C(x,t) = C_s \text{erfc}\left(\frac{x}{2\sqrt{Dt}}\right) = \frac{2C_s}{\sqrt{\pi}} \int_{x/2\sqrt{Dt}}^{\infty} e^{-v^2}\, dv \qquad (2.5.8)$$

where C_s is the surface concentration of dopant atoms. Note that

$$\text{erfc}(\eta) = 1 - \text{erf}(\eta) = 1 - \frac{2}{\sqrt{\pi}} \int_0^{\eta} \exp(-v^2)\, dv \qquad (2.5.9)$$

Thus, erfc $(0) = 1$, and the complementary error function decreases rapidly with increasing values of its argument η (Figure 2.19).

The complementary error function is a solution to Equation 2.5.7 for any method of dopant introduction that provides a constant value of dopant at the surface throughout the process $[C(0,t) = C_s]$. The combination of parameters $2\sqrt{Dt}$ used to normalize the x-axis of Figure 2.19 represents the characteristic diffusion length L associated with a particular diffusion cycle and describes the depth of penetration of the dopant. Note that the depth increases only as the square root of the diffusion time t.

If we know the surface concentration, the solid solubility and the diffusivity and diffusion time, we can calculate the impurity distribution from Equation 2.5.8. More important, for the deposition cycle we can find the total density N' of dopant atoms per unit surface area introduced by the diffusion. This quantity is calculated by integrating $C(x,t)$ over x to find

$$N' = \int_0^{\infty} C(x,t)\, dx = 2\sqrt{Dt/\pi}\; C_s \qquad (2.5.10)$$

for the complementary-error-function distribution. Thus, from Equation 2.5.10, the area density of impurities, like the diffusion depth, increases as the square root of the diffusion time t.

Dopant Redistribution. Dopant atoms are *redistributed* by subsequent heat treatments after deposition.

After ion implantation, dopants are Gaussian distributed (Equation 2.5.1). The subsequent *drive-in* diffusion broadens this initial distribution, reducing the peak concentration since the total number of dopant atoms is fixed. The new distribution is also Gaussian, but it is described by a new characteristic length that is a function of both the initial characteristic length $\sqrt{2}\,\Delta R_p$ and the further dopant spread described by the diffusion length $2\sqrt{Dt}$. These quantities are combined in a root-mean-square (*rms*) fashion so that the new characteristic length L' is

$$L' = \sqrt{2\,\Delta R_p^{\,2} + 4Dt} \qquad (2.5.11)$$

The new peak concentration is also found by replacing the initial characteristic length $\sqrt{2}\,\Delta R_p$ by L', so that the final dopant distribution is

$$C(x) = \frac{N'}{L'\sqrt{\pi}} \exp\left[-\left(\frac{x - R_p}{L'}\right)^2 \right]$$ (2.5.12)

Redistribution of dopant atoms added by gaseous deposition also leads to a Gaussian distribution. Since a typical drive-in diffusion produces an impurity distribution much deeper than that resulting from a gaseous deposition step, we often approximate the distribution at the beginning of the drive-in diffusion by a sheet of dopant at the semiconductor surface with a total concentration N' per unit area determined by the deposition cycle. We then assume that the drive-in diffusion simply redistributes this fixed amount of dopant impurity. With this boundary condition, the solution of the diffusion equation is again a Gaussian distribution

$$C(x,t) = \frac{N'}{\sqrt{\pi Dt}} \exp\left(-\frac{x^2}{4Dt} \right)$$ (2.5.13)

where the characteristic diffusion length $2\sqrt{Dt}$ is now determined by the temperature and the time of the drive-in diffusion.

When the dopant is successively redistributed by two or more diffusion steps, the over-all characteristic length is determined by an *rms* combination of the characteristic lengths associated with each process. For example, for m steps

$$L' = \left(\sum_{i=1}^{m} L_i^2 \right)^{1/2} = \left(\sum_{i=1}^{m} 4D_i t_i \right)^{1/2}$$ (2.5.14)

where the diffusivities are evaluated at the temperature corresponding to each heat cycle. (Note from Equation 2.5.14 that the squares of the diffusion lengths, *not the lengths themselves*, are summed, but that the individual $D_i t_i$ products are added.) In this manner, an approximation to the dopant distribution can be found after a number of different heat cycles such as those needed to carry out a complete *IC* fabrication process.

For a given area density of dopant atoms added to a wafer, the peak concentration is twice as large when a gaseous predeposition is used as when the atoms are added by ion implantation. This is the case because the dopant concentration is a maximum at the silicon surface for gaseous predeposition, permitting diffusion only into the wafer. The ion-implanted dopant, on the other hand, has a maximum concentration beneath the silicon surface so that diffusion proceeds both toward the surface and into the bulk which spreads the fixed number of dopant atoms throughout a larger volume.

Oxide Doping Sources. In an alternate method of dopant addition that is sometimes useful, a layer of oxide containing the dopant impurity can be deposited on the wafer surface. The dopant atoms are then diffused into the silicon from this glassy layer. One convenient method of forming the doped oxide layer is by *chemical vapor deposition (CVD)* of SiO_2 with a dopant species added to the oxide during its deposition. We discuss the CVD process in greater detail in Section 2.6. It is also possible to incorporate the dopant in particles of glass dispersed in an organic solvent. This material can be "spun-on" the wafer, dried, and the organic residue then driven off by heating the film to about 200°C. After the doped oxide has been deposited on the wafer by either of these two methods, the impurity must be diffused into the silicon with a drive-in diffusion step.

When a doped oxide is used, the concentration of dopant atoms at the silicon surface generally remains constant at a fixed fraction of the concentration of dopant in the oxide during the entire drive-in diffusion. This condition leads once again to a complementary-error-function distribution after the drive-in step (*cf* Equation 2.5.8 and Figure 2.19).

Diffusivity Variations. Because dopant diffusion occurs through the interaction of the diffusing species and point defects (chiefly silicon vacancies or interstitial silicon atoms), any factor that changes either the density of point-defects or the charges associated with them can modify the diffusion process. We briefly consider two important cases in which this occurs, (1) *oxidation-enhanced diffusion* and (2) *concentration-dependent diffusion* of a single species of dopant.

If the point-defect concentration varies with position in the crystal, the diffusivity also depends on position. In this case Equations 2.5.5 and 2.5.6 can be combined to express the diffusion equation as

$$\frac{\partial C}{\partial t} = \frac{\partial}{\partial x}\left(D\,\frac{\partial C}{\partial x}\right) \qquad (2.5.15)$$

instead of the simpler form of Fick's Law (Equation 2.5.7).

The solutions to Equation 2.5.15 are more complicated than the simple Gaussian or complementary-error-function distributions discussed previously. The exact profiles cannot generally be written in closed mathematical form and are found from Equation 2.5.15 using numerical techniques. A few qualitative comments can, however, provide useful insight into this more complex diffusion process.

A net localized charge is present in the crystal in the region of a silicon vacancy or interstitial. The charge state of these point defects depends on the position of the Fermi level in the crystal and is thus a function of the dopant concentration as well as the temperature. Since the diffusing dopant atoms move by interacting with charged, as well as, neutral point defects, this dependence on the Fermi level affects the dopant diffusivity. The diffusivity in this case can be written as the sum of components that account separately for interactions of the dopant atoms with different charge states of the point defects.[3] For example, for diffusion dominated by interaction with negatively charged vacancies, the effective diffusivity D_{eff} can be written

$$D_{eff} = h\left[D_i^0 + D_i^-\left(\frac{n}{n_i}\right) + D_i^=\left(\frac{n}{n_i}\right)^2\right] \qquad (2.5.16)$$

where h is a parameter that accounts for the effect of electric field on the diffusivities, and each D term is associated with a different charge state of a point-defect site.

From Equation 2.5.16 we see that the diffusivity becomes dependent on the dopant concentration when the terms involving the charged vacancies become comparable to the neutral vacancy term D_i^0. Because the various D_i values are of comparable magnitudes, this occurs when n/n_i becomes of order unity at the diffusion temperature. If, for example, we consider diffusion taking place at 1000°C, we may refer to Figure 2.10 to find $n_i \approx 9 \times 10^{18}$ cm^{-3}. Hence, for dopant densities above this value, we can expect to observe an increase in D_{eff}. Concentration-dependent diffusion can multiply the diffusivity by a factor of 10 or 20, greatly enhancing the diffusion rate in regions of high-dopant concentrations. The resultant dopant profile becomes distorted from the Gaussian or complementary error functions which apply to low-concentration doping profiles.

An example of concentration-dependent diffusion, often observed in bipolar integrated circuits, is termed the *emitter-push* effect. It occurs when the heavy phosphorus or arsenic doping used to form the emitter region pushes ahead the lightly doped transistor base region. The details of bipolar transistor design and operation are described in Chapter 6. Figure 2.21 is a Scanning Electron Microscope (SEM) picture showing the cross section of a transistor in which emitter push has

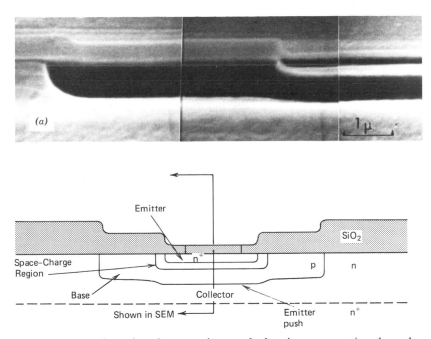

Figure 2.21 (*a*) Scanning electron micrograph showing cross section through a bipolar transistor, (*b*) sketch identifying the regions shown. The boron-doped base region has been pushed ahead (emitter push) by the concentration-dependent diffusion effects associated with heavy phosphorus doping in the emitter.[13]

occurred. Bulging dopant profiles caused by effects similar to emitter push can appear in other regions of some *ICs*.

In addition to concentration-dependent diffusion, interaction with point defects created by other mechanisms can also increase the diffusion rate. When silicon is oxidized, many bonds are broken at the surface, and point defects are generated. Many of these point defects migrate into the underlying silicon until they meet a different type of point defect that is capable of annihilating them. For example, a vacancy can recombine with (be annihilated by) an interstitial in similar fashion to the recombination of an electron and a hole. Before recombining, however, the point defects may migrate significant distances into the silicon, modifying the diffusion rate of any dopants in this region. This effect gives rise to *oxidation-enhanced diffusion* which is readily observed when portions of a silicon wafer are covered with a nonoxidizing layer while other regions are oxidized (Figure 2.22). Diffusion under the oxidizing regions of a wafer can be considerably deeper than that under portions that are not oxidized (that consequently do not contain large densities of point defects).

In addition to enhancement by high concentrations of point defects, diffusion can be greatly enhanced by lattice damage created during ion implantation. During a subsequent heat treatment the lattice damage is removed, but, at the same time, the dopant atoms diffuse. During the early stages of the annealing process, the

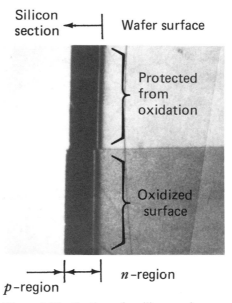

Figure 2.22 Section of a silicon wafer showing deeper diffused *n*–type region (dark area) under oxidized silicon surface (bottom) than underneath a surface protected from oxidation (top).[14]

lattice damage promotes rapid dopant migration, but the diffusion slows as the damage is removed. A quantitative description of the combined effects is extremely complex and must be treated by computer modeling. This interaction of lattice damage and dopant motion is especially severe when using rapid processing techniques in which the wafer is heated to a high temperature by radiation for only a few seconds to remove lattice damage and activate the implanted dopant.

Solid Solubility. The amount of dopant entering the silicon surface by gaseous deposition is limited by thermodynamics to the solid-solubility concentration (Figure 2.20). When dopant atoms are introduced by ion implantation, however, the impurity concentration can exceed the solid-solubility value because thermodynamic equilibrium is not involved. During the subsequent damage-removal and dopant-activation anneal, however, the amount of dopant entering substitutional sites is limited to the solid solubility at the annealing temperature. Any excess dopant forms clusters or precipitates. As seen in Figure 2.20, the solid solubility increases with temperature for typical dopants over temperature ranges of interest. Hence, a dopant concentration that is at its solid-solubility limit at high temperatures may exceed the limit at the temperature of a subsequent, lower-temperature anneal. If this is the case, the excess dopant will tend to move out of electrically active, substitutional sites. If the temperatures are sufficiently low (e.g. room temperature), however, expulsion of excessive dopant from the lattice can take a relatively long time. In practice, the doping concentrations can remain higher than the room-temperature solubility limit indefinitely if the dopant has been introduced at elevated temperatures.

Segregation. During thermal oxidation, silicon is consumed, and any dopant in the wafer must redistribute between the silicon and the growing oxide. At equilibrium, a constant ratio exists between the dopant concentrations on the two sides of the Si–SiO$_2$ interface. The *segregation coefficient* (*m*) describes this ratio

$$m = \frac{C_{Si}}{C_{SiO_2}} \tag{2.5.17}$$

The *n*–type dopants phosphorus and arsenic tend to segregate into the silicon ($m > 1$) and to be pushed ahead of a growing layer of SiO$_2$. Boron tends to deplete from the surface regions of silicon into the growing oxide layer ($m < 1$). However, because the boundary between silicon and SiO$_2$ is moving during thermal oxidation, equilibrium at the interface is only approached at very slow oxidation rates. For typical processing conditions, the amount of segregation differs significantly from that predicted by thermal-equilibrium considerations. Dopant segregation also depends on the diffusion rate of the dopant species in the oxide. Equilibrium is approached only if the dopant diffuses slowly in the oxide so that it is not exchanged at the oxide-gas interface. Another consideration is the transport of the dopant from the bulk to the interface by diffusion in the silicon. The ratio of the dopant concentration at the interface to its value in the bulk is primarily determined by the relative rates of oxidation and diffusion; the less able is the dopant to transfer between the bulk and the interface, the greater is the accumulation or depletion of dopant near the silicon surface.

Thus, the three important parameters which must be considered when determining the amount of surface segregation are (1) the segregation coefficient m, (2) the ratio of the oxidation rate to the square root of the dopant diffusivity in the silicon (which measures diffusion in the silicon), and (3) the ratio of the dopant diffusivities in silicon and in SiO_2.

2.6 Chemical Vapor Deposition

Although the basic elements of an integrated circuit can be formed by oxidation, lithography, and diffusion, more advanced structures require the flexibility of adding a semiconducting or insulating layer on top of a partially formed integrated circuit. Deposited insulators can be used to avoid high-temperature oxidation after dopant atoms have been introduced, while lightly doped, single-crystal silicon layers or polycrystalline-silicon films may be useful in other locations. These added layers can be formed by *chemical vapor deposition* (*CVD*). In *CVD* all constituents forming the deposited layer are introduced into the reactor; none come from the silicon wafer itself. The structure of a *CVD* film depends on the substrate on which it is deposited (amorphous or crystalline), and on the deposition conditions (mainly temperature, deposition rate, and gas pressure). Reactions that take place during the deposition by *CVD* occur over a wide temperature range. The reactions are usually promoted by heating the substrate, but energy may also be introduced into the system electrically by generating a plasma within the deposition chamber.

Epitaxy

We have described how impurities can be added to a wafer by ion implantation or by gaseous deposition and diffusion. These processes can reliably be employed to produce a layer at the surface that is of higher dopant concentration than had existed prior to the dopant addition. They do not, however, permit us to produce a layer that is less heavily doped near the surface than it is underneath. In theory this could be done by adding an approximately equal concentration of impurities of the opposite conductivity type, using compensation as described in Section 1.1. However, limited control of the accuracy of the diffusion process generally makes it impractical to attempt a nearly balanced compensation of dopants. An additional drawback to using a balanced compensation is that the carrier mobility in such a structure would be degraded since the mobility is affected by the total number of ionized impurities $N_d + N_a$ rather than by the net number of impurities $|N_d - N_a|$, which determines the carrier concentration.

We can fabricate the desired lightly doped layer above a heavily doped region by the process of *epitaxy* (Figure 2.23), which is the controlled growth of single-crystal silicon on a single-crystal wafer or *substrate*. To grow an epitaxial layer,

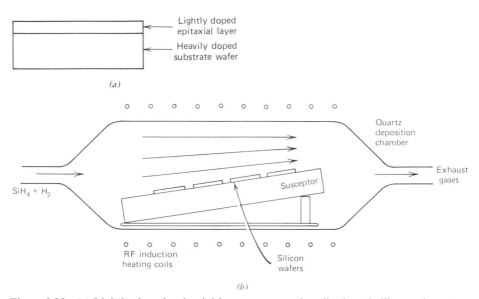

Figure 2.23 (a) Lightly doped epitaxial layer grown on heavily doped silicon substrate, (b) Epitaxial deposition system showing silicon wafers on susceptor that is heated by induction from the *RF* heating coils located outside the quartz deposition chamber.

the wafer is placed in a heated chamber where a gas such as silane SiH_4 or silicon tetrachloride $SiCl_4$ passes over its surface. The gas decomposes on the surface of the wafer and a layer of silicon is deposited there. Silane decomposes *pyrolytically*, that is, by the addition of heat alone.

$$SiH_4(g) \rightarrow Si(s) + 2H_2(g) \qquad (2.6.1)$$

while silicon tetrachloride requires reduction in a hydrogen ambient to form solid silicon.

$$SiCl_4(g) + 2H_2(g) \rightarrow Si(s) + 4HCl(g) \qquad (2.6.2)$$

Note that all the incoming species in reactions 2.6.1 and 2.6.2 are gases, which accounts for the name chemical *vapor* deposition (*CVD*).

To make the *CVD* of silicon an *epitaxial* deposition, it is necessary to heat the wafer sufficiently so that the depositing silicon atoms can move into a position to form covalent bonds to the substrate and thereby extend the single-crystal lattice before they become buried by subsequently arriving atoms. Single-crystal growth or epitaxy is usually carried out at temperatures between 900 and 1250°C. The epitaxial film may be more lightly doped than the substrate; hence, epitaxy provides a means for obtaining a low concentration of dopant above a high concentration. This capability is especially important in optimizing the structure of bipolar transistors. If desired, dopant atoms can be added to the growing film during its deposition by introducing dopant-containing gases such as arsine (AsH_3), phosphine (PH_3), or diborane (B_2H_6) into the reactor along with the silicon-containing gas.

Nonepitaxial Films

A number of other *CVD* films are useful for *IC* applications in addition to epitaxial silicon. For example, in any *IC*, conducting layers are needed to interconnect devices. These layers must, of course, be isolated from the substrate. Aluminum is very often used for these conducting layers, but its low melting point (660°C) and its reactivity with other elements generally precludes heating it higher than 500°C after it is deposited. If, as an alternative, a thin layer of silicon is used for an interconnecting path, subsequent heat treatments to above 1000°C may be carried out. This possibility is extremely important for a number of *IC* applications, and *CVD* silicon is often used in this way.

Polycrystalline Silicon. To deposit silicon for interconnections, methods similar to those used for epitaxy are often employed. However, because these layers are not deposited directly on the single-crystal silicon wafer, but usually over a SiO_2 layer, they cannot grow epitaxially. These *CVD* films are typically composed of many small crystallites (with roughly submicrometer dimensions) and are therefore called *polycrystalline silicon*, or simply *polysilicon*.

Polysilicon has a special importance in MOS processing where it is used as a transistor electrode in *silicon-gate* MOSFETs which are described in Chapter 9. Recently polysilicon has also been used in bipolar *IC*s, where it is formed over the single-crystal substrate by lowering the *CVD* temperature to about 600–700°C so that the depositing silicon atoms do not have enough energy to migrate on the surface and form an epitaxial layer.

Amorphous Silicon. When *CVD* silicon is deposited at still lower temperatures (below about 600°C), an amorphous film is formed regardless of the nature of the substrate. In amorphous silicon there is only very short-range order (typically over a small number of atomic spacings), and no crystalline regions can be observed. Amorphous silicon has not been as important as polysilicon in *IC*s, but it is under study for specialized applications.

Insulating Films. Insulating, as well as conducting films, can be formed by *CVD*. Especially important to *IC* processing are *CVD* films of silicon dioxide (SiO_2) and silicon nitride (Si_3N_4), formed by reacting a gas such as silane (SiH_4) with oxygen or nitrous oxide (N_2O) (for SiO_2) or with ammonia (NH_3) or nitrogen (for Si_3N_4). Although *CVD* oxides are not usually as pure as nor of equivalent electrical quality to thermally grown oxides, they do not require the high-temperature processing needed for thermal oxidation. For example, *CVD* oxide can be used above a poly-silicon interconnection layer before aluminum is deposited to avoid the dopant diffusion that would occur during a thermal oxidation. In another application,

CVD oxide can be used above an aluminum layer to protect the finished integrated circuit from subsequent contamination. In the latter case, the oxide is usually doped with phosphorus to impede the migration of any contaminant through the oxide to the circuit. A deposited oxide commonly used for this circuit protection is formed by the reaction of silane and oxygen at about 400°C. The reaction

$$SiH_4(g) + O_2(g) \rightarrow SiO_2(s) + 2H_2(g) \tag{2.6.3}$$

produces an oxide that is less dense and less chemically resistant than thermally grown oxide, but this oxide is useful when high temperatures must be avoided.

Since silicon-nitride (Si_3N_4) layers do not oxidize as readily as does silicon, such layers are useful for limiting the regions where thermal oxide is grown in a *local oxidation* (*LOCOS*) process used for device isolation in integrated circuits.[4] In the *LOCOS* process, a layer of silicon nitride is deposited on the silicon substrate and lithographically defined to retain the nitride in the device regions (Figure 2.24*a*). The nitride is removed from the area between the devices where a thick,

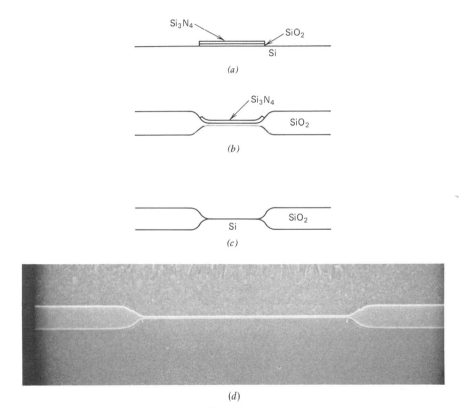

Figure 2.24 *LOC*al *O*xidation of *S*ilicon (*LOCOS*). (*a*) Defined pattern consisting of stress-relief oxide and Si_3N_4 covering the area over which further oxidation is not desired, (*b*) thick oxide layer grown over the bare silicon region, (*c*) stress-relief oxide and Si_3N_4 removed by etching to permit device fabrication, (*d*) SEM photo (5000 ×) showing *LOCOS*-processed wafer at step *b*).

isolating oxide layer is to be grown. A thin, *stress-relief* SiO_2 layer is usually inserted between the nitride and the silicon wafer to prevent stress from the nitride from inducing defects in the silicon wafer. After the nitride is defined, the wafer is inserted into an oxidation furnace, and a thick oxide is grown in the exposed silicon regions, usually in a steam or pyrogenic ($H_2:O_2$) ambient (Figure 2.24*b*). The nitride prevents oxidation of the device regions. A layer of SiO_2 more than one μm thick can be grown on exposed silicon while only a few tens of nm of Si_3N_4 are converted to SiO_2. After the oxidation, the thin oxide over the nitride is removed, and the nitride and thin stress-relief oxide are etched away to expose the bare silicon in the device regions (Figure 2.24*c*). Figure 2.24*d* shows a scanning-electron micrograph cross section of a device region formed by a *LOCOS* process. The oxide thickness is seen to taper gradually from the isolation region to the device region. This taper is an advantage in avoiding sharp edges that make continuous film coverage difficult. For devices made with very small dimensions, however, the taper becomes a disadvantage because its lateral extent cannot be reduced below about 1 μm which is excessive when 1 μm linewidths are being used.

Reaction Kinetics. Layers deposited by *CVD* are usually formed in open-flow reactors, as illustrated in (Figure 2.23). The gases flow continuously through the reaction chamber where the deposition takes place; gaseous byproducts are exhausted along with unused reactant gases. A *carrier gas* is often used to push the reactants through the chamber. The gases are usually mixed before entering the reaction chamber unless they react at low temperatures. At the surface of the wafers, there is a gas-phase boundary layer through which the reactants must diffuse (Figure 2.25*a*). This boundary layer is a transition region between the unrestricted flow region and the walls and fixtures in the chamber where the gas velocity is reduced by viscous forces which tend to pull against the moving gas (Figure 2.25*b*).

The reactants must pass through the boundary layer to reach the surface, where the reaction is promoted by heat. The deposition takes place on all heated surfaces reached by the gases. For example, if the walls of the chamber are hot, a generally undesired and bothersome film forms on them as well as on the wafers. Either the rate of diffusion through the boundary layer or the rate of reaction at the surface can limit the overall deposition rate. We discussed similar transfer-rate limitations in the case of thermal oxidation in Section 2.3. Unlike the case of thermal oxidation, however, the diffusion process for *CVD* occurs in the gas phase rather than in the solid SiO_2 layer. As in the case of oxidation, we write expressions for the fluxes of molecules diffusing through the boundary layer [$F(1)$] and reacting at the surface [$F(2)$] (Figure 2.26)

$$F(1) = D \frac{C_g - C_s}{\delta} \tag{2.6.4}$$

and

$$F(2) = k_s C_s \tag{2.6.5}$$

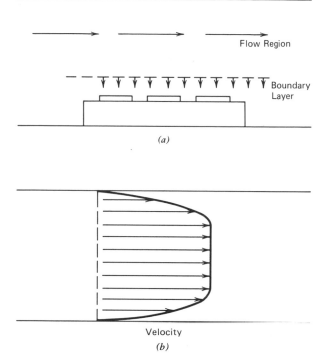

Flow Region

Boundary Layer

(a)

Velocity

(b)

Figure 2.25 *(a)* Section along horizontal, open-flow reactor showing gas flow parallel to the wafer surface and indicating the location of the boundary layer in which the gas flow is nearly perpendicular to the wafer surface, *(b)* representation of gas velocity distribution across the reactor tube.

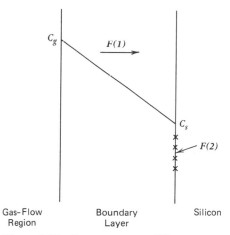

C_g

$F(1)$

C_s

$F(2)$

Gas–Flow Region

Boundary Layer

Silicon

Figure 2.26 Reactant gases diffuse through the boundary layer to the wafer surface at a rate $F(1)$ and react there at a rate $F(2)$.

In Equation 2.6.4, D is the gas-phase diffusivity of the reactant (only weakly temperature dependent), δ is the boundary-layer thickness, and the concentrations C_g and C_s occur at the outer edge of the gaseous boundary layer and at the surface, respectively. In Equation 2.6.5, k_s is the surface reaction-rate coefficient which is exponentially temperature dependent with an activation energy E_a $[k_s = k_{so} \exp(-E_a/kT)]$.

In steady state $F(1) = F(2) = F$, and the overall deposition rate R_d can be written

$$R_d = \frac{F}{N} = \frac{C_g/N}{\delta/D + 1/k_s} \qquad (2.6.6)$$

where N is the number of atoms per unit volume in the depositing film. The first term in the denominator of Equation 2.6.6 represents the impedance to gas-phase diffusion, while the second term is the impedance to the surface reaction. Reactions limited by surface processes depend strongly on k_s and typically dominate CVD processes at low temperatures. However, k_s increases rapidly with increasing temperature so that it no longer limits the overall process at higher temperatures, and gas-phase diffusion through the boundary layer becomes limiting. Typical temperature behavior is shown in Figure 2.27.

Choice of the proper region of operation is influenced by the geometry of the CVD reactor. In reactors operating at atmospheric pressure with the gas flow nearly parallel to the wafer surfaces, such as that shown in (Figure 2.23b), the gas-flow behavior is readily controlled, while the temperature is difficult to control. Consequently, the temperature-insensitive, gas-phase diffusion process is generally selected as the limitation of the overall CVD process. In the horizontal

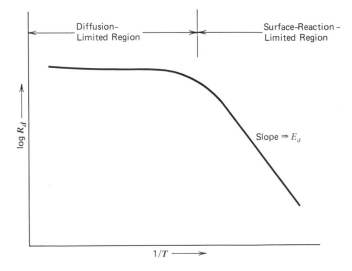

Figure 2.27 Typical dependence of overall deposition rate R_d as a function of reciprocal temperature $1/T$; the surface reaction rate limits the deposition at low temperatures, and gaseous diffusion (mass transport) limits at high temperatures.

reactor, however, throughput capacity is very limited since the wafers are placed flat in a single plane.

For high wafer capacity, vertical placement of closely spaced wafers, such as used in a diffusion or oxidation furnace (Figure 2.5) is desired. With this geometry, however, the gas flow is difficult to control since the gases first flow through the annular space surrounding the wafers and then into the narrow spaces between wafers (Figure 2.28a). Consequently, diffusion-limited-operation would lead to very nonuniform film thicknesses. On the other hand, the temperature in this type of furnace can be well controlled so that operation in the surface-reaction-limited region is feasible. Hence, high-capacity reactors with closely spaced, vertical wafers are generally operated at lower temperatures where the surface reaction limits the deposition rate. The diffusion limitation is further eased by operating at a reduced pressure. Because the gas-phase diffusivity is inversely proportional to the pressure, the first term in the denominator of Equation 2.6.6 then becomes even less of a limitation.

High-capacity, low-pressure *CVD* (*LPCVD*) reactors are now routinely used to deposit polysilicon, SiO_2 and Si_3N_4. The basic elements of an *LPCVD* system are shown in Figure 2.28b. This type of reactor has not been used for silicon epitaxy

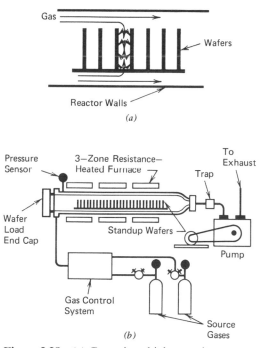

Figure 2.28   (a) Gases in a high-capacity reactor flow through the annular space between the wafers and reactor wall and then diffuse between the closely spaced wafers. (b) The basic elements of a *LPCVD* reactor.

because of the higher wafer temperatures needed to assure single-crystal growth. This requirement conflicts with the avoidance of a diffusion limitation as mentioned above.

Plasma-Enhanced CVD.[†] For some applications, the deposition process must occur at low temperatures. For example, the deposition of a passivating layer over aluminum interconnections already on the chip must take place below 500°C. Consequently, conventional, thermally activated *CVD* is limited, and an additional source of energy must be supplied to the wafer surface to allow the necessary chemical reactions to proceed. A plasma, which is an essentially neutral mixture of excited gaseous species, can be used to supply energy to the reactant gases and promote the chemical reactions. *Plasma enhanced CVD (PECVD)*, in which an *RF* field energizes a gas mixture, is assuming increasing importance in *IC* fabrication, being used for the deposition of silicon oxide, silicon nitride, and silicon. In general, *PECVD* does not produce equivalent quality in the deposited films to that achieved by thermal activation—the oxide and nitride may not be stoichiometric, and the silicon may have a poorly defined structure.

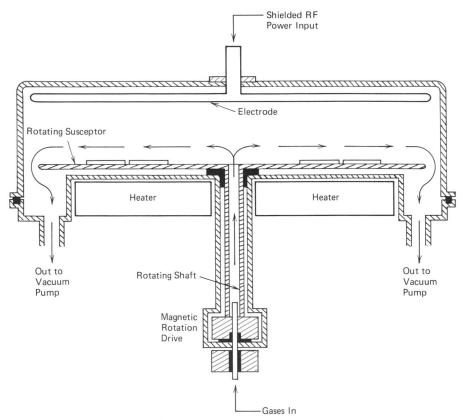

Figure 2.29   Cross section of a parallel-plate, plasma-enhanced *CVD* reactor.[15]

Figure 2.29 shows a cross-section of a parallel-plate plasma reactor. The nature of the plasma depends on many independent variables such as electron concentration, electron-energy distribution, gas density, and residence time of the excited species within the plasma. These *microscopic* variables are controlled by *macroscopic* parameters such as (1) the reactor geometry, (2) the intensity and frequency of the high-frequency power used to excite the plasma, (3) the pump speed, (4) the electrode temperature, and (5) the flow rates of the reactant and diluent gases. Since the relation between the macroscopic and microscopic parameters is not straightforward and there are many variables in a *PECVD* process, the deposition parameters are often determined empirically for best control and repeatability.

2.7 Interconnection and Final Assembly

To build an integrated circuit, the individual devices formed by the planar process must be interconnected by a conducting path. This procedure is usually called *interconnection* or *metallization*.

The simplest and most widely employed interconnection method is illustrated in Figure 2.30. First, the SiO_2 is removed from areas where a contact is to be made to the silicon. Then a layer of metal is deposited over the surface, typically by vaporizing a solid source using *electron-beam* (*EB*) bombardment in an evacuated chamber, or *sputtering* (ion bombardment) in a low-pressure chamber. The vaporized metal atoms travel to the wafers where they condense in a uniform film. The metal, usually aluminum or an aluminum alloy (such as aluminum with a low content of silicon or copper), is then removed from areas where it is not desired by lithography and etching operations similar to those already discussed. The aluminum is usually etched in phosphoric-acid solutions or by dry etching techniques.

As device dimensions are reduced, the requirements on the metallization continue to become more and more severe. We show in Chapter 10 for example, that the most usual technique for systematically reducing the feature sizes in an *IC* (scaling) results in an increased current density in the interconnections. If device dimensions on the surface are reduced by a factor K, the scaled current should also decrease by the same factor. However, the interconnection cross section decreases by K^2 so that the current density flowing through the interconnections increases by K. This increase results in a larger voltage drop in the interconnections

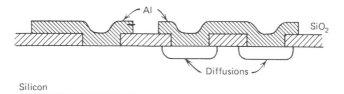

Figure 2.30   A thin layer of aluminum may be used to connect various doped regions of a semiconductor device.

so that a smaller fraction of an externally applied voltage is effective in activating an *IC* device. To minimize this effect, it is important to reduce the resistivity of the interconnection material.

Especially in MOS circuits, but at times also in bipolar circuits, one interconnection layer is composed of polycrystalline silicon, while another is made of aluminum. With these two interconnection layers plus a possible diffused interconnection line in the surface of the silicon wafer, current can be carried at three different vertical levels—an importance degree of freedom in circuits having many thousands of devices. In even more complex circuits, a second layer of metal may also be used.

The resistivity of polycrystalline silicon, however, is limited to about 500 $\mu\Omega$–cm so that significant voltage drops can occur across long polysilicon conductors. Perhaps more significant, however, the *RC* time constants associated with the resistance of a long polysilicon interconnection and its capacitance to the substrate can slow signal propagation through the *IC*. Therefore, alternative materials that are more conductive than polysilicon are beginning to be used for interconnections. Silicides of the refractory metals, such as tungsten silicide (WSi_2), molybdenum silicide ($MoSi_2$), tantalum silicide ($TaSi_2$), and titanium silicide ($TiSi_2$), as well as the refractory metals themselves, are being employed.

An interconnection phenomenon that poses an *IC* reliability problem is *electromigration*, which can cause a circuit to fail after a few hundred hours of successful operation by having an interconnection become discontinuous.

Electromigration refers to the movement of atoms of the conducting material as a result of momentum exchange between the mobile carriers and the atomic lattice. In aluminum the moving electrons collide with atoms and push them toward the positively biased electrode (Figure 2.31). As a result, aluminum piles up near this electrode and is depleted from other parts of the conductor, especially from the regions near the intersection of grain boundaries in the polycrystalline aluminum film. This transfer of material eventually causes voids in the film and a discontinuous interconnection. Electromigration occurs more rapidly at higher current densities and in severe temperature gradients. For aluminum, electromigration becomes a concern at current densities above 10^5 A cm^{-2}.

Electromigration can be reduced by adding small quantities of a second metal, such as copper, to the aluminum to inhibit the movement of aluminum atoms along the grain boundaries. The addition of 2–3 percent copper can increase the long-term, current-handling capability by two orders of magnitude without greatly increasing the resistivity of the film. Alternatively, higher-temperature metals such as tungsten may be used for metallization.

In addition to resistivity and electromigration, other points to consider in choosing an interconnection material include the following: (1) ability to make ohmic contacts to both *n*- and *p*-type silicon, (2) stability in contact with silicon after the circuit is completed, (3) adhesion to both silicon and silicon dioxide, (4) ability to be patterned using available lithography and etching (especially dry etching) techniques, (5) resistance to corrosion by reaction with the environment, (6) ability to be bonded to make connection to a suitable package, (7) coverage of

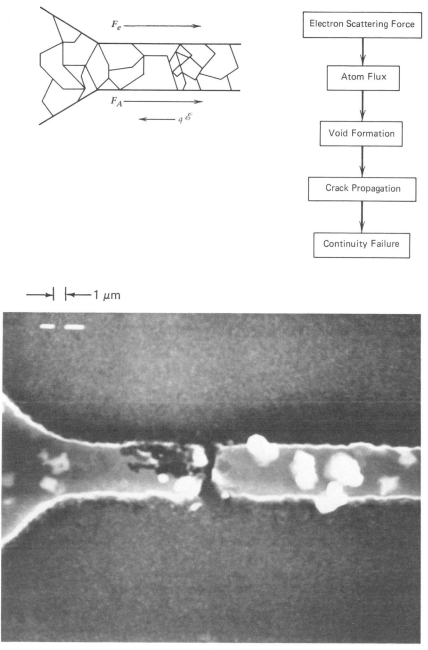

Figure 2.31 Electromigration mechanism in a conducting stripe. Directions of electron flux F_e, electrostatic force $q\mathscr{E}$ and resultant atomic flux F_A (upper left). *SEM* microphotograph showing void formation to the left of the break and accumulation of material in the form of hillocks to the right of the break (lower figure). The steps leading to electromigration failure are indicated at the upper right.[16]

steps in the *IC*, and (8) ability to be deposited without degrading the characteristics of devices already present in the *IC*. While no single interconnection material is optimum for all these requirements, aluminum and its alloys have satisfied enough criteria to become widely used. As the requirements on the interconnections become more severe, however, the limitations of aluminum (especially electromigration) are becoming more evident, and the search for alternate materials is continuing.

After the interconnection layer is deposited and defined, the wafer is placed in a low-temperature furnace at about 450°C to alloy the metal to the silicon, thus ensuring good ohmic contact. This heat treatment also improves the quality of the Si–SiO$_2$ interfaces. With the completion of the interconnection patterning, the processing of the *IC* wafer is complete. Several important steps are still necessary, however, before the *IC* is ready to be used by the customer.

Testing and Packaging

After the wafer fabrication process is complete, the *IC*s are electrically tested to determine which are functional so that only these need to be packaged. As the complexity of *IC*s increases, testing, which is typically done under computer control, becomes more and more difficult. Ease of electrical testing is an important consideration even in the initial circuit design and layout. The chip is often designed so that critical internal voltages can be accessed externally to determine whether the circuit is operating properly.

After this preliminary *functional testing* the wafer is *diced* into individual circuits or *chips*, often by fracturing the silicon along weak crystallographic planes after *scribing* the surface with a sharp, diamond-tipped instrument. Other techniques for dicing include sawing the wafer apart or melting part way through the wafer with a laser before breaking it. The back of each functional chip is then soldered to a package, and wires are connected or *bonded* from the leads on the package to the metal *pads* on the face of the semiconductor chip (Figure 2.32). Finally, the

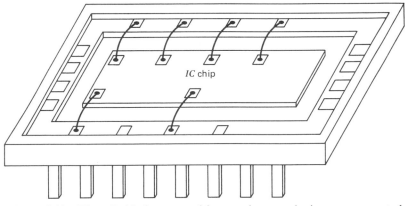

IC chip

Figure 2.32   The *IC* chip is mounted in a package, and wires are connected to the external leads.

package is sealed with a protective ceramic or metal cover or with plastic, and the circuit then undergoes further electrical testing.

As more complex systems containing many *IC* chips are designed, the number of interconnections required for communication between chips increases, and alternative packaging techniques are being explored. The requirement for more interconnections and greater reliability and packing density has led to the development of ceramic substrates containing several layers of metal interconnections. In some cases the *IC* chips are bonded face down on these substrates so that the metal pads on the *IC* chip are directly above corresponding pads on the ceramic. All leads are then simultaneously bonded by melting pre-formed *solder bumps* on the *IC* pads in what is called *flip-chip* bonding.

Power dissipation is an increasingly important factor in designing an *IC* or in choosing an assembly technique. Conventional packaging techniques limit power dissipated in a chip to approximately one watt, whereas more elaborate packaging and cooling techniques can increase this figure to several watts. As device dimensions decrease, the heat generated within a given size chip may remain constant or even increase. Power dissipation can, therefore, limit the complexity of an *IC* chip or lead to the choice of one type of circuit technology in preference to another. Bipolar circuits generally dissipate more power than do MOS circuits, and single-channel MOS circuits dissipate more than complementary MOS (CMOS) circuits, in which little *dc* power flows. These various *IC* technologies are discussed further in the following chapters.

2.8 Process Modeling

As integrated-circuit fabrication processes become more complex, analytical determination of the dopant profiles becomes increasingly difficult. At the same time, prediction of these variables is more important for the proper design of an *IC* process because their determination by experiment becomes extremely time consuming and difficult. Sophisticated computer programs have been developed to model the structures obtained by a given sequence of fabrication steps. Such simulation allows one to optimize a process theoretically before extensive experimental fabrication takes place, and can speed the development of process innovations and novel devices and circuits.

The modeling techniques can be strictly numerical or else they can combine numerical methods with analytical expressions. An analytical expression, when one can be used, often aids the designer by providing better visualization of the process model. More accurate formulations of fabrication processes [e.g., using the more complex diffusion equation (Equation 2.5.15) in place of the simpler Fick's-law expression (Equation 2.5.7)] typically leads to equations that can only be solved by numerical techniques.

Both one- and two-dimensional process-modeling programs have been formulated. While the one-dimensional programs are more well-developed, two-dimensional effects are becoming increasingly important as device features shrink and lateral interactions between device regions become more critical. The programs

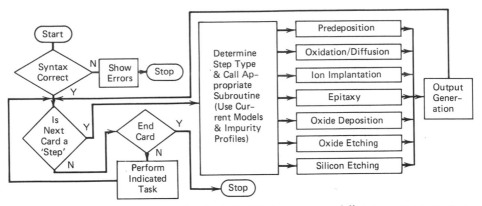

Figure 2.33　Block flow diagram for the *SUPREM* process-modeling computer program.

generally consider the individual process steps in the same sequence as the operations would physically be performed on a silicon wafer. In addition to predicting the final dopant profiles, some electrical parameters may be calculated directly by these programs, and the calculated dopant profiles can be used as the input of a device-modeling program to obtain more-detailed device parameters.

One widely used, one-dimensional modeling program is called *SUPREM* (for *Stanford University Process Engineering Modeling Program*).[5] The program input for *SUPREM* is a description of the processing schedule, specifying a sequence of times, temperatures, ambients and other parameters for diffusion, oxidation, implantation, deposition and etching. The output is the one-dimensional impurity profile in the bulk silicon and some overlying layers, such as SiO_2 or polysilicon. The basic structure of the *SUPREM* program, illustrated in Figure 2.33, is designed so that process steps can be simulated either individually or sequentially, with the dopant profile predicted after one operation used as the input for the next operation. This program includes detailed models for nonlinear diffusion, dopant segregation during oxidation, evaporation at the solid-gas interface, effects of the moving Si–SiO_2 boundary beneath a growing silicon-dioxide layer, impurity clustering during diffusion, concentration-enhanced oxidation, epitaxy, and ion-implantation, as well as several other models that go beyond the first-order considerations that can be treated analytically.

EXAMPLE　Process Modeling a Boron Implantation[6]

Use the modeling program *SUPREM* to investigate a silicon wafer that is (1) ion implanted with boron (B) and (2) subjected to several subsequent high-temperature process steps. Find the distribution of the boron after processing is completed.

Solution

The input to *SUPREM* includes several lines of computer code. First, a title such as the following is assigned to identify the analysis.

TITLE: BORON IMPLANT AND REDISTRIBUTION

Next, the silicon substrate is described. The dopant species and concentration are specified together with the crystal orientation.

SUBS ELEM = P, CONC = 2E15, ORNT = 100

In this example the substrate is doped with phosphorus (P) to a concetration $N_d = 2 \times 10^{15}$ cm^{-3} and has a (100) crystal orientation.

Parameters relating to the grid spacing in the vertical direction are then described. The vertical dimension in the silicon is divided into two portions, a high-resolution region just beneath the surface, and a lower-resolution region farther from the surface. The grid spacing, depth of the high-resolution region, and total depth to be simulated are specified by the user. There are typically 350 to 400 allowable grid points along the vertical dimension. Those points not specified for the high-resolution region are automatically distributed throughout the lower-resolution region.

GRID DYSI = 0.005, DPTH = 1.5, YMAX = 2.5

This line specifies the grid spacing ΔY to be 0.005 μm in the high-resolution region, which is 1.5 μm deep (thus using 300 points in this region). The total region to be simulated extends to 2.5 μm beneath the silicon surface and will therefore contain 50 to 100 grid points.

At this point, details of the output format for printing and plotting results can be entered. In addition, process models that are not included in the basic *SUPREM* program can be referenced. If there is no further specification, built-in models are used in the calculations.

After these initial parameters are entered, the actual implantation, diffusion, and oxidation operations are specified. A line is included for each operation or *step*.

STEP TYPE = IMPL, ELEM = B, DOSE = 3.2E13, AKEV = 380

In this example, boron is to be implanted with a dose of 3.2×10^{13} cm^{-2} at an implant energy of 380 keV. Subsequent heat-treatment steps are then specified. One step might be the following.

STEP TYPE = OXID, TEMP = 1000, TIME = 30, MODL = DRY1

Here an oxidation at 1000°C for 30 min is specified. The oxidation kinetics are described by the model called DRY1, which has been previously entered. It describes the dilute-oxygen ambient used during insertion of the wafers into the furnace. Other oxidation and inert-ambient heat treatments are then specified.

At the end of the sequence of steps, parameters calculated by the program are printed and the final dopant profile is plotted. In the case under consideration the

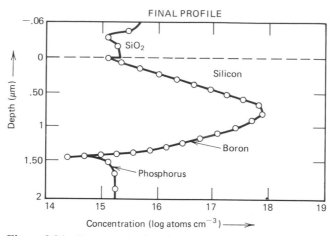

Figure 2.34   Boron and phosphorus profiles at the end of processing as calculated by *SUPREM*.

relevant parameters are the fraction of the boron dose remaining in the silicon, the portion in the oxide, and the dopant concentration at the surface. Figure 2.34 shows the *SUPREM* prediction for the impurity profile representing the boron subjected to the two process steps described above followed by four subsequent heat treatments similar to those used during the fabrication of an MOS transistor.

The models included in *SUPREM* and other process simulators are continually undergoing improvement, but many physical processes are still not completely understood. In addition, process models must often be simplified to complete the calculations in reasonable time. When many sequential operations are used, the small errors introduced in approximating the processes for each step typically compound so that the final predicted profile may be considerably in error. Consequently, measurements are essential to confirm computer predictions and to point the way to improved models for a given process. At present, computer simulations are useful guides, not replicas of reality.

2.9 Device: Integrated-Circuit Resistor

Resistors are simple electronic elements that are important in many integrated circuits. There are several different ways to fabricate them using the processing steps discussed in this chapter. We describe resistor fabrication techniques after a general discussion of the electrical behavior of diffused *IC* resistors.

In Chapter 1 we noted that the resistance of a bar of uniform conducting solid material is given by the equation

$$R = \frac{\rho L}{A} \qquad (2.9.1)$$

where the resistivity ρ is the reciprocal of the conductivity given by Equation 1.2.7.

$$\sigma = \frac{1}{\rho} = (q\mu_n n + q\mu_p p) \qquad (2.9.2)$$

A frequently used method for forming an integrated-circuit resistor is to define an opening in a protective SiO_2 layer above a uniformly doped silicon wafer and to introduce dopant impurities of the opposite conductivity type into the wafer as shown in Figure 2.35. In Chapter 4 we will see that the junction between two regions of opposite conductivity type presents a barrier to current flow. Therefore, if contacts are made near the two ends of the p-type region and a voltage is applied, a current will flow parallel to the surface in this region. It is not possible to use Equation 2.9.1, however, to calculate the resistance of the resistor because it is not a uniform bar. As shown in Figure 2.19, the dopant concentration resulting from the processing described above decreases from a maximum near the surface as one moves into the silicon. To calculate the resistance in this case, it is useful to work in terms of the conductance parallel to the surface.

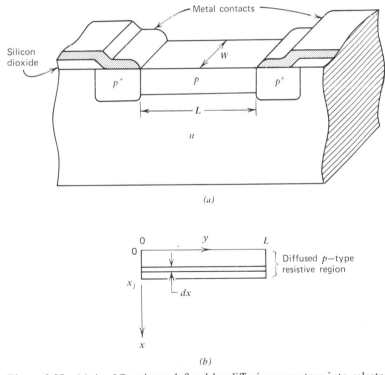

(a)

(b)

Figure 2.35 (a) An *IC* resistor defined by diffusing acceptors into selected regions of an n-type wafer. The p^+-regions are highly doped to assure good contact between the metal electrodes and the resistive p-type region. (b) The dimensions of a thin region in the resistor having conductance dG given by Equation 2.9.3.

Conductance. Consider a p-type resistor made by adding a p-type dopant into an n-type wafer as shown in Figure 2.35. The differential conductance dG of a thin layer of the p-type region of thickness dx parallel to the surface and at a depth x (shown in Figure 2.35b) is

$$dG(x) = q\mu_p p(x) \frac{W}{L} dx \tag{2.9.3}$$

We can find the conductance G of the entire p-type region by summing the conductance of each thin slab from the surface down to the bottom of the layer. This sum becomes an integral in the limit of many thin slabs.

$$G = \frac{W}{L} \int_0^{x_j} q\mu_p p(x)\, dx \tag{2.9.4}$$

where x_j is the depth at which the hole concentration becomes negligible (close to the point where $N_a = N_d$).

If the p-region has been formed in a gaseous deposition cycle followed by a drive-in diffusion, we can approximate the dopant profile $N_a(x)$ by a Gaussian distribution (Equation 2.5.13). As shown in Section 2.5, the total density of dopant N' per unit area diffused into the wafer is determined by the nature of the deposition process, whereas the characteristic length of the diffused region $2\sqrt{Dt}$ is associated with the drive-in diffusion. The hole concentration at any depth into the wafer is approximately given by the net concentration of p-type dopant atoms in excess of the original n-type dopant concentration of the starting wafer. In practice, most of the current in the diffused area is carried in the regions with the highest dopant concentration, which is usually of the order of 10^{18} cm^{-3} or greater. Since the starting wafer often has a dopant concentration of about 10^{15} cm^{-3}, we can neglect the background concentration and assume that $p(x) = N_a(x)$.* Substituting Equation 2.5.13 into Equation 2.9.4 we obtain

$$G = \frac{qN'_p}{\sqrt{\pi Dt}} \frac{W}{L} \int_0^{x_j} \mu_p \left[\exp\left(\frac{-x^2}{4Dt}\right)\right] dx \tag{2.9.5}$$

The mobility μ_p in Equation 2.9.5 is a function of the dopant concentration (Figure 1.15) and, therefore, a function of the distance from the surface. Consequently, we cannot move it outside the integral.

The most straightforward and accurate means of evaluating Equation 2.9.5 is by numerical integration using a computer. An alternative technique, which is often valuable, is to approximate various regions of the curve of Figure 1.15 by analytical relationships which may then be used in Equation 2.9.5. Both of these techniques produce results of high accuracy within the limits of the original assumptions. However, as described in Section 2.5, real diffused profiles may deviate from a simple Gaussian distribution. For this reason, detailed numerical analysis

---

* We have assumed here that the holes and the acceptor atoms have the same depth distribution. In Section 4.1 we show that this assumption is slightly inaccurate. In the present case, however, the small difference between N_a and p can be ignored.

is only warranted when the nature of the diffusion profile has been investigated experimentally for the particular dopant under consideration or more accurate process models such as those in *SUPREM* (Section 2.8) are available.

We can obtain an approximate value of the conductance by using an average value $\overline{\mu}_p$ for the mobility. Since much of the current is carried in a region with a dopant concentration close to the maximum, selecting a value of mobility corresponding to perhaps half the maximum dopant concentration is reasonable and consistent with the use of simplified diffusion theory. The expression for the conductance (Equation 2.9.4) then reduces to

$$G = N'q\overline{\mu}_p \frac{W}{L} = g\frac{W}{L} \tag{2.9.6}$$

where $g \equiv N'q\overline{\mu}_p$ is the conductance of a square resistor pattern ($L = W$). The conductance, in turn, is determined by the product of the average mobility $\overline{\mu}_p$ and the total dopant density per unit surface area N' (Equation 2.5.10). The resistance R is therefore

$$R = \frac{1}{G} = \frac{L}{W}\frac{1}{g} \tag{2.9.7}$$

Generally many resistors in an integrated circuit are fabricated simultaneously by defining different geometric patterns on the same mask. Since the same diffusion cycle is used for all of these resistors, it is convenient to separate the magnitude of the resistance into two parts; the ratio L/W determined by the mask dimensions, and $1/g$ determined by the diffusion process.

Sheet Resistance. Any resistor pattern on the mask can be divided into squares of dimension W on each side (Figure 2.36). The number of squares in any pattern is just equal to the ratio L/W. The value of a resistor is thus equal to the product of the number of *squares* into which it can be divided and the parameter $1/g$, which is usually denoted by the symbol $R_\square$, called the *sheet resistance*. The sheet resistance has units of ohms, but it is conventionally specified in units of ohms per square ($\Omega/\square$) to emphasize that the value of a resistor is given by the product of the number of squares times the sheet resistance. For example, a resistor 100 μm long and 5 μm wide contains 20 squares (20 $\square$). If the diffusion process used produces a diffused layer with a sheet resistance of 200 $\Omega/\square$, the value of the resistor is $20\ \square \times 200\ \Omega/\square = 4.0\ k\Omega$.

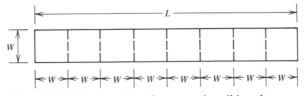

Figure 2.36 The number of squares describing the surface dimensions of a resistor is given by the ratio L/W.

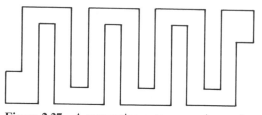

Figure 2.37 A serpentine pattern can be used when a long, high-valued resistor must be designed.

Attainable values of sheet resistance are such that resistors in the kilohm range and higher require patterns containing many squares. Since the width of the pattern is determined by the ability to mask and etch very narrow lines, the length of the resistor can become great in order to obtain the required number of squares. The large area needed for high-value resistors is a practical limitation in an integrated circuit, and circuits are usually designed to avoid large-value resistors. When a resistor containing a large number of squares must be used, it is generally designed to minimize area by using a serpentine pattern as shown in Figure 2.37. The current flow across the corner square of such a pattern is not uniform. The resistance contributed by a corner square can be estimated by taking it to be approximately 65% of the straight-path value.

For large-area resistors, the dimensions L and W are simply determined by the mask dimensions. However, for very narrow resistors (small W) the effective width W_{eff} may differ substantially from the mask dimension because impurities diffuse laterally under the oxide as well as vertically (Figure 2.38). If W is much greater than the diffusion depth x_j, we may neglect this effect. However, W is often made as small as masking tolerances allow so that a given number of squares fit into the smallest area. In this case x_j can be an appreciable fraction of W, and an effective value W_{eff} must be used for the width of the resistor to account for the lateral diffusion.

Figure 2.39 is a microphotograph of an integrated circuit (Type 741 Operational Amplifier) in which several diffused-resistor patterns can be identified. The resistance of the pattern indicated by the arrow is nominally 4kΩ.

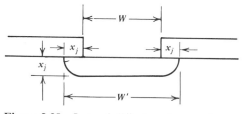

Figure 2.38 Lateral diffusion changes the dimensions of the resistor from the nominal mask dimensions.

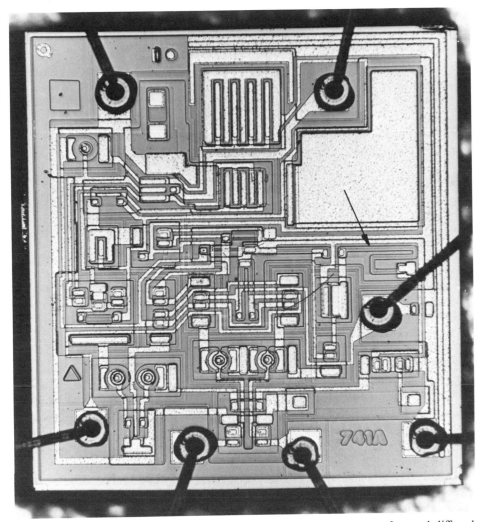

Figure 2.39 Microphotograph of an integrated circuit that makes use of several diffused resistor patters. A 4 kΩ resistor is indicated by the arrow. (*Courtesy Signetics Corp.*)

EXAMPLE   Diffused Resistor

A *p*-type resistor pattern is laid out for an *IC* with two highly conducting *p*-type regions contacting a resistive stripe that extends 24 μm between contacts and is 6 μm wide. The stripe has a junction depth x_j of 6 μm. The desired value of the resistance is 1kΩ.

Determine the sheet resistance and the average resistivity that is required to meet the specifications given. (Neglect lateral diffusion in this example.)

Solution

The number of squares in the resistor pattern is

$$L/W = 24/6 = 4$$

Hence, the sheet resistance R is

$$R = 1000/4 = 250\Omega/\square$$

From Equation 2.9.6 and the relationship $R_{\square} = (1/g)$, we can calculate the required dopant density per unit area N_a' as

$$N_a' = (q\overline{\mu_p}R_{\square})^{-1} \text{ dopant atoms cm}^{-2}.$$

The average volume density of impurity atoms $\overline{N_a}$ is related to the area density N_a' through the equation

$$\overline{N_a} = \frac{N_a'}{x_j}$$

and the average resistivity in the p-doped diffused resistor is (from Equation 2.9.2)

$$\overline{\rho} = (q\overline{\mu_p}\overline{N_a})^{-1} = \left[\frac{q\overline{\mu_p}N_a'}{x_j}\right]^{-1} = R_{\square}x_j = 250 \times 6 \times 10^{-4} = 0.5 \ \Omega\text{-cm}$$

Precision of Resistance Values. An important point should be made before leaving the subject of diffused resistors. As seen in Equation 2.9.7 the resistance is a function of two factors L/W, which is controlled by the lithography used, and the sheet resistance $1/g$ or $R_{\square}$, which depends on the dopant deposition and redistribution. Control of the sheet resistance is usually the most limiting in the design of a precision resistor. However, although the value of $R_{\square}$ obtained in a process may vary somewhat, it will be nearly constant over a typical die size. Thus, two nearby resistors in an *IC* can be expected to have the same value of sheet resistance, and the resistance ratio between them will be determined by their relative dimensions. With suitable processing care these dimensions can be controlled accurately, and significantly greater precision can be maintained in the ratios of paired resistors than in the resistance values themselves. For this reason, *IC*s are often designed so that their critical behavior depends on the ratio of two resistor values rather than on the absolute value of a specific resistor in the circuit.

Diffused resistors are commonly used in silicon *IC*s because of their compatibility with the remainder of the planar process. They are usually formed at the same time as other circuit elements and, therefore, do not add to the fabrication cost. Instead of diffusion, however, it is also possible to make resistors in *IC*s by forming patterns in epitaxial material which forms an isolating *pn*-junction with the underlying substrate. Bipolar transistor *IC*s typically are built with this type of epitaxial layer. Since the epitaxial material has the highest resistivity of the silicon used to form the circuit, one can obtain sheet resistances by this method that are roughly five or six times larger ($\sim 1000\ \Omega/\square$) than those normally attainable in p-type diffused resistors ($\sim 200\ \Omega/\square$). For even higher values of sheet resistance, a double-diffused structure is sometimes used in which an n-region is formed over the p-region to reduce the vertical dimension of the diffused resistor. Resistors of this type, known as *pinch-resistors*, can increase the sheet resistance by a factor of roughly 40 or 50. It is clearly more complicated to make a pinch-resistor, however, and reproducibility of the sheet resistance is typically poor.

If a particular circuit requires large or accurate values of resistance, alternative techniques may be needed. Instead of forming a resistor by adding dopant atoms to the wafer, a resistive film can be deposited on top of the insulating silicon dioxide that covers most of the circuit. This film is then defined by masking and connected to the rest of the circuit by aluminum interconnections. The use of deposited resistors provides flexibility in the design of circuits. Sometimes these resistors can be formed in a layer of material that is already included in an *IC* process. An example is the use of polysilicon that is deposited as part of the production process for silicon-gate MOS *ICs*. If a usable resistive layer is not a part of a given *IC* process, the added cost of depositing and patterning another layer of resistive material is typically only acceptable as a last resort.

In this discussion of *IC* resistors we have seen that particular constraints imposed by technology influence circuit design. We shall see this pattern repeated when we consider other *IC* devices in subsequent chapters.

Summary

The overwhelming importance of silicon to electronics is a result both of its advantageous material properties and also of the superb control that has been achieved over its technology. The ability to produce on silicon a highly insulating oxide with excellent, repeatable properties and with a well-controlled interface between the element and its oxide is not shared by any alternative set of materials. The *planar process* used to fabricate silicon integrated circuits is the basis for the accurate delineation of small-geometry devices. It also makes possible the simultaneous fabrication of many devices, and thus is the key to uniformity, reliability, and economy in *IC* production.

Large, single crystals of silicon are usually grown by the *Czochralski* method to provide the starting material for *IC* production. *Float-zone* refining is sometimes used when the oxygen content in finished ingots must be kept low. The single crystals are sliced into *wafers* before beginning the planar process. The oxidation of silicon to produce SiO_2 can be carried out in a dry ambient at temperatures near 1000°C or (at a significantly faster rate) in a steam ambient at similar temperatures. Production of the intricate patterns for an *IC* requires the definition of patterns in polymer *resist* films by lithographic techniques. *Selective* removal of oxide layers and other materials in *ICs* is accomplished using the patterned resist. Dopant impurities that can alter the conductivity (changing its magnitude as well as conductivity type) can then be added in patterns on the surface that conform to those in the original resist layer. Both *ion implantation* and *gaseous deposition* are used to deposit the dopant atoms in the desired areas of the *IC*. Subsequent *diffusion* of the dopants into the wafer can be modeled most simply by *Fick's Law*, which is a partial differential equation having two analytic solutions (*Gaussian* and *complementary error function*) that apply most frequently to *IC* diffusion processes. Fick's Law, however, does not account for several more complex aspects of diffusion, and computer modeling may be necessary to predict the diffusion process for more accurate *IC* design. *Chemical vapor deposition* is an important *IC* technology for producing single-crystal silicon on a single-crystal substrate (*epitaxy*) as well as depositing *polycrystalline silicon* and insulating films. Interconnection of the devices in an *IC* is an important step which places severe requirements on the conducting material used. The most frequently used conductor, aluminum, is becoming more susceptible to *electromigration* as device and conductor dimensions are reduced. Process-modeling using large computer programs such as *SUPREM* provides a useful tool for the engineer to investigate impurity profiles and to make predictions about the device behavior that can be expected for an *IC* design without carrying through all the steps of processing.

Diffused resistors are widely used in *IC*s. They are typically fabricated by employing the planar process to delineate a pattern that defines the current path for drifting carriers. Resistance values can be calculated from the *sheet resistance*, which is measured in ohms per square ($\Omega/\square$), and given the symbol $R_\square$. The sheet resistance represents the resistance across the opposite edges of a square pattern; hence, the resistance of a diffused resistor is obtained simply by multiplying $R_\square$ by the number of squares making up the resistor pattern.

References

1. J. F. Gibbons, W. S. Johnson, and S. W. Mylroie, *Projected Range Statistics*, 2nd ed. Dowden, Hutchinson, and Ross, New York (1975).
2. Research Triangle Institute, *Integrated Silicon Device Technology, Vol. IV, Diffusion* (*ASD-TDR-63-316*). Research Triangle Institute, Durham, N. C. (1964). Reprinted by permission of the publisher.
3. R. B. Fair, *Semiconductor Silicon 1977*. Electrochemical Society (1977), p. 968.
4. J. A. Appels, E. Kooi, M. M. Paffen, J. J. H. Schatorje, and W. H. C. G. Verkuylen, *Philips Research Rep.* **25**, 118 (1970).
5. D. A. Antoniadis and R. W. Dutton, *IEEE J. Solid-State Circuits* **SC-14**, 412 (April 1979).
6. R. D. Rung, "Silicon *IC* Technology Series: Computer Simulation of Silicon Processing," Videotape No. 90862. Hewlett-Packard Co. (1979).
7. R. R. Razouk, L. N. Lie, and B. E. Deal, *J. Electrochem. Soc.* **128**, 2214 (October 1981).
8. B. E. Deal and A. S. Grove, *J. Appl. Phys.* **36**, 3770 (1965).
9. C. Ho, J. D. Plummer, and J. D. Meindl, *J. Electrochem. Soc.* **125**, 665 (April 1978).
10. M. C. H. M. Wouters, H. M. Eijkman, and L. J. van Ruyven, *Philips Research Rep.* **31**, 278 (1976).

11. M. M. O'Toole, E. D. Liu, and G. W. Ray, *Hewlett-Packard J*. **33**, 5 (August 1982).
12. F. A. Trumbore, *Bell System Tech. J*. **39**, 205 (1960), G. Masetti, D. Nobili, and S. Solmi, *Semiconductor Silicon 1977*. Electrochemical Society (1977) p. 648, A. Armigliato, D. Nobili, P. Ostoja, M. Servidori, and S. Solmi, *Semiconductor Silicon 1977*. Electro-chemical Society (1977), p. 638, D. Nobili, A. Carabelas, G. Celotti, and S. Solmi, *J. Electrochem. Soc.*, **130**, 922 (April 1983), R. A. Craven, *Semiconductor Silicon 1981*. Electrochem. Soc. (1981), p. 254.
13. E. S. Meieran and T. I. Kamins, *Solid-State Electr*. **16**, 545 (1973).
14. B. Swaminathan, *Doctoral dissertation*, Department of EECS, Stanford University (April 1983).
15. R. S. Rosler, W. C. Benzing, and J. Baldo, *Solid-State Technology* **19**, p. 45 (June, 1976).
16. P. P. Merchant, *Hewlett-Packard Journal*, **33**, 28 (August 1982).
17. F. M. Smits, *Bell System Tech. J*. **37**, 711 (1958).

Textbooks

W. E. Beadle, J. C. C. Tsai, and R. D. Plummer (Editors), *Quick Reference Manual for Silicon Integrated-Circuit Technology*, Wiley-Interscience, New York (1985).
S. M. Sze (Ed.), *VLSI Technology*, McGraw-Hill, New York (1983).
R. A. Colclasser, *Microelectronics, Processing and Device Design*, Wiley, New York (1980).

Problems

2.1* A crystal of silicon is to be grown using the Czochralski technique. Prior to initiating the crystal growth, 1 mg of phosphorus is added to 10 kg of melted silicon in the crucible.

(a) What is the initial dopant concentration in the solid at the beginning of the crystal growth?

(b) What is the dopant concentration at the surface of the silicon crystal after 5 kg of the melt has solidified?

$$\left[\text{The segregation coefficient } \frac{C_{solid}}{C_{liquid}} \text{ for phosphorus in silicon is 0.3.} \right]$$

2.2 A Czochralski-grown silicon wafer is heated in a nitrogen ambient at a high tempera-ture to evaporate oxygen from the regions of the wafer near the surface. It is then heated at a low temperature to cause the remaining oxygen to precipitate in "clumps." Explain how and why this process improves the electrical properties of devices sub-sequently fabricated in the wafer.

2.3 A silicon wafer is oxidized several times during an *IC* process. Find the total thick-ness of silicon dioxide after each of the following steps which are carried out in sequence.

(a) 60 min at 1100°C in dry O_2 and HCl (enough HCl is added to enhance the oxida-tion rate by 10% over the rate in pure O_2).

(b) 2 h at 1000°C in pyrogenic steam (at 1 atm).

(c) 6 h at 1000°C in dry O_2.

2.4* A silicon wafer is covered with a 200-nm-thick layer of silicon dioxide. What is the added time required to grow an additional 100 nm of silicon dioxide in dry O_2 at 1200°C?

2.5 (a) How long does it take to grow 1 μm of silicon dioxide in steam at 1000°C and one atm? Consider (111) orientation for parts (a) through (d)].

    (b) How long does it take to grow 1 μm of silicon dioxide in steam at 1000°C and 10 atm?

    (c) How long does it take to grow 1 μm of silicon dioxide in steam at 800°C and 1 atm?

    (d) How long does it take to grow 1 μm of silicon dioxide in steam at 800°C and 10 atm?

    [This problem shows that a thick oxide (1 μm) can be grown at a reduced temperature (800°C) by using elevated pressures.]

2.6 Carry out a derivation of Equation 2.3.6.

2.7* In a *LOCOS* process (as described in Section 2.6), an 8-h oxidation in a steam ambient (at 1 atm and 1000°C) is carried out after a 50-nm layer of silicon nitride has been deposited and patterned. After the oxidation, the nitride layer is removed, exposing the original silicon surface. How far above the silicon surface is the top of the grown oxide layer? [About 24 nm of the silicon nitride is converted to silicon dioxide during the 8-h steam oxidation.]

2.8 Calculate the thickness of silicon dioxide produced on the surface of a silicon-nitride layer for every nm of silicon nitride that is oxidized in the *LOCOS* process.

2.9* A deep vertical groove 1 μm wide and several micrometers deep is etched in a silicon substrate. The grooved surface is bare silicon, but the plane silicon surface is covered with a thin layer of silicon nitride which serves as an oxidation mask (Figure P2.9). The wafer is then oxidized in steam at 1 atm and 1100°C to fill the groove with oxide.

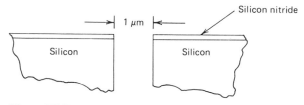

Figure P2.9

    (a) What is the width of the stripe of silicon dioxide that results when the groove is completely filled?

    (b) How long does it take to fill the groove with silicon dioxide?

    [*Hint*: Note, for part a that an oxide stripe x units wide is formed from $(x - 1)$ units of silicon. Do part b by applying Equation 2.3.6 with $\tau = 0$, and the data for (100) silicon in Figures 2.7 *a* and *b*.]

2.10† A dose of phosphorus ions equal to 3×10^{16} cm^{-2} is implanted into a silicon wafer at an energy of 50 keV ($R_p = 63$ nm, $\Delta R_p = 27$ nm) to form contacts to a transistor.

    (a) If the wafer is now oxidized, is it important to consider the effects of concentration-enhanced oxidation?

(b) The phosphorus is now diffused for 60 min at 1000°C prior to an oxidation step. Is concentration-enhanced oxidation important in this case?

(c) Reconsider parts a and b if phosphorus is implanted at 150 keV ($R_p = 180$ nm, $\Delta R_p = 64$ nm).

2.11 After a high concentration of boron is added near the surface of an n–type silicon wafer, a portion of the wafer is covered with a layer of polycrystalline silicon containing a high concentration of crystalline defects. The wafer is then oxidized, and the junction depth is found to be much greater in the regions which were *not* covered with polysilicon during the oxidation. Explain this result.

2.12 The linear coefficient of expansion of glass is 9×10^{-6} per°C. Assume that there is a 1°C increase in the mask temperature between two photolithography steps but that the wafer temperature stays the same. A wafer 10 cm in diameter is processed under these conditions with its center prefectly aligned. What is the minimum misalignment at the edges of the wafer?

2.13 Phosphorus is added to a silicon wafer from a gaseous source at 975°C for 30 min. Determine the junction depth for
(a) a 0.3 Ω-cm p–type substrate,
(b) a 20 Ω-cm p–type substrate.
Assume that the diffusion coefficient of phosphorus is 10^{-13} cm^2 s^{-1} and that its solid solubility is 10^{21} cm^{-3} at 975°C.

2.14[†] Boron atoms in an area concentration of 10^{15} cm^{-2} are introduced into an n-type wafer from a BCl$_3$ source in a carrier gas. The starting wafer has a uniform donor concentration of 5×10^{15} cm^{-3}. The subsequent drive-in diffusion is carried out at 1100°C in a nitrogen ambient. The desired junction depth is 2 μm.
(a) How long should the drive-in diffusion be continued?
(b) Sketch the resulting boron distribution on log N_a and linear N_a versus x plots.
(c) It is required to lay out a 500 Ω resistor. The minimum mask dimension is 4 μm. What is the length of the resistor as indicated on the mask? What is the silicon surface area used? An exact, closed form solution to this part of the problem is not possible, but the sketches of part b should indicate the method of solution. State your approximations.
(d) Qualitatively describe the effect of doubling the time of the drive-in diffusion. Note that the net number of acceptor impurities $N_a - N_d$ can contribute to the conduction. Also note that the mobility decreases as the impurity concentration increases above about 10^{16} cm^3 for Si.

2.15 A *four-point probe* can be used to measure the sheet resistance without making ohmic contact to the semiconductor. The four probes are spaced along a line (as shown in Figure P2.15) and a current is passed through the two end probes. The resulting voltage drop between the inner two probes is measured using a high-impedance meter which draws negligible current. The sheet resistance is given by

$$R_\square = \frac{\pi}{\ln 2} \times \frac{V}{I} = 4.53 \times \frac{V}{I}$$

if the probe spacing is large compared to the thickness of the sample, but small compared to its surface dimensions. [A correction factor can be used for small or thick samples.[17]]

If the four-point probe current $I = 1$ mA, what probe voltage V would be measured on a region in which a density $N' = 10^{12}$ phosphorus atoms cm^{-2} has been diffused into a very-high resistivity p-type wafer? Assume that the junction depth $x_j = 1$ μm and that $\bar{\mu}_n$ is the mobility associated with the average phosphorus density.

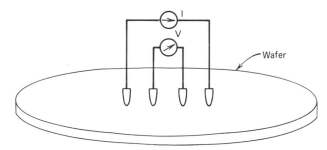

Figure P2.15

2.16 A dose of boron ions equal to 10^{12} cm^{-2} is implanted into a 5 Ω-cm n–type silicon wafer at 100 keV ($R_p = 290$ nm, $\Delta R_p = 70$ nm) and then diffused for 2 h at 1000°C ($D = 2 \times 10^{-14}$ cm^2 s^{-1}).
   (a) What is the peak boron concentration and how wide is the p–type region immediately after the implant?
   (b) What is the peak boron concentration after the subsequent diffusion?

2.17† A dose N' cm^{-2} of a dopant is implanted into silicon with a background dopant concentration C_B to create a pn junction.
   (a) Show that the vertical junction depth is

$$x_j = R_p + \Delta R_p \left[2 \ln \left(\frac{N'}{\sqrt{2\pi}\ \Delta R_p C_B} \right) \right]^{1/2}$$

   (b) Calculate x_j for a dose of 10^{15} As atoms cm^{-2} implanted at 60 keV into a wafer doped with $N_a = 10^{16}$ cm^{-3} boron atoms.

2.18* A polycrystalline silicon interconnection line having a resistivity of 500 μ Ω-cm is 5 μm wide and 0.5 μm thick. Current is supplied through a 1 mm length of this line to charge a capacitor that measures 0.1 $\times$ 0.5 mm^2 on its surface and which has plates spaced on either side of a silicon–dioxide layer that is 100 nm-thick. What is the RC time constant for the resulting resistor-capacitor series connection? [The resistivity for polysilicon given in this problem is about the minimum attainable; hence, this problem indicates a limitation to the use of polysilicon for interconnections in VLSI.]

2.19 Assume that uniform concentrations of 6.55 μm thickness can be imbedded in silicon of the opposite conductivity type to form an integrated-circuit resistor (Figure P2.19).

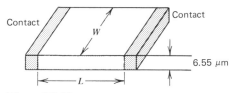

Figure P2.19

Make calculations for two cases: (i) for $N_d = 10^{16}$ cm^{-3} and (ii) for $N_a = 10^{16}$ cm^{-3} to obtain:

(a) The relationships between L and W for resistances of 100 Ω, 1 kΩ, and 10 kΩ between contacts at 25°C.

(b) The actual dimensions if the resistors should dissipate 10 mW of power each and the maximum power dissipation is 1 μW/μm^3.

(c) The temperature coefficients of the resistors (TCR) around 25°C. The TCR for a resistor R is defined by $(1/R)(\partial R/\partial T) \times 100$ (in % per degree).

2.20* Assume that the stepped doping profile shown in Figure P2.20a is attained at the surface of a silicon wafer.

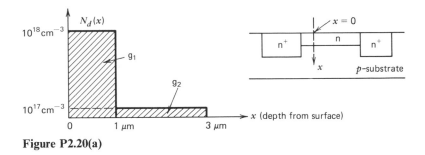

Figure P2.20(a)

(a) Calculate the sheet resistance without using an average mobility.
After some further processing steps, assume that the dopant profile along x changes to the shape shown in Figure P2.20b.

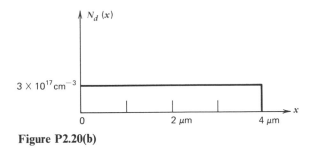

Figure P2.20(b)

(b) Assume that it is somehow possible to introduce extra dopant atoms uniformly between $x = 0$ and $x = 4$ μm. What type of dopant (donor or acceptor) and what concentration should be added to make the sheet resistance for a constant dopant profile (Figure P2.20b) equal to the sheet resistance obtained in part a?

2.21† An acceptor diffusion is carried out, introducing an acceptor profile $N_a = N_s \mathrm{erfc}\,(x/\lambda)$ with $N_s = 10^{18}$ cm^{-3} and $\lambda = 0.05$ μm into a lightly doped n-type silicon crystal. This diffusion will form a resistor in an integrated circuit.

(a) Show that the resistance across any square pattern on the surface of the resistor is approximately given by

$$R_\square \approx \left(q\mu \int_0^\infty N_s \mathrm{erfc}\left(\frac{x}{\lambda}\right) dx \right)^{-1}$$

State clearly where the approximation is made. This expression can be integrated by parts and the value for $R_\square$ written as

$$R_\square = \frac{\sqrt{\pi}}{q\mu N_s \lambda}$$

(b) Derive this form.
(c) What is the approximate resistance if the resistor is made $40\ \mu m$ wide and $2000\ \mu m$ long?
(d) What is the largest resistance that can be fabricated in a surface area that measures 200 by $70\ \mu m$ if $10\ \mu m$ line widths and $10\ \mu m$ spacings are the minimum dimensions? Sketch the resistor pattern. (Consider each corner square to be 65% effective.)

3

METAL-SEMICONDUCTOR CONTACTS

Most of the electronic devices that make up an integrated circuit are connected by means of metal-semiconductor contacts. Moreover, all integrated circuits communicate with the rest of an electrical system via metal-semiconductor contacts. As we shall see, the properties of these contacts can vary considerably, and it is necessary to consider several factors in order to understand them. We shall narrow our discussion whenever necessary to consider only metal contacts to silicon, but first let us consider in general the nature of the thermal equilibrium that is established when a metal and a semiconductor are in intimate contact. The concepts that are developed to understand this equilibrium are very important. They will prove useful many times in our discussion of devices because they underlie the basic properties of semiconductor *pn* junctions as well as the properties of interfaces between semiconductors and insulators and between metals and insulators.

Application of the equilibrium principles to metals and semiconductors provides a simple theory (Schottky theory) of ohmic and rectifying behavior in various metal-semiconductor systems. The Schottky theory is, however, not adequate in many cases: most notably not for metal-silicon systems without further consideration of the real nature of solid interfaces. The important influence of surface states and the origins of these states are therefore discussed. Several applications are then described, but special emphasis is given to the Schottky-diode clamp that has been so widely used in fast logic circuits.

3.1 Equilibrium in Electronic Systems

Metal-Semiconductor System

We shall find it very profitable to construct an energy-band representation for a metal-semiconductor contact. To do so we employ at first a general viewpoint. We consider the metal and the semiconductor as representing two systems of allowed electronic energy states. Using the concepts developed in Chapter 1, we recognize that these systems of allowed states can be regarded as being almost entirely populated at energies less than the Fermi energy, and nearly vacant at higher energies. When the metal and the semiconductor are remote from one another and therefore do not interact, both the systems of electronic states and their Fermi levels are independent. Let us designate the metal as state-system number 1 and the semiconductor as state-system number 2. Each system has a density of allowed states $g(E)$ per unit energy. Of these $g(E)$ states, $n(E)$ are full and $v(E)$ are empty. (The variables are all functions of energy; hence, the use of E in parentheses.) Our terminology is indicated in Table 3.1 and Figure 3.1. The Fermi-Dirac distribution functions

$$f_{D1,2} = \frac{1}{1 + \exp[(E - E_{f1,2})/kT]} \tag{3.1.1}$$

allow us to relate $v_{1,2}(E)$, $n_{1,2}(E)$, and $g_{1,2}(E)$.* The filled state density is

$$n_{1,2} = g_{1,2} \times f_{D1,2} \tag{3.1.2}$$

Table 3.1 **System Properties**

Material	Allowed Electron State Density	Filled State Density	Vacant State Density	Fermi-Dirac Distribution Function
1 (Metal)	$g_1(E)$	$n_1(E)$	$v_1(E)$	$f_{D1}(E,E_{f1})$
2 (Semiconductor)	$g_2(E)$	$n_2(E)$	$v_2(E)$	$f_{D2}(E,E_{f2})$

* A more complete discussion of Fermi–Dirac statistics would show that f_D may have slightly altered forms depending upon specific properties of the allowed electronic states. The alternate forms for f_D typically have factors of 2 or $\frac{1}{2}$ preceding the exponential in Equation 3.1.1. We shall not be concerned with this refinement.

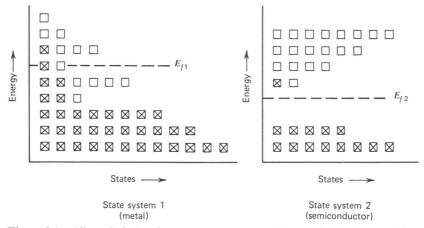

Figure 3.1 Allowed electronic-energy-state systems for two isolated materials. States marked with an x are filled; those unmarked are empty. System 1 is a qualitative representation of a metal; system 2 qualitatively represents a semiconductor.

and the vacant state density is

$$v_{1,2} = g_{1,2} \times (1 - f_{D1,2}) \qquad (3.1.3)$$

In Equations 3.1.1 through 3.1.3 we have omitted the unnecessary parentheses that have served only to emphasize the dependence on energy.

We now consider that the two remote systems are brought into intimate contact. The systems then become interactive and electron transfers take place between them. Equilibrium is reached when there is no net transfer of electrons at any energy. As we noted in Section 1.1, this does not mean a cessation of all processes; rather it means that every process and its inverse are occurring at the same rate. To make this point clear, we consider transfers in space of electrons which are at a given energy E_α. Equilibrium can be expressed mathematically by noting that the transfer probability, is proportional to the population of electrons $n(E_\alpha)$ available for transfer at a given energy and also proportional to the density of available states $v(E_\alpha)$ to which the electrons may transfer. Because of the Pauli exclusion principle, the available state density is not $g(E_\alpha)$, the state density, but $v(E_\alpha)$, the vacant-state density. The proportionality factor relating the transfer probability to these two densities is related to the detailed quantum nature of the states and is the same for transfers from system 1 to system 2 as it is for transfers from system 2 to system 1. Therefore, at thermal equilibrium,

$$n_1 \times v_2 = n_2 \times v_1 \qquad (3.1.4)$$

at any given energy.

Using Equations 3.1.2 and 3.1.3 in Equation 3.1.4, we have

$$f_{D1}g_1(1 - f_{D2})g_2 = f_{D2}g_2(1 - f_{D1})g_1$$

or

$$f_{D1}g_1g_2 = f_{D2}g_2g_1 \tag{3.1.5}$$

Equation 3.1.5 can only be true if $f_{D1} = f_{D2}$ or, by Equation 3.1.1, if $E_{f1} = E_{f2}$. We therefore have established the important property of any two systems in thermal equilibrium: they have the same Fermi energy. Note, from our derivation, that no constraints exist on the density-of-states functions g_1 or g_2. Regardless of the detailed nature of these functions, a statement that their Fermi levels are equal is equivalent to stating that the two systems are in thermal equilibrium. Generalization of this statement to more than two systems or its reduction to apply to one system of states is straightforward. *At thermal equilibrium the Fermi level is constant throughout a system.*

Inhomogeneously Doped Semiconductor

It is instructive to apply these thermal-equilibrium concepts to one additional example. Figure 3.2 is a sketch of the energy-band diagram at thermal equilibrium for an *n*-type semiconductor doped with N_{d1} donors cm^{-3} for $0 < x < a$ and with N_{d2} donors cm^{-3} for $x > a$. The figure represents what is effectively a single system of states because the dopant concentrations are much less than those of the host atoms in the crystal. To sketch the band diagram we first make use of the principle just derived and draw a constant Fermi level. Then, by applying Equation 1.1.21 to both regions of the crystal together with the knowledge that $n \approx N_{d1}$ for $0 < x < a$ and $n \approx N_{d2}$ for $x > a$, we can locate E_c, the conduction-band edge in each region. As we see in the figure, E_c is not constant but moves higher in energy in the lower-doped region. For a correct representation of the band picture close to the doping transition at $x = a$, we shall have to develop further theory. A smooth transition will be sketched for the present. Since the band gap is determined

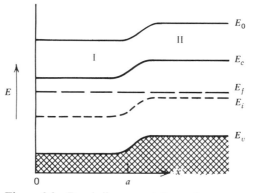

Figure 3.2   Band diagram at thermal equilibrium for an *n*-type semiconductor having doping N_{d1} for $0 < x < a$ and N_{d2} for $a < x$ ($N_{d1} > N_{d2}$).

by the properties of the silicon lattice, it is a constant of the system. Hence, E_v and E_i are drawn parallel to E_c in Figure 3.2 and only E_f is constant with position. The increase in the energies associated with the semiconductor band edges properly represents an increased potential energy for electrons in the less heavily doped region. We shall return to consider this and other points about Figure 3.2 in Chapter 4. For the present our emphasis is on Fermi-level constancy and on the use of this thermal-equilibrium principle to construct an energy-band diagram.

3.2 Idealized Metal-Semiconductor Junctions

Band Diagram

We now turn to the particular features of metal-semiconductor energy-band diagrams. For the purposes of our present discussion, the most distinctive characteristic of the electronic-energy states of the metal and the semiconductor is the relative positions of the Fermi levels within the densities of allowed states $g(E)$. In the metal, the Fermi level is immersed within a continuum of allowed states, while in a semiconductor, under usual circumstances, the density of states is negligible at the Fermi level. Plots of $g(E)$ versus energy for idealized metals and semiconductors are shown in Figures 3.3a and 3.3b.

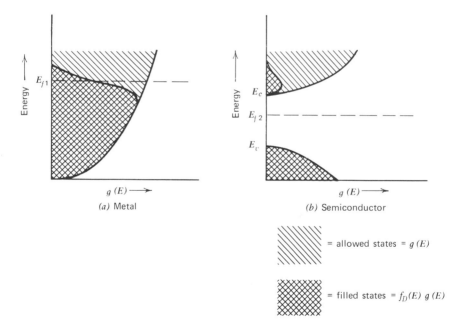

Figure 3.3 (a) Allowed electronic-energy states $g(E)$ for an ideal metal. The states indicated by cross-hatching are occupied. Note the Fermi level E_{f1} immersed in the continuum of allowed states. (b) Allowed electronic-energy states $g(E)$ for a semiconductor. The Fermi level E_{f2} is at an intermediate energy between that of the conduction-band edge and that of the valence-band edge.

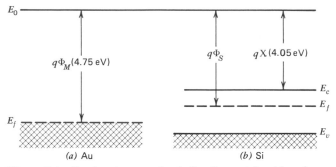

Figure 3.4 Pertinent energy levels for the metal gold and the semiconductor silicon. Only the work function is given for the metal, whereas the semiconductor is described by the work function $q\Phi_S$, the electron affinity qX_S, and the band gap $(E_c - E_v)$.

These points, which were made in general in our discussion of Figure 1.3, are apparent in the specific cases of the allowed energy states for gold and those for silicon shown in Figures 3.4a and 3.4b, respectively. Note that these sketches differ from those in Figure 3.3 in that they represent allowed energies in the bulk of the material versus position and not the density of allowed energy states versus energy. These two types of sketches are sometimes confused.

New quantities also appear in these figures. First, a convenient reference level for energy is taken, that is, the vacuum or free-electron energy E_0. This represents the energy that an electron would have if it were just free of the influence of the given material. The difference between E_0 and E_f is called the *work function*, usually given the symbol $q\Phi$ in energy units and often listed as Φ in volts for particular materials. In the case of the semiconductor the difference between E_0 and E_f is a function of the dopant concentration of the semiconductor, since E_f changes position within the gap separating E_v and E_c as the doping is varied. The difference between the vacuum level and the conduction-band edge is, however, a constant of the material. This quantity is called the *electron affinity*, and is conventionally denoted by qX in energy units. Tables of X in volts exist for many materials. The symbol X is a Greek capital letter *chi*.

The choice of E_0 as a common energy reference makes clear that if Φ_M is less than Φ_S and the materials do not interact, the electrons in the metal have, on the average, a total energy that is higher than the average total energy of electrons in the semiconductor. On the contrary, if Φ_M is greater than Φ_S, the average total energy of electrons in the semiconductor is higher than it is in the metal. For the sake of discussion, we consider the latter case where $\Phi_M > \Phi_S$. When an intimate contact is established, the disparity in the average energies can be expected to cause the transfer of electrons from the semiconductor into the metal.

Another way of looking at the establishment of equilibrium for this case where $\Phi_M > \Phi_S$ is to make use of the concepts developed in Section 3.1. The left-hand side of Equation 3.1.5 in that section is proportional to the flow of electrons from state-system 1 to state-system 2 while the right-hand side of the same equation is proportional to the inverse flow. The net flow is easily seen to be in the direction 2 to 1 if $f_{D2} > f_{D1}$ or, therefore, if $E_{f2} > E_{f1}$. The charge transfer will continue until equilibrium is obtained and a single Fermi level characterizes both the metal and the semiconductor. At equilibrium, the semiconductor, having lost electrons, will be charged positively with respect to the metal.

To construct a proper band diagram for the metal and the semiconductor in thermal equilibrium, we need to note two additional facts. The first of these is that the vacuum level E_0 must be drawn as a continuous curve. This is because E_0 represents the energy of a "just-free" electron and thus must be a continuous, single-valued function in space. If it were not such a function, one could conceive of means to extract work from an equilibrium situation by emitting electrons and then reabsorbing them an infinitesimal distance away where E_0 had changed value. Second, we note that electron affinity is a property associated with the crystal lattice like the forbidden-gap energy. Hence, it is a constant in a given material. Considering these three factors: constancy of E_f, continuity of E_0, and constancy of X in the semiconductor, we can sketch the general shape of the band diagram for the metal-semiconductor system. The sketch is given in Figure 3.5a for an n-type semiconductor for which $\Phi_M > \Phi_S$.

We see from Figure 3.5a that there is an abrupt discontinuity of allowed energy states at the interface. The magnitude of this step is $q\phi_B$ electron volts with

$$q\phi_B = q(\Phi_M - X) \tag{3.2.1}$$

Figure 3.5a indicates that electrons at the band edges (E_c and E_v) in the vicinity of the junction in the semiconductor are at higher energies than are those existing in more remote regions. This is a consequence of the transfer of negative charge from the semiconductor into the metal. Because of the charge exchange there is a field at the junction and a net increase in the potential energies of electrons within the band structure of the semiconductor. The free-electron population is thus depleted near the junction as indicated by the increased separation between E_c and E_f at the surface compared to that in the bulk.

Before considering the electrical properties of the metal-semiconductor junction we should note that our development thus far has relied on an important idealization. The idealization is that the basic band structures of the two materials are unchanged near the surface. We shall gain some useful results by working with this idealized model, but it will be necessary to consider conditions at the surface more carefully later, and then to construct a nonidealized band picture.

Charge, Depletion Region, and Capacitance

The charge and field diagrams for an ideal metal-semiconductor junction are sketched in Figures 3.5b and 3.5c. To the extent that the metal is a perfect conductor, the charge transferred to it from the semiconductor exists in a plane at the metal surface. In the idealized n-type semiconductor, positive charge can consist

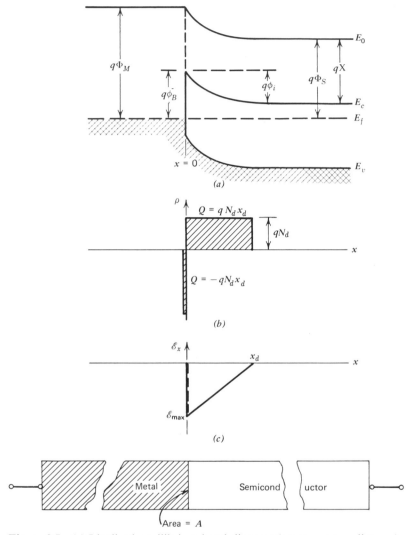

Figure 3.5 (*a*) Idealized equilibrium band diagram (energy versus distance) for a metal-semiconductor rectifying contact (Schottky barrier). The physical junction is at $x = 0$. (*b*) Charge at an idealized metal-semiconductor junction. The negative charge is approximately a delta function at the metal surface. The positive charge consists entirely of ionized donors (here assumed constant in space) in the depletion approximation. (*c*) Field at an idealized metal-semiconductor junction.

either of ionized donors or of free holes while electrons make up the negative charge. We have made several assumptions about the semiconductor charge in drawing Figures 3.5b and 3.5c. First, the free-hole population is assumed to be everywhere so small that it need not be considered; second, the electron density is much less than the donor density from the interface to a plane at $x = x_d$. Beyond x_d, the donor density N_d is taken to be equal to n. These assumptions make up what is usually called the *depletion approximation*. Although they are not precisely true, they are generally sufficiently valid to permit the development of very useful relationships. We shall reconsider the space-charge distributions at a junction in equilibrium in Chapter 4 and consider the depletion approximation in more detail.

Under the depletion approximation, the extent of the space-charge region is exactly x_d units, and the field (for this case of a constant doping in the semiconductor) is a decreasing linear function of position (Figure 3.5c). The maximum field $\mathscr{E}_{max}$ is located at the interface and, by Gauss' law, its value is given by

$$\mathscr{E}_{max} = \frac{-qN_dx_d}{\epsilon_s} \qquad (3.2.2)$$

where ϵ_s is the permittivity of the semiconductor. The voltage across the space-charge region, which is represented by the negative of the area under the field curve in Figure 3.5c, is given by

$$\phi_i = -\frac{1}{2}\mathscr{E}_{max}x_d = \frac{1}{2}\frac{qN_dx_d^2}{\epsilon_s} \qquad (3.2.3)$$

It is also frequently useful to express x_d in terms of ϕ_i. From Equation 3.2.3, we can write $x_d = \sqrt{2\phi_i\epsilon_s/qN_d}$. From Figure 3.5a, the built-in voltage ϕ_i is also seen to be $\Phi_M - \Phi_S = \Phi_M - X - (E_c - E_f)/q$. The space charge Q_s(per unit area) in the semiconductor is

$$Q_s = qN_dx_d = \sqrt{2q\epsilon_sN_d\phi_i} \qquad (3.2.4)$$

Applied Bias. Up to this point, we have been considering thermal equilibrium conditions at the metal-semiconductor junction. Now we take up the case of an applied voltage; that is, a nonequilibrium condition. We saw in Figure 3.5a that there is an abrupt step in allowed electron energies at the metal-semiconductor interface. This step makes it more difficult to cause a net transfer of free electrons from the metal into the semiconductor than it would be to obtain a net flow of electrons in the opposite direction. There is a barrier of $q\phi_B$ electron volts between electrons at the Fermi level in the metal and the conduction-band states in the semiconductor (Figure 3.5a). To first order this barrier height is independent of bias because no voltage can be sustained across the metal. If we refer to Figure 3.5c, we see that the voltage drop across the near delta function of space charge in the metal (equivalent to the area between the $\mathscr{E}$-field curve and the axis) is effectively zero in equilibrium. The total voltage drop across the space-charge region (ϕ_i) occurs within the semiconductor, as can be seen in Figure 3.5a. An applied voltage would similarly be sustained entirely within the semiconductor and would alter the equilibrium-band diagram (Figure 3.5a) by changing the total curvature of the bands, modifying the potential drop from ϕ_i. Thus, electrons in the bulk of the semiconductor at the conduction-band edge are impeded from transferring to the metal by a barrier that can be changed readily from its equilibrium value $q\phi_i$ by an applied bias. The barrier is reduced when the metal is biased positively with respect to the semiconductor, and it is increased under bias of the opposite polarity. Energy-band diagrams for two cases of bias are shown in Figures 3.6a and 3.6b. Since these diagrams correspond to nonequilibrium conditions, they are not drawn with a single Fermi level. The Fermi energy in the region from which electrons flow is higher than is the Fermi energy in the region into which electrons flow. Currents, of course, move in the direction opposite to the electron flow.

To investigate bias effects on the barrier, we shall consider the semiconductor to be grounded and take forward bias to correspond to the metal electrode being made positive. The applied voltage is called V_a, and the bias polarity is indicated in Figure 3.6c. Under reverse bias the metal is negatively biased ($V_a < 0$). If the metal-semiconductor junction is placed under reverse bias so that the voltage drop across the space-charge region increases to ($\phi_i - V_a$), then the space-charge density in the semiconductor increases from its equilibrium value Equation (3.2.4) to

$$Q_s = \sqrt{2q\epsilon_s N_d(\phi_i - V_a)} \tag{3.2.5}$$

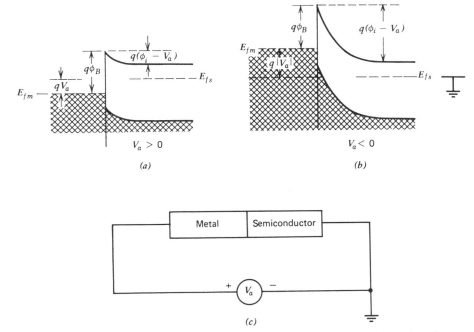

Figure 3.6 Idealized band diagram (energy versus distance) at a metal-semiconductor junction (a) under applied forward bias ($V_a > 0$) and (b) under applied reverse bias ($V_a < 0$). The semiconductor has been taken as the reference (voltage ground) as shown in (c). The vacuum levels for the two cases are not shown.

Under small-signal ac conditions, the junction shows a capacitive behavior that can be calculated by using Equation 3.2.5:

$$C = \left| \frac{\partial Q_s}{\partial V_a} \right| = \sqrt{\frac{q\epsilon_s N_d}{2(\phi_i - V_a)}} = \frac{\epsilon_s}{x_d} \qquad (3.2.6)$$

where C in Equation 3.2.6 represents capacitance per unit area. The last form of Equation 3.2.6 is a general result for small-signal capacitance C. Since C represents the ratio of a *differential* charge (hence a charge with differential linear extent) to differential voltage, it can always be expressed by the ratio of the permittivity to the total space-charge width. A further discussion of this result is given in Chapter 4.

If Equation 3.2.6 is solved for the total voltage across the junction, we obtain

$$(\phi_i - V_a) = \frac{q\epsilon_s N_d}{2C^2} \qquad (3.2.7)$$

The form of Equation 3.2.7 indicates that a plot of the square of the reciprocal of the small-signal capacitance versus the reverse bias voltage should be a straight line, as sketched in Figure 3.7. The slope of the straight line can be used to obtain the doping in the semiconductor, and the intercept of the straight line with the abscissa should equal ϕ_i. Measurements of small-signal capacitance as a function of dc bias and the subsequent construction of plots similar to Figure 3.7 from the data are often used to study semiconductors. In practice, the most serious inaccuracy comes from obtaining ϕ_i from the intercept with the voltage axis. The slope of the curve, however, usually gives an accurate indication of the semiconductor doping.

Measurements of the small-signal capacitance are useful even in cases where the semiconductor doping varies with distance. In that circumstance, the space-charge picture is altered from the one shown in Figure 3.5b to the configuration sketched in Figure 3.8. For a given dc (reverse) bias ($V_a < 0$), the space-charge

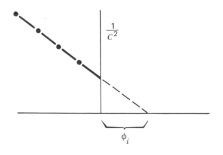

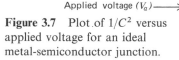

Applied voltage $(V_a) \longrightarrow$

Figure 3.7 Plot of $1/C^2$ versus applied voltage for an ideal metal-semiconductor junction.

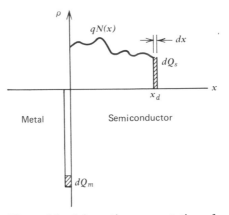

Figure 3.8 Schematic representation of space charge at a metal-semiconductor junction with nonuniform doping in the semiconductor.

layer is of width x_d. A small increase in the magnitude of V_a causes a small increase in Q_s where

$$Q_s = q \int_0^{x_d} N(x)\, dx \tag{3.2.8}$$

Consequently, an increment in voltage dV_a results in an increment in Q_s of value

$$dQ_s = qN(x_d)\, dx = -C\, dV_a$$

or

$$N(x_d) = -C/q(dx/dV_a) \tag{3.2.9}$$

where x_d is the depletion-region width that corresponds to an applied dc voltage V_a at which the small-signal capacitance C is to be measured. We can rewrite Equation 3.2.9 by noting that since the small-signal capacitance C is given by $C = \epsilon_s/x_d$, the derivative (dx/dV_a) can be written $dx/dV_a = (dx/dC)(dC/dV_a) = -(\epsilon_s/C^2)\, dC/dV_a$. Thus

$$N(x_d) = \frac{C^3}{q\epsilon_s(dC/dV_a)} \tag{3.2.10}$$

Equation 3.2.10 can be placed in a more useful form by recognizing that

$$\frac{d(1/C^2)}{dV_a} = -\left(\frac{2}{C^3}\right)\frac{dC}{dV_a}$$

so that

$$N(x_d) = \frac{-2}{q\epsilon_s[d(1/C^2)/dV_a]} \tag{3.2.11}$$

The result that we have derived in Equation 3.2.11 shows that the slope of a plot of $1/C^2$ versus reverse bias voltage directly indicates the doping concentration at the edge of the space-charge layer. This slope, when divided into $(2/q\epsilon_s)$, gives $N(x_d)$ directly. Several commercial "semiconductor profilers" make use of this fact; some have direct-reading outputs that convert the slope to its equivalent doping concentration.

EXAMPLE   Schottky Barrier Diode

A gold Schottky blocking contact is made to n-type silicon at the plane $x = 0$. The silicon has a surface layer extending from $x = 0$ to $x = x_1$ ($x_1 = 20$ nm) with doping $N_{do} = 5 \times 10^{14}$ cm^{-3}. Below this layer ($x > x_1$) the silicon is n-type with a higher doping density (N_{d1} cm^{-3}) as shown in the accompanying sketch. The built-in potential across the metal-silicon junction ϕ_i is 0.50 V. Assume that the work function for gold $q\Phi_M$ is 4.75 eV.

(a) What is the value of N_{d1}, the dopant density in the interior of the silicon?

(b) Sketch the charge density and the electric field in the diode at thermal equilibrium ($V_a = 0$).

(c) Calculate the maximum field assuming that the voltage ΔV_s across the thin surface layer is negligible. Using this calculated maximum field estimate the voltage between $x = 0$ and $x = x_1$ to determine if ΔV_s is negligible.

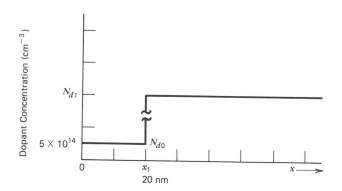

Solution

The surface layer is thin and very lightly doped. We assume that it is fully depleted and that the space-charge region extends into the bulk of the silicon crystal. If this assumption is incorrect our calculations will reveal it. From Figure 3.5,

$$q(\Phi_M - \Phi_S) = q\phi_i = 0.50\,\text{eV}$$

Hence,

$$q\Phi_S = 4.75 - 0.50 = 4.25 \text{ eV}$$

Again, from Figure 3.5,

$$q\Phi_S = qX + (E_c - E_f)$$

or

$$(E_c - E_f) = 4.25 - 4.05 = 0.20 \text{ eV}$$

and

$$(E_f - E_i) = (E_c - E_i) - (E_c - E_f)$$

so that,

$$(E_f - E_i) = 0.562 - 0.20 = 0.362 \text{ eV}$$

(a) Using Equation 1.1.26, we calculate

$$n = N_{d1} = n_i \exp\left(\frac{0.362}{0.0258}\right) = 1.8 \times 10^{16} \text{ cm}^{-3}$$

The calculated density N_{d1} is much higher than N_{d0}, supporting our assumption that the surface layer is fully depleted.

(b) The charge and field diagrams should appear as sketched below.

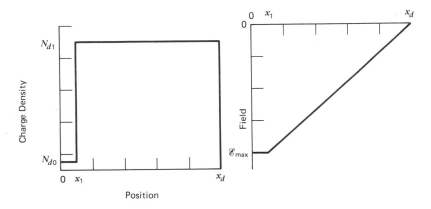

(c) If we neglect the voltage drop across the surface layer, we have, from Equation 3.2.3, $\mathscr{E}_{max} = -2\phi_i/x_d$ where $x_d = \sqrt{2\phi_i\epsilon_s/qN_d} = 190$ nm. Hence, $\mathscr{E}(x = 0) = \mathscr{E}_{max} = -5.27 \times 10^4$ V cm^{-1}. To estimate the voltage across the thin surface layer, we note that the change in the field in this region can be found from Gauss' Law.

$$\Delta\mathscr{E} = \frac{qN_{d0}}{\epsilon_s}\Delta x = \mathscr{E}(x = 0) - \mathscr{E}(x_1) = 154 \text{ V cm}^{-1}$$

Thus the field can be considered essentially constant across the surface layer, and we can calculate $\Delta V_s = -\mathscr{E}_{max} \times x_1 = 0.105$ V. Thus, ΔV_s is about 20%

of ϕ_i, sufficiently high to make it reasonable for the student to consider redoing the problem without assuming that the voltage drop is negligible across the surface layer.

Schottky Barrier Lowering.[†] We now reconsider our earlier statement that the barrier to electron flow from the metal into the semiconductor is "to first order" unchanged by an applied bias. The small dependence on applied voltage of this barrier height can be observed particularly under reverse bias. The dependence corresponds to an effect that was explained many years ago by Walter Schottky when electron emission into a vacuum was studied.

In our derivation we consider that within the semiconductor close to the interface it should be possible to approximate the electron energy by using the free-electron theory and by treating the metal as a plane conducting sheet. In this model we take account of the semiconductor in two ways: by assigning the electron an effective mass m_n^*, and by using a relative permittivity that differs from unity ($\epsilon_r = 11.7$ for Si). A sketch of the appropriate energy picture is given in Figure 3.9 under equilibrium conditions and also for the case of an applied field that tends to move electrons away from the metal surface. The function $E_1(x)$ that represents electron energy in the figure is derived by the classical technique; the conducting metal plane has the same effect on the electron as an image charge of opposite sign spaced equidistant behind the plane $x = 0$. In the presence of a field $-\mathcal{E}$ tending to move electrons away from the metal surface, the electron energy $E_2(x)$

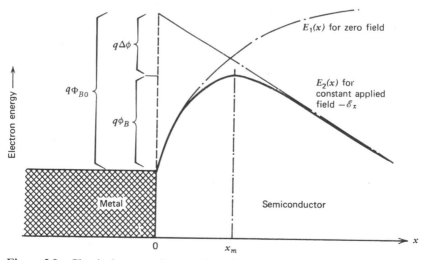

Figure 3.9 Classical energy diagram for a free electron near a plane metal surface at thermal equilibrium $[E_1(x)]$, and with an applied field $-\mathcal{E}_x[E_2(x)]$.

is

$$E_2 = \frac{-q^2}{16\pi\epsilon_s x} - q\mathscr{E}x \qquad (3.2.12)$$

From studies of metal-vacuum systems, Equation 3.2.12 has been shown to be accurate for distances greater than a few nanometers. The plane at which E_2 is a maximum is easily found, as is the energy $q\,\Delta\phi$ in Figure 3.9

$$q\,\Delta\phi = \sqrt{\frac{q^3\mathscr{E}}{4\pi\epsilon_s}} \qquad (3.2.13)$$

According to this model the height of the barrier to electron flow out of the metal $q\phi_B$ varies in the same way as does $q\,\Delta\phi$. The current flow is exponentially dependent on this height since only that fraction of the Boltzmann-distributed electrons in the metal with energies above the barrier maximum can pass across it. One therefore expects the emitted current from the metal under reverse bias to vary as

$$J = J_0 \exp\frac{\sqrt{q^3\mathscr{E}/4\pi\epsilon_s}}{kT} \qquad (3.2.14)$$

Using the depletion approximation, we can relate the field $\mathscr{E}$ to the bias V_a and built-in potential ϕ_i by

$$\mathscr{E} = \sqrt{\frac{2qN_d}{\epsilon_s}(\phi_i - V_a)} \qquad (3.2.15)$$

Equations 3.2.14 and 3.2.15 show that because of Schottky barrier lowering, the current emitted over a reverse-biased blocking contact depends exponentially on the fourth root of voltage at higher biases. Although this type of an exponential dependence is sometimes observed in practice, reverse currents that result from the generation of free carriers in the space-charge region may be larger than the component that we have considered here. In that case, the dependence on bias is more gradual. We shall consider generation as a source of reverse current when we discuss current flow in *pn* junctions in Chapter 5.

3.3 Current-Voltage Characteristics

The basic dependence of current on voltage in a Schottky-barrier diode can be deduced from qualitative arguments. These arguments provide fundamental insight into the nature of the equilibrium behavior of the metal-semiconductor system, and we therefore consider them before making a more rigorous calculation of the $I-V$ characteristic.

The starting point for the derivation is to consider the band diagram at thermal equilibrium shown in Figure 3.5. At equilibrium, the rate at which electrons cross over the barrier into the semiconductor from the metal is balanced by the rate at which electrons cross the barrier into the metal from the semiconductor. From the discussion of diffusion in Chapter 1, we know that free carriers in crystals are constantly in motion because of their thermal energies. In Problem 1.13, for example, this fact was used to deduce that a density n_0 of free carriers in thermal motion can be considered to cause a current density equal to $-qn_0v_{th}/4$ in an arbitrary direction. At thermal equilibrium, of course, this current density is balanced by an equal and opposite flow, and there is zero net current. Applying this concept to the boundary plane of the band diagram in Figure 3.5, we can consider that there is a tendency of electrons to flow from the semiconductor to the metal and an opposing balanced flux of electrons from the metal into the semiconductor. These currents are proportional to the density of electrons at the boundary. In the semiconductor, this density n_s, from Equation 1.1.21 is

$$n_s = N_c \exp\left(-\frac{q\phi_B}{kT}\right) \tag{3.3.1}$$

which can be written in terms of the bulk density $n = N_d$ by applying Equation 1.1.21 in the bulk of the semiconductor and noting from Figure 3.5 that $q\phi_B = q\phi_i + E_c - E_f$ in the semiconductor.

$$n_s = N_d \exp\left(-\frac{q\phi_i}{kT}\right) \tag{3.3.2}$$

Thus, equilibrium at the junction corresponds to

$$|J_{MS}| = |J_{SM}| = KN_d \exp\left(-\frac{q\phi_i}{kT}\right) \tag{3.3.3}$$

where J_{MS} and J_{SM} are the thermally induced current densities (current per area) directed from the metal toward the semiconductor and vice versa, and K is a proportionality constant.

When a bias V_a is applied to the junction as in Figure 3.6, the potential drop within the semiconductor is changed and we can expect the flux of electrons from the semiconductor toward the metal to be modified. If we assume that the surface and the bulk of the semiconductor remain nearly at thermal equilibrium under bias, then the equation for n_s will be modified from Equation 3.3.2 to

$$n_s = N_d \exp\left[-\frac{q(\phi_i - V_a)}{kT}\right] \tag{3.3.4}$$

The current from the metal to the semiconductor that results from the electron flow out of the semiconductor is therefore altered by the same factor. The flux of electrons from the metal to the semiconductor, however, is not affected by the applied bias because the barrier ($q\phi_B$) remains at its equilibrium value.

Subtracting these two components, we obtain an expression for the net current from the metal into the semiconductor under applied bias:

$$\begin{aligned} J &= J_{MS} - J_{SM} \\ &= KN_d \exp\left[-\frac{q(\phi_i - V_a)}{kT}\right] - KN_d \exp\left(\frac{-q\phi_i}{kT}\right) \end{aligned} \tag{3.3.5}$$

which can be written

$$J = J_0[\exp(qV_a/kT) - 1] \tag{3.3.6}$$

where $J_0 = KN_d \exp(-q\phi_i/kT)$ is a new constant.

Equation 3.3.6 is often called the ideal diode equation. It arises, as this derivation makes clear, when a barrier to electron flow affects the thermal flux of carriers asymmetrically. Although more detailed analysis will modify the current equation slightly, the essential dependence of current on voltage for metal-semiconductor barrier junctions is contained in Equation 3.3.6. The ideal diode equation predicts a saturation current $-J_0$ for V_a negative and a very steeply rising current when V_a is positive (Problem 3.10).

More detailed consideration of the $J-V$ characteristics at metal-semiconductor barrier junctions shows that the saturation current J_0 is not completely independent of applied voltage. Analyses that lead to this result are carried out in the remainder of this section.

Schottky Barrier[†]

The dependence of current on applied voltage in a metal-semiconductor junction can be obtained by integrating the equations for carrier diffusion and drift across the depletion region near the contact. This approach, which was first taken by Schottky,[1] assumes that the dimensions of this space-charge region are sufficiently large so that the use of a diffusion constant and a mobility value are meaningful—basically, that the width of the region is at least a few electronic mean-free paths and the field strength is less than that at which the drift velocity saturates. An alternative physical approach, adopted first by Bethe[2a,b] and based on carrier emission from the metal, is valid even if these constraints are not met, and it leads to the same $J-V$ dependence.[3]

If we refer to the metal-semiconductor contact under bias as shown in Figure 3.10 and consider a one-dimensional electron flow through the barrier region, then

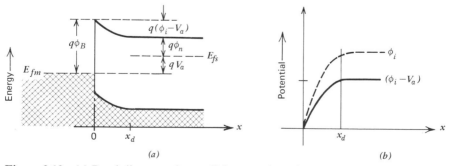

Figure 3.10   (*a*) Band diagram of a rectifying metal-semiconductor junction under forward bias. The applied voltage V_a displaces the Fermi levels: $qV_a = E_{fs} - E_{fm}$. (*b*) The potential across the surface depletion layer is decreased to $\phi_i - V_a$.

as shown in Chapter 1 (Equation 1.2.21), we can write

$$J_x = q\left[n\mu_n\mathscr{E}_x + D_n\frac{dn}{dx}\right] \tag{3.3.7}$$

If we denote potential in the region as ϕ, then by using $\mathscr{E}_x = -d\phi/dx$ and the Einstein relation (Equation 1.2.20), we can write

$$J_x = qD_n\left[\frac{-qn\,d\phi}{kT\,dx} + \frac{dn}{dx}\right] \tag{3.3.8}$$

A sketch of ϕ with the metal taken as ground is shown in Figure 3.10b.

We shall rewrite Equation 3.3.8 in an integrated form that can be evaluated at the two ends of the depletion region ($x = 0$ and $x = x_d$ in Figure 3.10). Multiplying both sides of Equation 3.3.8 by an integrating factor $\exp(-q\phi/kT)$ permits direct integration of the right-hand side. Using the limits of the depletion region, we obtain

$$J_x\int_0^{x_d}\exp\left(\frac{-q\phi}{kT}\right)dx = qD_n\left[n\exp\left(\frac{-q\phi}{kT}\right)\right]_0^{x_d} \tag{3.3.9}$$

In writing Equation 3.3.9 we have assumed that the particle current represented by J_x is not a function of position and may be placed outside the integral. This assumption has wide validity. Since the reference for potential is adjacent to the metal, the boundary conditions on voltage for use in Equation 3.3.9 are

$$\phi(0) = 0; \qquad \phi(x_d) = (\phi_i - V_a) \tag{3.3.10}$$

As we see from Figure 3.10a, $\phi(x_d)$ can also be written as

$$\phi(x_d) = (\phi_B - \phi_n - V_a) \tag{3.3.11}$$

where $q\phi_n$ is $(E_c - E_f)$ in the semiconductor bulk. Boundary conditions for n are also necessary in order to evaluate Equation 3.3.9. Using Equation 1.1.21, we can express these as

$$n(0) = N_c\exp\left(\frac{-q\phi_B}{kT}\right)$$

$$n(x_d) = N_d = N_c\exp\left(\frac{-q\phi_n}{kT}\right) \tag{3.3.12}$$

Substituting the boundary values into Equation 3.3.9, we find

$$J_x = qD_nN_c\exp\left(\frac{-q\phi_B}{kT}\right)\left[\exp\left(\frac{qV_a}{kT}\right) - 1\right]\bigg/\int_0^{x_d}\exp\left(\frac{-q\phi(x)}{kT}\right)dx \tag{3.3.13}$$

To obtain current as a function of voltage, the functional dependence of ϕ on x must be inserted into the integrand in the denominator of Equation 3.3.13 and the integration must be carried out. This functional dependence of ϕ on x is determined by the doping profile in the semiconductor near the contact.

The contact that we have been considering, a barrier to electron flow from a metal into a semiconductor of constant doping, is called a Schottky barrier in recognition of its initial analysis by Walter Schottky.[1] For the Schottky barrier we can use the depletion approximation, as in Section 3.2, to derive an expression for the potential in the depletion region.

$$\phi(x) = \frac{qN_d}{\epsilon_s} x \left(x_d - \frac{x}{2} \right) \qquad (0 < x < x_d) \qquad (3.3.14)$$

When Equation 3.3.14 is inserted into Equation 3.3.13 and the integration is performed, an explicit solution of J_x as a function of V_a is obtained.

$$J_x = J_S \left[\exp\left(\frac{qV_a}{kT} \right) - 1 \right] \qquad (3.3.15)$$

where

$$J_S = \frac{q^2 D_n N_c}{kT} \left[\frac{2q(\phi_i - V_a)N_d}{\epsilon_s} \right]^{1/2} \exp\left(\frac{-q\phi_B}{kT} \right) \qquad (3.3.16)$$

Equation 3.3.16 shows that J_S is not independent of voltage; hence, some of the voltage dependence in Equation 3.3.15 is implicit in J_S. The square-root dependence of J_s on voltage is, however, weak compared to the term that is exponential in voltage in Equation 3.3.15. It is, therefore, possible to approximate the current-voltage characteristic by writing, instead of Equation 3.3.15:

$$J_x = J_S' \left[\exp\left(\frac{qV_a}{nkT} \right) - 1 \right] \qquad (3.3.17)$$

where J_S' is independent of voltage and n is taken to be a constant having a value that is usually found experimentally to be between 1.02 and 1.15. (See Problem 3.9). The results of experimental measurements on a forward-biased, aluminum-silicon Schottky barrier are shown in Figure 3.11. The good fit between the measured data in Figure 3.11 and Equation 3.3.17 with n = 1.07 is typical.

The analyses that we have just carried out have made use of a number of assumptions, several of which have been explicitly noted. One important implicit assumption, however, should be mentioned before proceeding because it is very frequently relied upon for first-order device analysis. This is the assumption that the system is in *quasi-equilibrium*, that is, almost at thermal equilibrium even though currents are flowing. Quasi-equilibrium was implicitly invoked at a number of points in our analysis; for example, in writing Equation 3.3.4 and in using the Einstein relationship to write Equation 3.3.8. Logically we expect quasi-equilibrium to be more valid under low-bias conditions when currents are small—and, in fact, this is the case. Often, a quasi-equilibrium analysis suffices or else the analysis can be extended to cover all currents within the range of interest by making slight modifications to the theory. The ultimate test of any assumption is, of course, the agreement between measurements and predictions, as seen in Figure 3.11.

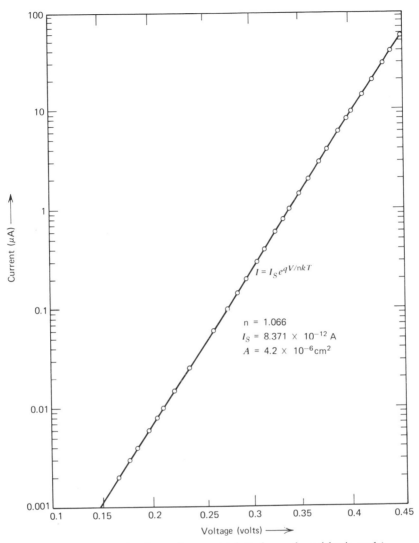

Figure 3.11 Measured values of current (plotted on a logarithmic scale) versus voltage for an aluminum-silicon Schottky barrier. Values for $I_S = J_S'A$ and n result from an empirical fit of the data to Equation 3.3.17).

Mott Barrier[†]

Our derivation thus far has been for a metal-semiconductor junction in which the semiconductor doping is constant throughout the space-charge region; that is, for a Schottky barrier. To derive the current-voltage characteristic for other dopant profiles, we can use the equations through 3.3.13, but we must modify the expression for voltage in the depletion region (Equation 3.3.14). One case of doping that is useful for some metal-semiconductor junctions is known as the Mott barrier after N.F. Mott who analyzed it in connection with studies of diodes composed of oxide compounds.[4]

In the Mott-barrier approximation, the semiconductor is characterized by an abrupt change in doping from a low value near the metal interface to a high value a very short distance from the surface. The distance is short in the sense that essentially no field lines terminate in it. (A shorthand description would be that the distance is much shorter than a Debye length L_D where L_D is discussed in Section 3.4.) An electron-energy diagram for this situation is sketched in Figure 3.12. Because the length of the lightly doped region is very short, the electric field can be assumed to be constant throughout it. The field lines are assumed not to penetrate into the highly doped region because the donor density there is so high. This condition might, for example, model a case in which the doping near the contact is altered during the fabrication of the junction. It might also represent the situation in which the metal contact is made to a thin, lightly doped, epitaxial

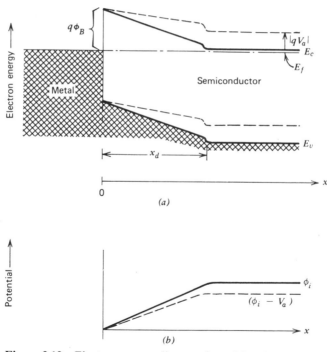

Figure 3.12 Electron-energy diagram for a Mott barrier (near-insulating region at the surface with a sharp transition at $x = x_d$ to a highly conducting region). The solid-line indicates thermal equilibrium; the dotted line, a forward bias of V_a volts with the metal held at ground potential. (b) Potential diagram for the Mott barrier.

film above a highly doped region in the crystal. We shall see that this latter case is encountered in the design of bipolar integrated circuits.

To obtain the dependence of current on voltage for the Mott barrier, we proceed as with the Schottky barrier, first expressing the potential ϕ as a function of x and then performing the integration indicated in Equation 3.3.13. From Figure 3.12 the dependence of ϕ on x is

$$\phi(x) = (\phi_i - V_a)\frac{x}{x_d} \qquad (0 < x < x_d) \tag{3.3.18}$$

Using Equation 3.3.18 in 3.3.13, we obtain a result that may be written in the form of Equation 3.3.15

$$J_x = J_M\left[\exp\left(\frac{qV_a}{kT}\right) - 1\right] \tag{3.3.19}$$

where J_M is more strongly dependent on V_a than is the parameter J_S, derived for the Schottky barrier, (Equation 3.3.16)

$$J_M = \frac{q^2 D_n N_c (\phi_i - V_a)\exp\left(\dfrac{-q\phi_B}{kT}\right)}{x_d kT\left\{1 - \exp\left[\dfrac{-q(\phi_i - V_a)}{kT}\right]\right\}} \tag{3.3.20}$$

Mott and Schottky barriers represent idealized metal-semiconductor rectifiers. In many cases, these idealizations suffice. To derive the J–V characteristic in cases for which they are not adequate, an accurate representation for potential throughout the barrier region must be found and used in place of Equations 3.3.14 or 3.3.18. In general, the major dependence on voltage is, however, contained in the exponential form that we derived both in Equations 3.3.15 and 3.3.19. An empirical fit to measured data is generally possible using a form similar to Equation 3.3.17 with a value of n that is nearly unity.

Both equations that have been derived for saturation currents, J_S in Equation 3.3.16 and J_M in Equation 3.3.20, become physically unreasonable as V_a approaches ϕ_i. In practice, if V_a were to equal ϕ_i, there would be no barrier at the junction and very large currents could then flow. When a large forward bias is applied to a practical diode, a fraction of the voltage is dropped across the resistance of the semiconductor regions in series with the junction. The actual forward voltage applied to the Schottky barrier is virtually never as large as the built-in voltage. A similar consideration will be given greater discussion when currents in *pn* junctions are described in Chapter 5.

3.4 Nonrectifying (Ohmic) Contacts

In our discussion of metal-semiconductor contacts, we have thus far considered cases in which the semiconductor near the metal is depleted of majority carriers relative to the bulk density and in which there is a barrier to electron transfer from the metal. In such cases any applied voltage is dropped mainly across the junction region, and currents are contact limited. *The inverse case, in which the contact itself offers negligible resistance to current flow when compared to the bulk, defines an ohmic contact.* Although this definition of an ohmic contact may sound awkward, it emphasizes this essential aspect: when voltage is applied across a device with ohmic contacts, the voltage dropped across the ohmic contacts is negligible compared to voltage drops elsewhere in the device. Thus, no power is dissipated in the contacts, and the ohmic contact can be described as being at thermal equilibrium even when currents are flowing. An important and useful consequence of this property is that all free-carrier densities at an ohmic contact are unchanged by current flow; the densities remain at their thermal-equilibrium values.

Tunnel Contacts

The metal-semiconductor contacts that we have considered in the previous section can, for example, be made ohmic if the effect of the barrier on carrier flow can be made negligible. In practice this is accomplished by heavily doping the semiconductor so that the barrier width x_d is reduced to a very small value. To see this, refer to Equation 3.2.4 and solve for x_d:

$$x_d = \sqrt{\frac{2\epsilon_s \phi_i}{qN_d}} \qquad (3.4.1)$$

The space-charge region, therefore, narrows as N_d is increased. When the barrier width approaches a few nanometers, a new transport phenomenon, *tunneling* through the barrier, can take place.

Figure 3.13*a* is a schematic illustration of the tunneling process through a very thin Schottky barrier. When the barrier is of the order of nanometers and the metal is biased negatively with respect to the semiconductor, electrons in the metal need not be energetic enough to surmount the barrier ($q\phi_B$ units above the Fermi

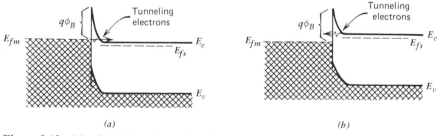

Figure 3.13 Metal-semiconductor barrier with a thin space-charge region through which electron tunneling can take place. (*a*) Tunneling from metal to semiconductor. (*b*) Tunneling from semiconductor to metal.

energy) to enter the semiconductor. Instead, they can tunnel through the barrier into the conduction-band states in the semiconductor. Likewise, when the semiconductor is biased negatively with respect to the metal, electrons from the semiconductor can tunnel into electronic states in the metal (Figure 3.13b). Many electrons are available to take part in these processes and currents rise very steeply as bias is applied. Hence, a metal-semiconductor contact at which tunneling is possible shows a very small resistance. It is virtually always an ohmic contact. Ohmic contacts are frequently made in this way in practice. To assure a very thin barrier, the semiconductor is often doped until it is degenerate (i.e., until the Fermi level enters either the valence or the conduction band).

Schottky Ohmic Contacts[†]

Another method to obtain an ohmic contact is to cause the majority carriers to be more numerous near the contact than they are in the bulk of the semiconductor. An ohmic contact of this type results if the semiconductor surface is not depleted when it comes into equilibrium with the metal, but rather has an enhanced majority-carrier concentration. Using the ideal Schottky theory that we described in Section 3.2, we see that this condition occurs in a metal-semiconductor junction between a metal and an n-type semiconductor with a larger work function than that of the metal. In this case, electrons are transferred to the surface of the semiconductor and the metal is left with a skin of positive charge.[*] The relevant energy diagram is sketched in Figures 3.14a, and the corresponding charge and field diagrams are given in Figure 3.14b and 3.14c. There is a qualitative similarity between these figures and Figures 3.5a to 3.5c that referred to a rectifying contact; the important distinction between the two situations is that the semiconductor charge consists of free electrons in the case of the ohmic contact, but it is fixed (on positive donor sites) in the barrier case. The charge distribution, field, and potential for such a contact can be calculated using techniques similar to those employed for the rectifying contact.[5] Details of this procedure are considered in Problem 3.11. The results are summarized as follows.

To find a solution for the distribution of the space charge in the semiconductor, we take a reference for potential ϕ in the metal. We assume that the excess electrons in the semiconductor surface region (denoted as n') are Boltzmann-distributed in energy so that $n' = n_s \exp(-q|\phi|/kT)$ where n_s represents the excess concentration in the semiconductor at the metal-semiconductor interface. We can write an integrable form of Poisson's equation if we represent the space-charge density in the semiconductor by qn' (which neglects the space charge contributed by donor ions). This approximation is reasonable for the major extent of the space-charge

---

[*] For an ohmic contact to a p-type semiconductor, the relative sizes of the work functions in the two regions need to be reversed to achieve a net positive charge in the semiconductor and, thereby, an enhanced hole density near the contact.

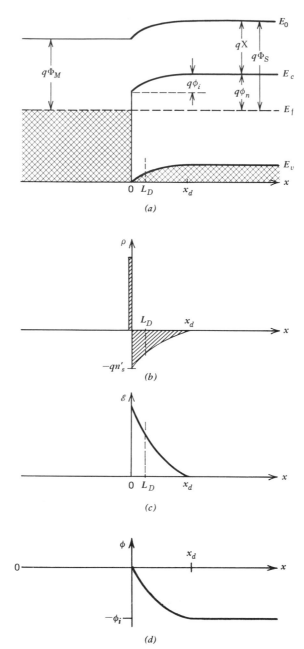

Figure 3.14 (a) Idealized equilibrium energy
diagram for a Schottky ohmic contact between a
metal and an n-type semiconductor. (b) Charge at
an ideal Schottky ohmic contact. A delta function
of positive charge at the metal surface couples to a
distributed excess-electron density $n'(x)$ in the semiconductor. (c) Field, and (d) potential at an idealized
Schottky ohmic contact. The Debye length L_D is a
characteristic measure of the extent of the charge
and field.

region (until the potential is within a few kT/q of the built-in value ϕ_i). Solving Poisson's equation, we find $\rho(x)$ the space-charge density.

$$\rho(x) = -qn_s \bigg/ \left(1 + \frac{x}{\sqrt{2}L_D}\right)^2 \tag{3.4.2}$$

where

$$L_D = \left(\frac{\epsilon_s kT}{q^2 n_s}\right)^{1/2} \tag{3.4.3}$$

is known as the *Debye length* at the surface. We shall discuss the Debye length in more detail after completing our analysis of the contact.

The field varies with distance according to the equation

$$\mathscr{E}_x = \frac{\sqrt{2}kT}{L_D q}\left(1 + \frac{x}{\sqrt{2}L_D}\right)^{-1} \tag{3.4.4}$$

and the space-charge layer extends into the semiconductor a distance

$$x_d = \sqrt{2}L_D \left[\exp\left(\frac{q|\phi_i|}{2kT}\right) - 1\right] \tag{3.4.5}$$

where the built-in voltage ϕ_i is seen from Figure 3.14a to be

$$|\phi_i| = \phi_n - (\Phi_M - X) \tag{3.4.6}$$

where $q\phi_n = (E_c - E_f)$ in the bulk. The condition $\Phi_M - X = \phi_n$ defines what can be called a neutral contact; that is, a contact with no built-in voltage and a surface that has the same density of free electrons as the bulk. In Problem 3.12 we show that a neutral contact can be considered to be "ohmic" for current levels less than $qn_0 v_{th}/4$ where n_0 is the electron density and v_{th} is the thermal electron velocity. For $\Phi_M - X \neq \phi_n$, the current limit for ohmic behavior at the contact differs from $qn_0 v_{th}/4$ by a factor $\exp\{q[(\Phi_M - X) - \phi_n]/kT\}$. Thus, contacts to *n*-type material are properly divided into *ohmic* behavior for cases in which the surface bands bend down and into *blocking* behavior for cases in which the bands bend up. The inverse conditions apply for contacts to *p*-material.

Summing up, we repeat the essential condition for ohmic behavior: an unimpeded transfer of majority carriers between the two materials forming the contact. At ohmic contacts there are generally built-in potentials. Unless penetrable barriers exist, majority carriers must be more numerous at an ohmic contact than they are in the bulk.

Debye Length. The expressions that we have derived for charge density (Equation 3.4.2), field (Equation 3.4.4), and space-charge-layer width (Equation 3.4.5) are all seen to contain the characteristic length L_D. Figures 3.14a through d show that L_D is an appropriate qualitative measure of the spatial extent of electrical effects at the boundary. The result of Problem 3.13 confirms this quantitatively by showing that 50% of the space charge in the semiconductor lies within $\sqrt{2}L_D$ units of the surface.

Considering a wider scope than this particular problem, we find that the solution of Poisson's equation in the presence of free charges always leads to a characteristic Debye length. If the charge configuration differs from the example given, the Debye length is still defined by Equation 3.4.3 except that n_s is replaced by the free-charge density appropriate to that configuration. We shall see an example of this in Chapter 4. The general result that L_D qualitatively measures the spatial extent of space charge is always true. Problem 3.14 shows that a relationship exists between L_D in a given region and the dielectric relaxation time in the same region. This relationship can be interpreted physically in terms of a balance of two transport mechanisms for free carriers: diffusion (thermal-induced motion), and drift (field-induced motion).

3.5 Surface Effects

When we began our consideration of metal-semiconductor contacts in Section 3.2 and made use of Fermi-level equality to infer an equilibrium energy-band picture, we made an important idealization. We considered that both the semiconductor and the metal had allowed energy states (energy bands) at the surface that were no different from those in the bulk. The real situation is more complicated, and we will need to develop the theory further to derive a more practical physical model. The most important correction needed is to account for the effects of surface states. Although these effects modify some conclusions, we shall be able to retain most of the ideas that were developed in previous sections.

Surface States

To begin, we should clarify the term *surface states*. These are extra allowed states for electrons that are present at the semiconductor surface, but not within the bulk. These extra states arise from a number of sources. First, let us consider an absolutely clean surface, that is, one composed purely of atoms of the host lattice. There will be extra states on this surface because the energy field of the crystal is one-sided; that is, electrons in the surface region are bonded only from the side directed toward the bulk (Figure 3.15). We would expect the characteristic energies for electrons at such sites to differ from the characteristic energies in the bulk.* Surface states of this type are called Tamm or Shockley states after original investigations by these scientists.[6,7] The density of the Shockley-Tamm states in a particular semiconductor is of the order of the density of atoms at the surface, or roughly $N_0^{2/3}$ (cm^{-2}) where N_0 is the bulk atomic density (atoms cm^{-3}). For silicon, N_0 is 5×10^{22} cm^{-3} (*cf* Table 1.3) and the density of the Tamm-Shockley states is about 10^{15} cm^{-2}. The energetic distribution of these states is not well

---

* In terms of quantum mechanics, the wave functions for the electrons are perturbed by the termination of the crystal potential; hence, the allowed energy states are different at the surface than in the bulk.

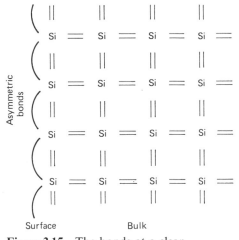

Figure 3.15   The bonds at a clean semiconductor surface are anisotropic and, consequently, differ from those in the bulk.

established although studies on the diamond lattice[8] indicate their density to be peaked about one-third of the forbidden-gap energy above the valence band, as sketched in Figure 3.16.

Bonded foreign atoms at the surface or crystal defects are sources for other types of surface states. An example is oxygen, which is virtually always found at a silicon surface. Oxygen can produce surface states spread over a range of energies depending upon the nature of the specific bonds it shares with silicon atoms. Other

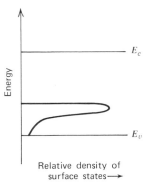

Figure 3.16   Approximate distribution of Tamm-Shockley states in the diamond lattice.[8] The distribution appears to peak sharply at an energy roughly one-third of the bandgap above E_v.

surface complexes of metals, hydroxyl ions, etc., can also be expected on any processed silicon surface. The net effect of all of these sources is a density of available electronic surface states that is not zero at any energy, although there may be pronounced peaks at specific energy values.

Besides varying in energy, surface states also vary in type and may be classified according to the charge they carry at equilibrium. For example, states that are neutral when occupied by electrons and positively charged when unoccupied are classified as donor states. States that are negative when occupied but neutral when empty are classified as acceptor states in a manner analogous to bulk dopant atoms, as discussed in Section 1.1.

One other classification of surface states arises from the properties of real interfaces between solids. On an atomic scale (a fraction of a nanometer), such interfaces are not plane but consist rather of zones of intermediate materials and impurities that are several to tens of atomic layers in thickness. Within these intermediate zones some surface states are physically close to the bulk semiconductor, and they remain in thermal equilibrium with bulk states even under fairly rapid changes in potential. These are called *fast surface states* because the electrons occupying them come into equilibrium quickly. In contrast to these are so-called *slow states*, which are states situated more remote from the bulk semiconductor within the intermediate layer. These states take relatively longer times to reach thermal-equilibrium conditions with the bulk states. Although the demarcation between these categories of states is vague, general usage sets the boundary between them at a response time corresponding to about 1 kHz.

Surface Effects on Metal-Semiconductor Contacts[†]

The presence of surface states can modify the contact theory that we have presented. If the semiconductor surface states are not neutral or if they change their charge state when the contact is formed, then the charge configurations that we obtain by applying Schottky theory are not valid. A theory for metal-semiconductor systems with surface states has been presented.[9] We shall not discuss this theory in detail, but rather give a summary of the viewpoints adopted and show the correspondence of theory with practical results.

To account for surface effects, the metal-semiconductor contact is treated as if it contained an intermediate region sandwiched between the two crystals. The intermediate region is thought to consist of an interfacial layer ranging from several to the order of ten atomic dimensions in width. This layer contains the impurities and added surface states. It is too thin to be an effective barrier to electron transfers (which can occur by tunneling), but it can sustain a voltage drop. Extra allowed electronic states are postulated to be distributed in energy at the assumed planar boundary between the interfacial layer and the semiconductor. Figure 3.17 is a sketch showing the band structure and the surface states. In the figure the states are assumed to be acceptor type and to have a density D_s states cm^{-2} eV^{-1}. Note in Figure 3.17 the thin layer of width δ, which sustains a voltage

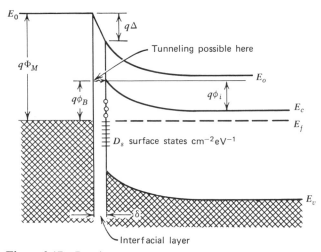

Figure 3.17 Band structure near a metal-semiconductor contact according to the model of Cowley and Sze.[9] The model presumes a thin interfacial layer of thickness δ that sustains a voltage Δ at equilibrium. Acceptor-type surface states distributed in energy are assumed to be described by a distribution function D_s states $cm^{-2} eV^{-1}$.

drop of Δ volts. Because this layer is thin enough for electron tunneling, the metal-semiconductor barrier height is measured between the Fermi level and the conduction band at the semiconductor surface. A surface layer should be associated with the semiconductor even without the metal being present (Figure 3.18). When acceptor surface states are present as in Figure 3.18, the n-type semiconductor is depleted of electrons near its surface, and negative charge exists in the acceptor surface states.

Consider the formation of a blocking contact to a semiconductor with surface states such as that shown in Figure 3.18. Such a contact would have an energy diagram similar to that in Figure 3.17. To draw this diagram we note that from Schottky theory a blocking contact would be formed if contact were made to a metal having Φ_M greater than Φ_S, because electrons would be transferred from the semiconductor into the metal. Transfer of electrons from the semiconductor bends the conduction band away from the Fermi level. For the semiconductor in Figure 3.18 this would remove charge from some of the surface states by lifting them above E_f. The larger is D_s, the greater is the charge removed for each incremental energy increase in E_c near the contact. If D_s is large, therefore, a negligible movement of the Fermi level at the semiconductor surface would transfer sufficient charge to equalize the Fermi levels. Under this condition, the Fermi level is said to be *pinned* by the high density of states. Note that pinning of the Fermi level is not exclusively associated with the surface states being acceptor type. Any electronic

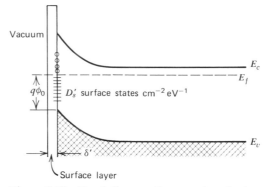

Figure 3.18 Band diagram for a semiconductor surface showing a thin surface layer containing acceptor-type surface states distributed in energy. A surface-depletion region is present because of charge in the surface states.

states clustered near the Fermi energy will cause the Fermi level to be pinned when the state density becomes very large because slight changes in Fermi energy would result in very sizable charge transfer.

When the Fermi level is pinned, the barrier height $q\phi_B$ becomes[9]

$$q\phi_B = (E_g - q\phi_0) \tag{3.5.1}$$

where $q\phi_0$ is $(E_f - E_v)$ at the surface when the semiconductor is not covered by metal as shown in Figure 3.18. As D_s approaches zero, the barrier height $q\phi_B$ approaches the height predicted by basic Schottky theory as given in Equation 3.2.1 and repeated here for reference.

$$q\phi_B = q(\Phi_M - X) \tag{3.5.2}$$

Whether a metal-semiconductor barrier height will be predicted by Equation 3.5.1 or by Equation 3.5.2 or by an intermediate value depends on the value of D_s at energies near the Fermi level and also on specific properties ascribed to the boundary layer, such as its precise thickness and permittivity.[9]

In practice most Schottky barrier heights for important semiconductors are predicted more accurately by Equation 3.5.1 than by Equation 3.5.2. There is thus only small dependence on the metal work function. This is true for silicon, germanium, and gallium arsenide, although cadmium sulfide is an exception. For silicon, as with germanium, gallium arsenide, and gallium phosphide, the quantity $q\phi_0$ is found experimentally to be about equal to $\frac{1}{3}E_g$ so that the barrier height $q\phi_B$ from Equation 3.5.1 is typically close to $\frac{2}{3}$ of the band gap or roughly 0.75 eV for silicon. The reason for this consistency may be the characteristic very high density of states that appears to be common to the diamond-type lattice and which pins the Fermi level at this energy (Figure 3.16). Some measured barriers for various metals contacting n- and p-type silicon crystals are shown in Table 3.2.

Table 3.2   Schottky Barriers to Silicon[10]
(qX for Silicon = 4.05 eV)

Silicon Type	Metal	$q\Phi_M$ (eV)	$q\phi_B$ (eV)
n	Al	4.1	0.69
p	Al	—	0.38
n	Pt	5.3	0.85
p	Pt	—	0.25
n	W	4.5	0.65
n	Au	4.75	0.79
p	Au	—	0.25

3.6 Metal-Semiconductor Devices: Schottky Diodes

The industrial use of metal-semiconductor barrier devices, generally designated as Schottky diodes, is widespread. The area of greatest application is that of digital logic circuits, that is, circuits that perform binary arithmetic. Schottky diodes are often used in these circuits as fast switches that can be made on integrated-circuit chips within very small dimensions on the surface. There is also an increasing interest in Schottky-diode power rectifiers since large-area devices with an excellent thermal path through the contact metal allow high currents to flow. These high currents are sustained at lower voltage drops than would exist in diffused *pn*-junction diodes.

Schottky diodes are also used as variable capacitors that can be operated efficiently in the microwave region of the spectrum. For variable capacitance applications, the diode is kept under reverse bias continuously. Voltage changes across it then are able to modulate the depletion-region width and capacitance in the manner discussed in Section 3.2.

Another application that makes use of modulation of the depletion width is the Schottky-barrier, field-effect transistor* also called the *MEtal-Semiconductor Field-Effect Transistor* (MESFET). This device, pictured in Figure 3.19, consists of a barrier junction at the input that acts as a control electrode (or *gate*), and two ohmic contacts through which output current flows (described as the *source* and *drain* electrodes). The output current varies when the cross section of the conducting path beneath the gate electrodes is changed. This device is a special form of a junction field-effect transistor (JFET), an amplifying device in which the control electrode is most usually made from a reverse-biased *pn* junction. Substitution of a Schottky barrier for the *pn* junction in JFETs has proven especially useful with semiconductor materials in which the fabrication of *pn* junctions is not practical. The major application of Schottky barriers to JFETs has been to make

---

* Since this is the first point at which we meet the term *transistor*, it is appropriate to point out that the word has its origin in an amalgamation of the descriptive terms *transfer resistor*. It was coined by W. Shockley and co-workers at the Bell Telephone Laboratories where the junction field-effect transistor and the bipolar transistor, to be discussed in Chapters 6 and 7, were both invented.

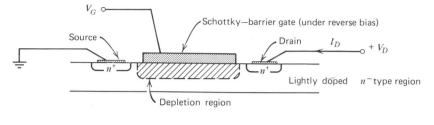

Figure 3.19   Schottky-barrier-gate, field-effect transistor. Current I_D flowing from drain to source is modulated by gate voltage V_G that controls the dimensions of the space-charge region. This, in turn, affects the conducting area for I_D. The source and drain contacts are ohmic because they are made to highly doped material.

high-frequency, gallium-arsenide devices. We shall discuss the operation of JFETs in more detail after we have considered *pn*-junction operation in Chapter 4.

Schottky Diodes in Integrated Circuits.   Two fortuitous designer's choices, made for reasons having nothing to do with Schottky barriers, combine to allow a very simple technology for Schottky diodes in silicon digital integrated circuits. These choices are the use of high resistivity *n*-type silicon in which to build *npn* bipolar transistors and the use of evaporated aluminum metal to form the "wire" inter-connections in integrated circuits. The aluminum forms a blocking contact to the lightly doped, *n*-type silicon provided that the silicon surface has been thoroughly cleaned. It is only necessary that the doping of the silicon be sufficiently low so that the barrier may not be penetrated by tunneling electrons, as was described in Section 3.4. Practically, this limits the doping to less than about 10^{17} cm^{-3} (Problem 3.7).

The barrier height between *n*-type silicon and aluminum is about 0.70 eV, and diodes made by evaporation of aluminum onto the silicon surface in a vacuum have characteristics that approximate theoretical predictions quite well (Figure 3.11). Because of the concentration of electric-field lines near the corners, however, reverse breakdown is not an abrupt function of voltage and occurs at a relatively low bias (~ 15 V). Several techniques, such as the use of diffused guard rings (Figure 3.20*a*) or field plates (Figure 3.20*b*), have been developed to improve the reverse characteristics. Because they complicate circuit processing, however, these techniques are avoided unless especially needed. Excellent Schottky diodes with high barriers can be made using refractory metals; platinum, in particular, has been employed in many instances. A technique of integrated-circuit metallization called *beam-leading* uses sputtered* platinum. If sputtering is included in the

---

* Sputtering is a technique for laying down thin films of materials in which a source, or target, is bombarded with high-energy gaseous ions. Sputtering is especially useful with refractory materials that are not easily evaporated.

Figure 3.20 Special processing techniques improve the performance of Schottky diodes. (*a*) The diffused guard ring leads to a uniform electric field and eliminates breakdown at the junction edge. (*b*) The field plate is an alternative means for achieving the same effect.

manufacturing process, an advantageous means of cleaning the silicon surface thoroughly by bombarding it with high-energy ions or *sputter-etching* can be employed. After sputter-etching, the subsequent sputter-deposition of platinum forms an excellent Schottky diode. Although Schottky diodes can be made to *p*-type silicon, the barrier heights are inherently smaller (roughly $\frac{1}{3}$ of the band gap or 0.36 V), and the yield and electrical performance are correspondingly poorer.

Designers who make use of Schottky diodes for digital logic circuits often call them *clamps* because they fix or clamp the voltage across one junction of a transistor and improve circuit performance. We shall see the reasons in Chapter 6, but at this point we merely assert that substantial improvement in the speed of a digital logic circuit is gained if clamps are put between the collector and base of a switching transistor to keep its collector-base junction from being forward biased. Schottky diodes are nearly ideal for this purpose.

To place this topic in context, we first discuss the concept of a *turn-on voltage*. This term refers to the forward drop that must be placed across a diode to "turn-it-on"; that is, to cause it to pass substantial current. The current-voltage equations for metal-semiconductor barriers (Equations 3.3.17 and 3.3.19) show, however, a continuous dependence of current on voltage with no abrupt change of characteristics to identify as a turn-on voltage. From an engineering point of view, however, there is a threshold for conduction that becomes apparent when we plot the current-voltage characteristics for a Schottky diode on linear scales (Figure 3.21). The linear scale permits only a small range of current to be plotted meaningfully. The strong dependence of current on voltage allows the data to be fit fairly well by two straight lines, one nearly horizontal at $J = 0$ and one nearly vertical. The intersection of the nearly vertical line with the voltage axis defines a turn-on voltage V_o. When digital circuits are designed, V_o is a good approximation to the voltage drop across any diode that is conducting. From Equation 3.3.17, at a given forward current density J_F, V_o is given by

$$V_o = \frac{nkT}{q} \ln \left(\frac{J_F}{J_S'} + 1 \right) \tag{3.6.1}$$

Thus, from the designer's point of view, V_o is most strongly dependent on J_S'. For diodes designed to pass currents in the milliamp range, aluminum Schottky diodes

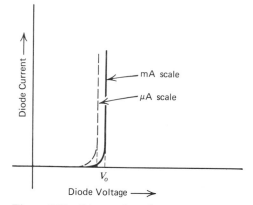

Figure 3.21   Linear plot of current versus voltage for a Schottky diode illustrating the concept of a diode "turn-on voltage."

fabricated on n-type silicon typically have a V_o of about 450 mV. Because this number is about 200 mV smaller than V_o for a comparable pn junction (Chapter 5), a Schottky diode placed in parallel with a pn-junction diode will not permit the forward bias to rise sufficiently for the junction diode to conduct. In the case of a Schottky-clamped collector junction, the bipolar transistor is held out of a condition called *saturation*, in which switching transients are inherently slowed. Thus, digital circuits using Schottky clamps are faster by several nanoseconds than are unclamped circuits.

The only penalty paid in electrical performance for using the Schottky diode is the small amount of extra capacitance with which the reverse-biased Schottky diode loads the circuit. The clamped transistor takes up only slightly more surface area on the silicon chip than does the unclamped transistor, a big advantage for an integrated-circuit device. There is, however, generally some loss in fabrication yield when Schottky processing is used because metallization and surface preparation are more critical than in circuits that do not make use of Schottky barriers

Summary

An ensemble of electrons at thermal equilibrium is characterized by a single *Fermi energy*. The Fermi energy is therefore the appropriate reference for a diagram of allowed energy states versus position. It is particularly useful in drawing the appropriate thermal-equilibrium diagram for an inhomogeneous material and for systems of materials in intimate contact. If two materials, characterized by different Fermi energies (and therefore not in thermal equilibrium with each other) are brought sufficiently close to interact with one another, electrons will be transferred from the material with the higher Fermi energy to the material with the lower

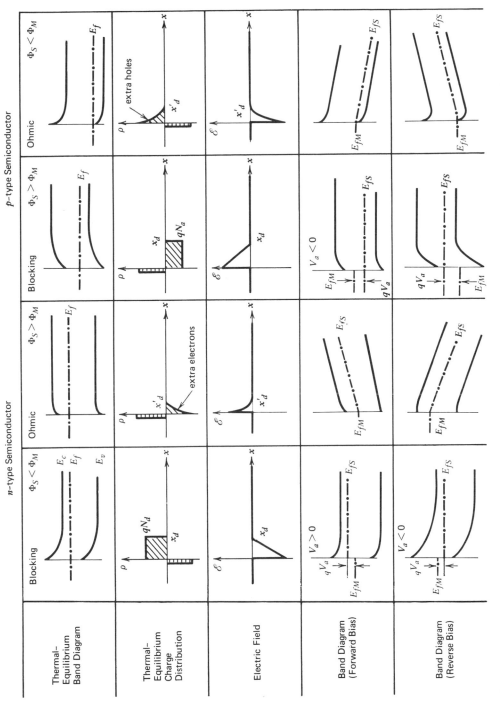

Figure 3.22 Diagrams for ideal metal-semiconductor Schottky diodes.

Fermi energy. Application of these principles to metals and semiconductors, under the idealized conditions that the bulk energy-state configuration continues to the surface and that the semiconductor is homogeneous, is the basis for Schottky, metal-semiconductor contact theory. This theory predicts blocking contacts and rectifying behavior for n-type semiconductors if the metal work function Φ_M exceeds the semiconductor work function Φ_S and ohmic behavior if Φ_S is greater than Φ_M. The inverse is true for metal contacts to p-type semiconductors. Diagrams for ideal Schottky contacts are shown in Figure 3.22. If the space-charge regions at a Schottky barrier become thin enough for substantial electron tunneling (which results from highly doping the semiconductor) the contacts are also ohmic.

To develop a theory of metal-semiconductor contacts, it is necessary to make use of Poisson's equation in conjunction with the thermal-equilibrium band diagram. Simplifying assumptions such as the *depletion approximation* and *quasi-equilibrium* make the theory tractable. Basic Schottky theory predicts a number of observed properties successfully. Among these are the major dependences of the current-voltage characteristics and the reverse-bias capacitance behavior of the Schottky barrier. Contacts to a semiconductor having a high-resistivity region on top of a low-resistivity bulk are analyzed by the same techniques to give results for a *Mott barrier*. A similar analysis scheme applied to the Schottky ohmic contact introduces the *Debye length*, a characteristic measure of the extent of electric-field penetration in a region having significant free charge. Although basic Schottky theory provides much useful information about metal-semiconductor contacts, it is unsuccessful in predicting barrier heights to silicon. A major inadequacy of the theory lies in the treatment of surface effects. Surface states originate from the termination of the lattice and from imperfect and impure surfaces. Theory for metallic blocking contacts to real silicon surfaces is based upon the assumption of thin interfacial layer of imprecise composition, but of well-specified electronic behavior. The applications of metal-semiconductor contacts in the form of Schottky diodes are widespread and growing. Schottky diodes have special relevance to integrated circuits because they are relatively easy to obtain using standard silicon planar technology.

References

1. W. Schottky, *Naturwissenschaften* **26**, 843 (1938).
2. (a) H. A. Bethe, *Theory of the Boundary Layer of Crystal Rectifiers*, MIT Radiation Laboratory Report 43–12 (1943).
   (b) S. M. Sze, *Physics of Semiconductor Devices*, 2nd Edition, Wiley-Interscience, New York (1981), pp. 255–258.
3. H. K. Henisch, *Rectifying Semiconductor Contacts*, Oxford at the Clarendon Press (1957), p. 172.
4. N. F. Mott, *Proceedings Cambridge Philosophical Society* **34**, 568 (1938).
5. A. Rose, *Concepts in Photoconductivity and Allied Problems*, Wiley-Interscience, New York (1963).

6. I. Tamm, *Phys. Z. Sowjetunion* **1**, *733* (1933).
7. W. Shockley, *Phys. Rev.* **56** 317 (1939).
8. D. Pugh, *Phys. Rev. Lett.* **12**, 390 (1964).
9. A. M. Cowley and S. M. Sze, *J. Appl. Phys.* **36**, 3212 (1965).
10. S. M. Sze, *Physics of Semiconductor Devices*, *2nd Edition*, Wiley-Interscience, New York (1981).

Textbooks

S. M. Sze, *Physics of Semiconductor Devices*, *2nd* Edition Wiley-Interscience, New York (1981).
A. G. Milnes, *Semiconductor Devices and Integrated Electronics*, Van Nostrand Reinhold, New York (1980).

Problems

3.1* Draw a diagram similar to Figure 3.2 for silicon in which the doping changes abruptly from 5×10^{18} donors cm^{-3} to 8×10^{15} donors cm^{-3}.
(a) What are the work functions associated with the two regions of the crystal?
(b) What is the potential difference between the two regions of silicon?

3.2 The energy-band diagrams for a metal and a semiconductor are sketched to the same scales in Figure P3.2a. Three possible band diagrams for a contact between these materials are sketched in Figures P3.2b to d. Assume thermal equilibrium.
(a) Explain why each figure is incorrect.
(b) Draw a correct diagram, assuming that basic Schottky theory applies.

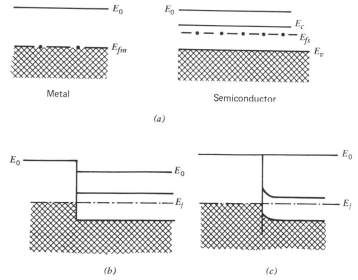

(a)

(b)   (c)

Figure P3.2

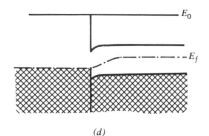

(d)

Figure P3.2   *(continued)*

3.3   Consider metal-semiconductor junctions that behave according to simple Schottky theory.

(a) Draw the theoretical energy-band diagram for copper (work function 4.5 eV) in contact with silicon having a work function of 4.25 eV.

(b) If light were to shine on this junction and create hole-electron pairs:
   (i) Which way would current flow within the device if the junction were connected into a circuit?
   (ii) What would be the maximum voltage that could be measured across the junction (zero output current)?

(c) Draw the energy-band diagram for copper in contact with silicon having a work function of 4.9 eV.

(d) Compare the electrical behavior of the metal-semiconductor systems described in a and c.

3.4   The accompanying data were obtained on metal contacts to silicon of equal area (Figure P3.4). If Schottky theory applies, which metal probably has the higher work function? Which data were taken on 1 Ω-cm silicon and which on 5 Ω-cm silicon? Justify your answers and explain the use of "probably" (Consider Section 3.5.)

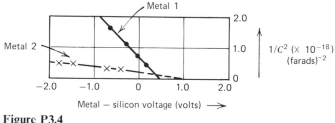

Figure P3.4

3.5* (a) Calculate the small-signal capacitance at zero dc bias and at 300 K for an ideal Schottky barrier between platinum (work function 5.3 eV) and silicon doped with $N_d = 10^{16}$ cm^{-3}. The area of the Schottky diode is 10^{-5} cm^2.

(b) Calculate the reverse bias at which the capacitance is reduced by 25% from its zero-bias value.

3.6† (a) Find the location x_m of the plane at which the barrier to emitted electrons [$E_2(x)$ in Figure 3.9] is a maximum and prove Equation 3.2.13.

(b) For an applied field of 10^5 V cm^{-1}, calculate x_m and $q\,\Delta\phi$.

3.7* Consider an aluminum Schottky barrier made to silicon having a constant donor density N_d. The barrier height $q\,\phi_B$ is 0.65 eV. The junction will pass high currents under reverse bias by tunneling from the metal if the barrier presented to the electrons is thin enough, as described in the following. We assume that the onset of efficient tunneling occurs when the Fermi level in the metal is equal to the edge of the conduction band (E_c) at a distance 10 nm into the semiconductor.
 (a) If this condition is reached at a total junction bias ($\phi_i - V_a$) of 5 V, what is the maximum value of N_d?
 (b) What limit does this place on the resistivity of the epitaxial layers used in Schottky-clamped circuits?
 (c) Draw a sketch of the energy-band diagram under the condition of efficient tunneling.

3.8† Carry through the steps needed to derive Equation 3.3.13.

3.9† Consider Equation 3.3.16 under conditions of low forward bias. Show that Equation 3.3.17 can be derived by using $(1 - V_a/\phi_i)^{1/2} = \exp[\frac{1}{2}\ln(1 - V_a/\phi_i)]$ and by approximating the resultant expression for J_S. This approach leads to Equation 3.3.17 with $n = (1 + kT/2q\phi_i)$, which is generally smaller than observed values. Other effects such as rounding of the barrier contribute to values for n in Equation 3.3.17 that are somewhat higher than those found from the expression derived in this problem.

3.10 Using linear scales, plot I versus V_a for a diode that obeys the ideal-diode law (Equation 3.3.6) under the condition that:
 (a) $I_0 = 1$ pA and $T = 150\ K$.
 (b) $I_0 = 1$ nA and $T = 300\ K$.
 (c) $I_0 = 1\ \mu A$ and $T = 450\ K$.
 (d) Considering the discussion in Section 3.6, state an appropriate value for the turn-on voltage V_o for each plot of I versus V_a.
 For clarity in the diagrams, use a scale change at $V = 0$. Show the forward characteristic through 5 mA.

3.11† Use the equations in Section 3.4 to represent the space charge in a Schottky ohmic contact between a metal and an n-type semiconductor and set up Poisson's equation. The equation form will be $d^2\phi/dx^2 = K\exp(\phi/V_t)$, where $V_t = kT/q$. It is convenient to convert this function of voltage ϕ and position x to a function of field $\mathscr{E}$ and voltage ϕ. This can be done by making use of $d^2\phi/dx^2 = \mathscr{E}\,d\mathscr{E}/d\phi$. The resultant equation can be solved to find $\mathscr{E} = \sqrt{2n_s kT/\epsilon_s}\,\exp(\phi/2V_t)$.
 (a) Carry through the steps which have been outlined in this problem.
 (b) Derive Equations 3.4.2, 3.4.4, and 3.4.5 by continuing with this analysis.

3.12† Draw the band diagram for a "neutral" contact, as described in Section 3.4. Consider the results of Problem 1.13, which express the random thermal flux of free electrons in a semiconductor as $qn_0 v_{th}/4$ where n_0 is the electron density and v_{th} is the thermal velocity. If currents drawn from the metal into the semiconductor are less than this value, the contact will not limit the flow and can be regarded as ohmic.
 (a) Show that a contact is ohmic for fields in the semiconductor less than $v_{th}/4\mu_n$.
 (b) Calculate the limiting ohmic current in a neutral contact made to a semiconductor having $N_d = 10^{16}$ cm^{-3} and $A = 10^{-5}$ cm^2. Take $v_{th} = 10^7$ cm s^{-1}.
 (c) What is the limiting ohmic current if the bands are bent such that $q(\Phi_M - X - \Phi_n) = 0.65$ eV?

3.13† Using Equations 3.4.2 and 3.4.5 show that half of the space charge in the Schottky ohmic contact exists within $\sqrt{2}L_D$ of the surface.

3.14 Show that the dielectric relaxation time $\tau_r = \epsilon_s/\sigma$ (discussed in Problem 1.12) can be related to the Debye length L_D by $L_D = (D\tau_r)^{1/2}$ where D is the diffusion constant in the material.

3.15 Assuming that basic Schottky theory applies, sketch the energy-band diagram for (a) an ohmic contact between p-type silicon and a metal at equilibrium, (b) a blocking contact between p-type silicon and a metal under 2 V reverse bias.

3.16* Both Schottky-barrier diodes and ohmic contacts are to be formed by depositing a metal on a silicon-integrated circuit. The metal has a work function of 4.5 eV. For ideal Schottky behavior, find the allowable doping range for each type of contact. Consider both p- and n-doped regions and comment on the practicality of processing the integrated circuit with the required doping.

3.17† A back-biased Schottky diode made to silicon is to be used as a tuning element for a broadcast-band radio receiver (550 to 1650 kHz). For ease of operation, it is desirable to have the resonant frequency $(1/2\pi\sqrt{LC})$ of a tuned circuit change linearly with voltage when a dc voltage applied to the circuit changes from 0 to 5 V. If the tuning inductance L were 2 mH, it is readily calculated that the capacitance at the two extremes of bias to achieve this behavior should be 41.8 and 4.65 pF, respectively. Consider that the diode area is 10^{-3} cm^2 and find the desired dopant variation for N_d. (Calculate the numerical values and sketch a semilogarithmic plot of the results.) *Hint.* To attack this problem note that

$$\frac{df}{dV} = -\frac{1}{4\pi\sqrt{LC}}\frac{1}{C}\frac{dC}{dV} = 0.22 \text{ MHz/V}$$

from the information given. Use Equation 3.2.10 together with $C = A\epsilon_s/x_d$ to find $N(x_d)$.

4

pn JUNCTIONS

We have seen in Chapter 3 that a system of electrons is characterized by a constant Fermi level at thermal equilibrium. This principle was used initially to deduce the energy-band diagram of a semiconductor having two doping levels. Systems that are initially not in thermal equilibrium approach equilibrium as electrons are transferred from regions with a higher Fermi level to regions with a lower Fermi level. The transferred charge causes the buildup of barriers against further electron flow, and the potential drop across these barriers increases to a value that just equalizes the Fermi levels. These concepts formed the bases for an extensive analysis of metal-semiconductor contacts.

In this chapter we consider similar phenomena in a single crystal of semiconductor material containing regions having different dopant concentrations. We shall find it useful to employ two important approximations that are frequently encountered in device analysis. One, the *depletion approximation*, has already been introduced in Chapter 3. The second, the *quasi-neutrality approximation*, serves the same simplifying purpose as the depletion approximation; it makes complicated problems tractable by focusing on the dominant physical effects in a given region of a device. The quasi-neutrality approximation is employed in a region of a semiconductor containing a slowly varying dopant concentration, while the depletion approximation is useful in the important case of a semiconductor containing adjacent *p* and *n*-type regions.

We discuss in some detail the transition region at a *pn* junction and the barrier associated with this transition region. We next consider the influence of an applied reverse voltage on the transition region and show that changes in this applied voltage lead to capacitive behavior. Limitations on the magnitude of the reverse bias imposed by two important breakdown mechanisms are then discussed. Since most semiconductor devices contain one or more *pn* junctions, our results are directly applicable to the analysis of practical devices. As an example of the use of the concepts developed for a reverse-biased *pn* junction, we conclude this chapter by discussing junction field-effect transistors.

Our focus in this chapter is on *pn* junctions at equilibrium and under reverse bias. We consider currents to be negligible except when junction fields are large enough to lead to breakdown. Current flow in *pn* junctions is discussed in Chapter 5.

4.1 Graded Impurity Distributions

In this section we consider equilibrium in a semiconductor with a dopant concentration that varies in an arbitrary manner with position (Figure 4.1*a*). We assume that initially the majority-carrier concentration equals the dopant concentration at every point in the material—a nonequilibrium condition. Then, we investigate the means by which the system approaches thermal equilibrium. From Equation 1.2.17 we note that a gradient in the mobile carrier concentration leads to diffusion of

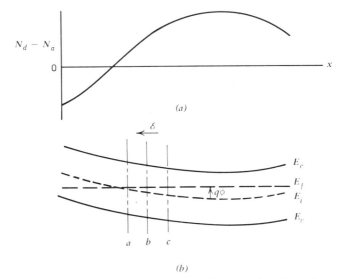

Figure 4.1 (*a*) Net dopant concentration as a function of position in an arbitrarily doped semiconductor. (*b*) Corresponding energy-band diagram versus position, indicating the potential ϕ. Locations *a*, *b*, and *c* are discussed in the text.

carriers from regions of higher concentration to regions of lower concentration. As the carriers move from their initial locations, they leave behind uncompensated, oppositely charged dopant ions. This separation of positive and negative charges creates a field that opposes the diffusion flow. Equilibrium is eventually reached when the tendency of the carriers to diffuse to regions of lower density is exactly balanced by the tendency to move in the opposite direction because of the electric field created by the charge separation. Thus, the equilibrium situation is characterized by a mobile carrier distribution that does not coincide exactly with the fixed dopant distribution, and also by a built-in electric field that keeps the two charge distributions from separating further. The space charge resulting from the mechanism just described is typically a small fraction of the dopant density,* but the field arising from it can significantly influence device behavior.

We can consider the implications of the built-in electric field in terms of the energy-band diagram. Since the dopant density and carrier concentration vary with position along the semiconductor, the separation between the Fermi level and the valence and conduction-band edges also varies with position, while the Fermi level itself remains constant throughout the system at thermal equilibrium. Figure 4.1b shows the energy-band diagram corresponding to the dopant distribution of Figure 4.1a. The separation between the Fermi level and the band edge is less in regions of high carrier density than in regions of lower density, and the intrinsic Fermi level E_i crosses the Fermi level E_f where the net dopant concentration $N_d - N_a$ is zero.

Potential. The presence of an electric field may be implied directly from this energy-band diagram, as well as from the particle model discussed above. Since Figure 4.1b represents the energy-band diagram of an electron, the energy of an electron is measured by its distance above the Fermi level on the band diagram. The separation of the Fermi level from the conduction-band edge may be taken to represent the potential energy of an electron while the energy above the conduction-band edge represents kinetic energy. Since the electric potential ϕ is related to the potential energy by the charge $-q$, the potential may be written

$$\phi_c = -\frac{1}{q}(E_c - E_f) = \frac{1}{q}(E_f - E_c) \tag{4.1.1}$$

where the subscript c implies reference to the conduction-band energy. The reference for potential energy is arbitrary, however, and we may shift it from E_c to E_i. Since E_i is usually used for the reference we will not subscript the symbol ϕ for potential which now is written

$$\phi = -\frac{1}{q}(E_i - E_f) = \frac{1}{q}(E_f - E_i) \tag{4.1.2}$$

as shown in Figure 4.1b. According to this definition the potential is positive for an n-type semiconductor and negative for p-type material.

* This assertion is considered further in an example given in Section 4.2.

Field. Since the electric field is the negative of the spatial gradient of the potential, the field $\mathscr{E}_x$ is found from Equation 4.1.2 to be

$$\mathscr{E}_x = \frac{-d\phi}{dx} = \frac{1}{q}\frac{dE_i}{dx} \tag{4.1.3}$$

Thus, a spatial variation of the band edges (hence, of the intrinsic Fermi level) implies the existence of an electric field in the semiconductor. At point b of Figure 4.1b, dE_i/dx is negative, and the field is directed toward the left. The resulting force on negatively charged electrons is toward the right; consequently, the field provides a force that opposes the tendency of electrons to diffuse from a high-concentration region at c to the low-concentration region at a. The situation in a p-type semiconductor is analogous, with the proper changes in signs and notation.

We now look at the problem quantitatively and find an expression for the electric field in terms of the graded impurity distribution. Once the system is in thermal equilibrium, there can be no current flowing at any point in the semiconductor. In particular, since thermal equilibrium requires that every process and its inverse are in balance, the electron current and the hole current must separately be zero at thermal equilibrium. In Section 1.2 we found that the total electron current is given by

$$J_n = q\mu_n n\mathscr{E}_x + qD_n \frac{dn}{dx} \tag{4.1.4}$$

This expression is applicable both in n-type material, where the electrons are majority carriers, and in a p-type semiconductor, where they are minority carriers.

The first term of Equation 4.1.4 represents the drift current, and the second, the diffusion current. When the total electron current is zero, the two terms are exactly balanced. No current actually flows, rather the two tendencies, drift and diffusion, are in equilibrium. Since $J_n = 0$, we may solve for the field in terms of the electron concentration and its gradient

$$\mathscr{E}_x = -\frac{D_n}{\mu_n}\frac{1}{n}\frac{dn}{dx} = -\frac{kT}{q}\frac{1}{n}\frac{dn}{dx} \tag{4.1.5}$$

where we have used the Einstein relation defined in Equation 1.2.20. Similarly, the field may be expressed in terms of the hole concentration by either considering the expression for hole current (Equation 1.2.22) or using the mass-action law (Equation 1.1.13) in Equation 4.1.5:

$$\mathscr{E}_x = \frac{kT}{q}\frac{1}{p}\frac{dp}{dx} \tag{4.1.6}$$

Equations 4.1.5 and 4.1.6 show that, if we could find the mobile carrier concentrations and their spatial derivatives, we could specify the fields in the semiconductor.

In developing this concept we can gain some physical insight by considering the relation between the electron density and the position of the band edge (or

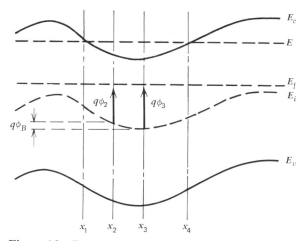

Figure 4.2 Energy-band diagram of an arbitrarily doped semiconductor, showing that an electron with energy E is constrained to remain in the region between x_1 and x_4 where $E > E_c$.

equivalently the intrinsic Fermi level) with respect to the Fermi level. Consider an electron at x_2 in Figure 4.2 with an energy E. A portion $E - E_c$ of this energy is kinetic energy; the remainder is potential energy. The electron can move freely in the region between x_1 and x_4 because it has energy greater than the potential energy associated with this region. The electron would require more potential energy than its total energy to enter the region to the left of x_1 or to the right of x_4. Consequently, it is forbidden to enter these regions, and the spatial variation of the energy bands indicates a *potential barrier* to the motion of the electrons.

Now let us relate the number of carriers at any two points in the material to the energy-band structure. We know that the electron density at x_2 is less than that at x_3 since the separation between the conduction-band edge and the Fermi level is greater at x_2. We may relate the carrier densities to the potential ϕ through the use of Equation 4.1.5. Since $\mathscr{E}_x = -d\phi/dx$, we have

$$d\phi = \frac{kT}{q}\frac{dn}{n} \tag{4.1.7}$$

at any point in the semiconductor. If we integrate this equation between any two points—for example, from x_2 to x_3—we obtain

$$\phi_3 - \phi_2 = \frac{kT}{q}\ln\frac{n_3}{n_2} \tag{4.1.8}$$

Rewriting Equation 4.1.8 in exponential form, we find

$$\frac{n_3}{n_2} = \exp\left[\frac{q}{kT}(\phi_3 - \phi_2)\right] \tag{4.1.9}$$

As might be expected, the carrier densities depend on the potential difference $\phi_B = \phi_3 - \phi_2$ between the two points. A physical interpretation of this result is that a fraction $\exp(-q\phi_B/kT)$ of the electrons at x_3 has enough energy to reach x_2.

Poisson's Equation. Equation 4.1.9 is often useful when the variation of carrier concentrations with position must be found. As is often the case, we start such an analysis by writing Poisson's equation:*

$$\frac{d^2\phi}{dx^2} = -\frac{\rho}{\epsilon_s} = -\frac{q}{\epsilon_s}(p - n + N_d - N_a) \qquad (4.1.10)$$

where ρ is the space-charge density. Using our definition of potential (Equation 4.1.2) in Equation 1.1.26, we may relate the carrier concentration n to the potential function ϕ:

$$n = n_i \exp\left(\frac{q\phi}{kT}\right) \qquad (4.1.11)$$

Poisson's equation can then be rewritten in the form

$$\frac{d^2\phi}{dx^2} = \frac{q}{\epsilon_s}\left(2n_i \sinh\frac{q\phi}{kT} + N_a - N_d\right) \qquad (4.1.12)$$

Equation 4.1.12 is the differential equation for the potential distribution in an arbitrarily doped semiconductor material. Unfortunately, this equation cannot be solved in the general case, and approximations must be made to obtain solutions appropriate to specific situations.

To proceed, we consider two special cases. In the first, the dopant concentration varies gradually with position (small gradient case) as, for example, the donor distribution within a diffused n-type region. The second case is just the opposite and involves abrupt spatial variations of dopant concentration (large gradient case), as found, for example, in the junction between p-type and n-type semiconductor regions.

Quasi-Neutrality. For the small gradient case we consider n-type silicon in which the dopant concentration may vary from about 10^{18} to 10^{16} cm^{-3} in several hundred nanometers after a typical dopant diffusion. This change in dopant concentration corresponds to a potential difference of about 0.1 V and a field (Equation 4.1.5) of the order of 10^4 V cm^{-1} or less. If we take a specific case in which the field varies from zero to 10^4 V cm^{-1} in 0.5 μm, the average field gradient is 2×10^8 V cm^{-2}. Using this value in Poisson's equation (Equation 4.1.10) and neglecting the minority-carrier density p, we find that the difference between n and N_d must be less than about 10^{15} cm^{-3}. Since this number is only a small fraction

---

* Since much of semiconductor device analysis is concerned with the spatial variations of carriers and potentials from one region of a device to another, Poisson's equation is often encountered, as we have already seen in Chapter 3. Together with approximations that place it in mathematically tractable form, it is one of the most useful principles in semiconductor device analysis.

of the donor concentration over most of the region under consideration, it is reasonable to approximate n by N_d in order to proceed with the analysis. In essence the approximation means that the semiconductor region under consideration is nearly neutral or *quasi-neutral*; that is, the majority-carrier distribution does not differ much from the donor distribution. The *quasi-neutrality approximation* is the more valid the smaller is the gradient in dopant density. Under the assumption of quasi-neutrality, the field in the n-type semiconductor is found directly from the donor concentration by using Equation 4.1.5:

$$\mathscr{E}_x = -\frac{kT}{q}\frac{1}{N_d}\frac{dN_d}{dx} \qquad (4.1.13)$$

In a p-type semiconductor under quasi-neutral conditions, the field is similarly

$$\mathscr{E}_x = \frac{kT}{q}\frac{1}{N_a}\frac{dN_a}{dx} \qquad (4.1.14)$$

Since Equations 4.1.13 and 4.1.14 depend on the quasi-neutrality approximation, they are not valid if the dopant concentration has a steep gradient.

One frequently considered case in which the quasi-neutrality approximation is useful is that of an exponential dopant distribution

$$N_d = N_0 \exp\left(\frac{-x}{\lambda}\right) \qquad (4.1.15)$$

where λ is the characteristic length describing the decrease of donor atoms away from the semiconductor surface at $x = 0$. Typical Gaussian or complementary-error-function diffusion profiles (Section 2.5) are often approximated by exponential distributions for mathematical simplicity. Because of the relationship of an exponential function and its derivative, the field has a constant value $kT/q\lambda$ throughout the region in which exponential doping may be assumed. As we shall see in Chapter 6, the approximation of a constant field simplifies some useful cases of device analysis. The exponential approximation may, however, obscure important implications of real diffusions that become apparent when more detailed dopant distributions are considered.

4.2 The *pn* Junction

In the analysis of the previous section we restricted our discussion to material of one conductivity type with carrier and dopant concentrations that had a gradual variation with position. These limitations permitted a solution for the electric field in a quasi-neutral region having a spatially varying dopant concentration. Now, we consider the other extreme: a semiconductor with a dopant concentration that has a steep gradient. In this case there can be significant departures from charge neutrality in localized regions of the semiconductor. In particular, we consider the junction between a p-type and an n-type semiconductor and find that we may treat the transition region as if it were depleted of mobile carriers. This *depletion approximation* is the opposite extreme of the *quasi-neutrality* approximation. When analyzing device structures, it is frequently very useful to divide them into regions that are assumed to be quasi-neutral and other regions that are considered to be completely depleted of mobile carriers. Although an idealization, this simplification is adequate for many calculations.

To build a model for the *pn* junction, we begin by considering initially separated *n*- and *p*-type semiconductor crystals (Figure 4.3*a*). When these are brought into

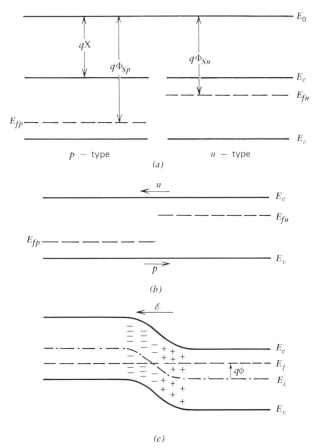

Figure 4.3 (*a*) *n*-type and *p*-type semiconductor regions separated and not in thermal equilibrium. (*b*) The two regions brought into intimate contact allowing diffusion of holes from the *p*-region and electrons from the *n*-region. (*c*) Transfer of free carriers leaves uncompensated dopant ions; these cause a field that opposes and balances the diffusion tendencies of holes and electrons.

intimate contact as shown in Figure 4.3*b*, the large difference in electron concentrations between the two materials causes electrons to flow from the *n*-type semiconductor into the *p*-type semiconductor. Similarly, holes flow from the *p*-type region into the *n*-type region. As these mobile carriers move into the oppositely doped material, they leave behind uncompensated dopant atoms near the junction so that an electric field is built up. The field lines extend from the donor ions on the *n*-type side of the junction to the acceptor ions on the *p*-type side (Figure 4.3*c*). The presence of this field causes a potential barrier between the two types of material. When equilibrium is reached, the magnitude of the field is such that the tendency of electrons to diffuse from the *n*-type region into the *p*-type region is exactly balanced by the tendency of the electrons to drift in the opposite direction under the influence of the built-in field.

Potential Barrier. The magnitude of the potential barrier associated with this field may be found by considering the difference in the Fermi levels of the initially separated materials (Figure 4.3*a*) as was done for the metal-semiconductor system in Chapter 3. When the two sections of semiconductor are at equilibrium, the Fermi level must be constant throughout the entire system. Consequently, the potential barrier that forms between the two materials must be just equal to the initial difference between the Fermi levels in the separated pieces of semiconductor. This is equivalent to the difference in work functions of the separated semiconductor regions, since the work function of a semiconductor is defined as

$$q\Phi_s \equiv qX + (E_c - E_f) \tag{4.2.1}$$

where qX is the electron affinity.

Far from the junction, the carrier concentrations remain at the values they had in the isolated semiconductor crystals. The electron concentration in the *n*-type material is equal to the donor concentration, and the hole density is given by n_i^2/N_d. Similarly, in the *p*-type material far from the junction, the hole concentration is equal to N_a and the electron concentration is given by n_i^2/N_a. Since the carrier concentrations are known far from the junction, we can solve for the potential defined in Equation 4.1.2 by making use of Equation 4.1.11. Close to the junction, we cannot specifically give the free-carrier concentration without further analysis. In this region, however, we can see qualitatively from considering the sketch of potential shown in Figure 4.3*c* together with Equation 4.1.11 that the carrier concentrations are much less than the concentrations in the respective neutral materials since $|\phi|$ is small. Thus, the transition region is often referred to as a depletion region in which the space charge is overwhelmingly made up of dopant ions.

In the *depletion approximation* we assume that the semiconductor may be divided into distinct zones that are either neutral or completely depleted of mobile carriers. These zones join each other at the edges of the depletion or space-charge region, where the majority-carrier density is assumed to change abruptly from the dopant concentration to zero. The depletion approximation appreciably simplifies the solution of Poisson's equation (Equation 4.1.10).

Since the carrier concentrations are assumed to be much less than the net ionized dopant density in the depletion region, the second derivative of the potential is proportional to the net dopant concentration in the depletion region:

$$\frac{d^2\phi}{dx^2} = \frac{-q}{\epsilon_s}(N_d - N_a) \qquad (4.2.2)$$

In the general case N_d and N_a may be functions of position, and we cannot solve Equation 4.2.2 explicitly.

EXAMPLE   The Quasi-Neutrality Approximation

Investigate the assumption of *quasi-neutrality* for *n*-type silicon with nonuniform doping by considering the density of space charge present in a region where the dopant density $N_d(x)$ changes from 10^{16} to 10^{18} cm^{-3} over a length $\lambda = 1\,\mu$m. Assume that the dopant varies as

$$N_d(x) = 10^{16} \times \exp\left\{\ln(100)\left[\frac{x}{\lambda} - \frac{1}{2\pi}\sin\left(\frac{2\pi}{\lambda}x\right)\right]\right\} \qquad \left(0 < \frac{x}{\lambda} < 1\right)$$

(This mathematical form is a smooth differentiable function with the proper end values.)[8]

Calculate the field and charge distribution in the region $0 < \frac{x}{\lambda} < 1$.

Solution

Because the donor-dopant density increases with x, we expect the free-electron density to be characterized by a positive gradient. Hence, electrons will tend to diffuse in the negative x direction. At thermal equilibrium, this diffusion tendency must be balanced by an electron drift tendency toward positive x; hence, there is a built-in field in the negative x direction. Associated with this field is a space charge so that the graded-dopant region is not truly charge neutral. In this example we calculate the density of charge present to investigate quantitatively the departure from charge neutrality.

Using Equation 4.1.8 we calculate the total potential difference across the region $\Delta\varphi$ as

$$\Delta\varphi = |\mathscr{E}_{avg}|\lambda = \frac{kT}{q}\ln(100) = 0.12 \text{ V}.$$

where $\mathscr{E}_{avg} = -1.19 \times 10^3$ V cm^{-1} is the average field in the variable doping region. The field as a function of x is (from Equation 4.1.13)

$$\mathscr{E} = -\frac{kT}{q}\frac{1}{N_d}\frac{dN_d}{dx} = \mathscr{E}_{avg}\left(1 - \cos\frac{2\pi}{\lambda}x\right)$$

To find the charge density we use Poisson's equation $\rho(x) = \epsilon_s \, d\mathscr{E}/dx$. Normalizing $\rho(x)$ to the electronic charge, we have

$$\frac{\rho(x)}{q} = \frac{\epsilon_s \mathscr{E}_{avg}}{q\lambda}\left[2\pi \sin\left(\frac{2\pi}{\lambda}x\right)\right]$$

The maximum charge density ($|\rho/q|$), occurring at $x/\lambda = 1/4$ and $x/\lambda = 3/4$, is 4.8×10^{14} cm^{-3}, which is appreciably smaller than the minimum dopant density ($N_d = 10^{16}$ cm^{-3}). Hence, we conclude that *quasi neutrality* does prevail in the region over which the dopant density varies. The accompanying figures show the dopant density, field, and charge density in the region of the doping gradient.

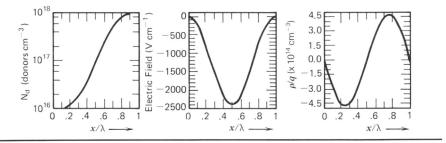

Step Junction

There are two idealized dopant configurations that provide useful insight into the behavior of real *pn*-junctions. One configuration, called the *linearly graded junction*, corresponds to a continuous gradient in dopant between the *n*-type and *p*-type regions. We consider linearly graded junctions later, however, after first discussing a second idealized configuration, the *step junction*. Step junctions (sometimes called *abrupt junctions*), are characterized by a constant *n*-type dopant density that changes with position in stepwise fashion to a constant *p*-type dopant density. An abrupt junction can be formed, for example, by epitaxial deposition of a constant-doped *n*-type region on a *p*-type substrate (as was described in Chapter 2 for bipolar processing).

Approximate Analysis. We can solve Equation 4.2.2 for the step junction shown in Figure 4.4, where the dopant concentration changes abruptly from N_a to N_d at $x = 0$. Using the depletion approximation, we assume that the region between $-x_p$ and x_n is totally depleted of mobile carriers, as shown in Figure 4.4b, and that the mobile majority-carrier densities abruptly become equal to the respective dopant concentrations at the edges of the depletion region. The charge density is, therefore, zero everywhere except in the depletion region where it takes the value of the ionized dopant concentration (Figure 4.4c). In the *n*-type material ($x > 0$) Equation 4.2.2 becomes

$$\frac{d^2\phi}{dx^2} = -\frac{d\mathscr{E}}{dx} = -\frac{qN_d}{\epsilon_s} \tag{4.2.3}$$

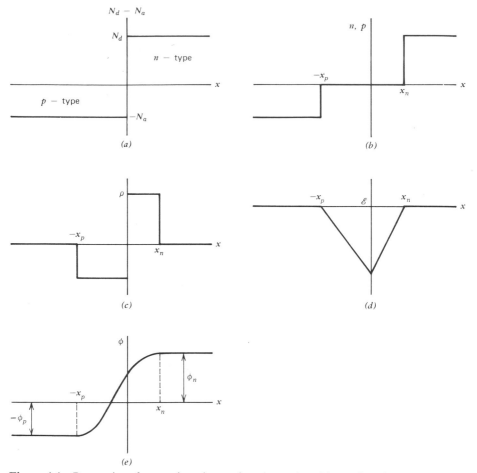

Figure 4.4 Properties of a step junction as functions of position using the complete-depletion approximation: (*a*) net dopant concentration, (*b*) carrer densities, (*c*) space charge used in Poisson's equation, (*d*) electric field found from first integration of Poisson's equation, and (*e*) potential obtained from second integration.

which may easily be integrated from an arbitrary point in the *n*-type depletion region to the edge of the depletion region at x_n, where the material becomes neutral and the field vanishes. Carrying through this integration, we find the field to be

$$\mathscr{E}(x) = -\frac{qN_d}{\epsilon_s}(x_n - x) \qquad 0 < x < x_n \qquad (4.2.4)$$

The field is negative throughout the depletion region and varies linearly with *x*, having its maximum magnitude at $x = 0$ (Figure 4.4*d*). The direction of the field toward the left is physically reasonable since the force it exerts must balance the tendency of the negatively charged electrons to diffuse toward the left out of the neutral *n*-type material. The field in the *p*-type region is similarly found to be

$$\mathscr{E}(x) = -\frac{q}{\epsilon_s} N_a(x + x_p) \qquad -x_p < x < 0 \tag{4.2.5}$$

The field in the *p*-type region is also negative in order to oppose the tendency of the positively charged holes to diffuse toward the right.

At $x = 0$, the field must be continuous, so that

$$N_a x_p = N_d x_n \tag{4.2.6}$$

Thus, the width of the depleted region on each side of the junction varies inversely with the magnitude of the dopant concentration; the higher the dopant concentration, the narrower the space-charge region. In a highly asymmetrical junction where the dopant concentration on one side of a junction is much higher than that on the other side, the depletion region penetrates primarily into the lightly doped material, and the width of the depletion region in the heavily doped material can often be neglected. The charge, field, and potential at this type of junction, called a *one-sided step junction*, are identical to the results obtained for the ideal Schottky barrier, and the sketches of these quantities shown in Figure 4.4 reduce to the equivalent sketches in Figure 3.5. The properties of one-sided, abrupt, planar junctions are so frequently referred to that it is worthwhile to collect them in the useful nomograph form of Table 4.1.

The expressions for the field may be integrated again to obtain the potential variation across the junction. In the *n*-type material

$$\phi(x) = \phi_n - \frac{qN_d}{2\epsilon_s}(x_n - x)^2 \qquad 0 < x < x_n \tag{4.2.7}$$

as shown in Figure 4.4e, where ϕ_n is the potential at the neutral edge of the depletion region as obtained by solving Equation 4.1.11

$$\phi_n = \frac{kT}{q} \ln \frac{N_d}{n_i} \tag{4.2.8}$$

Similarly, in the *p*-type material

$$\phi(x) = \phi_p + \frac{qN_a}{2\epsilon_s}(x + x_p)^2 \qquad -x_p < x < 0 \tag{4.2.9a}$$

$$\phi_p = \frac{-kT}{q} \ln \frac{N_a}{n_i} \tag{4.2.9b}$$

where $\phi_p < 0$ is the potential at the neutral edge of the depletion region in the *p*-type material.

The total potential change ϕ_i from the neutral *p*-type region to the neutral *n*-type region is $\phi_n - \phi_p$ ($\phi_p < 0$) and is dependent on the dopant concentration

Table 4.1 Nomograph for Silicon Uniformly Doped, One-Sided, Step Junctions (300 K). (See Figure 4.13 to correct for junction curvature.) (*Courtesy AT&T Bell Laboratories*).

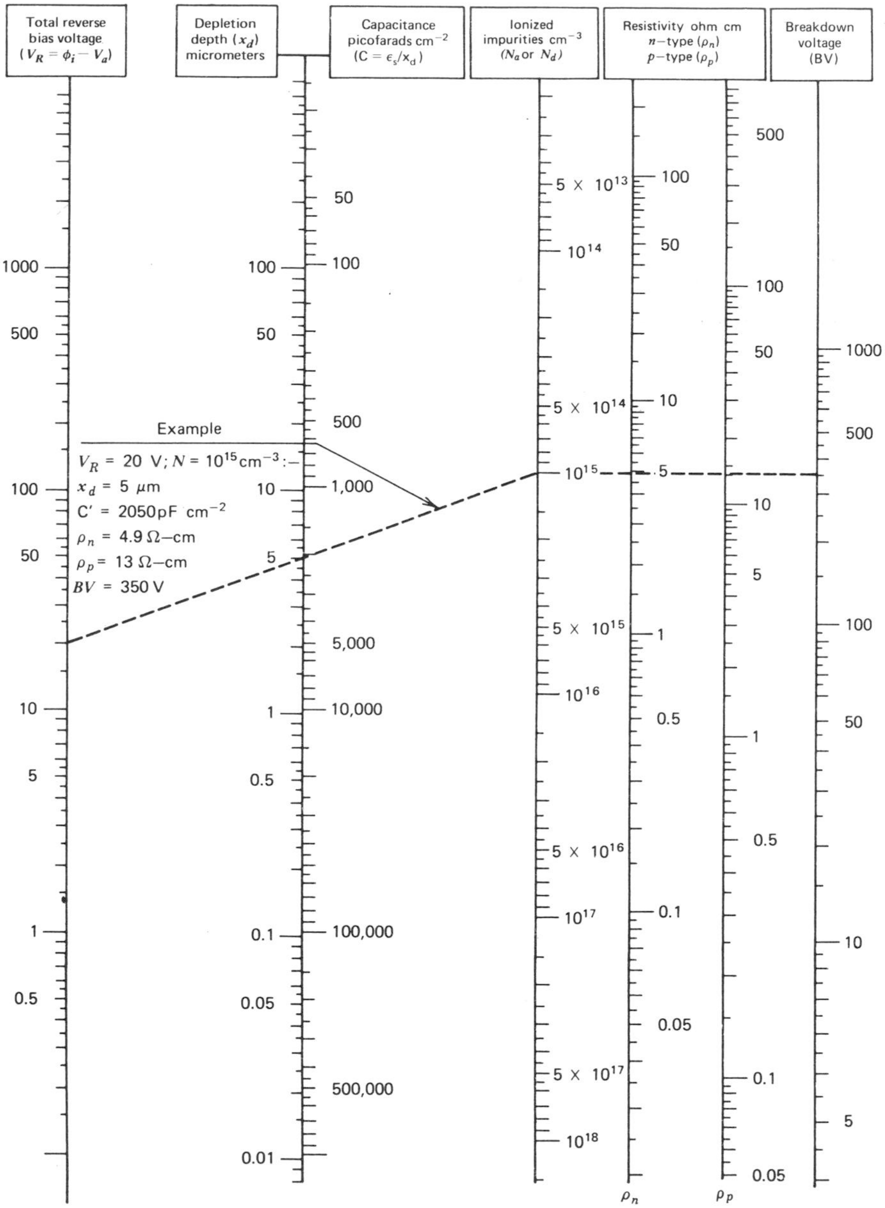
Total reverse bias voltage ($V_R = \phi_i - V_a$)
Depletion depth (x_d) micrometers
Capacitance picofarads cm^{-2} ($C = \epsilon_s/x_d$)
Ionized impurities cm^{-3} (N_a or N_d)
Resistivity ohm cm n–type (ρ_n) p–type (ρ_p)
Breakdown voltage (BV)
1000
500
100
50
10
5
1
0.5
100
50
10
5
1
0.5
0.1
0.05
0.01
50
100
500
1,000
5,000
10,000
100,000
500,000
5 × 10^{13}
10^{14}
5 × 10^{14}
10^{15}
5 × 10^{15}
10^{16}
5 × 10^{16}
10^{17}
5 × 10^{17}
10^{18}
100
50
10
5
1
0.5
0.1
0.05
500
100
50
10
5
1
0.5
0.1
0.05
1000
500
100
50
10
5
ρ_n
ρ_p
Example
V_R = 20 V; $N = 10^{15}$cm^{-3}:–
x_d = 5 μm
C′ = 2050pF cm^{-2}
ρ_n = 4.9 Ω–cm
ρ_p = 13 Ω–cm
BV = 350 V

in each region:

$$\phi_i = \phi_n - \phi_p = \frac{kT}{q} \ln \frac{N_d}{n_i} + \frac{kT}{q} \ln \frac{N_a}{n_i} = \frac{kT}{q} \ln \frac{N_d N_a}{n_i^2} \qquad (4.2.10)$$

Note that the *built-in potential* ϕ_i is positive, that is, the *n*-side is at a higher potential than the *p*-side, which is proper to obtain a balance between drift and diffusion across the junction. The major portion of the potential change occurs in the region with the lower dopant concentration, and the depletion region is wider in the same region. Note that the potential at the junction plane ($x = 0$) is not exactly zero unless the junction is symmetrical (i.e., $N_a = N_d$).

EXAMPLE　Built-In Voltage at a *pn* Junction

A lightly doped, *n*-type sample of silicon has a resistivity of 4 Ω-cm. It is used to make a *pn* junction with a region in which the dopant density N_a is 1000 times higher than that in the *n*-type silicon. What is the built-in voltage ϕ_i at the junction?

Solution

Equation 4.2.10 expresses ϕ_i as a function of the dopant densities N_a and N_d on either side of the junction. From Figure 1.14, for a resistivity of 4 Ω-cm, $N_d = 10^{15}$ cm^{-3}. Since the *p*-region dopant density is 1000 times higher, $N_a = 10^{18}$ cm^{-3}.

$$\phi_i = \frac{kT}{q} \ln \left(\frac{N_a N_d}{n_i^2} \right) = \frac{kT}{q} \ln \left(\frac{10^{18} \times 10^{15}}{(1.45 \times 10^{10})^2} \right)$$

$$\phi_i = 0.753 \text{ V}$$

This voltage, equivalent to roughly 2/3 of E_g/q, is a typical value for ϕ_i in many IC *pn* junctions.

At very high dopant concentrations, Equation 4.2.10 is no longer valid because it is based on the use of Maxwell-Boltzmann statistics (*cf* Equation 4.2.8). When the dopant density approaches the effective density of states N_c or N_v (essentially for concentrations of order 10^{19} cm^{-3}), Fermi-Dirac statistics should be used in any derivations. When calculating the potential at a *pn* junction, however, we do not need to consider Fermi-Dirac statistics in detail because the practical result of heavy doping is that the Fermi level comes very near the band edge, and the potential in the doped silicon is approximately equal to $E_g/2q$ or 0.56 V. Thus, the built-in potential across a *pn* junction made between heavily doped *p*-type material (usually denoted as p^+ silicon) and lightly doped *n*-type silicon is

$$|\phi_i| = 0.56 + \frac{kT}{q} \ln \left(\frac{N_d}{n_i} \right) \qquad (\text{heavily doped})$$

A similar result with N_a in place of N_d applies to junctions between n^+ silicon and lightly doped *p*-type silicon.

For the step *pn* junction with arbitrary dopant concentrations, the total depletion-region width is found from Equations 4.2.6 through 4.2.10 to be

$$x_n + x_p = \left[2 \frac{\epsilon_s}{q} \phi_i \left(\frac{1}{N_a} + \frac{1}{N_d} \right) \right]^{1/2} \qquad (4.2.11)$$

From Equation 4.2.11, we see that the depletion-region width depends most strongly on the material with the lighter doping, and varies approximately as the inverse square root of the smaller dopant concentration.

EXAMPLE   Step Junction

Consider a *pn* junction with constant doping concentrations N_a on the *p*-type side and N_d on the *n*-type side.

Derive an expression for the percentage P_n of the total reverse-bias voltage that is dropped across the *n*-type region if an external voltage $V_a = -5\text{V}$ is applied. Evaluate P_n for $N_a = 10^{17} \text{ cm}^{-3}$ in the following three cases: (a) $N_d = 10^{-1} N_a$, (b) $N_d = 10^{-2} N_a$, and (c) $N_d = 10^{-3} N_a$.

Use a modified form of Equation 4.2.11 to obtain the total depletion-layer width for each of these cases. Compare the calculated depletion-layer width x_d to values obtained by using Table 4.1 for a one-sided step junction having a dopant density N_d in each of these three cases.

Solution

A plot of the field for this problem is shown in Figure 4.4d. If we designate the maximum field by $\mathscr{E}_{\max}$, the voltage V_p dropped across the *p*-region is

$$V_p = \frac{1}{2} \mathscr{E}_{\max} x_p,$$

while that dropped across the *n*-region V_n is

$$V_n = \frac{1}{2} \mathscr{E}_{\max} x_n$$

From these two equations, we have $V_p/V_n = x_p/x_n$. The total reverse-bias voltage is $V_R = V_p + V_n$. Therefore, the required percentage P_n is

$$P_n = \frac{V_n}{V_n + V_p} \times 100 = \frac{1}{1 + V_p/V_n} \times 100$$

$$= \frac{1}{1 + x_p/x_n} \times 100.$$

Using Equation 4.2.6, we have therefore

$$P_n = \frac{1}{1 + N_d/N_a} \times 100$$

The percentages (P_N values) for the three junctions described above are

(a) $N_d = 10^{16}$ $P_n = 91\%$
(b) $N_d = 10^{15}$ $P_n = 99\%$
(c) $N_d = 10^{14}$ $P_n = 99.9\%$

Hence, we see that an order-of-magnitude ratio in dopant density between the two sides of a step junction results in more than 90% of the total reverse bias being dropped across the more lightly doped region. As the dopant ratio increases, the junctions becomes more and more one-sided.

In the second part of this example, we consider the total depletion-layer width when the junction is under 5V reverse bias. To apply Equation 4.2.11 we replace ϕ_i, the built-in potential, with the total reverse voltage $V_R = \phi_i + |V_a|$ (measured in the *p*-region with respect to the *n*-region). From Equation 4.2.10, we find ϕ_i for cases (a), (b) and (c) to be

(a) $\phi_i = 0.753$ V for $N_a = 10^{17}$, $N_d = 10^{16}$ cm^{-3}
(b) $\phi_i = 0.694$ V for $N_a = 10^{17}$, $N_d - 10^{15}$ cm^{-3}
(c) $\phi_i = 0.634$ V for $N_a = 10^{17}$, $N_d = 10^{14}$ cm^{-3}

Since $V_a = -5$V, $V_R = 5.753$, 5.694, and 5.634 V for cases (a), (b) and (c), respectively. Using these values in the modified form of Equation 4.2.11, and also in Table 4.1, we find for x_d:

	Equation 4.2.11	Table 4.1
(a)	0.91 μm	0.85 μm
(b)	2.73 μm	2.7 μm
(c)	8.54 μm	8.6 μm

From this example, it is apparent that one-sided behavior is observed in a reverse-biased *pn* junction with relatively small ratios of the dopant densities. Table 4.1 can be useful for rough estimates of the depletion-layer width even when dopant densities differ by only one order of magnitude.

The analysis of the abrupt *pn* junction has employed the depletion approximation to solve Poisson's equation and to specify the boundary conditions. Before proceeding further we discuss briefly the more exact solution and consider the consequences of the depletion approximation. We again look at a junction with a step-function change in dopant concentration at $x = 0$ (Figure 4.5a). However, we no longer assume an abrupt change from neutral regions to completely depleted regions at x_n and $-x_p$; instead, we consider transition regions that are only partially depleted near these boundaries (Figure 4.5b). Since the net charge densities in the

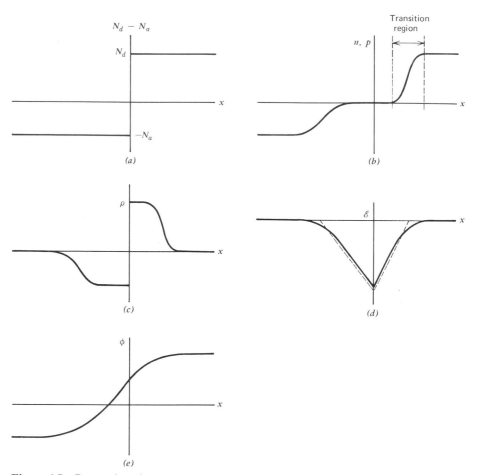

Figure 4.5 Properties of a step junction as functions of position considering a gradual transition between neutral and depleted regions: (*a*) net dopant concentration, (*b*) carrier densities, (*c*) space charge, (*d*) electric field, (*e*) potential. (Compare with Figure 4.4).

transition regions (Figure 4.5*c*) are less than the dopant concentrations, the fields change more gradually than predicted by the depletion approximation. The solid line of Figure 4.5*d* indicates the field found from the more exact solution, while the dashed line represents the field found using the depletion approximation. In the more exact analysis the field extends further into the semiconductor interior and the space-charge region is widened.

Debye Length.[†] Although an exact solution of the step junction can be obtained without making use of the depletion approximation, we will not carry out this more detailed analysis, but rather consider an analytical approximation for the potential near the edges of the space-charge region (near x_n and $-x_p$) in order to investigate the validity of the depletion approximation. If we consider only small variations of potential from ϕ_n near $x = x_n$, we may rewrite Equation 4.1.10 by neglecting the minority-carrier concentration p and letting $\phi' = \phi_n - \phi$ in Equation 4.1.11

$$\frac{d^2\phi'}{dx^2} = \frac{q}{\epsilon_s}(N_d - n) = \frac{q}{\epsilon_s}\left[N_d - n_i \exp\left(\frac{q(\phi_n - \phi')}{kT}\right)\right]$$

$$= \frac{q}{\epsilon_s}N_d\left[1 - \exp\left(-\frac{q\phi'}{kT}\right)\right] \qquad (4.2.12)$$

Since we are restricting ϕ' to be small, we may expand the exponential term in Equation 4.2.12 in a Taylor series, retaining only the first two terms so that the equation reduces to

$$\frac{d^2\phi'}{dx^2} = \frac{q}{\epsilon_s}N_d\frac{q\phi'}{kT} = \frac{\phi'}{L_D^2} \qquad (4.2.13)$$

where L_D, the extrinsic *Debye length* given by

$$L_D = \left[\frac{\epsilon_s kT}{q^2 N_d}\right]^{1/2} \qquad (4.2.14)$$

is a characteristic length associated with the spatial variations of potential.* We met a similar form for the Debye length in Equation 3.4.3 in our discussion of Schottky ohmic contacts. The solution of Equation 4.2.13 is of the form $\phi' = B \exp x/L_D$ with B a constant of integration. Hence, the potential ϕ' varies exponentially with distance near the edges of the space-charge region with a characteristic length equal to the extrinsic Debye length. Since the carrier concentration itself depends exponentially on the potential, the carrier concentration changes rapidly from the dopant concentration to essentially zero within a few Debye lengths. Therefore, the depletion approximation is questionable only within a few extrinsic Debye lengths of the edges of the space-charge region, x_n and $-x_p$. For symmetric junctions with typical dopant densities of 10^{16} cm^{-3}, L_D equals 40 nm while x_n is found from simultaneous solutions of Equations 4.2.6 and 4.2.11 to be

---

* More rigorously, the Debye length L_D describes the screening of electric fields by the rearrangement of mobile carriers and depends on the total (hole and electron) free-carrier concentrations in the region.

$$L_D = \left[\frac{\epsilon_s kT}{q^2(n + p)}\right]^{1/2}$$

For an extrinsic *n*-type semiconductor, the minority-carrier density is negligible and L_D takes the form of Equation 4.2.14. The Debye length increases as carrier density decreases. The maximum length is the intrinsic Debye length L_{Di} given by the expression

$$L_{Di} = \left[\frac{\epsilon_s kT}{2q^2 n_i}\right]^{1/2}$$

At room temperature, $L_{Di} = 24$ μm.

approximately 210 nm. Consequently, for this case the depletion approximation is reasonable, but clearly still an approximation.

Most usually the step or abrupt junction is not an adequate representation for a *pn* junction made by diffusion. It is especially inapplicable to most practical cases of *double-diffused junctions*, that is, junctions formed by two successive diffusions of opposite type dopant atoms. A general analytical solution of Poisson's equation for double-diffused junctions (Equation 4.1.12) is not possible, and specific cases are usually considered only approximately. If more accurate results are needed, numerical techniques are used.

Linearly Graded Junction

In addition to the abrupt junction there is another doping profile that can be treated exactly and that gives useful results for approximating real *pn* junctions. In a *linearly graded* junction the net dopant concentration varies linearly from the *p*-type material to the *n*-type material. This type of junction is characterized by a constant *a*, which is the gradient of the net dopant concentration and thus has units of cm^{-4}. The net dopant concentration can be written

$$N_d - N_a = ax \qquad (4.2.15)$$

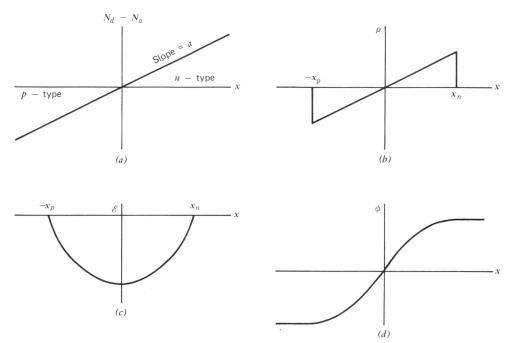

Figure 4.6 Properties of a linearly graded junction using the depletion approximation: (*a*) net dopant concentration: $N_d - N_a = ax$, (*b*) space charge, (*c*) electric field, (*d*) potential.

throughout the space-charge region (Figure 4.6a). The field and potential are readily found from Poisson's equation by using the depletion approximation. Since the space charge varies linearly with position in the depletion region, the field varies quadratically and the potential varies as the third power of position in the space-charge region (Figure 4.6). (The details of the analysis are considered in Problem 4.6.)

Although linearly graded junctions are not realized physically, many practical cases can be approximated by a linearly graded junction over at least a limited voltage range. If an abrupt junction is heated so that the dopant atoms diffuse across the junction, the junction becomes less abrupt and may be approximated by a linearly graded junction as long as the space-charge region is narrow compared to the diffusion length of the impurity atoms. Even diffused junctions are sometimes approximated by linearly graded junctions over a limited distance as shown in Figure 4.7a for an *n*-type diffusion into a *p*-type wafer. On the other hand, a diffusion into a uniformly doped wafer may be approximated by a one-sided step junction if the diffusion length is short compared to the width of the space-charge region (Figure 4.7b).

Exponential Doping. Although more realistic junctions cannot be treated analytically, some qualitative comments about their behavior may be made. We may approximate either a complementary-error-function or Gaussian diffusion profile by an exponential function over a considerable distance. Within this approximation the net dopant concentration after an *n*-type diffusion into a uniformly doped *p*-type wafer may be written

$$N_d - N_a = N_0 e^{-x/\lambda} - N_a \qquad (4.2.16)$$

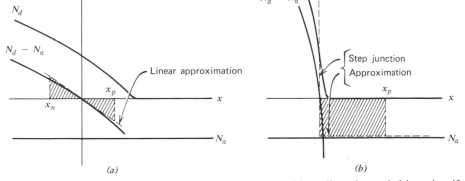

(a)   (b)

Figure 4.7 (a) A diffused junction can be approximated by a linearly graded junction if the diffusion length $2\sqrt{Dt}$ is much longer than the space-charge region. (b) A "one-sided" step junction is more appropriate if the space-charge region is much greater than the diffusion length $2\sqrt{Dt}$.

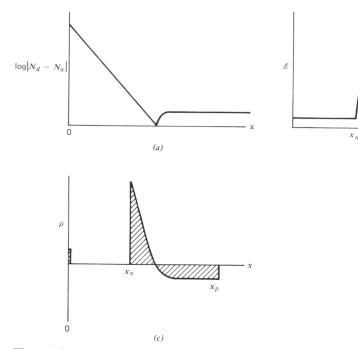

Figure 4.8   Properties of an exponential junction as functions of position: (*a*) net dopant concentration (semilogarithmic scale), (*b*) electric field, (*c*) space charge.

as shown in Figure 4.8*a*. In Equation 4.2.16, λ is the characteristic length associated with the diffusion, and N_a is the dopant concentration in the *p*-type wafer. As we saw in Section 4.1, there is an electric field in the exponentially doped, quasi-neutral, *n*-type side of the junction to balance the tendency of carriers to diffuse to regions of lower carrier density. There is no field in the uniformly doped *p*-type neutral region. The absence of mobile carriers in the depletion region on the *n*-type side of the junction requires that the field there be higher than in the quasi-neutral *n*-type region; the field throughout the structure is indicated schematically in Figure 4.8*b*. The space charge is found by differentiating the field. Since the field is constant in the quasi-neutral region, the space charge there is zero, although there is a sheet of charge at the surface (Figure 4.8*c*). The remainder of the space charge is confined to the depletion region.

4.3 Reverse-Biased *pn* Junctions

In Section 4.2, we used the depletion approximation to simplify our consideration of electrical effects at *pn* junctions in thermal equilibrium. To discuss the junction under bias, we again make use of the depletion approximation together with several assumptions about the applied bias. We assume that ohmic contacts connect the p and n regions to the external voltage source so that negligible voltage is dropped at the contacts. We also consider that small currents flow through the neutral regions and cause only small voltage drops. We consider the n-region to be grounded and voltage V_a to be applied to the p-region. With these assumptions, the entire applied voltage appears across the junction. Furthermore, under these assumptions, the solutions of Poisson's equation that were found for thermal-equilibrium conditions in the last section apply also to the junction under bias. Only the total potential across the junction changes from the built-in value ϕ_i to $\phi_i - V_a$.

If V_a is positive, the barrier to majority carriers at the junction is reduced and the depletion region is narrowed. A conceptionally useful way of visualizing the reduction in depletion-layer width is that the applied voltage moves majority carriers toward the edges of the depletion region where they neutralize some of the space charge. This reduces the overall depletion-region width. The total potential across the junction is $\phi_i - V_a$ with $V_a > 0$, and the junction is *forward biased*. Under forward bias, appreciable currents can flow even for small values of V_a. We shall defer further consideration of forward bias until Chapter 5.

If negative voltage is applied to the p-region, the barrier to majority flow increases. Again the total potential drop may be expressed as $\phi_i - V_a$ except that now V_a is negative and the junction is under *reverse bias*. Under reverse bias, majority carriers are pulled away from the edges of the depletion region, which therefore widens. There is very little current flow since the bias polarity aids the transfer of electrons from the p-side to the n-side and holes from the n-side to the p-side. Since these are minority carriers in each region, they are low in density. We shall discuss further the currents that do flow in a reverse-biased pn junction in Chapter 5.

Depletion Width, Maximum Field. For an abrupt pn junction we may find the depletion-region width as a function of voltage by replacing the built-in potential ϕ_i in Equation 4.2.11 by $\phi_i - V_a$ so that

$$x_d = x_n + x_p = \left[\frac{2\epsilon_s}{q}\left(\frac{1}{N_a} + \frac{1}{N_d}\right)(\phi_i - V_a)\right]^{1/2} \tag{4.3.1}$$

As V_a becomes considerably larger than ϕ_i, the depletion region begins to vary as the square root of the reverse-bias voltage in a step junction. Expressions for x_d for other dopant profiles are similarly modified from the thermal-equilibrium result. For example, the depletion-region width is given by

$$x_d = \left[\frac{12\epsilon_s(\phi_i - V_a)}{qa}\right]^{1/3} \qquad \text{linear} \tag{4.3.2}$$

in a linearly graded junction with a doping gradient a (Equation 4.2.15).

It is often important to know the maximum field at the junction $\mathscr{E}_{max}$ and its relationship to applied voltage. The step-junction case is particularly simple because the field varies linearly with distance (Equations 4.2.4 and 4.2.5). Hence, the area under the field curve that represents the potential may be written by inspection as one-half the maximum field times the depletion-layer width. Formally

$$\tfrac{1}{2}\mathscr{E}_{max}x_d = (\phi_i - V_a)$$

so that

$$\mathscr{E}_{max} = \frac{2(\phi_i - V_a)}{x_d} \tag{4.3.3}$$

in a step junction. For the linearly graded junction, the results of Problem 4.5 can be used to find that the maximum field is

$$\mathscr{E}_{max} = \frac{3(\phi_i - V_a)}{2x_d} \qquad \text{linear} \tag{4.3.4}$$

Capacitance. In Section 3.2, we analyzed capacitive behavior in a metal-semiconductor junction resulting from modulation of the stored charge in the junction depletion region as a function of a varying applied voltage. Analogous behavior occurs in *pn* junctions except that the depletion region extends with bias in two directions. As the dopant concentration on one side of a *pn* junction increases, the depletion region on that side of the junction narrows. Ultimately, the width of the depletion region on the heavily doped side will become a negligible part of the total space-charge-layer width, and the field- and space-charge-configurations for the *pn* junction become very similar to those at a rectifying metal-semiconductor contact.*

For a step junction with dopant concentrations N_a and N_d on its two sides, we may find the small-signal capacitance per unit area by using the expression for the charge Q_s (per unit area) in the depletion region on either side of the junction

$$Q_s = qN_d x_n = qN_a x_p \tag{4.3.5}$$

in the definition of the small-signal capacitance C (per unit area)

$$C = \frac{dQ}{dV_a} = qN_a \frac{dx_n}{dV_a} = qN_a \frac{dx_p}{dV_a} \tag{4.3.6}$$

Since $x_p = (N_d/N_a)x_n$ and $x_d = x_n + x_p$, we find from Equation 4.3.1 that

$$\frac{dx_n}{dV_a} = \frac{1}{N_d}\left[\frac{\epsilon_s}{2q\left(\dfrac{1}{N_a} + \dfrac{1}{N_d}\right)(\phi_i - V_a)}\right]^{1/2} \tag{4.3.7}$$

and

$$C = \left[\frac{q\epsilon_s}{2\left(\dfrac{1}{N_a} + \dfrac{1}{N_d}\right)(\phi_i - V_a)}\right]^{1/2} \tag{4.3.8}$$

---

* Looked at another way, as charge becomes more concentrated and finally approaches a δ function in space, fields are terminated more and more abruptly. Thus, the heavily doped semiconductor assumes the properties associated with idealized metals; in particular, all charge is confined to the surface.

Thus, for $|V_a|$ greater than ϕ_i, the capacitance of a step *pn* junction decreases approximately inversely with the square root of the reverse bias. Using Equation 4.3.1 in Equation 4.3.8, we obtain the expression $C = \epsilon_s/x_d$, the general relationship for small-signal capacitance.

Although we presented a heuristic proof for the general equation $C = \epsilon_s/x_d$ in Chapter 3, the meaning of the result is important enough to be reinforced at this point by making a careful derivation. We consider an arbitrarily doped *pn* junction having a depletion region that extends from $-x_p$ to x_n (Figure 4.9*a*). The charge per unit area Q stored between a point x and the edge of the depletion region at x_n is just

$$Q = q \int_x^{x_n} N \, dx \tag{4.3.9}$$

where $N = N_d - N_a$ is the net dopant density. Since $\mathscr{E}_x(x_n) = 0$, the electric field

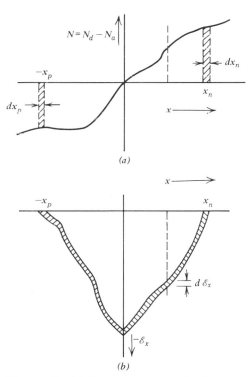

(a)

(b)

Figure 4.9 (*a*) Dopant concentration in an arbitrarily doped junction showing modulation of the carrier densities at the edges of the space-charge region by an applied voltage. (*b*) Electric-field distributions for two slightly different applied voltages.

at x is found from Gauss' law to be

$$-\mathscr{E}_x(x) = \frac{1}{\epsilon_s} \int_x^{x_n} qN \, dx = \frac{Q}{\epsilon_s} \tag{4.3.10}$$

and is illustrated in Figure 4.9*b*. As the applied voltage V_a is changed by a small amount dV_a, the width of the *n*-type side of the depletion region changes by dx_n and the charge stored between x and the edge of the depletion region changes by

$$dQ = qN(x_n) \, dx_n \tag{4.3.11}$$

Consequently, the field at x changes by

$$-d\mathscr{E}_x = \frac{dQ}{\epsilon_s} = \frac{q}{\epsilon_s} N(x_n) \, dx_n \tag{4.3.12}$$

as the voltage changes. Since the area under the $\mathscr{E}_x$ versus x curve corresponds to the total potential $\phi_i - V_a$, the differential change in voltage corresponds to the change in area under the curve (the shaded region of Figure 4.9*b*).

$$dV_a \approx -x_d \, d\mathscr{E}_x = \frac{x_d}{\epsilon_s} \, dQ \tag{4.3.13}$$

Using Equation 4.3.13 in the definition of the small-signal capacitance (Equation 4.3.6), we find

$$C = \frac{dQ}{dV_a} = \frac{\epsilon_s}{x_d} \tag{4.3.14}$$

Thus, we have proven this simple relationship for an arbitrarily doped junction. The voltage dependence of the depletion-region width x_d and of the capacitance depends, of course, on the actual dopant profile.

As in the case of the metal-semiconductor junction, measurements of the variation with applied voltage of the small-signal capacitance of a *pn*-junction may be used to determine the dopant concentration as a function of position. However, an additional complication arises in the *pn* junction since the depletion region extends in both directions from the junction plane, that is, the plane where $N = N_d - N_a = 0$. To investigate this further, we differentiate Equation 4.3.14 and write $x_d = x_n + x_p$ to find

$$\frac{dC}{dx_n} = -\frac{\epsilon_s}{(x_n + x_p)^2} \left(1 + \frac{dx_p}{dx_n} \right) \tag{4.3.15}$$

Since the changes in charge on either side of the junction are equal in magnitude,

$$|dQ| = |qN(-x_p) \, dx_p| = qN(x_n) \, dx_n \tag{4.3.16}$$

and

$$\frac{dC}{dx_n} = -\frac{C^2}{\epsilon_s} \left(1 + \frac{N(x_n)}{|N(-x_p)|} \right) \tag{4.3.17}$$

Combining this expression with Equation 4.3.11 and the definition of the capacitance, we obtain

$$N(x_n) = -\frac{C^3}{\epsilon_s q(dC/dV_a)}\left(1 + \frac{N(x_n)}{|N(-x_p)|}\right) \qquad (4.3.18)$$

If the p-type side of the junction is much more heavily doped than the n-type side, the factor on the right of Equation 4.3.18 reduces to unity, and the equation simplifies to the same form as Equation 3.2.10 which we obtained for the metal-semiconductor junction. Physically, this is reasonable since a depletion region scarcely extends into a highly doped semiconductor, just as it extends a negligible amount into the metal side of a metal-semiconductor junction. In the more general case, however, the presence of the factor $[1 + N(x_n)/N(-x_p)]$ in the expression for $N(x_n)$ complicates the interpretation of capacitance measurements in terms of dopant concentrations. The use of iterative techniques and measurements of other quantities are then required to obtain the dopant distributions.

4.4 Junction Breakdown

We have seen that the depletion-region width and the maximum electric field in a *pn* junction increase as reverse bias increases. Intuitively, we know that there must be physical limitations to these increases. At high voltages, some of the materials making up the device structure, such as the insulating layers of silicon dioxide or packaging materials, may rupture, or else the current through the *pn* junction itself may increase rapidly. In the first case, irreversible damage usually occurs, and the device is destroyed. The second case—breakdown of the barrier to current flow within the junction itself—is generally not destructive (unless the high currents involved melt a portion of the junction) and is of practical importance. The voltage at which breakdown occurs depends on the structure of the junction and the dopant concentrations in a reasonably well-defined manner and junctions can be constructed with predictable breakdown characteristics.

At high fields in a semiconductor, one of two breakdown processes can occur. In one process, the field may be so high that it exerts sufficient force on a covalently bound electron to free it. This creates two carriers, a hole and an electron to contribute to the current. In terms of an energy-band picture, in this breakdown process an electron makes a transition from the valence band to the conduction band without the interaction of any other particles. This type of breakdown is called *Zener breakdown* and involves electron tunneling through energy barriers, a phenomenon that was introduced in our discussion of ohmic contacts in Section 3.4. In the second breakdown process, free carriers are able to gain enough energy from the field between collisions for them to break covalent bonds in the lattice when they collide with it. In this process, called *avalanche breakdown*, every carrier interacting with the lattice as described above creates two additional carriers. All three carriers can then participate in further avalanching collisions, leading to a sudden multiplication of carriers in the space-charge region when the maximum field becomes large enough to cause avalanche.

One or the other of these two breakdown processes predominates in a given *pn* junction. The factors that determine which process occurs will be more evident after details of the breakdown processes have been considered. We discuss first avalanche breakdown, which is most frequently observed. Then we consider Zener breakdown and, finally, we contrast these two processes.

Avalanche Breakdown[†]

We consider an electron traveling in the space-charge region of a reverse-biased *pn* junction. The electron travels, on the average, a distance *l*, its mean free path, before interacting with an atom in the lattice and losing energy. The energy ΔE gained from the field $\mathscr{E}$ by the moving electron between collisions is

$$\Delta E = q \int_0^l \mathscr{E} \cdot dx \tag{4.4.1}$$

The boldface letters in Equation 4.4.1 denote a vector product. If the electron gains sufficient energy from the field before colliding with an atom, it can break the bond between the atom core and one of the bound electrons during a collision so that three carriers—the initial electron and the hole and electron created by the collision—are free to leave the region of the collision. This process is indicated schematically in Figure 4.10. If all three carriers are assumed to have equal mass, conservation of energy and momentum requires that the original electron possess kinetic energy of at least $\frac{3}{2}E_g$ in order to break the bond (Problem 4.11).

For simplicity, we assume an abrupt junction in which the *n*-type region is much more heavily doped than the *p*-type region. In this case, as will be shown in Chapter 5, most of the carriers entering the depletion region under moderate reverse bias are electrons from the *p*-type region. The few holes entering from the *n*-type region may be neglected in our analysis. Near the edges of the space-charge region, the electric field is low and essentially no carriers can gain enough energy

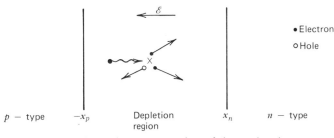

Figure 4.10 Schematic representation of the avalanche process. An incident electron (wavy arrow) gains enough energy from the field to excite an electron out of a silicon-silicon bond during a lattice collision. This creates an additional electron-hole pair.

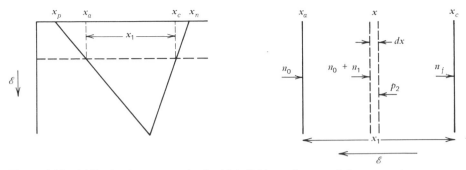

Figure 4.11 (*a*) Ionization occurs in the high-field portion x_1 of the space-charge region. (*b*) Carrier pairs are created in the region dx at x by electrons flowing from the left and by holes flowing from the right.

from the field to create a hole-electron pair before they lose their kinetic energy in a collision with the lattice. Consequently, avalanche is confined to the central portion of the space-charge region where the field is sizable. This central region has been labeled x_1 in the plot of field versus distance for this case shown in Figure 4.11*a*.

We consider the number of carriers created by avalanche in a small volume of width dx located within x_1 at x (Figure 4.11*b*). Let the density of electrons entering x_1 from the left at x_a be n_0. Avalanching will increase this density between x_a and x so that the electrons entering the volume $A\,dx$ at x from the left have a density $n_0 + n_1$. The probability that electrons create electron-hole pairs while traveling through dx is given by the product of a proportionality factor α_n, called the *ionization coefficient*, times the length dx. Since the electrons gain energy more rapidly when the field is higher, we expect the ionization coefficient to be a function of electric field and, hence, of position. The added density of electrons (and, therefore, the added density of holes) created in dx by the electrons entering from the left is

$$dn' = dp = \alpha_n n\,dx = \alpha_n(n_0 + n_1)\,dx \tag{4.4.2}$$

Since we have assumed that no holes enter the ionization region at x_c, any holes entering the infinitesimal volume at x from the right were created between x and x_c. We designate the density of these holes as p_2. The holes will also be avalanche-multiplied within dx to create added hole and electron densities.

$$dn'' = dp = \alpha_p p\,dx = \alpha_p p_2\,dx \tag{4.4.3}$$

where α_p is the ionization coefficient for holes. The total increase in electron density created within dx is equal to the sum $(dn' + dn'')$ (Equations 4.4.2 and (4.4.3)

$$dn = \alpha_n(n_0 + n_1)\,dx + \alpha_p p_2\,dx \tag{4.4.4}$$

If we let n_f be the density of electrons that reaches x_c,

$$n_f = n_0 + n_1 + n_2 \tag{4.4.5}$$

where n_2 is the density of electrons created between x and x_c, and $n_2 = p_2$ since electrons and holes are created in pairs. We may then write

$$\frac{dn}{dx} = (\alpha_n - \alpha_p)(n_0 + n_1) + \alpha_p n_f \qquad (4.4.6)$$

To proceed further, it is necessary to have additional information about the ionization coefficients α_n and α_p. We can gain some useful perspective about avalanche breakdown by taking $\alpha_n = \alpha_p$, although this is only approximately true. Making that assumption and defining $\alpha \equiv \alpha_n = \alpha_p$, we can integrate Equation 4.4.6 subject to the boundary conditions $n(x_a) = n_0$ and $n(x_c) = n_f$. The result of this integration is

$$n_f - n_0 = n_f \int_{x_a}^{x_c} \alpha \, dx \qquad (4.4.7)$$

We denote the ratio of the density of electrons n_f leaving the space-charge region to the density n_0 entering as the multiplication factor M:

$$M = \frac{n_f}{n_0} = \frac{1}{1 - \int_{x_a}^{x_c} \alpha \, dx} \qquad (4.4.8)$$

As the integral in the denominator of Equation 4.4.8 approaches unity, the multiplication factor increases without bound. Hence, avalanche breakdown can be defined to occur when

$$\int_{x_a}^{x_c} \alpha \, dx = 1 \qquad (4.4.9)$$

Had we not taken $\alpha_n = \alpha_p$, there would be a more complicated integral than that in Equation 4.4.9, but a form similar to Equation 4.4.8 in which the denominator vanishes could be derived.

As stated earlier, the ionization coefficient is a strong function of the electric field since the energy necessary for an ionizing collision is imparted to the carrier by the field. The exact field dependence of the ionization coefficient is complicated, but an expression of the form

$$\alpha = K\mathscr{E} \exp\left(-\frac{B}{\mathscr{E}}\right) \qquad (4.4.10)$$

is often used. That Equation 4.4.10 is a plausible expression for α may be seen by the following arguments. The density of ionizing collisions at x is proportional to n^*, the density of excited electrons arriving at x with sufficient energy to create electron-hole pairs. The density n^*, in turn, is just the total electron density n times the probability that an electron has not collided in a distance d necessary to gain adequate energy, that is,

$$n^* = n \exp\left(-\frac{d}{l}\right) \qquad (4.4.11)$$

where l is the mean-free path. The length d may be found from Equation 4.4.1 by letting E_1 be the minimum energy necessary for an ionizing collision and $\mathscr{E}$ be the average field that accelerates the electron:

$$d = \frac{E_1}{q\mathscr{E}} \tag{4.4.12}$$

The number of ionizing collisions in the distance dx is also proportional to dx/d if we assume that the electron collides soon after gaining sufficient energy to ionize an atom. Then

$$dn = A'n^* \frac{dx}{d} = \frac{A'q\mathscr{E}}{E_1}\left[\exp\left(-\frac{E_1}{lq\mathscr{E}}\right)\right]n\,dx \tag{4.4.13}$$

where A' is a proportionality constant. Comparing Equations 4.4.2 and 4.4.13, we see that the dependence of α on field (in Equation 4.4.10) is at least reasonable. Since the ionization coefficient depends strongly on the electric field, the multiplication factor also increases rapidly with field. Thus, a small increase in field as the voltage approaches breakdown causes a sharp increase in current, a behavior that is dramatically confirmed in practical diodes.

Not only does the ionization coefficient vary with field, and consequently with position, in the space-charge region, but the width of the space-charge region also varies with voltage. Therefore, the evaluation of M in Equation 4.4.8 is difficult, and an empirical approximation of the form

$$M = \frac{1}{1 - (|V_R|/BV)^n} \qquad (2 < n < 6) \tag{4.4.14}$$

is often used, where $V_R < 0$ is the applied (reverse) bias and BV is the breakdown voltage, at which the current is observed to increase rapidly.

To relate the breakdown voltage to the material parameters, we consider a one-sided step junction with $N_a \ll N_d$ and assume that breakdown occurs when the maximum field in the junction reaches a critical value $\mathscr{E}_1$ that causes the integral in Equation 4.4.8 to approach unity. Since the maximum field is approximately given by

$$\mathscr{E}_{max} = \left(\frac{2qN_a|V_R|}{\epsilon_s}\right)^{1/2} \tag{4.4.15}$$

the breakdown voltage has the approximate form

$$BV = \frac{\epsilon_s\mathscr{E}_1{}^2}{2qN_a} \tag{4.4.16}$$

This equation illustrates the important fact that the breakdown voltage decreases for more heavily doped material, although the decrease is not quite as rapid as indicated. In practical diodes the breakdown voltage generally varies with doping as $N^{-2/3}$. The more gradual variation with doping in the practical case is a consequence of the more efficient avalanche multiplication in junctions having wider

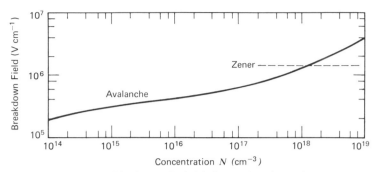

Figure 4.12   The critical electric fields for avalanche and Zener
breakdown in silicon as functions of dopant concentration.[1,2,3]

depletion regions. At higher dopant concentrations, a slightly higher critical field
$\mathscr{E}_1$ is needed (Figure 4.12).

All comments thus far have been in terms of one-dimensional geometry. In a
practical planar junction, an important consideration arises because of the finite
lateral dimension of the junction. In Section 2.1 we considered a planar junction
formed by diffusion through an opening in a silicon dioxide layer. The impurities
diffuse laterally beneath the Si–SiO$_2$ interface as well, as vertically into the silicon,

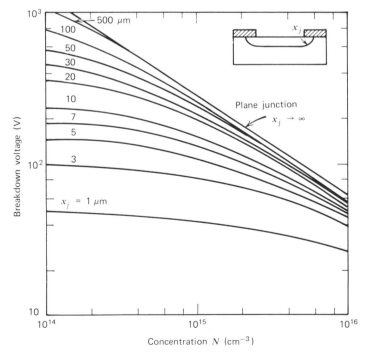

Figure 4.13   Breakdown voltage of one-sided, plane, silicon step
junction showing the effect of junction curvature.[4,5]

creating a rounded region of the junction beneath the edge of the SiO_2 (Figure 2.2d). The field in this corner region can be markedly higher than the field in the remainder of the junction, causing breakdown to occur there at unexpectedly low voltages. The reduction in breakdown voltage is especially severe for a shallow junction with a small radius of curvature. It becomes less serious as the junction depth x_j increases. The influence of junction curvature on the breakdown voltage is shown in Figure 4.13 for a one-sided silicon planar step junction.

EXAMPLE *pn*-**Junction-Diode Breakdown**

Diodes are to be made in a boron-doped silicon wafer that has a resistivity $\rho = 1.5$ Ω-cm. The processing sequence consists of

1. A predeposition of phosphorus atoms on the surface with a concentration $N' = 5 \times 10^{15}$ cm^{-2}.
2. A drive-in diffusion at 1000°C for 27 min.
   (a) What value is expected for the breakdown voltage BV if the entire silicon surface is exposed to the phosphorus doping?
   (b) What value is expected for the breakdown voltage BV if the wafer is oxidized and *windows* are opened through the oxide to the silicon surface at selected areas before the phosphorus is deposited so that separated diode regions are formed?

Solution

The process described will produce *np*–junction diodes. For part a, a large diode will result with the junction consisting of a single plane. For part b, there will be diffused regions each having a cross section similar to that shown in Figure 2.2d. The diodes in part b will therefore have enhanced fields in the corner regions and we can expect them to have a lower value of BV than the diode formed in part a.

The 27 min. diffusion at 1000°C redistributes a fixed total quantity of phosphorus, and therefore results in a Gaussian distribution of the n-type dopant. From Figure 2.18, the diffusion coefficient of phosphorus at 1000°C is 1.6×10^{-13} cm^2 s^{-1}. Hence, $\sqrt{Dt} = 0.16$ μm.

To decide whether or not the diodes can be considered to be one-sided, we determine the density of phosphorus after redistribution. The surface concentration for this Gaussian distribution is obtained from Equation 2.5.13 at $x = 0$ as $C_s = N'/(\sqrt{\pi} \sqrt{Dt}) = 1.76 \times 10^{20}$ cm^{-3}. We find the doping density N_a in the 1.5 Ω-cm wafer to be 10^{16} cm^{-3}, either from Figure 1.14 or Table 4.1. The ratio N_a/C_s is therefore 5.7×10^{-5}, and the diodes can be considered to be one-sided step junctions.

The junction depth x_j can be determined from Figure 2.19 by finding x/L such that $C/C_s = N_a/C_s$. Using the results for a Gaussian in Figure 2.19, we find $x_j = 3.1 \times 2\sqrt{Dt} = 1.0$ μm.

(a) The breakdown voltage *BV* for the plane junction is now found either from Figure 4.13 or Table 4.1 to be 63 V.
(b) From Figure 4.13, the individual diffused diodes have a breakdown voltage *BV* equal to 26 V.

This example shows that field concentration at the corners of diffused junctions can have important practical consequences. From Figure 4.13 we see that *BV* increases rapidly as the junction is diffused deeper into the wafer.

Zener Breakdown[†]

As the dopant concentration increases, the width of the depletion region decreases and the critical field at which avalanche occurs also increases. At high dopant concentrations, the field required for avalanche breakdown to occur exceeds the field necessary for Zener breakdown and the latter becomes more probable. As stated earlier, Zener breakdown occurs when the force exerted by the applied field is strong enough to rip an electron from its covalent bond to create an electron-hole pair directly. Figure 4.14 is an energy-band picture that shows Zener breakdown schematically.* As the figure indicates, a large number of electrons in the valence band on the *p*-type side of the junction are separated by the narrow depletion region from empty allowed states at the same energy in the conduction band of the *n*-type material. As the dopant concentration in the semiconductor increases, the width of the depletion region at a given reverse bias decreases, and the energy bands in the depletion region are bent more steeply. Because of the wave

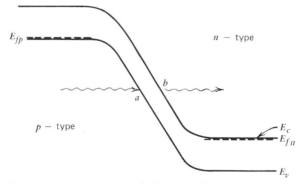

Figure 4.14 Energy-band diagram of a reverse-biased junction that has a high dopant concentration on both sides. Tunneling or Zener breakdown is likely in this type of junction.

---

* For this discussion it is useful to consider electrons in both the conduction band and the valence band rather than employing the concept that only holes are important in the valence band.

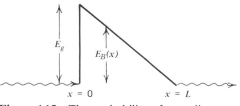

Figure 4.15 The probability of tunneling across the junction can be approximated by considering tunneling through a triangular barrier.

nature of the electron, there is a finite probability that an electron in the valence band of the p-type semiconductor approaching the forbidden gap can *tunnel* through the forbidden region and appear at the same energy in the conduction band of the n-type semiconductor. Since the probability of transmission of an electron through a barrier is a strong function of the thickness of the barrier, tunneling is only significant in highly doped material in which the fields are high and the depletion region is narrow.

To investigate the probability of tunneling from the valence band to the conduction band, we may approximate the barrier that the electron sees by a triangular barrier (Figure 4.15). The height of the energy barrier E_B decreases linearly from E_g at $x = 0$ to 0 at $x = L$, and the average field is $\mathscr{E} = E_g/qL$. The probability of tunneling Θ can be approximated by using the barrier height in the equation*

$$\Theta \approx \exp\left[-2\int_0^L \sqrt{\frac{2m^*E_B}{\hbar^2}}\,dx\right] \qquad (4.4.17)$$

Carrying through the integration, we find the tunneling probability to be

$$\Theta = \exp\left(-\frac{B}{\mathscr{E}}\right) = \exp\left(-\frac{qBL}{E_g}\right) \qquad (4.4.18)$$

where

$$B = \frac{4\sqrt{2m^*}\,E_g^{3/2}}{3q\hbar} \qquad (4.4.19)$$

Thus, the probability of tunneling decreases rapidly as the electric field decreases or the tunneling distance increases.

We can roughly estimate the field necessary to obtain appreciable tunneling from the above approximations. The current is the product of the area, the electron charge, the number of valence-band electrons in the p-region arriving at the barrier per second that "see" empty states across the barrier, and the probability that each electron tunnels through the barrier. The number of electrons arriving at the

---

* This "WKB" approximation is discussed in most quantum-mechanics text books; see, for example, reference 6.

barrier can be expressed as the density $\mathcal{N}$ of electrons in the valence band times their velocity, and the current may be written as

$$I = qA\mathcal{N}v\Theta \tag{4.4.20}$$

For appreciable tunneling, a current of, for example, 10 mA may flow across a junction 10^{-5} cm^2 in area. The density of electrons in the valence band that are at energies corresponding to empty allowed states in the conduction band across the barrier is comparable to the atomic density of about 10^{22} cm^{-3}. We assume that the electrons are moving with their thermal velocity of about 10^7 cm s^{-1} so that 10^{29} electrons cm^{-2} s^{-1} strike the barrier. The tunneling probability corresponding to this current is found from Equation 4.4.18 to be about 10^{-7}. Using this value and a semiconductor band gap of 1 eV, we find the corresponding tunneling distance and electric field to be about 4 nm and 10^6 V cm^{-1}, respectively. That is, for appreciable current to flow by tunneling or Zener breakdown, the barrier that the electrons "see" must be less than about 4 nm wide and the electric field in the depletion region must be greater than about 10^6 V cm^{-1}. These values are consistent in order-of-magnitude with observations.

As the dopant concentrations decrease, the width of the space-charge region increases and the probability of tunneling decreases rapidly. Avalanche breakdown then becomes more likely than Zener breakdown. Thus, Zener breakdown is only expected for the most heavily doped junctions, while more lightly doped junctions break down by the avalanche mechanism.

Devices exhibiting Zener breakdown generally have breakdown voltages lower than those that break down by avalanching. In silicon, pure Zener breakdown is usually found in diodes having $BV < 5$ V. At higher voltages, avalanche breakdown predominates most often. Commercially available diodes with well-defined breakdown characteristics are generally called *Zener diodes* regardless of their breakdown mechanism.

It is possible to determine whether avalanche or Zener breakdown is occurring in a junction by noting the temperature sensitivity of BV, the breakdown voltage. Although not large, the temperature variation of the two types of breakdown is of opposite sign. In the case of Zener breakdown, BV decreases with increasing temperature because the flux of valence-band electrons available for tunneling increases as temperature rises. The effect of temperature on BV for avalanching junctions is just the opposite. The breakdown voltage increases as temperature increases because the mean-free path of energetic electrons (l in Equation 4.4.11) decreases. For BV values in the range of $\sim$5 to 6 V, both avalanche- and tunnel-breakdown can occur simultaneously so that the net temperature variation is very slight. This characteristic is useful for establishing a voltage reference in some *IC*s.

4.5 Device: The Junction Field-Effect Transistor

We have seen that the depletion-layer width of a *pn* junction can be varied by modulating a reverse-bias voltage applied to the junction. In this section, we consider a device that makes use of this mechanism to control the current through a

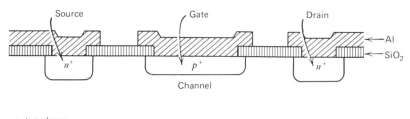

Figure 4.16 Basic structure of an *n*-channel, junction field-effect transistor. A lightly-doped *n*-type channel layer is bounded by the *p*-type substrate and a *p*-type gate diffusion.

region bounded by one or more *pn* junctions. Since little current flows into a reverse-biased *pn* junction, only a small amount of power is consumed at the control electrode, while substantially more power can be delivered by the controlled current. The device, which is called a *junction field-effect transistor* (often abbreviated JFET) can therefore be used as a power amplifier.

Consider the structure shown in Figure 4.16, which consists of a lightly doped *n*-type layer on top of a *p*-type substrate. For the reasons outlined in Section 2.6, it is usual to obtain this structure by growing an *n*-type epitaxial layer on a *p*-type substrate. The *n*-type region is then relatively uniform, and the dopant concentration is well controlled. Alternatively, a well-controlled *n*-type layer can be introduced by ion-implantation techniques.

After the uniform lightly doped *n*-type layer is formed, two heavily doped *n*-type regions (denoted n^+) are added by diffusion, as shown in Figure 4.16, so that good ohmic contact can be obtained (as discussed in Section 3.4) to the lightly doped region known as the *channel*. Note that the channel region is similar to the resistor structure discussed in Section 2.9 except for the presence of the p^+ layer. The n^+ electrodes are called the *source* and *drain* electrodes. The source is the electrode that supplies majority carriers to the channel. Consequently, conventional current flows from drain to source in an *n*-channel JFET. If the doping type were changed in all regions of the structure shown in Figure 4.16, the sketch would represent a *p*-channel JFET. In the *p*-channel device, conventional current flows from source to drain.

The *pn* junction above the channel in Figure 4.16 serves as the control element when reverse bias is applied to it and is called the *gate*. The channel is defined from above by the depletion region at the gate and from below by the depletion region at the substrate *pn* junction. The substrate is usually at ground potential. If the drain is biased positively, current flows from it to the source through the channel. If we now ground the source and apply a negative voltage to the p^+ electrode, the junction depletion region widens and the channel narrows. As the channel narrows, its resistance increases, and less current flows from drain to source. Consequently, a signal applied to the gate controls the current flowing through the channel.

Having seen qualitatively the basis for JFET operation, we will find it straight-forward to develop a quantitative theory for the device. We shall see that many of the ideas that we develop will be useful in our later discussion of the metal-oxide-semiconductor field-effect transistor (MOSFET).

Device Analysis. To analyze the JFET, we consider first a small bias V_D applied to the drain electrode while the source is grounded. Under this condition the gate-channel bias and therefore the width of the gate depletion region is uniform along the entire channel. The voltage at the gate is V_G. An expanded view of the channel region is shown in Figure 4.17. We assume a one-dimensional structure with a gate length L between the source and drain regions and a width W perpendicular to the plane of the paper. (Usually $W \gg L$.) Drain current flows along the dimension y. We assume a one-sided step junction at the gate with N_a in the p-region much greater than N_d in the channel. Therefore, the depletion layer extends primarily into the n-channel. The distance between the p-type gate and the substrate is t, the thickness of the gate depletion region in the n-type channel is x_d, and the thickness of the neutral portion of the channel is x_w. To focus on the role of the gate, we assume that the depletion zone at the substrate junction extends primarily into the substrate so that $x_w \simeq (t - x_d)$. This is approximately true in practice.

The resistance of the channel region can be written

$$R = \frac{\rho L}{x_w W} \tag{4.5.1}$$

where $\rho = (q\mu N_d)^{-1}$ is the resistivity of the channel. Hence, the drain current is

$$I_D = \frac{V_D}{R} = \left(\frac{W}{L}\right)(q\mu_n N_d x_w V_D) \tag{4.5.2}$$

The dependence on gate voltage is incorporated in Equation 4.5.2 by expressing $x_w = t - x_d$, where x_d (from Equation 4.3.1) is

$$x_d = \left[\frac{2\epsilon_s}{qN_d}(\phi_i - V_G)\right]^{1/2} \tag{4.5.3}$$

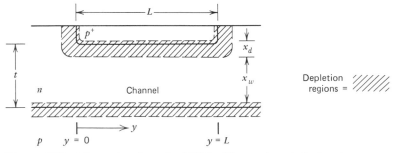

Figure 4.17 Channel region of a JFET with gate length L showing depletion regions. Typical dimensions might be $L \approx 8 \ \mu$m and $t \approx 1 \ \mu$m.

and ϕ_i is the built-in potential. The current can now be written as a function of the gate and drain voltages.

$$I_D = \frac{W}{L} q\mu_n N_d t \left\{ 1 - \left[\frac{2\epsilon_s}{qN_d t^2} (\phi_i - V_G) \right]^{1/2} \right\} V_D \qquad (4.5.4)$$

The factor in front of the bracketed terms is equal to the conductance G_0 of the n-region when it is completely undepleted (the so-called metallurgical channel). In terms of G_0, Equation 4.5.4 may be rewritten as

$$I_D = G_0 \left\{ 1 - \left[\frac{2\epsilon_s}{qN_d t^2} (\phi_i - V_G) \right]^{1/2} \right\} V_D \qquad (4.5.5)$$

Hence, at a given gate voltage, we find a linear relationship between I_D and V_D. This is a consequence of our having assumed small applied drain voltages. The square-root dependence on gate voltage in Equation 4.5.5 arises from our assumption of an abrupt gate-channel junction. From Equation 4.5.5, we see that the current is maximum at zero applied gate voltage and decreases as $|V_G|$ increases. The equation predicts zero current when the gate voltage is large enough to deplete the entire channel region.

EXAMPLE   Source and Drain Resistance in a JFET

As seen in the JFET cross section in Figure 4.16, the heavily doped source and drain electrodes are typically set apart from the channel region. This separation adds undesirable resistance in series with the JFET channel. Consider an n-channel JFET with $L = 5$ μm and $W = 10$ μm in which the channel region is separated from the source and drain diffusions by 5 μm. The JFET channel is formed by diffusing a p-type gate region 0.5 μm into a 1.5 μm-thick, n-type, epitaxial layer doped with $N_d = 5 \times 10^{15}$ cm^{-3}. Assume that the gate diffusion forms a step junction with a dopant concentration $N_a = 1 \times 10^{19}$ cm^{-3}, and neglect the depletion layer at the junction between the channel and the substrate.

Find the percentage increase in resistance between the source and drain contacts (for the JFET in the linear region of operation) caused by the separation from the channel over the resistance presented by the channel alone.

Solution

The conductance of the channel is given by Equation 4.5.4. We use Equation 4.2.10 to calculate $\phi_i = 0.854$ V. Either from Table 4.1 or from Figure 1.14, we find $\rho_n = 1$ Ω-cm in the channel region, where $\rho_n = (q\mu N_d)^{-1}$. Using these values in Equation 4.5.4, together with $W/L = 2$ and $t = 1$ μm, we calculate the channel conductance G_c

$$G_c = (2.08 \times 10^{-4})(1 - 0.472) = 1.1 \times 10^{-4} \ \Omega^{-1} = 0.11 \text{ mS}$$

where the conductance unit S is a Siemen.

The conductance of the series resistance near the source and drain regions $G_{s,d}$ is

$$G_{s,d} = \frac{W}{L} q\mu N_d t_e = 3.1 \times 10^{-4} \ \Omega^{-1} = 0.31 \ \text{mS}$$

where t_e, the epitaxial layer thickness, is 1.5 μm.

The channel resistance is $1/G_c = 9.1$ kΩ. The source and drain series resistors are each $1/G_{s,d} = 3.2$ kΩ, so the total resistance is 15.5 kΩ. The percentage increase is therefore $6.4/9.1 \times 100 = 70\%$.

The extra resistance is *parasitic*, that is, it reduces the achievable gain of the JFET because it is connected in series with the output voltage. The parasitic resistors have a maximum effect under the conditions of this problem (with no gate voltage applied). Their effect is reduced as the gate becomes reverse biased causing G_c to decrease.

Now that we can see the physics behind the device, we remove the restriction of small drain voltages and consider the problem for arbitrary V_D and V_G values (with the restriction that the gate must always remain reverse biased). With V_D arbitrary, the voltage between the channel and the gate is a function of position y. Consequently, the depletion-region width and therefore the channel cross section also vary with position. The voltage across the depletion region is higher near the drain than near the source in this n-channel device. Therefore, the depletion region is wider near the drain as shown in Figure 4.18.

We now make use of what has been called the *gradual-channel approximation*. This approximation assumes that the channel- and depletion-layer widths vary slowly from source to drain so that the depletion region is influenced only by

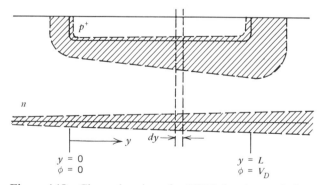

Figure 4.18 Channel region of a JFET showing variation of the width of depletion regions along the channel when the drain voltage is significantly higher than the source voltage.

fields in the vertical dimension and not by fields extending from drain to source. In other words, the field in the y-direction is much less than that in the x-direction in the depletion regions, and we may find the depletion-region width from a one-dimensional analysis.

Within this approximation, we can write an expression for the increment of voltage across a small section of the channel of length dy at y as

$$d\phi = I_D \, dR = \frac{I_D \, dy}{Wq\mu_n N_d(t - x_d)} \tag{4.5.6}$$

The width x_d of the depletion region is now controlled by the voltage $\phi_i - V_G + \phi(y)$ where $\phi(y)$ is the potential in the channel at point y, so that

$$x_d = \left[\frac{2\epsilon_s}{qN_d}(\phi_i - V_G + \phi(y))\right]^{1/2} \tag{4.5.7}$$

This expression may be used in Equation 4.5.6, which is then integrated from source to drain to obtain the current-voltage relationship for the JFET

$$\frac{I_D \int_0^L dy}{Wq\mu_n N_d} = \int_0^{V_D} \left\{t - \left[\frac{2\epsilon_s}{qN_d}(\phi_i - V_G + \phi)\right]^{1/2}\right\} d\phi \tag{4.5.8}$$

After integrating and rearranging, we find

$$I_D = G_0\left\{V_D - \frac{2}{3}\left(\frac{2\epsilon_s}{qN_d t^2}\right)^{1/2}\left[(\phi_i - V_G + V_D)^{3/2} - (\phi_i - V_G)^{3/2}\right]\right\} \tag{4.5.9}$$

At low drain voltages, Equation 4.5.9 reduces to the simpler expression of Equation 4.5.5, and the current increases linearly with drain voltage. However, at higher drain voltages, the current increases more gradually.

At large enough drain voltages, Equation 4.5.9 indicates that the current reaches a maximum and begins decreasing with increasing drain voltage, but this maximum corresponds to the limit of validity of our analysis. Indeed, from Figure 4.19 we see that, as the drain voltage increases, the width of the conducting channel near the drain decreases, until finally the channel is completely depleted in this region (Figure 4.19b). When this occurs, Equation 4.5.6 becomes indeterminate ($x_d \to t$). The equations are, therefore, only valid for V_D below the drain voltage that *pinches off* the channel. Current continues to flow when the channel has been pinched off because there is no barrier to the transfer of electrons traveling down the channel toward the drain. As they arrive at the edge of the pinched-off zone, they are pulled across it by the field directed from the drain toward the source. If the drain bias is increased further, any additional voltage is dropped across a depleted, high-field region near the drain electrode, and the point at which the channel is entirely depleted moves slightly toward the source (Figure 4.19c). If this slight movement is neglected, the drain current remains constant (saturates) as the drain voltage is increased further, and the bias condition is referred to as *saturation*. The drain voltage at which the channel is entirely depleted near the drain

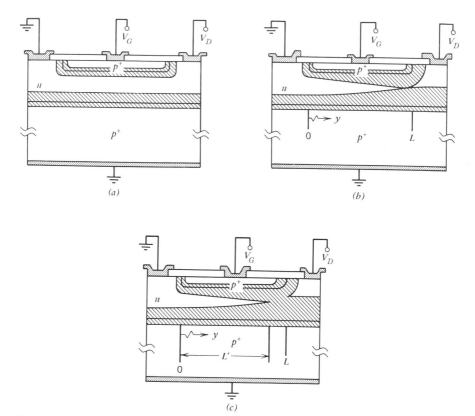

Figure 4.19 Behavior of the depletion regions in a JFET. (*a*) For small drain voltage, the channel is nearly an equipotential and the dimensions of the depletion regions are uniform. (*b*) When V_D is increased to $V_{D\text{sat}}$, the depletion regions on both sides of the channel meet at the pinch-off point (at $y = L$). (*c*) When $V_D > V_{D\text{sat}}$, the pinch-off point (at $y = L'$) moves slightly closer to the source.[7]

electrode is found from Equation 4.5.7 to be

$$V_{D\,\text{sat}} = \frac{qN_dt^2}{2\epsilon_s} - (\phi_i - V_G) \tag{4.5.10}$$

and the corresponding drain current is

$$I_{D\,\text{sat}} = G_0\left[\frac{qN_dt^2}{6\epsilon_s} - (\phi_i - V_G)\left\{1 - \frac{2}{3}\left[\frac{2\epsilon_s(\phi_i - V_G)}{qN_dt^2}\right]^{1/2}\right\}\right] \tag{4.5.11}$$

Based on our analysis, we may divide the drain current-drain voltage characteristic into three regions (Figure 4.20): (1) the linear region at low drain voltages, (2) a region with less than linear increase of current with drain voltage, and (3) a saturation region where the current remains relatively constant as the drain voltage is further increased. As expected from the physics of the device, Equation 4.5.11 predicts the current to be maximum for zero gate bias and to decrease as

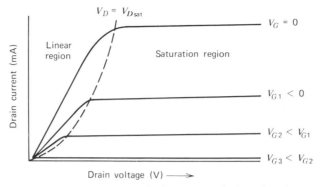

Figure 4.20 The output (drain-current, drain-voltage) characteristics of a JFET as a function of the gate voltage. Operation in the "linear" region (to the left of the $V_D = V_{Dsat}$ curve) corresponds to Figure 4.19a. In saturation, (to the right of $V_D = V_{Dsat}$) Figure 4.19c applies.

negative gate voltage is applied. As the gate voltage becomes more negative, the drain saturation voltage and the corresponding current decrease so that a family of curves may be generated (Figure 4.20), each curve showing the drain current versus drain voltage characteristic for a particular value of gate voltage. At a sufficiently negative value of gate voltage, the saturation drain current becomes zero. This turn-off voltage V_T is found from Equation 4.5.11 to be

$$V_T = \phi_i - \frac{qN_dt^2}{2\epsilon_s} \tag{4.5.12}$$

The drain current actually increases slightly as the drain voltage is increased beyond V_{Dsat} because the end point for the integration in Equation 4.5.8 now becomes L' rather than L, where L' is the point at which the channel becomes completely depleted [$\phi(L') = V_{Dsat}$] (Figure 4.19). For $V_D > V_{Dsat}$, the expression for I_{Dsat} in Equation 4.5.11 is multiplied by the ratio L/L' (which is greater than unity). Although most JFETs exhibit characteristics with well-defined saturation regions as shown in Figure 4.20, significant departures can be seen in devices with short channel lengths. In a device with a donor density of 10^{16} cm^{-3} in the channel region, an excess of 5 V beyond V_{Dsat} depletes an additional 1.0 μm. Thus, for a typical channel length of 8 μm, the ratio $L/L' \approx \frac{8}{7}$, and the deviation from simple theory is not severe. For devices with channel lengths of the order of 2 μm, however, important deviations occur.*

Field-effect transistors are often operated in the saturation region where the output current is not appreciably affected by the output (drain) voltage but only by the input (gate) voltage. For this bias condition, the JFET is almost an ideal current source controlled by an input voltage. The transconductance g_m of the

---

* Channel-length modulation is discussed further in connection with metal-oxide-semiconductor field-effect transistors in Chapter 10 along with other short-channel effects.

transistor expresses the effectiveness of the control of the drain current by the gate voltage. It is defined by the relation

$$g_m \equiv \left. \frac{\partial I_D}{\partial V_G} \right|_{V_D = \text{const}} \tag{4.5.13}$$

and may be found by differentiating Equation 4.5.9

$$g_m = G_0 \left(\frac{2\epsilon_s}{qN_d t^2} \right)^{1/2} \left[(\phi_i - V_G + V_D)^{1/2} - (\phi_i - V_G)^{1/2} \right] \tag{4.5.14}$$

In the saturation region, g_m reaches a maximum value, which is found from Equation 4.5.14

$$g_{m\,\text{sat}} = G_0 \left[1 - \left(\frac{2\epsilon_s}{qN_d t^2} (\phi_i - V_G) \right)^{1/2} \right] \tag{4.5.15}$$

Our analysis has included several simplifying assumptions. In practical devices, however, some of these assumptions may not be sufficiently valid to obtain a good match between theory and experiment. One such assumption is that the depletion-layer width is controlled by the gate-channel junction and not by the channel-substrate junction. There will be a variation of the potential across the channel-substrate junction along the channel, with the maximum potential and depletion-layer thickness near the drain. Consequently, the channel becomes completely depleted at lower drain voltage than is indicated by Equation 4.5.10. A bias applied between the source and substrate is sometimes gainfully used to control the characteristics of a JFET. The effect of such a bias can be readily calculated by including the effect of substrate bias in the expression for the undepleted channel thickness in Equation 4.5.6.

The presence of lightly doped regions between the active channel and the heavily doped source and drain contacts can be more troublesome. Because of photomasking limitations and breakdown-voltage requirements, the n^+ contact diffusions generally are separated from the p-type gate diffusion. As we saw in the example, the series resistance of the intermediate regions may cause deviations from the ideal characteristics, especially at high current levels, and must be considered when analyzing or designing practical devices.

We have seen that the characteristics of the JFET are highly sensitive to the thickness of the channel region t and to its dopant concentration. The n-type region in which the channel is formed can be made with excellent control by epitaxial deposition. The more crucial fabrication step is diffusion of the p-type gate. This diffusion can introduce troublesome variations in the effective channel thickness. To improve control, ion implantation is used to introduce the gate impurities in some processes. JFETs can also be fabricated using two implantations, one for the n-type channel dopant, and one for the p-type gate. This eliminates the need for epitaxial deposition in cases where it is not required for other devices in the same *IC*.

Figure 4.21 shows an integrated circuit in which several JFETs are used (type 355 operational amplifier). The interdigitated structures on the lower edge of the

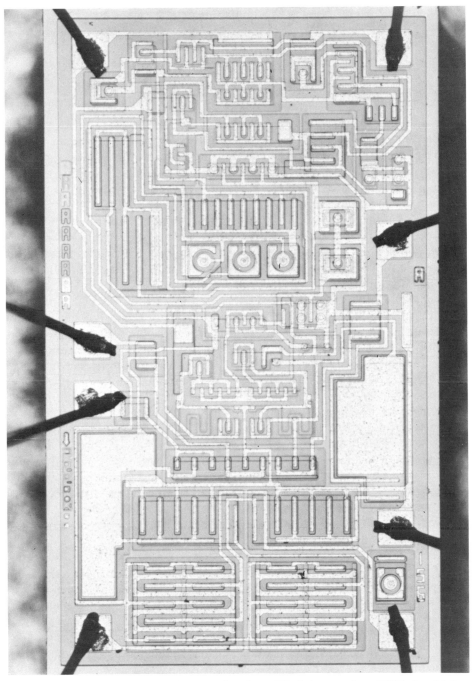

Figure 4.21 An integrated circuit (operational amplifier) that employs JFETs to obtain high input resistance. (*Courtesy National Semiconductor Corp.*)

circuit are large input JFETs. The sources and drains are made in a comblike pattern. The gate electrode snakes back and forth between the comb electrodes that contact the source and drain.

Summary

As was the case for metal-semiconductor contacts, a basic understanding of some important properties of inhomogeneously doped semiconductors can be gained by applying the principles of thermal equilibrium. If the doping in a semiconductor is nonuniform but of one type, there will be a built-in electric field that balances the diffusion tendency of free carriers with an opposing drift tendency. Often, the space charge associated with this field is small and the semiconductor can be treated as quasi-neutral so that the net dopant concentration equals the majority-carrier concentration. The *quasi-neutral approximation* becomes less valid as the gradient of the dopant concentration becomes larger. At *pn* junctions, dopant gradients are generally large and quasi-neutrality does not apply.

Effects at a *pn* junction are usually analyzed by making use of the *depletion approximation*. In the depletion approximation, the space charge is assumed to terminate abruptly and is made up of uncompensated dopant ions. The use of this approximation leads to predictions for field and potential that are inaccurate within a region that is roughly an extrinsic Debye length from each respective neutral boundary. There is a built-in voltage at a *pn* junction in equilibrium just as in a metal-semiconductor junction. The built-in voltage equals the difference that the Fermi levels of the *p*- and *n*-regions would have if they were isolated.

Little current flows when the voltage applied to the junction has a polarity that increases the built-in potential, that is, if a positive voltage is connected to the *n*-type region with the *p*-type region grounded. This bias polarity, called reverse bias, causes the junction space-charge region to widen. Depletion-region widening can be detected readily by small-signal capacitance measurements made at different dc reverse biases. The specific behavior of a series of capacitance-voltage measurements can provide useful information about the dopant concentration in the junction region.

At high fields, semiconductors can suddenly become highly conductive because of the internal generation of extra free carriers. When this occurs in a reverse-biased junction there is a sudden increase in current. Therefore, the phenomenon is referred to as breakdown. One of two mechanisms may be responsible for breakdown: (1) *avalanche* or (2) *tunneling* (*Zener breakdown*). One use of reverse-biased *pn* junctions in integrated circuits is for the gate of a *junction field-effect transistor* (JFET). Operation of a JFET depends directly on the modulation of the depletion-layer width x_d in a reverse-biased *pn* junction. This reverse-bias voltage can modulate a current if that current is made to flow through a region having a cross section that depends on x_d.

References

1. A. S. Grove, *Physics and Technology of Semiconductor Devices*, Wiley, New York, 1967, p. 193. Reprinted by permission of the publisher.
2. A. G. Chynoweth, W. L. Feldman, C. A. Lee, R. A. Logan, G. L. Pearson, and P. Aigrain, *Phys. Rev.* **118**, 425 (1960).
3. S. L. Miller, *Phys. Rev.* **105**, 1246 (1957).
4. A. S. Grove, *ibid.*, p. 197.
5. H. L. Armstrong, *IRE Trans. Electron Devices* **ED-4**, 15 (1957). Reprinted by permission of the publisher.
6. E. Merzbacher, *Quantum Mechanics*, 2nd edition, Wiley, New York, 1970.
7. A. S. Grove, *ibid.*, p. 244.
8. Courtesy W. G. Oldham.

Textbooks

D. A. Fraser, *The Physics of Semiconductor Devices*, 2nd Edition, Oxford at the Clarendon Press, 1979.

G. W. Neudeck, *The PN Junction Diode*, Vol II Modular Series on Solid-State Devices, Addison-Wesley, Reading, Mass., 1983.

Problems

4.1* An abrupt silicon *pn* junction has dopant concentrations of $N_a = 1 \times 10^{15}$ cm^{-3} and $N_d = 2 \times 10^{17}$ cm^{-3}.

   (a) Evaluate the built-in potential ϕ_i at room temperature.

   (b) Using the depletion approximation, calculate the width of the space-charge layer and the peak electric field for junction voltages V_a equal to 0 V and -10 V.

4.2 Consider abrupt silicon *pn* junctions that are very heavily doped on one side and have dopant concentrations of (a) 10^{15} cm^{-3}, (b) 10^{16} cm^{-3}, (c) 10^{17} cm^{-3}, and (d) 10^{18} cm^{-3} on the less heavily doped side. Find as a function of dopant concentration the length that can be depleted of mobile carriers before the maximum electric field reaches the breakdown field shown in Figure 4.12. What are the corresponding applied voltages?

4.3* Calculate the magnitude of the built-in field in the quasi-neutral region of an exponential impurity distribution:

$$N = N_0 \exp\left(-\frac{x}{\lambda}\right)$$

Let the surface dopant concentration be 10^{18} cm^{-3} and $\lambda = 0.4$ μm. Compare this field to the maximum field in the depletion region of an abrupt *pn* junction with acceptor and donor concentrations of 10^{18} cm^{-3} and 10^{15} cm^{-3}, respectively, on the two sides of the junction.

4.4 Find an expression for the potential in the example of Section 4.2 as a function of position within the region of variable doping. Assume that the doping is constant on either side of the transition region and make a plot of the potential and a sketch of the energy bands.

4.5 (a) Find and sketch the built-in field and potential for a silicon *pin* junction with the doping profile shown in Figure P4.5. Indicate the length of each depletion region. (The symbol *i* represents a very lightly doped or nearly intrinsic region.)

(b) Compare the maximum field to the field in a *pn* junction that contains no lightly doped intermediate region, but has the same dopant concentrations as in part a in the other regions.

(c) Explain physically what is happening in the intrinsic region. (That is, what does the depletion approximation mean here?)

(d) Discuss how the depletion capacitance for this structure varies with voltage, comparing it with the depletion capacitance of a structure with no intrinsic region but with the same dopant concentrations in the other regions. Sketch $1/C^2$ versus applied reverse bias for the two cases; use the same axes so that the two cases may be directly compared.

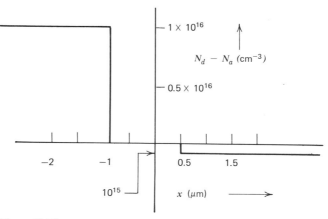

Figure P4.5

4.6 Use the *depletion approximation* to study a linearly graded junction with $(N_d - N_a) = ax$ throughout the depletion region. Assume that the equilibrium space-charge region is x_{d0} units wide. In terms of the parameters given, derive expressions for (a) the built-in potential, (b) the electric field as a function of applied voltage and distance, and (c) the depletion capacitance as a function of voltage.

4.7[†] We know that the capacitance of an abrupt *pn* junction varies as $V_a^{-1/2}$ for $V_a \gg \phi_i$, where ϕ_i is the built-in potential and V_a is the reverse bias applied across the junction. The capacitance of a linearly graded junction varies as $V_a^{-1/3}$. In a TV tuning circuit we need a capacitance that varies as V_a^{-1} for $V_a \gg \phi_i$. Qualitatively, discuss the general form of doping profile needed, indicating in each of the three cases the variation of the depletion-region width with voltage.

4.8[†*] Assume that the dopant distributions in a piece of silicon are as indicated in Figure P4.8.

(a) If a *pn* junction is desired at $x_0 = 1\ \mu m$, what should be the value of the surface dopant density N_{d0}?

(b) Assume the depletion approximation and make a sketch of the space charge near to the junction. Approximate this space charge as if this were a linearly graded junction. Choose an appropriate value for the doping gradient a.

(c) Under the approximation of part b, take $\phi_i = 0.7$ V and use Equations 4.3.2 and 4.3.4 to calculate $\mathscr{E}_{max}$ at thermal equilibrium. Then, sketch the fields at thermal equilibrium throughout the regions. (Take $N_{a0} = 10^{18}$ cm^{-3}, $x_0 = 10^{-4}$ cm, $\lambda_a = 10^{-4}$ cm, and $\lambda_d = 2 \times 10^{-4}$ cm.)

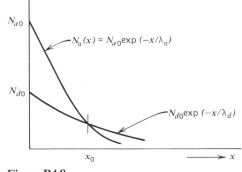

Figure P4.8

4.9 The small-signal capacitance C_d of a *pn* junction diode with area 10^{-5} cm^2 is measured. A plot of $(1/C_d^2)$ *vs.* the applied voltage V_a is shown in Figure P4.9.

(a) If the diode is considered as a one-sided step junction, find the indicated doping level on the lower conductivity side (use the slope of the curve).

(b) Sketch the doping density on the low conductivity side of the junction. Calculate the location of any point at which the dopant density changes.

(c) Use the intercept on the $(1/C_d^2)$ plot to find the doping density on the highly doped side.

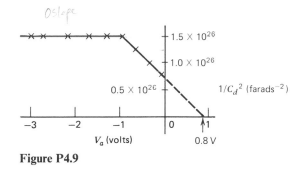

Figure P4.9

4.10* All parts of the following question refer to the system shown in Figure P4.10. The sketch shows the cross section of a silicon wafer fabricated by a planar process. It consists of a p substrate of resistivity 10 Ω-cm and an epitaxial layer 2.5 μm thick having 5×10^{15} donors cm^{-3}. At A, a platinum (Pt) contact is made directly to the epitaxial Si surface. At B, an n^+ contact is diffused 1.5 μm deep into the epitaxial material. The donor density throughout the n^+ region can be assumed to be 3×10^{18} cm^{-3}. Ignore all edge effects in the following considerations.

(a) What is the energy interval between the Fermi level and the mid-gap energy (E_i):
(i) In the epitaxial n-region? (ii) In the substrate p-region?

(b) The barrier height (energy difference between the conduction-band edge and the Fermi level) at the Pt-Si junction is measured to be 0.85 eV.
(i) What is the built-in voltage for the Pt-Si junction?
(ii) Is the 0.85 eV barrier consistent with idealized Schottky theory? Comment intelligently.

(c) Show whether or not it would be possible to deplete fully the n^- layer below the Pt contact without reaching a breakdown field of 3×10^5 V cm^{-1}. (Consider that contact B is shorted to contact C and that both are held at ground while voltage is applied to contact A.) What voltage would be required to accomplish this depletion?

(d) Sketch the equilibrium energy-band diagram along axis $1 - 1$ (through the Pt-Si junction and into the substrate). Make the diagram qualitatively correct; show the vacuum level and assume that ideal Schottky theory *does* apply.

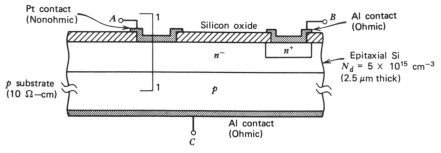

Figure P4.10

4.11 To treat avalanche from first principles, consider that an incident electron collides with the lattice and frees a hole-electron pair. Assume that after the interaction, the three particles each have equal kinetic energies. Also assume that all three have equal masses. Use conservation of energy and momentum principles to find that the threshold for avalanche occurs when the incident electron possesses $(3/2) E_g$ units of kinetic energy. (Despite the many approximations made in this problem, the energy is a useful first-order measure of what is found in practice.)

4.12 Is Zener breakdown more likely to occur in a reverse-biased silicon or germanium pn-junction diode if the peak electric field is the same in both diodes? Discuss. (Consider the size of the bandgap of each material.)

4.13[†] Carry through the analysis following Equation 4.4.20 and show that the tunneling distance and field derived are the proper values. The coefficient B given in Equation 4.4.19 is 7.87×10^7 V cm^{-1} for a bandgap of 1.1 eV if m^* is taken to be the rest mass of the electron. Use the conductivity effective mass in the calculations.

4.14 For a JFET derive an expression for the major temperature variation of the conductance in the linear region at a fixed gate voltage. Assume that the mobility varies as $T^{-3/2}$.

4.15[†] For a JFET derive an expression for the drain conductance $g = \partial I_D/\partial V_D$ at a given gate voltage in the saturation region. Assume that this conductance results from the widening of the depletion region near the drain and approximate the latter by a one-dimensional step junction with very heavy doping in the drain region.

4.16† Figure P4.16 shows a JFET made in an annular geometry. The junctions shown can be approximated as being linearly graded with a doping gradient $dN/dx = a$. Series resistance at the source and drain is negligible.

(a) What is the value of the gate turn-off voltage V_T in terms of the properties of the device?

(b) Write a differential equation that can be integrated to find the dependence of I_D on V_D, V_G, and device properties. Do not solve this differential equation, but place it in a form that involves definite integrals.

(c) How does the transconductance $(\partial I/\partial V_G)$ for this structure depend upon the radii r_1 and r_2? Specifically, if a device is made with $r_1 = 10\ \mu m$ and $r_2 = 40\ \mu m$ and it has a g_m value of 10 mS, what value of g_m would be expected for a device made with all parameters the same except that r_2 were made 60 μm in length?

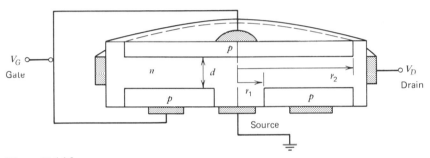

Figure P 4.16

5

CURRENTS IN *pn* JUNCTIONS

Under reverse bias, little current flows across a *pn* junction unless the applied voltage exceeds a breakdown value. Thus far, our consideration has been of reverse-biased *pn* junctions; we have therefore neglected consideration of current flow at low bias values. Instead, we have focused attention on the barrier to majority-carrier transfers at the *pn* junction and on changes in the depletion-region width as the applied voltage was varied.

The major topic in this chapter is current flow across a *pn* junction under both forward and reverse bias. A detailed understanding of this topic is important not only to gain insight into junction-diode behavior but, of even greater significance, to grasp the basis for junction-transistor operation. As a first step toward analyzing current flow, we derive a continuity equation for free carriers, that is, an equation that takes account of the various mechanisms affecting the population of carriers in an infinitesimal volume inside a semiconductor. To formulate some significant terms in this equation, those which account for the generation and recombination process, it will be necessary to consider several basic physical processes in more detail than was done in Chapter 1. After deriving a continuity equation that treats generation and recombination, it will be possible to characterize the minority-carrier distributions in the quasi-neutral regions of a *pn* junction under bias. We shall consider in detail the solutions for two especially simple cases of *pn* junctions and carry out what is known as the *ideal-diode analysis*. Then, in order to relate this result to real silicon diodes, it will be necessary to discuss generation and recombination in the space-charge region. The physical model developed for the steady-state current-voltage relationship will help in our consideration of charge storage and diode transients. Finally, we assess the role of *pn* junctions in integrated circuits and give a practical perspective to the earlier theory in a concluding section.

5.1 Continuity Equation

To discuss current flow in a *pn* junction, it is useful to write an equation that is basically an accounting for the flux of free carriers into and out of an infinitesimal volume in space. A *continuity equation* such as this can be written for both majority and minority carriers in semiconductors. We shall find that solutions of the minority-carrier continuity equation in semiconductors have special importance in many device applications.

To derive a one-dimensional continuity equation for electrons, we consider an infinitesimal slice of thickness dx located at x (Figure 5.1). The number of electrons in the slice may increase because of net flow into the volume and from net carrier generation in the slab. The overall rate of electron increase equals the algebraic sum of

(1) the number of electrons flowing into the slab, minus
(2) the number flowing out, plus

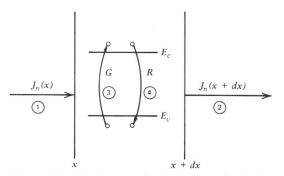

Figure 5.1 The increase in the electron density in an infinitesimal slice of thickness dx is related to the net flow of electrons into the slice and the excess of generation over recombination.

(3) the rate at which electrons are generated, minus

(4) the rate at which they recombine.

The first two components are found by dividing the currents at each side of the slice by the charge on an electron; for the present we symbolize the last two by G and R, respectively. The rate of change in the number of electrons in the slab is then

$$\frac{\partial n}{\partial t} A \, dx = \left(\frac{J_n(x)}{-q} - \frac{J_n(x + dx)}{-q} \right) A + (G_n - R_n) A \, dx \tag{5.1.1}$$

where A is the cross-sectional area of the slice and G_n and R_n represent the generation and recombination rates per unit volume for electrons. Expanding the second term on the right-hand side in a Taylor series,

$$J_n(x + dx) = J_n(x) + \frac{\partial J_n}{\partial x} \, dx + \cdots \tag{5.1.2}$$

we derive the basic continuity equation for electrons

$$\frac{\partial n}{\partial t} = \frac{1}{q} \frac{\partial J_n}{\partial x} + (G_n - R_n) \tag{5.1.3a}$$

A similar continuity equation applies to holes except that the sign of the first term on the right-hand side of Equation 5.1.3a is changed because of the charge associated with a hole

$$\frac{\partial p}{\partial t} = -\frac{1}{q} \frac{\partial J_p}{\partial x} + (G_p - R_p) \tag{5.1.3b}$$

To obtain equations that can be solved, we must relate the quantities on the right-hand side of Equations 5.1.3 to the carrier densities n and p. This is straightforward for the current terms, since J_n and J_p have been written in terms of carrier densities in Equations 1.2.21 and 1.2.22. When these equations are inserted, we obtain

$$\frac{\partial n}{\partial t} = \mu_n n(x) \frac{\partial \mathscr{E}(x)}{\partial x} + \mu_n \mathscr{E}(x) \frac{\partial n(x)}{\partial x} + D_n \frac{\partial^2 n(x)}{\partial x^2} + (G_n - R_n) \tag{5.1.4a}$$

and

$$\frac{\partial p}{\partial t} = -\mu_p p(x) \frac{\partial \mathscr{E}(x)}{\partial x} - \mu_p \mathscr{E}(x) \frac{\partial p(x)}{\partial x} + D_p \frac{\partial^2 p(x)}{\partial x^2} + (G_p - R_p) \tag{5.1.4b}$$

Note that we have assumed that mobility μ and diffusion constant D are not functions of x. Although this assumption is not valid in a number of important cases, the major physical effects are included in Equations 5.1.4 and more exact formulations are seldom considered.

If the electric field is zero or negligible in the region under consideration, the first two terms on the right-hand side of Equations 5.1.4 can be neglected and the analysis is greatly simplified. Even if the field is not negligible, some of the terms in Equations 5.1.4 may be unimportant. For example, if the field is constant, the first term in each equation drops out. As we saw in Section 4.1, a constant field is present in an exponentially graded semiconductor. Rarely is it necessary to deal with the full complexity of Equations 5.1.4.

It is worthwhile to recall some basic facts of calculus before we seek specific solutions to the continuity equations. The continuity equations (Equations 5.1.4) are partial differential equations because they are functions of time and position. Thus, they have an infinity of solutions, one of which applies to a given problem because it matches boundary and initial conditions. The continuity equations simplify to ordinary differential equations if, for example, one is interested in steady-state solutions. In that case, the time dependence on the left-hand side of the equations falls away and only position-dependent derivatives remain.

5.2 Generation and Recombination

To formulate correct expressions for generation and recombination in terms of free-carrier densities, it is necessary to develop the physics of semiconductors beyond the discussion of Chapter 1. In Chapter 1, we noted that an electron could be excited by thermal energy from the valence to the conduction band, leaving behind a hole in the valence band. Both the hole and the electron could contribute to conduction. In thermal equilibrium, this generation rate equals the rate at which the inverse process—the direct transfer of an electron into a valence-band site— occurs. These processes do describe one means for free-carrier generation and recombination. There are, however, other processes by which generation-recombination can take place.

Although direct transitions between the valence band and the conduction band occur in all semiconductors,* a complication of the crystal structures makes them unlikely in silicon and germanium except when very high densities of holes and electrons are present. In these materials, electrons at the lowest energy in the conduction band have a nonzero property that is equivalent to classical momentum. Since the holes at the valence-band edge do have a zero "momentum," a direct transition that conserves both energy and momentum is impossible without a lattice (phonon) interaction occurring simultaneously. Thus, in silicon or germanium, direct transitions across the forbidden-energy gap correspond to a simultaneous interaction of three particles: the electron, the hole, and a phonon that represents the lattice interaction (refer to Section 1.2).

Three-particle interactions are far less likely than are two-particle interactions, such as those between a free carrier and a phonon, that can take place if there are localized allowed energy states into which electrons or holes can make transitions. In practice, localized states at energies between E_v and E_c are always present because of lattice imperfections caused by misplaced atoms in the crystal or, more

---

* Direct transitions are the most important generation-recombination process in gallium arsenide and gallium-arsenide-phosphide, two semiconductors that are used for light-emitting diodes (LEDs).

usually, because of impurity atoms. Furthermore, they are always present in sufficient numbers to dominate the generation-recombination process in silicon and
germanium. These localized states act as stepping stones. In a recombination event,
for example, an electron falls from the conduction band to a state that we logically
call a *recombination center*,* and then it falls further into a vacant state in the
valence band, thus recombining with a hole.

Localized States: Capture and Emission

The four processes through which free carriers can interact with localized states
are indicated in Figure 5.2. The illustration shows a density N_t of states at an
energy E_t within the forbidden gap. The states shown are acceptor type—that is,
neutral when empty and negative when full—but the processes described apply
also to donor-type states.

In the first process, *electron capture*, an electron falls from the conduction band
into an empty localized state. The rate at which this process occurs is proportional
to the density of electrons n in the conduction band, the density of empty localized
states, and the probability that an electron passes near a state and is captured by
it. The density of empty localized states is given by their total density N_t times
one minus the probability $f(E_t)$ that they are occupied. When thermal equilibrium
applies, f is just f_D, the Fermi function as given by Equation 1.1.18. In the nonequilibrium case, f differs from f_D, but we need not specify it further for this
discussion.

The probability per unit time that an electron is captured by a localized state
is given by the product of the electron thermal velocity v_{th} and a parameter σ_n
called the *capture cross section*. The capture cross section describes the effectiveness

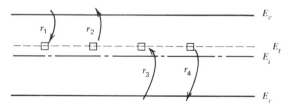

Figure 5.2   Free carriers can interact with localized
states by four processes: r_1 electron capture, r_2
electron emission, r_3 hole capture, and r_4 hole
emission. The localized state shown is acceptor type
and at energy E_t within the forbidden-energy gap.

---

* Because the states behave symmetrically as interim sites either for generation or recombination of
free carriers, they are properly termed generation-recombination centers. For brevity, however, this is
usually shortened to "recombination centers."

of the localized state in capturing an electron. This product $v_{th}\sigma_n$ may be visualized as the volume swept out per unit time by a particle with cross section σ_n. If the localized state lies within this volume, the electron is captured by it. The capture cross section is generally determined experimentally for a given type of localized state. A typical size for an effective recombination center is about 10^{-15} cm^2 for gold or iron.[1] An abnormally large cross section, 10^{-10} cm^2, is associated with beryllium.[2] Combining the factors discussed above, we may write the total rate of capture of electrons by the localized states as

$$r_1 = n\{N_t[1 - f(E_t)]\}v_{th}\sigma_n \qquad (5.2.1)$$

The second process is the inverse of electron capture: that is, *electron emission.* The emission of an electron from the localized state into the conduction band occurs at a rate given by the product of the density of states occupied by electrons $N_t f(E_t)$ times the probability e_n that the electron makes this jump.

$$r_2 = [N_t f(E_t)]e_n \qquad (5.2.2)$$

The emission probability can be expressed in terms of the quantities already defined in Equation 5.2.1 by considering the capture and emission rates in the limiting case of thermal equilibrium. At thermal equilibrium, the rates of capture and emission of carriers must be equal and the probability function $f(E)$ is given by the Fermi function $f_D(E)$ (Equation 1.1.18). Thus, we can write

$$r_1 = r_2 = nN_t[1 - f_D(E_t)]v_{th}\sigma_n = N_t f_D(E_t)e_n \qquad (5.2.3)$$

and

$$e_n = v_{th}\sigma_n n_i \exp\left(\frac{E_t - E_i}{kT}\right) \qquad (5.2.4)$$

The right-hand side of Equation 5.2.4 can be used to replace e_n in the general case described by Equation 5.2.2. From Equation 5.2.4, we see that electron emission from the localized state becomes more probable when its energy is closer to the conduction band since $E_t - E_i$ is then greater.

Corresponding relationships describe the interactions between the localized states and the valence band. For example, the third process, *hole capture,* is proportional to the density of localized states occupied by electrons $N_t f(E_t)$, the density of holes and a transition probability. This probability may then be described by the product of the hole thermal velocity v_{th} and the capture cross section σ_p of a hole by the localized state. Thus,

$$r_3 = [N_t f(E_t)]pv_{th}\sigma_p \qquad (5.2.5)$$

The fourth process, *hole emission,* describes the excitation of an electron from the valence band into the empty localized state. By arguments similar to those for electron emission, hole emission is given by

$$r_4 = \{N_t[1 - f(E_t)]\}e_p \qquad (5.2.6)$$

The emission probability e_p of a hole may be written in terms of σ_p by considering the thermal-equilibrium case, for which $r_3 = r_4$, and is given by

$$e_p = v_{th}\sigma_p n_i \exp\left(\frac{E_i - E_t}{kT}\right) \tag{5.2.7}$$

Analogously to Equation 5.2.4, the probability of emission of a hole from the localized state to the valence band becomes much greater as the energy of the state approaches the valence-band edge.

Before making use of Equations 5.2.1, 5.2.2, 5.2.5, and 5.2.6 for the kinetics of interactions between the valence and conduction bands via the localized states at E_t, it is worthwhile to consider qualitatively the physics that they represent. First, we recognize that at thermal equilibrium $r_1 = r_2$ and $r_3 = r_4$ since thermal equilibrium requires every process to be balanced by its inverse. When we have a nonequilibrium situation, $r_1 \neq r_2$ and $r_3 \neq r_4$. To gain insight about these rates, imagine specifically that holes in an *n*-type semiconductor are suddenly increased in number above their thermal-equilibrium value. This would cause r_3 to increase. The effect of this rate increase would be to increase r_4 and r_1 (both of which eliminate holes at E_t). If most of the holes disappear from E_t via r_1, they will remove electrons, and the localized state will be an effective recombination center. If the holes are removed from the level at E_t predominantly by an increase in r_4, they will return to the valence band, and the site will be effective as a *hole trap*. A given localized state will generally be effective in only one way: either as a *trap* or as a *recombination center*. Our interest for the present is in recombination centers.

Shockley-Hall-Read Recombination[†]

The equations describing generation and recombination through localized states or recombination centers were originally derived by Shockley and Read[3] and by Hall,[4] and the process is referred to frequently as Shockley-Hall-Read (or *SHR*) recombination. According to the *SHR* model, when nonequilibrium occurs in a semiconductor, the overall population of the recombination centers is not greatly affected. The reason for this is that they quickly capture majority carriers (there are so many of them around) but have to wait for the arrival of a minority carrier. Thus, the states are nearly always full of majority carriers whether under thermal-equilibrium conditions or in nonequilibrium.

To clarify this behavior, consider a typical example: acceptor-like recombination centers in an *n*-type semiconductor. At thermal equilibrium, the Fermi level is near E_c and, therefore, above the energy of the recombination centers. Hence, they are virtually all filled with electrons and r_1 and r_2 are both much greater than r_3 or r_4. When equilibrium is disturbed by low-level excitation (in which the hole population is changed greatly while the electron population is hardly affected), r_1 will have to exceed r_2 by only a very slight amount in order to accommodate the increased rate of hole capture represented by r_3. Thus, the population of the localized states remains constant, and the net rate of electron capture $r_1 - r_2$ equals the net rate of hole capture by the states. These net rates are, in turn, just the net rate of recombination that we define by the symbol U.

$$U \equiv R_{sp} - G_{sp} = r_1 - r_2 = r_3 - r_4 \qquad (5.2.8)$$

where the subscript sp stands for *spontaneous*, that is, recombination and generation that responds only to deviation from thermal equilibrium.* Inserting the expressions for r_1 through r_4 into Equation 5.2.8, we can eliminate f and solve for
U to obtain

$$U = \frac{N_t v_{th} \sigma_n \sigma_p (pn - n_i^2)}{\sigma_p \left[p + n_i \exp\left(\dfrac{E_i - E_t}{kT} \right) \right] + \sigma_n \left[n + n_i \exp\left(\dfrac{E_t - E_i}{kT} \right) \right]} \qquad (5.2.9a)$$

$$= \frac{(pn - n_i^2)}{\tau_{no} \left[p + n_i \exp\left(\dfrac{E_i - E_t}{kT} \right) \right] + \tau_{po} \left[n + n_i \exp\left(\dfrac{E_t - E_i}{kT} \right) \right]} \qquad (5.2.9b)$$

where $\tau_{no} = (N_t v_{th} \sigma_n)^{-1}$ and $\tau_{po} = (N_t v_{th} \sigma_p)^{-1}$.

Equation 5.2.9 shows that U is positive, and therefore there is net recombination if the pn product exceeds n_i^2. The sign changes and there is net generation if
the pn product is less than n_i^2. The term $(pn - n_i^2)$ represents the restoring "force"
for free-carrier populations in a nonequilibrium condition.

The dependence of U on the energy level of the recombination centers in Equation 5.2.9 can be more easily grasped if we consider the case of equal electron
and hole capture cross sections. For the case $\sigma_p = \sigma_n \equiv \sigma_o$, we can define $\tau_o \equiv$
$(N_t v_{th} \sigma_o)^{-1}$ and, therefore,

$$U = \frac{(pn - n_i^2)}{\left[p + n + 2n_i \cosh\left(\dfrac{E_t - E_i}{kT} \right) \right] \tau_0} \qquad (5.2.10)$$

The dependence on the energy level of the recombination center is contained in
the hyperbolic cosine term in Equation 5.2.10. This term is symmetric around
$E_t = E_i$, which reflects a symmetry in the capture of holes and electrons by the
center. The denominator is at its minimum value at $E_t = E_i$, so that U is a maximum value for recombination centers having energies near the middle of the gap.
In Figure 5.3 (solid curve) U normalized to its maximum value has been plotted
versus $(E_t - E_i)/kT$ from Equation 5.2.10 for a reasonable case of recombination
in an n-type semiconductor. The conditions used for the figure are: $p < n$,
$n = 10^{16}$ cm^{-3}, $(pn - n_i^2) = 1.5 \times 10^{31}$ cm^{-6}, and $\tau_0 = 10^{-7}$ s. We shall find
shortly that these values are appropriate in the quasi-neutral region near a forward-
biased pn junction. As a second illustration of Equation 5.2.10, Figure 5.3 (dotted
curve) shows the dependence of U (normalized to its maximum value) on
$(E_t - E_i)/kT$ for a case involving generation. Again, reasonable values have been
used for the other terms in Equation 5.2.10: $(pn - n_i^2) = -2.1 \times 10^{20}$ cm^{-6}, p and

---

* In contrast to the spontaneous recombination and generation transitions are those caused by
the stimulation, for example, of a radiative source.

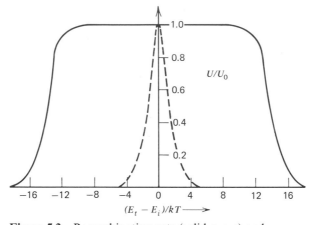

Figure 5.3   Recombination rate (solid curve) and generation rate (dotted curve) as functions of the difference between the energy of the recombination center E_t and the intrinsic Fermi energy E_i. The curves are normalized to the rates for $E_t = E_i$, and have been drawn using Equation 5.2.10 with parameter values listed in the text.

n being much less than n_i, and $\tau_0 = 10^{-7}$ s. These conditions might fit the region near the center of a _pn_ junction depletion zone under reverse bias. The results, sketched in Figure 5.3, show that the dependence of the rate U on recombination-center energy is more pronounced for the case of generation in a depleted semiconductor than for recombination in an undepleted region. The reason for this is that the energy level plays a greater role in equalizing the rates of transfer between the recombination center and the conduction and valence bands when all carrier densities are small. In either generation or recombination, however, the most effective recombination centers will have E_t close to E_i. As practical examples, gold and copper give rise to two effective recombination centers. The values of $(E_t - E_i)$ in silicon for these elements are 0.03 and 0.01 eV, respectively.[5]

Equations 5.2.9 and 5.2.10 are the major results of the _SHR_ recombination analysis. They show that U, the net recombination rate through a recombination center, will be a function of the free-carrier densities as well as specific properties of the recombination site. The equations usually can be simplified for a particular problem. The predictions of material and device behavior based upon these equations have been generally corroborated by experiments.

Excess-Carrier Lifetime

To understand the physical significance of the net recombination rate U, we may consider a semiconductor with no current flow in which thermal equilibrium is disturbed by the sudden creation of equal numbers of excess electrons and holes.

These excess carriers then decay spontaneously as the semiconductor returns to thermal equilibrium. Solutions of the continuity equations (Equations 5.1.3) for this case give the excess electron density as a function of time. Let us consider this problem under an assumption that is frequently satisfied in practice; we assume that the disturbance of equilibrium corresponds to *low-level injection*. In this condition the external disturbance does not appreciably change the total free-carrier density from its equilibrium value. For the case we are considering, if we call the extra injected electron density n' and the extra hole density p', then low level injection implies that n' and p' are both much less than $(n_o + p_o)$ where n_o and p_o represent the thermal-equilibrium densities of carriers in the semiconductor. From these definitions, $n' \equiv n - n_o$ and $p' \equiv p - p_o$, where $n' = p'$.*

If $\sigma_n = \sigma_p$, then the recombination rate U is given by Equation 5.2.10 and hence the continuity equation 5.1.3a can be written

$$\frac{dn'}{dt} = G - R = -U = \frac{-(n_o + p_o)n'}{\left(n_o + p_o + 2n_i \cosh\left[\dfrac{E_t - E_i}{kT}\right]\right)\tau_0} \qquad (5.2.11)$$

Solving for n', we find that the excess carrier density decays exponentially with time

$$n'(t) = n'(0)\exp(-t/\tau_n) \qquad (5.2.12)$$

where the *lifetime* τ_n is given by

$$\tau_n = \left[\frac{n_o + p_o + 2n_i \cosh\left(\dfrac{E_t - E_i}{kT}\right)}{(n_o + p_o)}\right]\tau_0 \qquad (5.2.13)$$

As we noted in the previous section, for recombination centers to be effective, the term $(E_t - E_i)$ is relatively small, and, therefore, the third term in the numerator of Equation 5.2.13 is negligible compared to the sum of the first two terms. Equation 5.2.13 then reduces to

$$\tau_n = \tau_0 = \frac{1}{N_t v_{th}\sigma_o} \qquad (5.2.14)$$

and

$$U = \frac{n'}{\tau_n} \qquad (5.2.15)$$

Equation 5.2.14 shows that the excess-carrier lifetime is independent of the majority-carrier concentration for recombination through recombination centers under low-level injection. We may understand this behavior physically by considering the kinetics of the recombination process. For example, in a *p*-type semiconductor most of the recombination centers are empty of electrons since $E_f < E_t$

---

* Since electron and hole densities increase or decrease at the same rate, the excess densities n' and p' remain equal to one another.

(assuming traps near midgap). The recombination process is therefore limited by the capture of electrons from the conduction band. Once an electron is captured by a recombination center, one of the many holes in the valence band is quickly captured. Thus, the rate-limiting step in the recombination process is the capture of a minority carrier by the recombination center; this is insensitive to the majority-carrier population.

Minority-carrier lifetimes can vary widely, dependent upon the density and type of recombination centers in the semiconductor. For devices such as radiation detectors that depend upon long lifetime for minority carriers, special care can yield silicon having millisecond lifetimes or greater. For integrated circuits, typical values range from a fraction of a microsecond to hundreds of microseconds.

Auger Recombination.[†] When excess carriers recombine in a region that has a high dopant concentration, the probability of direct recombination between holes and electrons may not be negligible compared to the probability of recombination through traps (_SHR_ recombination). This direct recombination process is called _Auger_[*] recombination. In Auger recombination, three free carriers interact, either two electrons and a hole, or two holes and an electron. Two of the carriers recombine, and the third carries away the momentum of the incoming carriers and the energy released by the recombination event. Auger recombination is exactly the inverse of avalanche pair production, discussed in Section 4.4, in which the energy and momentum of an incoming carrier create a hole and an electron. Because of the need for the simultaneous interaction of three carriers, we can expect that Auger recombination is highly unlikely except in heavily doped material, such as that forming the emitter region of a bipolar transistor (Chapter 6).

An expression for the Auger recombination rate U_A is

$$U_A = R_A - G_A = \Gamma_n n(pn - n_i^2) + \Gamma_p p(pn - n_i^2) \tag{5.2.16}$$

where Γ_n is the coefficient representing interactions in which the remaining carrier is an electron, and Γ_p represents recombination events in which the remaining particle is a hole. Typical values for these coefficients are about 1 or 2×10^{-31} cm^6 s^{-1}.

As an example of the use of Equation 5.2.16, we consider the recombination of electrons in heavily doped _p_-type material. If the Γ values are comparable, the first term is negligible ($p \gg n$), and the lifetime τ_A of excess electrons n' for pure Auger recombination is given by

$$\tau_A = \frac{n'}{U_A} \approx \frac{1}{\Gamma_p N_a^2}$$

where we have assumed that the electron density is approximately equal to n', and $n' \ll N_a$. From this result, using $\Gamma_p = 10^{-31}$, we see that at a dopant concentration of 10^{19} cm^{-3}, τ_A would be of order 100 ns.

---

[*] Named for physicist P. Auger and pronounced "oh-zhay".

If the lifetime for *SHR* recombination is given by τ_n, an overall expression for the lifetime τ_{nA} when both types of recombination take place is calculated from

$$\frac{1}{\tau_{nA}} = \frac{1}{\tau_n} + \frac{1}{\tau_A} \tag{5.2.17}$$

Surface Recombination.[†] Thus far, we have considered generation-recombination centers that are uniformly distributed throughout the bulk of the semiconductor material. As discussed in Section 3.5, a semiconductor surface is the location of an abundance of extra localized states having energies within the forbidden gap. The presence of a passivating layer of silicon dioxide over the semiconductor surface, as is usual in devices made by the planar process, ties up many of the bonds that would otherwise contribute to surface states and protects the surface from foreign atoms. A passivating oxide can reduce the density of surface states from about 10^{15} to less than 10^{11} cm^{-2}. Even with passivated surfaces, however, surface states provide additional generation-recombination centers over those present in the bulk. Since the properties of many practical semiconductor devices are affected by generation and recombination at the surface, a brief discussion is in order.

The kinetics of generation-recombination at the surface are similar to those considered for bulk centers with one significant exception. While we considered the volume density N_t(cm^{-3}) of bulk centers, we must discuss the area density N_{st}(cm^{-2}) of surface centers.* Although the N_{st} surface centers are actually distributed over a thickness of many atoms, the poorly defined microscopic structure near the semiconductor surface makes useful a description in terms of an equivalent number of states located at the surface. We may write an expression for the recombination rate U per unit area at the surface analogous to Equation 5.2.9:

$$U_s = \frac{N_{st} v_{th} \sigma_n \sigma_p (p_s n_s - n_i^2)}{\sigma_p \left[p_s + n_i \exp\left(\dfrac{E_i - E_{st}}{kT}\right)\right] + \sigma_n \left[n_s + n_i \exp\left(\dfrac{E_{st} - E_i}{kT}\right)\right]} \tag{5.2.18}$$

where the subscript s denotes concentrations and conditions near the surface and E_{st} is the energy of the surface generation-recombination centers. To stress the physical significance of surface recombination, we again simplify the mathematics by considering the most efficient centers, which are located near midgap, and equal capture cross sections for electrons and holes. With these assumptions, Equation 5.2.18 reduces to

$$U_s = N_{st} v_{th} \sigma \frac{(p_s n_s - n_i^2)}{p_s + n_s + 2n_i \cosh\left(\dfrac{E_{st} - E_i}{kT}\right)} \tag{5.2.19}$$

---

* The density N_{st} represents interface trapping states that are active as generation-recombination sites. These states are discussed further in Section 8.5.

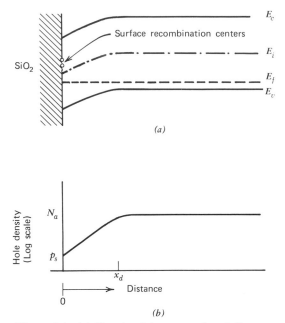

Figure 5.4 (a) Sketch of the energy-band diagram near the surface of p-type silicon that has been covered with a passivating oxide. (b) Hole density near the surface.

We saw in Chapter 3 that the surface of a semiconductor is often at a different potential than the bulk so that surface carrier concentrations may differ from their values in the neutral bulk region. Even in the case of oxide-passivated surfaces, there is generally a space-charge region near the surface of the semiconductor as shown in Figure 5.4 for p-type silicon. As described in Chapter 2, dopant segregation at the oxide-silicon interface causes silicon surfaces to be less strongly p-type or more n-type than the bulk. If we assume that the pn product remains constant throughout the space-charge region, the product at the surface $p_s n_s$ may be expressed in terms of quantities at the neutral edge of the space-charge region:

$$p_s n_s = p_p(x_d) n_p(x_d) \approx N_a n_p(x_d) \tag{5.2.20}$$

in a p-type semiconductor.* Equation 5.2.19 may then be written

$$U_s = N_{st} v_{th} \sigma \frac{N_a[n_p(x_d) - n_{po}]}{(p_s + n_s + 2n_i)} = N_{st} v_{th} \sigma \frac{N_a}{(p_s + n_s + 2n_i)} n_p'(x_d) \tag{5.2.21}$$

where we have assumed $E_{st} \approx E_i$. In Equation 5.2.21 we have expressed the surface recombination rate U_s in terms of the deviation n_p' of the minority-carrier

---

* The subscripts p and n are used generally to denote carrier concentrations in p and n-type material, respectively.

concentration from its equilibrium value at the interior boundary of the surface space-charge region.

The coefficient of n_p' on the right side of Equation 5.2.21 is usually defined as a parameter s, which describes the characteristics of the surface recombination process:

$$s = N_{st}v_{th}\sigma \frac{N_a}{(p_s + n_s + 2n_i)} \qquad (5.2.22)$$

The value of s depends on the physical nature and density of the surface generation-recombination centers as well as on the potential at the surface. If the surface region is depleted of mobile carriers, n_s and p_s are small, and s is large. If the surface is neutral, $p_s \sim N_a$; s is small and given by

$$s = s_o = N_{st}v_{th}\sigma \qquad (5.2.23)$$

where the subscript o denotes that the surface and bulk are at the same potential; that is, the surface region is neutral. The dependence of s on surface potential is of practical significance.

The dimensions of s are cm s^{-1}, and s is consequently called the *surface recombination velocity*, although it is not directly related to an actual velocity. A physical interpretation of s can be obtained by comparing Equation 5.2.23 to Equation 5.2.14 for the minority-carrier lifetime; s is related to the rate at which excess carriers recombine at the surface just as $1/\tau$ is related to the rate at which they recombine in the bulk.

EXAMPLE   Surface Recombination Velocity

An n-type silicon wafer with resistivity $\rho_n = 0.025$ Ω-cm at room temperature is illuminated through a passivating layer of SiO$_2$ so that the hole and electron densities exceed their thermal-equilibrium values. Assume that the surface electron density $n_s = 10^{16}$ cm^{-3} at thermal equilibrium, the photogeneration rate of carriers is 10^{14} cm^{-2} s^{-1}, and the surface hole density under illumination increases to $p_s = 10^{10}$ cm^{-3}.

(a) Find a value for s, the surface recombination velocity such that 50% of the excess carriers recombine at the surface.

(b) If the surface recombination is due to the presence of a density of surface recombination centers $N_{st} = 10^{11}$ cm^{-2}, determine the interaction cross section σ giving rise to the recombination rate described in part a.

Solution

We first find $N_d = 10^{18}$ cm^{-3} for $\rho_n = 0.025$ Ω-cm using either Figure 1.14 or Table 4.1. Since $n_s = 10^{16}$ cm^{-3}, the surface of the wafer is slightly depleted of majority carriers even under illumination. Using Equation 1.1.13, we find the thermal-equilibrium density of holes at the surface

$$p_s = n_i^2/n_s = 2.1 \times 10^4 \text{ cm}^{-3}$$

Hence, the excess density of holes at the surface

$$p_s' = 10^{10} - 2.1 \times 10^4 \approx 10^{10} \text{ cm}^{-3}$$

The rate of recombination at the surface is just half the generation rate. Hence,

$$U_s = sp_s' = s \times 10^{10} = 0.5 \times 10^{14} \text{ cm}^{-2} \text{ s}^{-1}$$

Therefore, $s = 5000$ cm s^{-1}, which is the answer required in part a. For part b, we use a form of Equation 5.2.22 appropriate for *n*-type silicon together with Equation 5.2.23 to calculate

$$s_o = s \frac{(p_s + n_s + 2n_i)}{N_d} \approx 5000 \frac{10^{16}}{10^{18}} = 50 \text{ cm s}^{-1}$$

Taking $N_{st} = 10^{11}$ cm^{-2} and $v_{th} \approx 10^7$ cm s^{-1} (Section 1.2), we find $\sigma = 5 \times 10^{-17}$ cm^2.

5.3 Current-Voltage Characteristics of *pn* Junctions

Through use of the continuity equations (Equations 5.1.3) together with the concept of excess-carrier lifetime as derived from the *SHR* generation-recombination model, we are in a position to find expressions for current in a *pn* junction under bias. Solutions of the continuity equations in the quasi-neutral regions will give carrier densities in terms of position and time. Expressions for current are then obtained directly by using Equations 1.2.21 and 1.2.22, which define carrier flow in terms of carrier densities. The total current consists, in general, of the sum of four components: hole and electron drift currents and hole and electron diffusion currents.

We consider a *pn*-junction diode that is connected to a voltage source with the *n*-region grounded and the *p*-region at V_a volts relative to ground. The diode structure is of constant cross-sectional area A, and it has the longitudinal dimensions sketched in Figure 5.5. The junction is not illuminated, and carrier densities within the diode are only influenced by the applied voltage. The applied voltage V_a is dropped partially across the quasi-neutral regions and partially across the junction itself. Since the voltage drops across the quasi-neutral regions are ohmic (current times resistance), they are typically small in *IC* devices at low and moderate currents. However, for small device cross-sections such as those used in some VLSI design, ohmic voltage drops can limit device performance.

For our analysis we neglect ohmic drops and assume that V_a is sustained entirely at the junction. Then, the total junction voltage will be $\phi_i - V_a$ under bias where ϕ_i is the built-in voltage. If V_a is positive, that is, for *forward bias*, the applied voltage reduces the barrier to the diffusion flow of majority carriers at the junction.

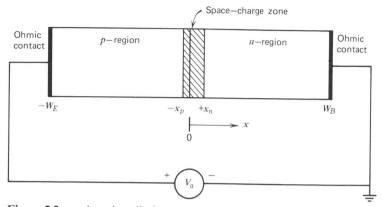

Figure 5.5 *pn*-junction diode structure used in the discussion of currents. The sketch shows the dimensions (cross-sectional area A is assumed to be uniform) and the bias convention.

The reduced barrier, in turn, permits a net transfer of holes from the *p*-side into the *n*-side and of electrons from the *n*-side into the *p*-side. When these transferred carriers enter the quasi-neutral regions, they are minority carriers and they are quickly neutralized by majority carriers that enter the quasi-neutral regions from the ohmic contacts at the ends. This neutralization of injected carriers is just the dielectric-relaxation process that we considered in Chapter 1. Once the minority carriers have been injected across the space-charge region, they tend to diffuse into the neutral interior.

If V_a is negative, that is, for *reverse bias*, the barrier height to diffusing majority carriers is increased. Since equilibrium is disturbed, minority carriers near the junction space-charge region tend to be depleted. The majority-carrier concentration is likewise reduced by the process of dielectric relaxation.

From these few remarks, we can see that the minority-carrier densities deserve special attention because they really determine what currents flow in a *pn* junction. The majority carriers act only as suppliers of the injected minority-carrier current or as charge neutralizers in the quasi-neutral regions. It is helpful to consider the majority carriers as willing "slaves" of the minority carriers. Accordingly, we shall seek solutions of the continuity equations for the minority-carrier densities in each of the quasi-neutral regions.

Boundary Values of Minority-Carrier Densities

To write these solutions in a useful fashion, however, it will be necessary to relate the boundary values of the minority-carrier densities to the applied voltage V_a. The most straightforward way of doing this is to make two additional assumptions: first, that the applied bias leads to low-level injection and, second, that the applied bias is small enough so that the detailed balance between majority and minority populations across the junction regions is not appreciably disturbed. The first of these assumptions, low-level injection, has been considered in Section 5.2. Briefly, it implies that the majority-carrier populations are negligibly changed at the edge of the quasi-neutral regions by the applied bias. The assumption that detailed balance nearly applies allows the use of Equation 4.1.9 across the junction where the potential difference is known to be $\phi_i - V_a$.

Both of these assumptions are certainly valid when V_a is small ($|V_a| \ll \phi_i$). Their validity at higher biases deserves more careful consideration. As with our previous assumptions we shall defer this consideration until the analysis is completed.

By the assumption of low-level injection, the electron density at the quasi-neutral boundary in the *n*-region next to the *pn* junction is equal to the dopant density whether at equilibrium or under bias. As in Chapter 4 we call this position x_n (Figure 5.5), and we denote the thermal equilibrium density with an extra subscript *o*.

Likewise, the boundary of the quasi-neutral *p*-region is at $-x_p$ and the hole density there is equal to the acceptor dopant density at equilibrium as well as under bias. Summarizing these statements with equations, we have

$$n_{po}(-x_p) = n_{no}(x_n) \exp\left(\frac{-q\phi_i}{kT}\right)$$

$$= N_d(x_n) \exp\left(\frac{-q\phi_i}{kT}\right) \tag{5.3.1}$$

$$p_{no}(x_n) = p_{po}(-x_p) \exp\left(\frac{-q\phi_i}{kT}\right)$$

$$= N_a(-x_p) \exp\left(\frac{-q\phi_i}{kT}\right) \tag{5.3.2}$$

$$n_p(-x_p) = N_d(x_n) \exp\left[\frac{-q(\phi_i - V_a)}{kT}\right] \tag{5.3.3}$$

and

$$p_n(x_n) = N_a(-x_p) \exp\left[\frac{-q(\phi_i - V_a)}{kT}\right] \tag{5.3.4}$$

These four equations can be combined to express the excess minority-carrier densities at the boundaries in terms of their thermal-equilibrium values. We define the excess densities by

$$n' \equiv n - n_o \tag{5.3.5}$$

and

$$p' \equiv p - p_o \tag{5.3.6}$$

Then

$$n'_p(-x_p) = n_{po}(-x_p)\left[\exp\left(\frac{qV_a}{kT}\right) - 1\right] \tag{5.3.7}$$

and

$$p_n'(x_n) = p_{no}(x_n) \left[\exp\left(\frac{qV_a}{kT}\right) - 1 \right]$$ (5.3.8)

Equations 5.3.7 and 5.3.8 are extremely important results that we shall use to define specific solutions for the continuity equations for minority carriers in the quasi-neutral regions near a *pn* junction. The equations show that the minority-carrier density is an exponential function of applied bias while the majority-carrier density has been assumed to be insensitive (to first order) to it. Since the minority-carrier density at thermal equilibrium is typically 11 or 12 orders of magnitude below the majority-carrier density, Equations 5.3.7 and 5.3.8 will not be in conflict with the low-level injection assumption until the exponential factor is typically of the order 10^{11} or 10^{12}. We shall look into the second assumption, that of quasi-equilibrium of carrier fluxes, after we have obtained the voltage dependence of currents.

Ideal-Diode Analysis

Having reviewed the reasons for our focus on the minority carriers and having obtained the bias dependence of excess minority carriers, we are ready to consider solutions of the continuity equations (Equation 5.1.4) in the quasi-neutral regions. Let us first do this under a series of idealizations that comprise what has been called the *ideal-diode analysis*.

First, we consider excess holes injected into the *n*-regions, where bulk recombination through generation-recombination centers is dominant. Thus, the term $(G_p - R_p)$ in Equation 5.1.4 can be expressed through Equation 5.2.9. Because of our assumption of low level injection, the arguments used in the discussion of excess-carrier lifetime that led to Equation 5.2.15 apply also to this case. Therefore, the continuity equation in the quasi-neutral approximation discussed in Section 5.1 becomes

$$\frac{\partial p_n}{\partial t} = D_p \frac{\partial^2 p_n}{\partial x^2} - \frac{p_n - p_{no}}{\tau_p}$$ (5.3.9)

where the subscripts *n* emphasize that the holes are in the *n*-region.

Fortunately, the simplest case of impurity doping, that of a constant donor density along *x*, is frequently encountered in practice. Taking this case and considering steady state ($\partial p/\partial t = 0$), we can rewrite Equation 5.3.9 as a total differential equation in terms of the excess density p', which was defined in Equation 5.3.6. This equation

$$0 = D_p \frac{d^2 p_n'}{dx^2} - \frac{p_n'}{\tau_p}$$ (5.3.10)

has the simple exponential solution

$$p_n'(x) = A \exp\left(-\frac{x - x_n}{\sqrt{D_p \tau_p}}\right) + B \exp\left(\frac{x - x_n}{\sqrt{D_p \tau_p}}\right)$$ (5.3.11)

where A and B are constants determined by the boundary conditions of the problem. The characteristic length $\sqrt{D_p \tau_p}$ in Equation 5.3.11 is called the diffusion length and denoted by L_p. (The diffusion length of an electron in a p-type region is denoted by L_n.) For specific applications of the solution to the continuity equation given in Equation 5.3.11, let us consider two limiting cases based upon the length W_B of the n-region from the junction to the ohmic contact (Figure 5.5).

Long-Base Diode. First, if W_B is long compared to the diffusion length L_p, essentially all of the injected holes recombine before traveling completely across it. This case is popularly known as the _long-base diode_ for reasons that will become clear later. For the long-base diode, L_p is the average distance traveled in the neutral zone before an injected hole recombines (Problem 5.8). Since p_n' must decrease with increasing x, the constant B in Equation 5.3.11 must be zero. The constant A in the solution is determined by applying Equation 5.3.8, which specifies $p_n'(x_n)$ as a function of applied voltage. The complete solution is therefore

$$p_n'(x) = p_{no}(e^{qV_a/kT} - 1)e^{-(x-x_n)/L_p} \qquad (5.3.12)$$

as shown in Figure 5.6. Having derived in Equation 5.3.12 an expression for the excess hole density, we obtain in a straightforward manner an expression for hole current. The hole current flows only by diffusion, since we have assumed that the field is negligible in the neutral regions; therefore, from Equation 1.2.22

$$J_p(x) = -qD_p\frac{dp_n}{dx} = qD_p\frac{p_{no}}{L_p}(e^{qV_a/kT} - 1)e^{-(x-x_n)/L_p}$$

$$= qD_p\frac{n_i^2}{N_dL_p}(e^{qV_a/kT} - 1)e^{-(x-x_n)/L_p} \qquad (5.3.13)$$

The hole current is therefore a maximum at $x = x_n$ and decreases away from the junction (Figure 5.7) because the hole gradient decreases as carriers are lost by

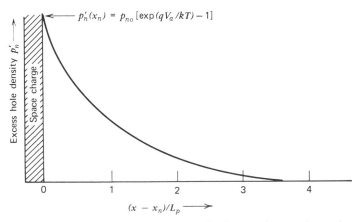

Figure 5.6 Spatial variation of holes in the quasi-neutral n-region of a long-base diode under forward bias V_a. The excess density p_n' has been calculated from Equation 5.3.12.

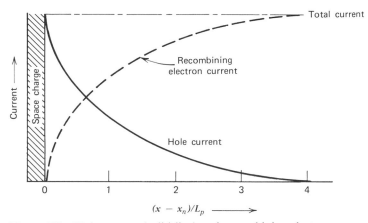

Figure 5.7 Hole current (solid line) and recombining electron current (dotted line) in the quasi-neutral *n*-region of the long-base diode of Figure 5.5. The sum of the two currents *J* (dot-dash line) is constant. The hole current is calculated from Equation 5.3.13.

recombination. Since the total current must remain constant with distance from the junction in the steady-state case, the electron current must increase as we move away from the junction. This electron current supplies the electrons with which the holes are recombining.

The total current is entirely carried by electrons at the ohmic contact at W_B. Moving toward the junction, the electron current decreases as recombination occurs with the injected holes. At the junction, the only electron current flowing is the one injected into the *p*-region. The electrons injected into the *p*-region constitute the minority-carrier current there. Thus, we see that the the total current is obtained by summing the two minority-carrier injection currents: holes into the *n*-side plus electrons into the *p*-side.

The minority-carrier electron current that is injected into the *p*-region can be found by an analogous treatment to the analysis used to obtain Equation 5.3.13. If the ohmic contact is at $-W_E$ where $W_E \gg L_n \equiv \sqrt{D_n \tau_n}$, then

$$J_n = qD_n \frac{n_i^2}{N_a L_n} (e^{qV_a/kT} - 1) e^{(x + x_p)/L_n} \tag{5.3.14}$$

Because we have chosen the origin for *x* at the physical junction (Figure 5.5), *x* is a negative number throughout the *p*-region. Hence, J_n decreases away from the junction as did J_p in the *n*-region. To obtain an expression for the total current J_t, we sum the minority-carrier components at $-x_p$ and $+x_n$, respectively, as expressed in Equations 5.3.13 and 5.3.14:

$$J_t = J_p(x_n) + J_n(-x_p) = qn_i^2 \left(\frac{D_p}{N_d L_p} + \frac{D_n}{N_a L_n} \right) (e^{qV_a/kT} - 1)$$

$$= J_0(e^{qV_a/kT} - 1) \tag{5.3.15}$$

where J_0 is the magnitude of a saturation current density predicted by this theory when a negative bias equal to a few kT/q volts is applied. Equation 5.3.15 is identical in form to Equation 3.3.6, which was derived for a metal-semiconductor Schottky-barrier diode. The similar dependence on voltage arises from the application in both cases of quasi-equilibrium assumptions, which led to Equations 5.3.7 and 5.3.8.

Short-Base Diode. A second limiting case occurs when the lengths W_B and W_E of the *n*- and *p*-type regions are much shorter than the diffusion lengths L_p and L_n. In this case, there is little recombination in the bulk of the quasi-neutral regions. In the limit, all injected minority carriers recombine at the ohmic contacts at either end of the diode structure. We can obtain solutions most easily in this case by approximating the exponentials in Equation 5.3.11 by the first two terms of a Taylor series expansion so that

$$p_n'(x) = A' + B' \frac{(x - x_n)}{L_p} \tag{5.3.16}$$

Because of the ohmic contact at $x = W_B$, $p_n'(W_B) = 0$. The boundary condition at $x = x_n$ is given by Equation 5.3.8 as before, so that the solution for the excess hole density in the *n*-region becomes

$$p_n'(x) = p_{no}(e^{qV_a/kT} - 1)\left(1 - \frac{x - x_n}{W_B'}\right) \tag{5.3.17}$$

where $W_B' = W_B - x_n$ is the length of the quasi-neutral *n*-type region (Figure 5.5). From Equation 5.3.17, we see that the excess hole density decreases linearly with distance across the *n*-type region (Figure 5.8). The approximation $L_p \gg W_B$ is equivalent to stating that all the holes diffuse across the *n*-type region before

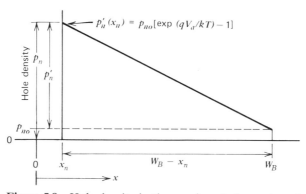

Figure 5.8   Hole density in the quasi-neutral *n*-region of an ideal short-base diode under forward bias of V_a volts. The excess hole density p_n' is calculated from Equation 5.3.17.

recombining. The assumption that no recombination occurs in the *n*-type region could also have been made by letting the lifetime τ_p approach infinity in Equation 5.3.10. The differential equation that results has a linear solution.

A linearly varying concentration indicates that the hole current remains constant throughout the *n*-type region, and that no electron current is needed to compensate for recombining holes. Therefore,

$$J_p = -qD_p\frac{dp}{dx} = qD_p\frac{p_{no}}{W_B'}(e^{qV_a/kT}-1) = qD_p\frac{n_i^2}{N_dW_B'}(e^{qV_a/kT}-1) \quad (5.3.18)$$

Comparing Equations 5.3.18 and 5.3.13 for the hole currents in the short- and long-base diodes, we see that the solutions obtained are similar except for the characteristic length appropriate to each geometry. In the long-base diode, the characteristic length is the minority-carrier diffusion length. In the short-base diode, it is the length of the quasi-neutral region. As in the long-base diode, the total current in the short-base diode is made up of electrons injected into the *p*-region and of holes injected into the *n*-region. Hence, for the short-base diode with $W_E \ll L_n = \sqrt{D_n\tau_n}$ and $W_B \ll L_p = \sqrt{D_p\tau_p}$,

$$J_t - qn_i^2\left(\frac{D_p}{N_dW_B'} + \frac{D_n}{N_aW_E'}\right)(e^{qV_a/kT}-1) \quad (5.3.19)$$

Of course, a given diode may be approximated by a combination of these two limiting cases; that is, it may be short-base in the *p*-region and long-base in the *n*-region or vice versa. Handling these cases is straightforward.

Both diode cases, Equations 5.3.15 and 5.3.19 predict a large current flow under forward bias and a small saturation current in reverse bias. Physically, this marked nonuniformity results because forward bias aids the injection of majority carriers from each region across the junction. These carriers are replaced at the end ohmic contacts. Under reverse bias, the net flow across the junction is composed of minority carriers from each region. These are few in number and not replaced at the end contacts. Hence, only a small saturation current flows under reverse bias (Problem 5.7).

We see from Equations 5.3.15 and 5.3.19 that the most lightly doped side of the junction determines the reverse saturation current. For example, if the dopant concentration on the *n*-side of the junction is much less than that on the *p*-side, the hole current injected across the junction into the *n*-side is much greater than the electron current injected into the *p*-region. Such a situation would characterize a heavy *p*-type diffusion into a lightly doped *n*-type wafer.

EXAMPLE   Long- and Short-Base Diodes

(a) A silicon diode is formed by diffusing a high concentration of boron into a 350 μm-thick, phosphorus-doped wafer having resistivity $\rho = 4.5$ Ω-cm. Contacts are made to the diffused boron region and to the backside of the wafer. The _pn_ junction, 0.5 μm below the surface with a plane area $= 10^{-4}$ cm^2, can be considered to be a one-sided step junction. Find the saturation current in the diode if the hole lifetime τ_{ps} in the substrate is 1 μs.

(b) In a different process, a similar diffusion schedule is used to form a p^+n junction in a 2 μm-thick, phosphorus-doped, epitaxial layer having a resistivity $p = 4.5$ Ω-cm. The epitaxial layer is deposited on a heavily doped ($N_a > 10^{19}$ cm^{-3}) _n_-type substrate. The lifetime τ_{pe} in the epitaxial layer 1 μs. In the substrate the lifetime is short ($\tau_{ps} \approx 100$ ps). The _pn_ junction, formed 0.5 μm below the surface with a plane area $= 10^{-4}$ cm^2, can be considered to be a one-sided step junction. Find the saturation current in the diode.

Solution

(a) Either from Figure 1.14 or Table 4.1, we find that for 4.5 Ω-cm _n_-type material, $N_d = 1 \times 10^{15}$ cm^{-3}. The diffusion constant D_p is found from Figure 1.15 to be 12 cm^2 s^{-1}. Therefore, the diffusion length $L_p = \sqrt{D_p \tau_p}$ is 34.6 μm. Since L_n is much less than the thickness of the silicon wafer (350 μm), the device is a long-base diode. The saturation-current density J_0 is therefore given by Equation 5.3.15, and the saturation current $I_0 = J_0 A = 1.2 \times 10^{-14}$ A.

(b) In process b, the _pn_ junction is formed in an epitaxial layer having a thickness much smaller than L_p. Therefore, almost all injected holes diffuse to the substrate where they rapidly recombine. This structure is therefore a short-base diode, and J_0 is given by Equation 5.3.18. The effective base width is 1.5 μm (the thickness of the epitaxial layer minus the junction depth). Using Equation 5.3.18, we calculate the diode saturation current $I_0 = J_0 A = 2.7 \times 10^{-13}$ A.

The short base is seen in this example to increase the saturation current because the gradient of the minority-carrier density is increased.

Validity of Approximations. Before taking a more searching look at the physics of _pn_-junction diodes, we return to consider two of the initial assumptions used to carry out the analysis. These assumptions are first, that ohmic drops in the quasi-neutral regions are small so that V_a is sustained entirely across the junction space-charge region and, second, that applied bias does not greatly alter the detailed balance between drift and diffusion tendencies that exists at thermal equilibrium.

To consider these assumptions, let us take the fairly typical case of $N_d = 5 \times 10^{15}$ and $N_a = 5 \times 10^{18}$. If we take first a short-base diode with $W_B = W_E = 3$ μm, then from Equation 5.3.19 we calculate a saturation current of approximately 10^{-9} A cm^{-2}. For a typical cross-sectional area of 10^{-5} cm^2 and a typical forward-bias voltage of ~ 0.65 V, this would lead to a current of ~ 1 mA or 100 A cm^{-2}.

This current would consist almost entirely of holes injected into the *n*-side since the first term in the saturation current of Equation 5.3.19 far exceeds the second. Thus, the ohmic drop would only exist in the highly conducting *p*-region that has a resistivity of 0.03 Ω-cm. The field in this region is just 3 V cm^{-1} and, hence, the potential dropped in the 3 μm length is approximately 1 mV, which is certainly negligible compared to the applied 0.65 volts. Note that there is essentially no ohmic drop in the *n*-region for this case because there is only a negligible flow of electron current. If we take a long-base diode with lower doping on either side, the assumption is on shakier grounds. To see this, assume that $N_a = N_d = 10^{16}$ and that we are considering a long-base diode of cross-sectional area 10^{-5} cm^2 with 0.65 V applied bias. If $L_n \simeq L_p = 30$ μm, using Equation 5.3.15, we calculate that such a diode would pass approximately 32×10^{-6} A. If the *p* and *n* regions were each 100 μm long there would be ohmic voltage drops of about 0.03 V in the *p*-region and 0.02 V in the *n*-region. Again, this can be considered negligible to first order as compared to the 0.65 V applied. These results are typical of many practical cases, and we conclude that it is reasonable to assume that the entire applied voltage changes the height of the potential barrier at the *pn* junction for low-to-moderate currents.

If the applied forward voltage is nearly as large as the built-in potential, however, the barrier to majority-carrier flow is substantially reduced and large currents may flow. In that case a significant portion of the applied voltage is dropped across the neutral regions, and the series resistance of the neutral regions must be considered. The effect of series resistance can easily be included using circuit analysis, and we shall not consider it further here.

To determine the validity of the quasi-equilibrium assumption we must compare the magnitude of typical currents to the balanced drift and diffusion tendencies at thermal equilibrium. For a typical integrated-circuit *pn* junction in which the hole concentration changes from 10^{18} to 10^4 cm^{-3} across a depletion region about 10^{-5} cm wide, the hole diffusion current for the average gradient is of the order of 10^5 A cm^{-2}. As we saw in Chapter 4, at thermal equilibrium this tendency of holes to diffuse is exactly balanced by an opposite tendency to drift under the influence of the electric field in the depletion region. We have already argued that typical forward-bias diode currents are roughly 10^2 A cm^{-2}—only 0.1% of the two current tendencies in balance at thermal equilibrium. Thus, it is reasonable to treat the case of small and moderate biases by considering only slight deviations from thermal equilibrium. This means that we may relate the carrier concentrations on either side of the junction space-charge region by considering the effective barrier height to be $(\phi_i - V_a)$ and by using Equations 5.3.7 and 5.3.8. These two equations depend also on the validity of the low-level injection assumption. The current tendencies at *pn* junctions that are in detailed balance at thermal equilibrium are so large relative to currents that flow in practice that a more general relationship than Equations 5.3.7 and 5.3.8 is valid.

The more general relationship can be deduced by referring to the quasi-Fermi levels defined in Equations 1.1.28 and 1.1.29. The quasi-Fermi levels (ϕ_{fn} and ϕ_{fp}), we recall, express the densities of free carriers under nonequilibrium conditions. At thermal equilibrium, there is a single Fermi level, valid both for electron and hole densities, while under nonequilibrium conditions, a different quasi-Fermi level represents each carrier density.

From our discussion of the *pn* junction under forward bias we expect that the two quasi-Fermi levels will be separated in the region near the biased junction where injection markedly changes the minority-carrier densities, while far from the junction, ϕ_{fn} and ϕ_{fp} merge with one another. Since the majority-carrier densities are only negligibly affected (for the low and moderate bias condition discussed thus far), the quasi-Fermi levels for the majority-carrier densities are essentially unchanged, while those for the minority-carrier densities are modified.* The applied bias V_a is therefore equal to the offset $\phi_{fp}(x = -x_p) - \phi_{fn}(x = +x_n)$, that is, the difference in the quasi-Fermi level for holes on the *p*–side and that for electrons on the *n*–side. Applying Equation 1.1.30 to the space-charge region of a *pn* junction $(-x_p \leq x \leq +x_n)$, we find, therefore

$$pn = n_i^2 \exp\frac{q(\phi_p - \phi_n)}{kT} = n_i^2 \exp\frac{qV_a}{kT} \qquad (5.3.20)$$

Equation 5.3.20 is consistent with the boundary values for low-level injection derived for minority-carrier densities in Equations 5.3.3 and 5.3.4. It is also useful in cases where injected carrier densities approach the thermal-equilibrium concentrations of majority carriers (Chapter 7).

Space-Charge-Region Currents[†]

The analysis of *pn* junctions that led to the diode equations (Equations 5.3.15 and 5.3.19) was based on events in the quasi-neutral regions. The space-charge region was treated solely as a barrier to the diffusion of majority carriers, and it played a role only in the establishment of minority-carrier densities at its boundaries (Equations 5.3.7 and 5.3.8). This is a reasonable first-order description of events and the equations derived from it (Equations 5.3.15 and 5.3.19) are called the *ideal-diode* equations. Over a significant range of useful biases, however, the ideal-diode equations are quite inaccurate, especially for silicon *pn* junctions. It is necessary to consider corrections to these equations that arise from a more complete treatment of events in the space-charge region at the junction.

As we noted in Chapter 4, the space-charge region typically has dimensions of the order of 10^{-4} cm. Like the quasi-neutral zones of the diode, it contains generation-recombination centers; unlike the quasi-neutral zones, it is a region of steep impurity gradients and rapidly changing populations of holes and electrons. Since the injected carriers, under forward bias, must pass through this region, some carriers may be lost by recombination. Under reverse bias, generation of carriers

---

* Another way to visualize the near constancy of the quasi-Fermi levels for the majority carriers is through Equations 1.2.25 and 1.2.26 which represent total currents in terms of the products of the carrier densities and the gradients of the quasi-Fermi levels. For a given current, the higher the carrier density, the lower is the necessary gradient—hence, the gradients of the quasi-Fermi levels for minority carriers are much larger than those for majority carriers.

in the region leads to excess current above the saturation value predicted by the ideal-diode equations.

To find expressions for generation-recombination in the space-charge region we make use of the Shockley–Hall–Read theory. For simplicity, we take the case of equal hole and electron capture cross sections (Equation 5.2.10) and evaluate it in the space-charge region with an applied bias V_a. The *pn* product there is given by Equation 5.3.20 and thus the overall recombination-generation rate $U = -dn/dt = -dp/dt$ is

$$U = \frac{n_i^2(e^{qV_a/kT} - 1)}{\left(p + n + 2n_i \cosh \dfrac{(E_t - E_i)}{kT}\right)\tau_0} \tag{5.3.21}$$

The recombination rate is thus positive for forward bias and negative under reverse bias. The total current arising from generation and recombination in the space-charge region is given by the integral of the recombination rate across the space-charge region:

$$J_r = q \int_{-x_p}^{x_n} U \, dx \tag{5.3.22}$$

Although this integral is not easily evaluated, a qualitative discussion permits us to deduce the dependence of current on voltage. Consider first, forward bias and, as in Section 5.2, note that recombination centers located near the middle of the forbidden gap are the most effective (Figure 5.3). Thus, we consider $E_t \approx E_i$ in Equation 5.3.21. From this equation, it is evident that the recombination rate is maximized when the sum $p + n$ is minimized. If we consider this sum as a variable function in p and n that is subject to the product of p and n being given by Equation 5.3.20, we can easily show (Problem 5.15) that U is maximized when

$$p = n = n_i \exp\left(\frac{qV_a}{2kT}\right) \tag{5.3.23}$$

For typical forward-bias conditions, the sum $p + n$ is much greater than n_i within the space-charge region. If the dominant contribution to the integral in Equation 5.3.22 is given by this maximum value of U extending across a portion x' of the space-charge region, the recombination current may be expressed as

$$J_r = \frac{qx'n_i^2(e^{qV_a/kT} - 1)}{2n_i(e^{qV_a/2kT} + 1)\tau_0}$$

$$\approx \frac{qx'n_i}{2\tau_0} \exp\left(\frac{qV_a}{2kT}\right) \tag{5.3.24}$$

where $\tau_0 = 1/N_t\sigma v_{th}$ is again the lifetime associated with the recombination of excess carriers in a region with a density N_t of recombination centers. Thus, unlike current arising from carrier diffusion and recombination in the quasi-neutral regions, the current resulting from recombination in the space-charge region varies with applied voltage as $\exp(qV_a/2kT)$ (assuming x' is only a weak function of voltage). This different exponential behavior can be observed in real diodes, especially at low current. Because terms like τ_0 are not known with high precision, space-charge-region recombination is usually not given more elaborate analysis than our treatment and x', in particular, is often approximated by the entire space-charge-region width x_d. If we make this approximation and express the ratio between the ideal-diode current J_t (Equation 5.3.15) and the space-charge-region recombination-current J_r under forward bias, we find

$$\frac{J_t}{J_r} = \frac{2n_i}{x_d}\left[\frac{L_n}{N_a} + \frac{L_p}{N_d}\right]\exp\left(\frac{qV_a}{2kT}\right) \qquad (5.3.25)$$

Thus, space-charge recombination current J_r becomes less significant relative to ideal-diode current as bias increases. Also, the more defect-free is the material, the longer are the diffusion lengths and the more dominant is J_t over J_r. Taking typical (one-sided) silicon diode values of $L_n = 60$ μm, $x_d = 0.25$ μm, and $N_a = 10^{16}$, it is easily seen that J_t exceeds J_r for V_a greater than about 0.375 V.

Under reverse bias, the numerator of Equation 5.3.21 approaches $-n_i^2$. Therefore U is negative, indicating net generation in the space-charge region. From Equation 5.3.20 the pn product becomes vanishingly small. As before, the maximum value of U occurs when both p and n are equal; in this case both are much less than n_i. It would take more elaborate analysis than we carry out here to show that p and n are both substantially less than n_i over a portion of the space-charge region x_i bounded by the points at which the intrinsic Fermi level E_i crosses the quasi-Fermi levels.[7] This region may be considerably smaller than the total space-charge region width x_d. Outside x_i either p or n is greater than n_i, and the generation rate falls precipitously. The net generation rate in the space-charge region can be approximated by the product of the maximum generation rate and the width x_i:

$$J_g = \frac{qn_ix_i}{2\tau_0} \qquad (5.3.26)$$

where we have again assumed that the most effective centers are located at E_i. For a one-sided pn junction with a heavily doped p-type side, virtually all of the space-charge region extends into the lightly doped n-type semiconductor, and we may solve for the width of the total space-charge region x_d and for the region x_i with maximum generation rate:

$$x_d = \left[\frac{2\epsilon_s}{qN_d}(\phi_i - V_a)\right]^{1/2} \qquad (5.3.27)$$

$$= \left[\frac{2\epsilon_s kT}{q^2 N_d}\left(\ln\frac{N_d N_a}{n_i^2} - \frac{qV_a}{kT}\right)\right]^{1/2}$$

and

$$x_i = \left(\frac{2\epsilon_s kT}{q^2 N_d}\right)^{1/2}\left[\left(\ln\frac{N_d}{n_i} - \frac{qV_a}{kT}\right)^{1/2} - \left(\ln\frac{N_d}{n_i}\right)^{1/2}\right] \qquad (5.3.28)$$

where V_a is the applied voltage ($V_a < 0$ for reverse bias). Both x_d and x_i depend on the square root of the applied voltage for large reverse bias, and the difference

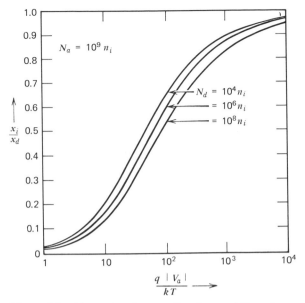

Figure 5.9   The ratio of generation-layer width x_i to space–charge–region width x_d as a function of reverse voltage for several donor concentrations in a one-sided step junction.[7]

between the two quantities becomes small at high reverse bias. Figure 5.9 shows the ratio x_i/x_d as a function of voltage for several donor concentrations in a one-sided step junction. Because the density of recombination centers in practical diodes may vary with position and is generally not well known, it is often not worthwhile to distinguish between x_i and x_d, and the latter is frequently used in Equation 5.3.26.

From Equations 5.3.26 and 5.3.28, we see that the generation current in the space-charge region is only a weak function of the reverse bias, varying roughly as the square root of applied voltage. We may estimate the relative importance of the contributions to the reverse current from the quasi-neutral region and from the space-charge region for a long-base diode by taking the ratio of Equation 5.3.15 under reverse bias to Equation 5.3.26:

$$\frac{J_t}{J_g} = \frac{2n_i}{x_i}\left[\frac{L_n}{N_a} + \frac{L_p}{N_d}\right] \tag{5.3.29}$$

Values of the parameters for practical diodes are such that J_t/J_g is much less than unity. Hence, the current in reverse-biased silicon diodes is generated dominantly in the space-charge region. Generation-recombination centers in the space-charge region are also responsible for excess currents in reverse-biased Schottky diodes (Chapter 3). In that case, however, the dominant currents are associated with recombination centers at the metal-semiconductor interface.

EXAMPLE **Generation Currents in a Reverse-Biased Diode**

(a) Compare the importance of neutral-region generation and space-charge-region generation in a reverse-biased, diffused silicon diode at room temperature. Assume that the diode is formed by introducing a high concentration of boron atoms into a thick, 5 Ω-cm, *n*-type wafer. The lifetime in both the neutral and space-charge regions is 10 μs, and 5 V of reverse bias is applied.

(b) Determine the relative importance of the two current components at 125°C. Assume that the lifetime and diffusion constant vary with temperature as $T^{-1/2}$ and $T^{-2.2}$ (Figure 1.16), respectively.

Solution

(a) Either from Figure 1.14 or Table 4.1, we find that $N_d = 9 \times 10^{14}$ cm^{-3} for 5 Ω-cm *n*–type silicon. Using Figure 1.15, we find $D_p = 12$ cm^2 s^{-1}. Hence, the diffusion length at 300 K, $L_p = \sqrt{D_p \tau_p} = 110$ μm. For a p^+n junction, the generation current in the charge-neutral region J_t is (from Equation 5.3.15)

$$J_t \approx q n_i^2 \frac{D_p}{N_d L_p} = 4.1 \times 10^{-11} \text{ A cm}^{-2} = 41 \text{ pA cm}^{-2}.$$

To calculate the current due to generation in the space-charge region, we first use Equation 5.3.28 to determine the width x_i over which generation takes place. For the values given, we find $x_i = 2.12$ μm. From Equation 5.3.26 we calculate $J_g = 24.6$ nA cm^{-2}. Comparing these results, we see that almost 600 times as much current is produced by generation in the space-charge region as is produced in the charge-neutral regions.

(b) At 125°C, the diffusion constant D_p is reduced

$$D_p(398 \text{ K}) = D_p(300 \text{ K}) \times \left(\frac{300}{398}\right)^{2.2} = 12 \times 0.534 = 6.44 \text{ cm}^2 \text{ s}^{-1}.$$

The lifetime is also reduced to 8.68 μs. Using the formula in Table 1.4, we calculate $n_i = 6.5 \times 10^{12}$ cm^{-3} at 125°C. With these values, we find $L_p = 75$ μm and $J_t = 6.45$ μA cm^{-2} for generation in the charge-neutral region. The generation width in the space-charge region x_i is 2.24 μm, and J_g is 13.4 μA cm^{-2}. Hence, at 125°C, the ratio of the two currents $J_g/J_t = 2.1$. The two currents have become almost equal because one increases as n_i^2 while the other varies directly with n_i.

Summary. Our analysis of current in *pn* junctions has touched on several points and has become rather lengthy. It is worthwhile, therefore, to summarize the results at this point, and to make several comments about them.

We have considered solutions of the continuity equation for two cases of uniformly doped step-junction diodes. Both of these led to a dependence of current

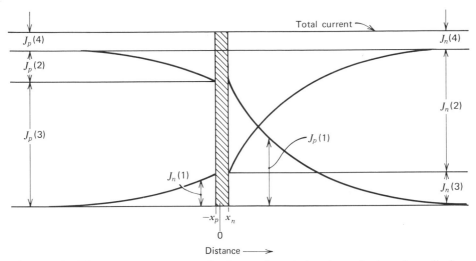

Figure 5.10 The current components in the quasi-neutral regions of a long-base diode under moderate forward bias: $J(1)$ injected minority-carrier current, $J(2)$ majority-carrier current recombining with $J(1)$, $J(3)$ majority-carrier current injected across the junction, $J(4)$ space-charge-region recombination current.

on voltage (Equations 5.3.15 and 5.3.19) which is the same as was found earlier for Schottky-barrier diodes, and which is generally known as the ideal-diode law [i.e., $J = J_0 \, (\exp(qV_a/kT) - 1)$]. In practical diodes, currents that result from events in the space-charge region must be added to this ideal law to obtain the overall current-voltage relationship. These added currents dominate in silicon diodes under reverse bias and under low forward bias. A sketch of the currents flowing in a long-base diode at moderate forward bias is shown in Figure 5.10. The total current close to the junction in each of the quasi-neutral regions consists of (1) injected minority carriers diffusing away from the junction, (2) majority carriers drifting toward the junction to recombine with minority carriers injected into the quasi-neutral regions, (3) majority carriers drifting toward the junction to be injected into the opposite quasi-neutral region, (4) majority carriers drifting toward the junction space-charge region where they recombine with injected carriers entering the junction region from the opposite side. Far from the junction in either region, the entire current is carried by drifting majority carriers.

A next logical step is to analyze *pn* junctions in which the dopant densities are nonuniform in the quasi-neutral regions. We can expect nonuniform dopant distributions in many diffused junctions. Nonuniform doping is, however, most gainfully discussed by using approximation techniques—in particular, by methods that we shall develop when we discuss transistors. We therefore defer direct considerations of this situation to Chapter 6. We shall see that the ideal-diode law is again obtained in practical cases of nonuniform doping, but the saturation current J_0 has a different formulation than our earlier results. As with the ideal diode, generation-recombination in the space-charge region leads to added current components. The observed steady-state dependence of current on voltage is made up of the superposition of the ideal-diode current and these components.

5.4 Charge Storage and Diode Transients

In the previous section we saw that forward bias across a _pn_ junction causes the injection of electrons from the _n_-type region into the _p_-type region and of holes in the opposite direction. After injection across the junction, these minority carriers move into the quasi-neutral regions. The resulting distribution of minority carriers leads both to current flow and to charge storage in the junction diode. In this section we consider the stored charge, its relation to the current, and its effect on the transient response of the _pn_ junction to a change in diode bias conditions.

On a fundamental basis, the time-dependent behavior of minority carriers is included in the continuity equations (Equations 5.1.4). Since these partial differential equations are functions of time and position, they can be solved for various forcing functions and initial conditions to yield the transient behavior of minority carriers. This is generally not done in practice for several reasons. First, explicit solutions can be obtained only in special cases and for idealized forcing functions that can only approximate the real circuit conditions. It therefore makes little sense to carry out the exact mathematics of a solution when that solution will only apply approximately to the real problem. A second reason for not carrying out the solution of the partial differential equation is that the diode transient is not governed only by the minority charges stored in the quasi-neutral regions, but is also a function of charge storage in the depletion region. Thus, one must assess the changes in both charge stores simultaneously. Overall, the best way to look at the transient problem is to consider the physical behavior of these charges directly.

Minority-Carrier Storage

The total injected minority-carrier charge per unit area stored in the quasi-neutral _n_-type region can be found by integrating the excess hole distribution across the quasi-neutral region.

$$Q_p = q \int_{x_n}^{W_B} p_n'(x) \, dx \tag{5.4.1}$$

Since the region is quasi-neutral, the _n_-region contains Q_p/q extra electrons per unit area above the thermal-equilibrium value to balance the $+Q_p$ charge on the holes.

We first consider the long-base diode, in which all of the injected minority carriers recombine before they reach the end of the _n_-type region. For simplicity, we also assume that each region of the semiconductor is uniformly doped. From

Equation 5.3.12 we know that the minority-carrier distribution decays exponentially as the holes diffuse further into the n-type region. Inserting Equation 5.3.12 into Equation 5.4.1 and integrating from the edge of the space-charge region across the quasi-neutral region, we find the charge Q_p resulting from holes stored in the n-type region to be

$$Q_p = qL_p p_{no}(e^{qV_a/kT} - 1) \tag{5.4.2}$$

The stored charge, like the current, depends exponentially on applied bias. This behavior could be expected since the exponential term arises from the magnitude of the excess minority-carrier density at the edge of the space-charge region for both the current and the stored charge. We can gain further insight into the physical behavior by expressing the stored charge in terms of the corresponding hole current. By using the expression for the hole current (Equation 5.3.13) at $x = x_n$ in Equation 5.4.2, we find the simple equation

$$Q_p = \frac{L_p^2}{D_p} J_p(x_n) = \tau_p J_p(x_n) \tag{5.4.3}$$

Thus, the stored charge is the product of the current and the lifetime for this case of the ideal long-base diode. This relationship is reasonable, since the injected carriers diffuse farther into the n-type region before recombining if their lifetime is greater; more holes are then stored.

The situation in the ideal short-base diode is slightly different. To treat it we use the expression for the hole distribution (Equation 5.3.17) in Equation 5.4.1 to obtain the stored hole charge as

$$Q_p = \frac{q(W_B - x_n)}{2} p_{no}(e^{qV_a/kT} - 1) \tag{5.4.4}$$

where W_B is the length of the n-type region (Figure 5.5). Again, we can express the stored hole charge in terms of the hole current (Equation 5.3.18)

$$Q_p = \frac{(W_B - x_n)^2}{2D_p} J_p \tag{5.4.5}$$

Although the group of constants in front of J_p in Equation 5.4.5 has the dimensions of time, it is not the lifetime, as was found in Equation 5.4.3 for the long-base diode. To see the physical significance of this group of constants, we rewrite Equation 5.4.5 as

$$\frac{Q_p}{J_p} = \frac{(W_B - x_n)^2}{2D_p} \tag{5.4.6}$$

The amount of stored charge divided by the rate at which the charge enters or leaves the n-type region is just equal to the time an average carrier spends in this region. Therefore, the right-hand side of Equation 5.4.6 is equal to the average *transit time* τ_{tr} of a hole moving through the n-type region of the short-base diode.

Transient-Behavior of Minority-Carrier Storage.[†] We can extend the physical picture of minority-carrier storage to discuss qualitatively the nature of the transient buildup and decay of Q_p. Consider the behavior of holes in the n-region of an initially uncharged, ideal, long-base diode to which a positive constant-current source is suddenly applied. Before steady state can be reached, holes must be transported into the quasi-neutral n-region to establish a particular hole distribution. Theoretically, at $t = 0^+$ a near-infinite gradient of holes is possible because the n-region is nearly free of holes at that time. The actual gradient is set, however, by the size of the current source. The stored hole charge Q_p builds with time as holes are supplied, and the voltage across the junction rises to reflect the boundary value of the hole density. A sketch of the transient buildup for this situation is shown in Figure 5.11. The time required to reach the steady-state hole distribution is given by the ratio of the steady-state stored hole charge to the size of the current source (if we neglect recombination of the holes and charges stored on the space–charge–region edges).

The turn-off time of the diode is limited by the speed with which stored holes can be removed from the quasi-neutral region. When a reverse bias is suddenly applied across the forward-biased junction, the current can reverse direction quickly since the gradient near the edge of the space-charge region can change with only a small change in the number of stored holes (curve 1 of Figure 5.12a). As long as sufficient minority carriers are present at the edge of the space-charge region, the diode will be able to conduct a large amount of current in the reverse direction. The junction will, in fact, remain forward biased until the injected minority carriers near the edge of the depletion region are removed. This means that a plot of current versus time (Figure 5.12b) will initially be nearly constant (and determined by the

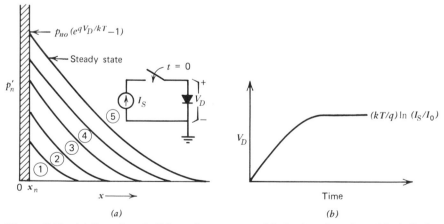

(a) (b)

Figure 5.11 (a) Transient buildup of excess stored holes in a long-base ideal diode. The case shown is one in which a constant current drive is supplied starting at time zero with the diode initially unbiased. Note the constant gradient at $x = x_n$ as time increases from (1) through (5), which indicates a constant injected hole current. (Circuit shown in inset). (b) Diode voltage V_D versus time.

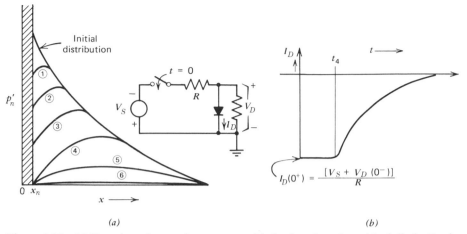

$$I_D(0^+) = \frac{[V_S + V_D(0^-)]}{R}$$

(a)   (b)

Figure 5.12 (a) Transient decay of excess stored holes in a long-base ideal diode. In the case shown, a diode with an initial forward bias is subjected at time zero to an abrupt reverse bias supplied through a series resistor. (Circuit shown in inset). (b) Diode current I_D versus time.

external circuit), and will remain in this condition until t_4, at which time $p_n'(x_n)$ goes to zero. Thereafter, (curves 5 and 6 in Figure 5.12a) the rate at which holes can be delivered across the space-charge region becomes more and more limited. The current then decays on a time scale that is determined largely by carrier lifetime in the quasi-neutral region.

Since the switching time of a diode depends on the amount of stored charge that must be injected or removed, we may decrease the switching time by reducing the stored charge. From Equation 5.4.3 we see that this reduction in stored charge may be achieved either by restricting the forward current or by decreasing the lifetime of minority carriers. Consequently, optimal design for a given diode may not maximize the minority-carrier lifetime by forming the purest material possible. In fact, in many applications where devices must switch rapidly, recombination centers are purposely introduced into the semiconductor to reduce the lifetime so that little charge is stored. The critical concern is, of course, to introduce these "lifetime-killing" impurities in a controlled manner so that the amount of charge storage is reduced without increasing the reverse leakage current ($-J_0$ in Equation 5.3.15) to unacceptable levels. For example, gold may be diffused into a semiconductor wafer at a precisely controlled temperature.

The forward current through a *pn* junction can be restricted by the Schottky-diode clamping technique discussed in Section 3.6. Since a Schottky diode has a lower turn-on voltage than a *pn* junction, the voltage across a *pn* junction and the current through it may be limited by placing a Schottky diode in parallel with the *pn* junction. Charge storage in the *pn* junction is then decreased, and the turn-off time is reduced. Additional current flows through the Schottky diode, of course, but to first order no minority charge is stored there.

Thus far, we have discussed hole storage in the quasi-neutral *n*-type side of the *pn* junction. There is, of couse, a similar storage of electrons in the *p*-region. Expressions analogous to Equations 5.4.2 and 5.4.4 can be derived to express this stored electron charge. The total minority-carrier charge stored in the diode is the sum of the two components.

We shall find these concepts of stored minority charge, transit time, and the interrelationship between them to be useful in discussing transients in *pn* junctions. To treat the transient problem, we can neglect changes in the majority-carrier concentrations since majority carriers respond to changes in fields within a dielectric relaxation time $\tau_r = \epsilon_s/\sigma$, which is of the order of 10^{-12} s (one picosecond) for 1 Ω-cm silicon. This time is much shorter than the lifetime (Equation 5.4.3) or the transit time (Equation 5.4.6) of minority carriers. For integrated-circuit devices these times are typically between 10^{-10} and 10^{-5} s.

Total Junction Storage. In Section 4.3 we examined one mechanism of charge storage in a *pn* junction. We saw that majority carriers near the edges of the depletion region move as the depletion region expands or contracts in response to a changing reverse bias. We saw that this charge storage in the depletion region could be modeled by a small-signal capacitance. Capacitance owing to charge storage in the depletion region is usually symbolized by C_j. An example is the abrupt-junction depletion capacitance derived in Equation 4.3.8.

Similarly, the variation of stored minority-carrier charge in the quasi-neutral regions under forward bias can be modeled by another small-signal capacitance. This capacitance is usually called the diffusion capacitance (denoted C_d), since in the ideal-diode case the minority carriers move across the quasi-neutral region by diffusion. We can find the contribution to C_d from the stored holes in the *n*-region by applying the definition $C_d = dQ_p/dV_a$ to Equation 5.4.2 for the long-base ideal diode or to Equation 5.4.4 for the short-base diode. (Because Q_p represents charge per unit area, C_d denotes capacitance per unit area.) Since both diodes have the same voltage dependence, we find that in general

$$C_d = \frac{qQ_{po}e^{qV_a/kT}}{kT} \tag{5.4.7}$$

where

$$Q_{po} = J_{p0}\tau_p = qp_{no}L_p \tag{5.4.8}$$

for the long-base diode (where $J_{p0} = qD_pp_{no}/L_p$), and

$$Q_{po} = J_{p0}\tau_{tr} = qp_{no}(W_B - x_n)/2 \tag{5.4.9}$$

for the short-base diode (where $J_{p0} = qD_pp_{no}/(W_B - x_n)$).

Again, it is straightforward to add components to C_d, which represent electron storage in the *p*-region in cases where that storage is significant. As should be expected, Equation 5.4.7 shows that C_d is inconsequential under reverse bias because the minority-carrier storage is negligible. Under forward bias, both Q_p and C_d rise exponentially with voltage.

From the foregoing we can see that the relative significance of charge storage in the space-charge region (as represented by C_j) and charge storage in the quasi-neutral regions depends strongly on the actual junction voltage. Under reverse bias, storage in the quasi-neutral regions is negligible and the storage represented by the junction-capacitance term will dominate. Under forward bias, although C_j increases (because x_d decreases), the exponential factor in the formula for C_d generally makes diffusion capacitance and its associated charge storage dominant. For accurate solution of practical problems, however, it is frequently necessary to account for both types of charge storage in the forward-bias region.

EXAMPLE   Junction and Free-Carrier Storage

An abrupt, pn-junction, long-base diode in which $L_p = 20$ μm is doped in the p-region with $N_a = 10^{19}$ cm^{-3} and in the n-region with $N_d = 10^{16}$ cm^{-3}. Plot the charge Q_p as a function of applied voltage V_a for $-3 < V_a < 0.6$ V. On the same axes plot Q_V, the charge added to the n-region to change the depletion-region charge storage when the junction is biased.

Solution

For this p^+n diode we neglect electron injection into the heavily doped p-region and use Equation 5.4.8 to calculate $Q_{po} = 6.7 \times 10^{-18}$ C cm^{-2}. Q_p is obtained by multiplying by the diode factor

$$Q_p = Q_{po}\left(\exp\frac{qV_a}{kT} - 1\right)$$

Q_V is just the negative of the difference in the charge stored in the depletion region at $V = V_a$ and the charge stored there at $V = 0$. Hence, using Equation 4.3.1,

$$Q_V = -qN_d(x_d - x_{do}) = -\sqrt{2\epsilon_s qN_d}\left[(\phi_i - V_a)^{1/2} - \phi_i^{1/2}\right]$$

where we find $\phi_i = 0.872$ V from Equation 4.2.10.

We use these two expressions to calculate the values of Q_p and Q_V plotted in the accompanying figure. For all values of negative bias, we find Q_p is negligible compared to Q_V. Both charge storages are negative. At a forward bias between 0.5 and 0.6 V, the two charge densities become equal. At 0.6 V, Q_p is 84.3 nC cm^{-2}, about 3.4 times as large as Q_V, and at higher voltages Q_p becomes far larger than Q_V. This example emphasizes that under reverse bias, charge storage at the depleted junction (Q_V) is most important, while under forward bias, minority-carrier storage (Q_p) becomes dominant.

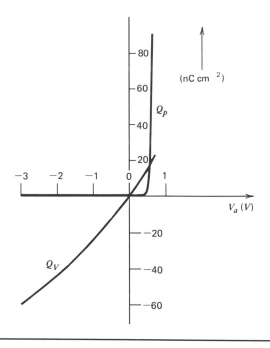

Circuit Modeling

The nature of charge storage at the *pn* junction strongly complicates the use of equivalent circuits for hand calculations involving transient problems. For example, the most frequent use of diodes in integrated circuits is as switches, which are successively forward and reverse-biased during circuit operation. In this application, the dominant charge storage shifts during the transient itself between the diffusion and depletion-layer charge. To make hand calculations in such situations, the most successful technique is, first, to determine the total charge storage at the beginning and end of the transient. Then, the time required for switching is calculated by dividing the increment in stored charge ΔQ by the driving current I that causes the diode to switch (Problem 5.18).

This technique does not, of course, give an accurate picture of the current and voltage of the diode as functions of time. For many applications, such detail is not of interest, but if it has importance, the diode can be *piecewise-linear* approximated. This phrase means that the nonlinear charge-storage effects can be approximated to first order by linear elements over a small voltage range. If the voltage increment is made small enough, this approximation will be accurate, and an arbitrary precision can be maintained by joining together a sufficient number of piecewise-linear approximations to represent the entire voltage variation in a given transient problem.

Although we did not label them as such, the small-signal capacitances C_j and C_d in Equations 4.3.8 and 5.4.7 are piecewise-linear approximations because they represent an overall nonlinear charge-storage effect in terms of linear circuit elements (capacitors). A complete piecewise-linear model for a *pn* junction must include a conductance as well as a capacitance to represent the real current through the device. The conductance (per unit area) in the ideal diode is (from Equation 5.3.15)

$$g_d = \frac{dJ}{dV} = \frac{q}{kT} J_0(e^{qV_a/kT}) = \frac{q}{kT}(J + J_0) \qquad (5.4.10)$$

The total piecewise-linear circuit for the diode consists of g_d in parallel with C_j and C_d (Figure 5.13). Since we have discussed only basic junction processes, the circuit in Figure 5.13 does not include elements that are inherent to the planar technology used for *IC* fabrication. We shall refer to these in Section 5.5.

For specialized applications such as the low-voltage sinusoidal excitation of a diode that is biased quiescently (at dc) to some operating point, analysis with the circuit shown in Figure 5.13 provides adequate accuracy. It is, therefore, called the *diode small-signal equivalent circuit*. The diode small-signal equivalent circuit is especially useful for analysis of a wide range of communication and linear amplification circuits.

Hand calculations using the small-signal equivalent circuit in piecewise-linear analysis quickly become cumbersome. The complexity is such that the problem is gratefully and rapidly given over to a computer in all cases except exercises for university students. To carry out the analysis on a computer, an initial state is specified in which starting values of the currents, voltages, and capacitances are found.

As an example to clarify these comments, we consider the circuit in Figure 5.14. The initial state might correspond to an initially unbiased circuit ($V_S = 0$ in Figure 5.14). At time $t = 0$, the source voltage V_S in Figure 5.14 is changed. Since the voltage across a capacitor cannot change instantaneously, the size of the capacitor and the value of the current source that represents the diode can be calculated from the initial state (at $t = 0^-$). The elements in the circuit are assumed to hold these values for a time increment Δt. At the end of Δt, the voltage across the diode

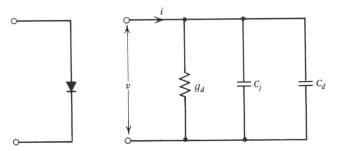

Figure 5.13 Diode small-signal equivalent circuit—a piecewise–linear representation for the nonlinear current-voltage relationship of a junction diode. Element values (per unit area) for a long-base ideal diode are given by Eqs. 4.3.8 for C_j, 5.4.7 for C_d, and 5.4.10 for g_d.

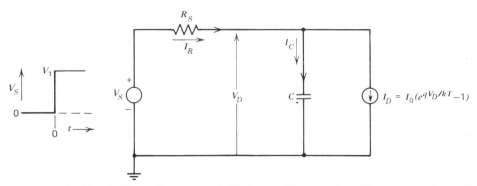

Figure 5.14 Circuit illustrating a typical diode switching problem. The source voltage V_S changes from zero to V_1 at $t = 0$. The analysis of this problem is discussed in the text.

will have changed to a new value V_D. This new voltage V_D is then used to compute new values for the components of the equivalent circuit, and the procedure is repeated in an iterative manner until the diode reaches its final state.

If the circuit of Figure 5.14 were analyzed by a computer program in a piece-wise-linear manner, the sequence of operations might be the following:

Computer Operation	*Parameter Values*
Initialize I_D, C, and V_D, for steady state with $V_S = 0$.	$V_D = 0$, $I_D = 0$, $C = C_0$
Set time $t = 0$.	$t = 0$
Set V_S to new value.	$V_S = V_1$
→ Increment time.	$t = t + \Delta t$
Compute currents I_R and I_C.	$I_R = \dfrac{V_S - V_D}{R_S}$
	$I_C = I_R - I_D$
Compute change in voltage V_D during Δt.	$\Delta V_D = \dfrac{I_C \Delta t}{C} = \dfrac{(I_R - I_D)}{C} \Delta t$
Calculate new voltage.	$V_D = V_D + \Delta V_D = V_D + \dfrac{I_R - I_D}{C} \Delta t$
Calculate new diode current.	$I_D = I_0(e^{qV_D/kT} - 1)$
Test whether I_D has reached its steady-state value.	
No. Continue.	
Yes. Exit from loop.	
Calculate new capacitance. (An abrupt junction is assumed).	$C = \dfrac{C_0}{\sqrt{1 - V_D/\phi_i}} + \dfrac{q}{kT}\tau I_D$
Go to beginning of loop for next time increment.	
→ STOP	

The analysis continues until the diode current approaches to within some predetermined increment of the steady-state value. The computer instructions may also include output commands so that any variable, such as voltage or current, can be plotted to give the user an indication of the circuit behavior.

This piecewise-linear numerical approach can be utilized in any nonlinear circuit where computer simulation is the only reasonable means for studying the behavior of the total circuit. It is one of several alternative computational approaches (or algorithms) that enable the computer solution of nonlinear problems. Models for diodes and other nonlinear circuit elements based either on piecewise-linear approximations or on other techniques are typically included as subroutines in more elaborate computer-analysis programs for integrated circuits. The use of these computer-simulation programs is an invaluable aid to the design and analysis of integrated circuits.

5.5 Devices: Integrated-Circuit Diodes

In order to form a *pn*-junction diode using integrated-circuit technology, it is only necessary to diffuse a *p*-type region into an *n*-type wafer and make contact to the front and back of the wafer (Figure 5.15). If the diode is part of an integrated circuit, however, many *pn* junctions are typically formed on the same wafer, and simple diffusion as described above results in all of the *n*-regions being electrically connected. This, of course, is generally unacceptable. Interconnections between diodes, or between other devices in an integrated circuit, can be avoided by surrounding them with *pn* junctions that are kept under reverse bias at all times. This technique of *junction isolation* for integrated-circuit devices is not the only isolation method, but it is frequently used. To illustrate some aspects of diode structures

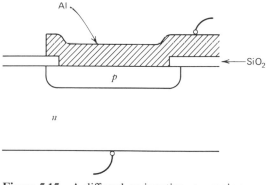

Figure 5.15   A diffused *pn* junction, a *p*-region is diffused into an *n*-type wafer; the surface is passivated with silicon dioxide, and ohmic contacts are made to the front and back of the wafer.

in integrated circuits, we consider that an array of diodes is desired in an integrated circuit.

To achieve a junction-isolated array of diodes, the most usual procedure would be to start with a *p*-type wafer on which a thin *n*-type epitaxial silicon layer is grown (Figure 5.16*a*). A typical "epi layer" may be from 1 to 5 *μ*m thick for a digital logic switching circuit and as thick as 10 to 20 *μ*m for a linear circuit. The *pn* junction at the epi-substrate interface provides electrical isolation in the vertical direction as long as it is reverse biased. The substrate is usually relatively lightly doped *p*-type material ($N_a \approx 2 \times 10^{15}$ cm^{-3}) in order to reduce undesired or "parasitic" junction capacitance between the active region and the substrate and to maintain a high breakdown voltage.

The following sequence of steps includes oxide growth, photomasking, and etching to define the web of isolating *pn* junctions. The acceptor dopant diffused into the *n*-epitaxial region to form these junctions must penetrate completely through the epitaxial layer in order to surround the *n*-regions with *p*-type material (Figure 5.16*b*). The resulting isolated *n*-regions are commonly and picturesquely called "tubs" in the integrated-circuit industry. The *p*-isolation diffusion, because of its depth, typically takes the longest of any diffusion time used in the *IC* process. Because of the lateral diffusion under the oxide mask that takes place only a little slower than the vertical diffusion toward the substrate, the designer must allow for a considerable surface area to be consumed by the isolation junctions. This is

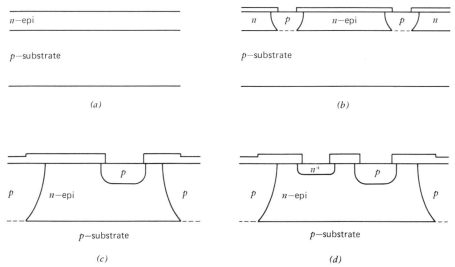

Figure 5.16 Planar-process steps in the production of junction diodes for integrated circuits: (*a*) substrate junction formed by *n*-type epitaxial layer on *p*-type substrate, (*b*) *p*-type isolation diffusion through *n*-type epitaxial layer, (*c*) shallower *p*-type diode diffusion into epitaxial layer, (*d*) highly doped, *n*-type diffusion to provide ohmic contact to *n*-type region. (Vertical scale expanded in parts *c* and *d*.)

a major disadvantage of junction isolation. The greatest advantage of this isolation technique is its simplicity and compatibility with other *IC* processing steps.

After isolation, the next step in producing a junction diode is to introduce a second diffused *p*-region—this time keeping the diffusion time relatively short so that a *pn* junction is formed within the *n*-type tub (Figure 5.16*c*). It is important at this point to keep the two space-charge regions, one from the diode junction and one from the substrate isolation junction, from coming into contact. Proper design must insure that they do not touch, regardless of the voltage conditions that may be imposed on the diode and substrate junctions. Since higher reverse bias widens space-charge regions, the design focus is on the highest reverse-bias conditions. If the two space-charge regions do meet, the condition is described by the term *punch-through* since the quasi-neutral *n*-region separating the two *p*-regions has been "punched through." Under this condition the *p*-region for the diode and the *p*-type substrate are not isolated and large currents can flow between them (Problem 5.20). To avoid punch-through to the sidewalls, the edge of the diode diffusion must also be located sufficiently far from the edge of the isolation diffusion.

The next step is to make electrical connection to both sides of the diode from the top of the wafer. First, a highly doped n^+ contact diffusion is added to the *n*-type region (Figure 5.16*d*) in order to achieve good ohmic contact. As we saw in Chapter 3, aluminum does not make ohmic contact to lightly doped *n*-type silicon. After contact areas are opened, the junction-isolated diode is completed by covering the wafer with aluminum, defining the aluminum pattern by selective etching, and alloying it. In Figure 5.17, the diode structure is drawn with a single scale for both horizontal and vertical dimensions. In Figure 5.16, as in most sketches of *IC* cross sections, the vertical dimension has been exaggerated for clarity.

An integrated-circuit diode made by the process we have outlined above suffers from a serious disadvantage, however, and in practice some modification of the structure is almost always necessary. To obtain a relatively large current-carrying area it is desirable to have the current flow vertically through the diode. However, this requires that majority-carrier electrons must flow to the junction region through a path that includes the long, thin *n*-region bordered by the two *pn* junctions. This is a path very like the pinch resistor described at the end of Chapter 2. For a typical epitaxial-layer resistivity of 0.5 Ω-cm and typical integrated-circuit dimensions, the series resistance introduced by this path would be several kilohms.

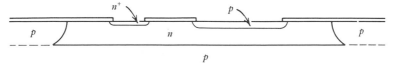

Figure 5.17 The diode of Figure 5.16 sketched with the same scale for both *x* and *y* axes to give a proper perspective. The relative dimensions shown are typical for switching diodes.

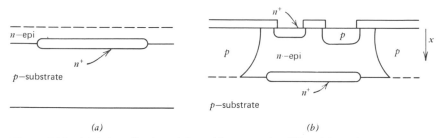

Figure 5.18   (*a*) A heavily doped *buried layer* can be diffused into the *p*-type substrate before the epitaxial layer is grown; this reduces resistance in series with the diode. (*b*) An integrated-circuit diode incorporating a buried layer.

Such a high series resistance would produce sufficient ohmic drop, even at relatively low forward currents, to cause the *pn* junction bias to vary along its length. Because current is exponentially related to the junction voltage under forward bias, there would be *current crowding* toward the regions of the junction that are nearest the n^+ contact. The effective area of the diode would be reduced from its geometric area, leading to many undesirable effects.

To eliminate this problem, a heavily doped *n*-type region is formed on the substrate before growing the epitaxial layer. This diffused region is usually included beneath each junction region (Figure 5.18*a*). After the epitaxial layer has been grown and the diffusions discussed above have been carried out, the structure appears as shown in Figure 5.18*b*. By using this *buried layer* of *n*-type material, the series resistance is reduced to a few ohms, which is generally quite acceptable. An added advantage of the buried layer is that it reduces the possibility of punch-through between the *p*-region at the surface and the substrate. This is true because the added donors in the buried layer inhibit the extension of the space-charge region into that zone. Although punch-through is inhibited, the breakdown voltage of the junction diode may be reduced by the presence of the buried layer because it acts to confine the space-charge region and thus to increase the maximum field.

The incorporation of a buried layer does add some fabrication complications to integrated-circuit production. First, since the diffusion of the buried-layer impurities is carried out before the epitaxial-growth step, these impurities go through the temperature cycling needed for the growth and also for all the subsequent processing steps. This means that a slow-diffusing, relatively nonvolatile impurity must be used. The slow diffusion is required to maintain the desired dimensions for the devices; nonvolatility is necessary to avoid *autodoping* the epitaxial layer during its growth. In practice, relatively good results are obtained by using arsenic or antimony for the buried-layer dopant. The *p*-type diffusions are then carried out with boron, which diffuses much faster. Phosphorus is used for subsequent *n*-type diffusions that must consist of any appreciable fraction of the epitaxial-layer width. The lack of a *p*-type dopant with the desirable slow-diffusing properties of arsenic or antimony has made it difficult to fabricate practical *p*-type buried layers.

The n-type buried layer, by narrowing the space-charge region between the n-type region and the p-type substrate, does add to the parasitic capacitance between the epitaxial layer and the substrate. This is one reason that the buried-layer region is usually introduced only beneath the active regions. One unavoidable disadvantage of the incorporation of a buried layer is that the strain associated with its high impurity density generally reduces the lifetime and mobility in the epitaxial region above it. Since this is precisely the critical region with respect to device performance, the drawback can be severe, but it is less with arsenic than with antimony. Optimum design requires choosing a buried-layer doping that balances these disadvantages with the advantages that motivated the inclusion of a buried layer.

As a final topic concerned with the practical aspects of junction diodes for integrated-circuits, we give some perspective to the widespread use of tables and nomographs for design and analysis. Properties of pn junctions formed by (1) a Gaussian diffusion profile into a constant background doping and (2) a complementary-error function profile into a constant background doping have been published by Lawrence and Warner.[8] These authors obtained their results by finding exact solutions of the junction equations using computer routines. Their calculations permit the determination of junction capacitance, total depletion-layer thickness, and the extent of the depletion layer on either side of the junction. Reference 8 presents the data in especially useful fashion. To make use of the Lawrence and Warner results, it is necessary to know the background dopant density of the region of constant doping as well as the density of the dopant at the surface from which diffusion is taking place. These data specify a family of curves that are plots of the output data as functions of total junction voltage. Because it is a closely related calculation, the Lawrence-Warner analysis also provides curves of the maximum junction field versus total voltage across the junctions. Reference to these results can provide a major, time-saving aid to the designer of an integrated circuit.

Tables of other parameters for typical IC processing conditions are also consulted frequently in the design of circuits. These tables relate to specific processes that are employed by a manufacturer to produce integrated circuits. An example is included at the end of Chapter 6. In addition to making use of published tables and graphs, designers more and more frequently carry out their own computer analyses. A process simulation program such as $SUPREM$ (discussed in Chapter 2) is used to find the dopant profiles resulting from a given process. These profiles are then used with numerical solutions of many of the equations that we have presented in this chapter to obtain the device characteristics required for a circuit design.

Finally we note that although the pn-junction isolation technique that we have described in this section is frequently employed in IC design, newer techniques are often used for small-geometry circuits. For example, the local oxidation ($LOCOS$) process described in Section 2.6 produces regions of insulating silicon dioxide between devices. This technology greatly reduces the area needed for the isolation regions and decreases some parasitic capacitances. A cross section of an

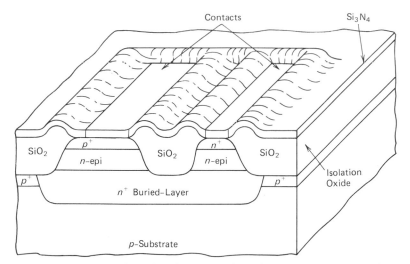

Figure 5.19   Three-dimensional cross section of an _IC_ diode produced using _LOCOS_ oxide isolation.

oxide-isolated diode is shown in Figure 5.19. The area reduction is possible because diode and contact diffusions can extend to the oxide-isolation regions. The advantages of a _LOCOS_ structure are gained at a considerable increase in processing complexity, however, and the epitaxial-layer thickness that can be isolated by the _LOCOS_ technique is limited to about 1 μm.

Summary

The continuity equation for free carriers, a partial differential equation that accounts for the mechanisms causing an increase or a decrease in carrier densities, is a powerful tool for semiconductor device analysis. By solving the continuity equation for minority carriers in the quasi-neutral regions near a _pn_ junction, we can obtain the carrier densities. Boundary values for minority-carrier densities at the edges of the quasi-neutral regions may be written as functions of applied junction bias. Using these boundary values and the defining relationships for current as a function of free-carrier density, it is possible to obtain current-voltage characteristics in the steady state for several simple cases with important practical applications. These solutions lead to the _ideal-diode equation_ $J = J_0[\exp(qV_a/kT) - 1]$.

Solutions to the time-dependent continuity equation can be carried out in simple cases to give meaning to the concept of the lifetime of excess carriers. In most cases of interest, the recombination rate, which determines the carrier _lifetime_, is a function of the properties of recombination centers. These recombination centers consist of localized electronic states having energies within the forbidden gap, typically close to the intrinsic Fermi level. The theory for recombination through localized centers is called *Shockley–Hall–Read (SHR)* recombination theory. Generation and recombination through recombination centers located inside the space-charge region are responsible for significant currents, especially under reverse and low forward bias, in silicon junction diodes. *Auger recombination* occurs when there are very high densities of carriers present. In Auger recombination the energy and momentum released by recombination are carried away by a third free particle. Because there are extra localized states at a silicon surface and also because a space-charge region is usually present in addition, surface recombination can be an added important effect. Surface recombination is usually described by a parameter *s* known as the *surface-recombination velocity*.

The transient behavior of junction diodes is governed by the charge and discharge of stored minority carriers, as well as by the variation of stored depletion-layer charge. These components of charge are nonlinear functions of voltage, and linear circuit representation is only approximately valid for small bias variations. A *small-signal equivalent circuit* can be developed with capacitors representing the charge storage. This circuit can be used in piecewise fashion to make accurate calculations for large-signal transients. Any practical problems are usually carried out by computer analysis.

The *pn*-junction diode is a basic building block for integrated circuits. There is widespread use of *pn* junctions as isolating elements to avoid unintentional interactions between the devices making up an integrated circuit. Series resistance between surface contacts and forward-biased *pn* junctions can be reduced by adding buried layers of donor dopant atoms below diffused junctions. Many of the properties of *pn* junctions for integrated circuits can be obtained by reference to curves of published results obtained from numerical solutions of the junction equations. Alternatives to junction isolation, such as oxide isolation, have advantages in area and parasitic element reduction, but they require a more complicated technology.

References

1. K. Thiessen and G. Zech, *Phys. Stat. Sol.* (a) **10**, K133 (1972).
2. W. Fahrner and A. Goetzberger, *J. Appl. Phys.* **44**, 725 (1973).
3. W. Shockley and W. T. Read, *Phys. Rev.* **87**, 835 (1952).
4. R. N. Hall, *Phys. Rev.* **87**, 387 (1952).
5. E. M. Conwell, *Proc. IRE* **46**, 1281 (1958).
6. J. L. Moll, *Physics of Semiconductors*, McGraw-Hill, New York, 1964, p. 117.
7. P. U. Calzolari and S. Graffi, *Solid-State Electron.* **15**, 1003 (1972).
8. H. Lawrence and R. M. Warner, Jr., *Bell Syst. Tec. J.* **34**, 105 (1955).
9. P. E. Gray, D. DeWitt, A. R. Boothroyd, and J. F. Gibbons, *Physical Electronics and Circuit Models of Transistors*, Wiley, New York, 1964, p. 75.

Textbook

G. W. Neudeck, *The PN-Junction Diode*, Volume II Modular Series on Solid-State Devices, Addison-Wesley Inc, Reading, MA, 1983.

Problems

5.1* A 0.6 Ω-cm, n-type silicon sample contains 10^{15} cm^{-3} generation-recombination centers located at the intrinsic Fermi level with $\sigma_n = \sigma_p = 10^{-15}$ cm^2. Assume $v_{th} = 10^7$ cm s^{-1}
   (a) Calculate the generation rate if the region is depleted of mobile carriers.
   (b) Calculate the generation rate in a region where only the minority-carrier concentration has been reduced appreciably below its equilibrium value.

5.2* Light is incident on a silicon bar doped with 10^{16} cm^{-3} donors, creating 10^{21} cm^{-3} s^{-1} electron-hole pairs uniformly throughout the sample. There are 10^{15} cm^{-3} bulk recombination centers at E_i with electron and hole capture cross sections of 10^{-14} cm^2.
   (a) Calculate the steady-state hole and electron concentrations with the light turned on.
   (b) At time $t = 0$ the light is turned off. Calculate the time response of the total hole density and find the lifetime. The thermal velocity is 10^7 cm s^{-1}, and there is no current flowing.

5.3 (This problem is intended to provide perspective on the concept of charge neutrality in semiconductors.) Assume that the distribution of free electrons in Figure P5.3a could be maintained in a block of silicon without any neutralizing charges. Consider that the electric field associated with the charge vanishes at $x = W$ and take $n_1 = 10^{18}$ cm^{-3}, $W = 1$ μm, $D_n = 7$ cm^2 s^{-1}, and $T = 300$ K.

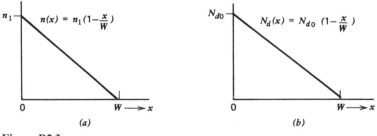

Figure P5.3

(a) Calculate the field and current at $x = 0$ that is required to sustain this distribution of free charge. (Note that both a drift and diffusion component will be present.)
(b) In light of the results of (a), is the charge configuration reasonable?
(c) If a current density of approximately 10^5 A cm^{-2} can be sustained before power dissipation in the silicon causes irreversible damage, what is the maximum field that can exist in silicon having a uniform electron density equal to n_1? (The

current densities in most operating *IC* devices are two or more orders of magnitude lower than this value.)

(d) If the compensating positive charge density shown in Figure P5.3*b* exists in the silicon, calculate the value of N_{d0} such that the field of part c will be present. Express this as a fraction of the free-electron density n_1, and note thereby how close to electrical neutrality the block of silicon will be. (Use values at $x = 0$).

5.4[†] If recombination of electrons takes place through a donorlike recombination center, the capture cross section σ_n for the site can be crudely estimated by considering that the free electron enters a region where its thermal energy $\frac{3}{2}kT$ is less than the energy associated with Coulombic attraction; this means that the carrier is drawn to the center. The radius at which this occurs describes an area that may be taken to be equal to σ_n.

(a) Obtain an expression for σ_n based upon this model and evaluate it for silicon at 300 K.

(b) Considering the dependence on temperature of σ_n from this model together with other temperature-dependent terms, how would the electron capture rate for the center depend on temperature? (Assume that the center exists in extrinsic silicon.)

5.5[†] The analysis of recombination through recombination centers in Section 5.2 assumed low level injection with no current flowing so that we could find the characteristic time in which excess carriers decayed. Following a similar analysis, find the "effective lifetime" in the opposite case of high level injection in an *n*-type semiconductor, assuming that charge neutrality holds. Let $n' = p' \gg n_0$, p_0, and compare the decay time or effective lifetime to that found in the case of low level injection for both $\sigma_n \neq \sigma_p$ and $\sigma_n = \sigma_p$.

5.6 Consider the continuity equation for minority-carrier holes that are injected across a *pn* junction into an *n*-type region that is short so that recombination of the holes takes place essentially only at the contact at $x = W_B$. Show by the direct use of Equation 5.3.10 that the holes are distributed linearly along *x* as was found in Equation 5.3.17.

5.7 An ideal *pn*-junction diode has long *p*- and *n*-regions and negligible generation or recombination in the space-charge region, so that the current-voltage characteristics are described by Equation 5.3.15. Show that the limiting reverse-bias current predicted by Equation 5.3.15 corresponds to the integrated generation rate of minority carriers on either side of the *pn* junction. *Hint. Use Equation 5.3.12 and its equivalent for electrons and consider that* $G_p - R_p = -p_n'/\tau_p$.

5.8 Holes are injected across a forward-biased *pn* junction into an *n*-region that is much longer than L_p, the hole diffusion length. Show, by using Equation 5.3.12 that L_p is the average length that a hole diffuses before it recombines in the *n*-region.

5.9* For forward–biased current in an ideal *pn*-junction diode show that the ratio of hole current to total current can be controlled by varying the relative doping on the two sides of the junction. If we call the ratio of hole current to total current γ, express γ as a function of N_a/N_d. Calculate γ for a silicon *pn* junction for which resistivity of the *n*-side is 0.001 Ω-cm and that of the *p*-side is 1 Ω-cm. Assume that $\tau_p = 0.1\,\tau_n$ and that both neutral regions are much longer than the minority-carrier diffusion lengths in them.

5.10[†] A *pn*-junction diode has the configuration shown in Figure P5.10. Assume that
1. $N_a = N_d = N_o \gg n_i$.

2. $W \ll L$, the minority-carrier diffusion length; $L_1 \gg L$.
3. All the diffusion constants $= D$, all lifetimes $= \tau$.
4. The space-charge-region width $\ll W$.
5. The external applied voltage is V_B, where $V_B \gg \phi_i$, the built-in voltage.
6. The recombination velocity is infinite at $x = +W$.

If a light beam that produces G_0 hole-electron pairs per unit area per unit time in a slab of negligible width along the x-direction is incident on the diode at the plane $x = -W/2$ on the p-type side

(a) Assuming low level injection, find and sketch the minority-carrier concentrations in the neutral regions of the diode.
(b) Calculate the current that flows in the illuminated diode in terms of G_0 and the diode properties.
(c) In terms of the constants associated with the diode, what current flows when the light beam is removed (steady-state)?[9]

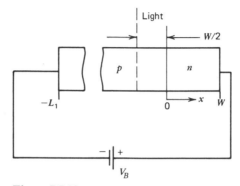

Figure P5.10

5.11* Consider an ideal, long-base, silicon abrupt pn-junction diode with uniform cross section and constant doping on either side of the junction. The diode is made from 1 Ω-cm p-type and 0.2 Ω-cm n-type materials in which the minority-carrier lifetimes are $\tau_n = 10^{-6}$ s and $\tau_p = 10^{-8}$ s, respectively. ("Ideal" implies that effects within the space-charge region are negligible and that minority carriers flow only by diffusion mechanisms in the charge neutral regions.)

(a) What is the value of the built-in voltage?
(b) Calculate the density of the minority carriers at the edge of the space-charge region when the applied voltage is 0.589 V (which is $23 \times kT/q$).
(c) Sketch the majority- and minority-carrier currents as functions of distance from the junction on both sides of the junction, under the applied bias voltage of part b.
(d) Calculate the location(s) of the plane or planes at which the minority-carrier and majority-carrier currents are equal in magnitude for the applied bias voltage of part b.

5.12 Construct a sketch showing the distribution of hole and electron currents in the neutral regions of a pn-junction diode (with forward bias) having _significant_ recombination in the space-charge layer. Assume that

(a) The injected hole current is twice the injected electron current.

(b) The net rate at which pairs recombine in the space-charge layer is equal to half of the net rate at which electrons recombine in the p-type region.

Demonstrate that the total diode current is given by adding a term that describes the space-charge-region recombination current to the sum of the diffusion currents at the edges of the space-charge region.

5.13 Consider the integrated-circuit structures shown in cross section in Problem 4.10.

(a) Ignoring series resistance but considering all junctions, sketch the two sets of I-V characteristics that you would expect to see for measurements of current when (i) voltage V_C (both polarities) is applied to contact C and contact A is grounded while contact B is left open, (ii) voltage V_B (both polarities) is applied to contact B and contact A is grounded while contact C is left open.

(b)† If it were possible to deplete the n^- layer fully under the conditions of part $a(i)$ above, what would you expect to happen to the current flowing through contact A? Explain the reasons for your conclusions.

5.14† A silicon diode of junction area 10^{-5} cm^2 is made with a uniformly doped p-type region having 5×10^{18} acceptors cm^{-3} and a step junction to an n-type region doped with 10^{15} donors cm^{-3}. The diode is found to be adequately described by the simplest theory; that is, one-dimensional flow with dominance of bulk recombination and transport of minority carriers under forward bias by pure diffusion into the neutral regions from the edge of the space-charge region. Recombination generation in the space-charge region is negligible, and the minority-carrier lifetime in each region is 100 ns.

This junction diode is to be used in a circuit that requires a rectification ratio between forward current I_F and reverse current $I_R (|I_F/I_R|)$ of 10^4 at 0.5 V and a maximum reverse saturation current of 100 nA. What is the maximum temperature at which the diode will perform adequately? Consider only the major temperature dependences (i.e., neglect changes in D, μ, τ_p, N_c, and N_v with temperature).

5.15 Prove that the recombination rate in the space-charge region of a pn junction in which $\sigma_n = \sigma_p$ is maximized when $p = n$, as given in Equation 5.3.23.

5.16† Use the values for a silicon junction diode given in the paragraph following Equation 5.3.25 together with the equation itself to make a semilogarithmic plot of J_t/J_r versus applied voltage V_a under forward bias. The diode consists of a step junction to a highly doped n-region (i.e., $N_a \ll N_d$).

5.17 Consider an abrupt diode biased so that $(\phi_i - V_a) = 5$ V. The diode has the following properties: $N_a = 10^{17}$ cm^{-3}, $\tau_n = 10^{-6}$ s, $N_d = 10^{18}$ cm^{-3}, and $\tau_p = 10^{-8}$ s.

(a) Use Equation 5.3.29 to determine the ratio J_t/J_g (take $x_i = x_d$).

(b) Discuss the expected temperature dependence of this ratio.

5.18† An ideal short-base diode has been built with an abrupt junction in which $N_d \gg N_a$ and $N_a = 10^{17}$ cm^{-3}. (Assume that the n-region is degenerately doped.) The width of the p-region between the edge of the space-charge region and the contact at which all recombination takes place is 3 μm. The area of the diode is 10^{-5} cm^2.

(a) Calculate the charge stored in the neutral p-region if 0.5 mA flows through the diode.

(b) Determine the charge stored in the narrowed space-charge layer under forward bias.

(c) How much time would it take a current source of 0.5 mA to cause the diode to switch from an "off" condition (with $V_a = 0$) to the steady state of 0.5 mA of current flow?

5.19   Consider a small-signal equivalent circuit for a Schottky diode under forward bias. Compare the circuit to that for a _pn_-junction diode as sketched in Figure 5.13. Discuss the differences and similarities in the two circuits.

5.20[†] Consider qualitatively the effects of punch-through on the structure shown in Figure 5.16_d_ as follows: (a) sketch a band diagram at thermal equilibrium along an axis that passes through the middle of the diffused _p_-region and runs through to the substrate; (b) with the substrate grounded, consider the imposition of sufficient reverse bias at the upper _pn_ junction so that the junction between the epitaxial _n_-region and the _p_-substrate becomes forward biased. When this occurs, the holes in the substrate can be injected into the upper _p_-region. (In steady-state, a _space-charge-limited current_ of holes will flow.) Sketch a band diagram for this condition.

5.21* Calculate the small-signal, incremental resistance and capacitance for an ideal, long-base silicon diode using the following parameters:

$$N_d = 10^{18} \text{ cm}^{-3}, \qquad N_a = 10^{16} \text{ cm}^{-3}, \qquad \tau_n = \tau_p = 10^{-8} \text{ s},$$
$$A = 10^{-4} \text{ cm}^2, \qquad T = 300 \text{ K}$$

(a) For 0.1, 0.5, and 0.7 V forward bias.
(b) For 0, 5, and 20 V reverse bias.
(c) What is the series resistance of the _p_-type quasi-neutral region if this region is 0.1 cm long. This resistance must be added to the elements modeling the ideal diode when considering the response of a real diode.

5.22   A _pn_ junction is initially off. A step of current is applied to the device with a polarity that turns the device on. Explain physically why it takes much less time for the width of the space-charge region to change than for the minority-carrier distributions to reach steady state in the quasi-neutral regions. (That is, compare the physical processes in the two situations.)

5.23   Consider the silicon integrated-circuit diode shown in Figure P5.23. Can injected holes flowing from the _p_-type diffusion to the n^+ buried layer best be described by the long-base or short-base approximation? What about carriers flowing sideways from the edge of the _p_-type diffusion to the n^+ contact region? The hole lifetime is 1 μsec in the _n_-type epitaxial layer, and the n^+ buried layer is an effective sink for excess minority carriers. Justify you answers. (Note that the figure is not to scale.)

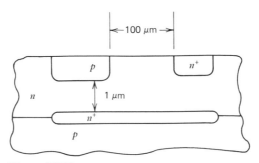

Figure P5.23

5.24 At high current densities, a significant fraction of the voltage applied across a diode can be dropped across the neutral regions of the device. Consider the current-voltage relation for a one-sided step junction with N_d donors on the lightly doped side of the junction. Find the current and applied voltage V_a at which 10% of V_a appears across the n-type neutral region for a typical integrated-circuit diode. Assume that the cross-sectional area is 10^{-5} cm^2, the length of the neutral region in the n-type silicon is 10 μm, $N_d = 5 \times 10^{15}$ cm^{-3}, and $\tau_p = 1$ ns.

6

BIPOLAR TRANSISTORS I: BASIC PROPERTIES

In Chapter 5, we saw that a *pn* junction, under bias such that the *p*-region is positive with respect to the *n*-region, conducts current because holes are injected from the *p*-region and electrons are injected from the *n*-region. These are majority carriers in each region, and their supply is plentiful near the junction. Consequently, current rises rapidly as voltage is increased. Currents are much smaller under reverse bias because they are carried only by minority carriers that are generated either in the junction space-charge region or nearby. The current passed by a reverse-biased junction will, however, increase if the supply of minority carriers in the vicinity of the junction is enhanced. This enhancement can, for example, be the result of the incident radiation of energetic particles in diode photodetectors or radiation sensors.

Another means of enhancing the minority-carrier population in the vicinity of a reverse-biased *pn* junction is simply to locate a forward-biased *pn* junction very close to it. This method is especially advantageous because it brings the minority-carrier population under electrical control, that is, under the control of the bias applied to the nearby forward-biased junction.

The modulation of the current flow in one *pn* junction by changing the bias on a nearby junction is called *bipolar transistor action*. It is one of the most significant ideas in the history of device electronics, and the investigations that led to it were honored with a Nobel Prize in Physics for the inventors of the bipolar junction transistor (or *BJT*), William Shockley, John Bardeen, and Walter Brattain.

In this chapter we formulate the basic physical description of bipolar transistor action. From the formulation we discover that transistor operation may be described more simply when various specific bias ranges are considered separately. We concentrate first on the so-called *active region* in which transistors act as amplifiers, and we consider the other bias regions in terms of transistor switching. We introduce the Ebers-Moll model, which provides an extremely useful and informative means of describing basic transistor operation in each of the bias ranges. The relationship of this model to our discussion of transistor action at the beginning of the chapter is then made clear. In a final section, we consider the design of planar diffused bipolar amplifying and switching transistors.

6.1 Transistor Action

A simple realization of a structure showing transistor action is sketched in Figure 6.1. Two *pn* junctions are shown spaced a distance W units apart in a single bar of semiconductor material. The bar has a uniform cross section of area A. The

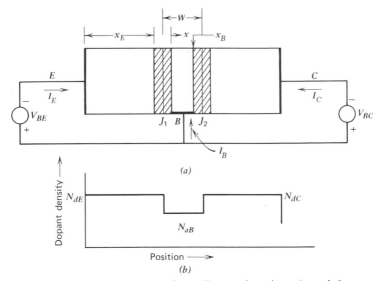

Figure 6.1   (*a*) Prototype transistor. Two *pn* junctions J_1 and J_2 are spaced W units apart. (*b*) Each region has a constant doping density. The doping changes abruptly at J_1 and J_2. The quasi-neutral portion of the middle *p*-region is bounded by the edges of space-charge regions at $x = 0$ and $x = x_B$, respectively. The contacts at E, B, and C are ohmic.

junctions are located "close" to one another so that electrons that are injected across junction 1 when V_{BE} is positive are in the vicinity of junction 2. From our discussion in Chapter 5, we recognize that the qualitative term "close" means that the junction spacing W is small enough so that the loss of electrons by recombination in the middle p-region is minimal. The middle region is commonly called the *base* of the transistor. For our initial discussion, we assume that electron recombination or generation in the base region is not significant; this assumption is reconsidered later.

An important observation about current in a bipolar transistor such as that shown in Figure 6.1 is that there is a negligible flow of holes (base majority carriers) between junctions 1 and 2 or in the reverse direction. This is true under all bias conditions because there can be only a vanishingly small flow of holes from either n-region into the p-type base. If we call the longitudinal dimension x and write the expression for hole current in the x direction (assuming no recombination) that was derived in Equation (1.2.22), we have

$$J_p = 0 = q\mu_p p \mathscr{E}_x - qD_p \frac{dp}{dx} \tag{6.1.1}$$

and

$$\mathscr{E}_x = \frac{D_p}{\mu_p} \frac{1}{p} \frac{dp}{dx}$$
$$= \frac{kT}{q} \frac{1}{p} \frac{dp}{dx} \tag{6.1.2}$$

Thus, the condition of zero hole current in the base has led to an expression (Equation 6.1.2) for the longitudinal electric field.* The field is seen to be dependent both on the magnitude of the majority-carrier (hole) density and on its gradient. If, for example, the base is homogeneously doped, then dp/dx is zero and there is no electric field. In general, however, there will be a field in the base that is proportional to the ratio between the majority-carrier (hole) density gradient and the majority-carrier (hole) density itself.

In contrast to a current of holes, however, a current of *electrons* flowing between the two junctions is possible since either junction can freely supply electrons from the n-regions to the middle p-region if that junction is forward biased. The electron current (from Equation 1.2.21) is

$$J_n = q\mu_n n \mathscr{E} + qD_n \frac{dn}{dx} \tag{6.1.3}$$

$$= kT\mu_n \frac{n}{p} \frac{dp}{dx} + qD_n \frac{dn}{dx}$$

---

*Strictly, Equations 6.1.1 and 6.1.2 must be considered as approximations because second order effects (to be described later) can account for small quantities of x-directed hole current.

and

$$J_n = \frac{qD_n}{p}\left(n\frac{dp}{dx} + p\frac{dn}{dx}\right) = \frac{qD_n}{p}\frac{d(pn)}{dx} \tag{6.1.4}$$

It is useful to write an integrated form of Equation 6.1.4 with arbitrary limits x and x' and to treat recombination as negligible so that J_n can be removed from the integral:

$$J_n \int_x^{x'} \frac{p\,dx}{q\,D_n} = \int_x^{x'} \frac{d(pn)}{dx}\,dx$$
$$= p(x')n(x') - p(x)n(x) \tag{6.1.5}$$

where we have used the fact that the right-hand side of Equation 6.1.4 is a perfect differential. Equation 6.1.5 shows the minority-carrier (electron) current in the base to depend on the difference in the hole-electron products across a region divided by the integrated majority-carrier density in the same region. If we use the junctions themselves as the two boundaries of the region, then $x = 0$ becomes the lower limit and $x' = x_B$, the upper limit for the integrals in Equation 6.1.5. From Equation 5.3.20, the pn products at the boundaries can then be related to the junction voltages.

$$p(0)n(0) = n_i^2 \exp\left(\frac{qV_{BE}}{kT}\right)$$

$$p(x_B)n(x_B) = n_i^2 \exp\left(\frac{qV_{BC}}{kT}\right) \tag{6.1.6}$$

Thus, the longitudinal electron current in the base can be expressed as a function of the junction voltages V_{BE} and V_{BC}.

$$J_n = \frac{qn_i^2\left[\exp\left(\dfrac{qV_{BC}}{kT}\right) - \exp\left(\dfrac{qV_{BE}}{kT}\right)\right]}{\displaystyle\int_0^{x_B} \frac{p\,dx}{D_n}} \tag{6.1.7}$$

Before we discuss the many implications of Equation 6.1.7, we shall modify it slightly. First, we note that D_n is frequently a weak function of position in the base and can therefore be expressed by an average value $\tilde{D}_n$ and removed from the integral in the denominator of Equation 6.1.7.* With D_n removed, the integral expresses the total majority-carrier density per unit area in the base. The charge associated with this density is called Q_B

$$Q_B = q \int_0^{x_B} p\,dx \tag{6.1.8}$$

*Strictly, $\tilde{D}_n$ is not a direct spatial average, but rather

$$\tilde{D}_n \equiv \int_0^{x_B} p\,dx \Big/ \int_0^{x_B} (p/D_n)\,dx$$

With these changes, we can express the density of electron current flowing from the first junction to the second junction.

$$J_n = J_S \left[\exp\left(\frac{qV_{BC}}{kT}\right) - \exp\left(\frac{qV_{BE}}{kT}\right) \right] \qquad (6.1.9)$$

where

$$J_S = \frac{q^2 n_i^2 \tilde{D}_n}{Q_B} \qquad (6.1.10)$$

From Equation 6.1.9 we see that the current J_n can be switched on and off through the action of the junction voltages. If both V_{BC} and V_{BE} are negative and significantly greater than kT/q, J_n will be insignificant. If, on the other hand, either V_{BE} or V_{BC} is positive and greater than kT/q, J_n will be a sensitive function of the most positive voltage.

Equation 6.1.9 is an elegant representation of the physics of transistor action, and it is capable of describing more phenomena than just switching action in the device. Before we explore these phenomena more fully, however, let us consider the physical basis for transistor action.

Prototype Transistor

The equation that we have derived to represent transistor action (Equation 6.1.7) is not a function of the base doping profile; rather it depends only on the integrated base majority charge. We shall see that this feature makes the equation extremely useful. For further insight into transistor action, however, it is worthwhile to consider currents in a very simple transistor from a different point of view. The simple transistor consists of the structure in Figure 6.1 with abrupt junctions and constant base doping. We call this the *prototype transistor* because it resembles closely the device described by W. Shockley in his original discussion of the bipolar junction transistor.[1]

In Figure 6.2 we sketch the energy-band diagrams and electron densities (base minority carriers) for the prototype transistor at equilibrium and under various bias conditions. Figure 6.2a refers to the transistor under equilibrium (zero bias) conditions. There are only a few electrons in the base region, and transfer of electrons from either end region is inhibited by energy barriers. Negative bias applied to both junctions tends to increase these barriers and to deplete the base region of the few electrons present under equilibrium conditions. The band diagram and a sketch of the electron population for this case are shown in Figure 6.2b. Positive bias applied to both junctions injects electrons into the base region by lowering the built-in barrier height, and the resulting vast increase in the base electron density allows the ready flow of current between the two junctions (Figure 6.2c)

Active Bias. In another bias condition one junction is forward biased and the other is reverse biased. This arrangement holds the greatest interest because, as we shall see, it makes signal amplification possible. Diagrams for this bias condition,

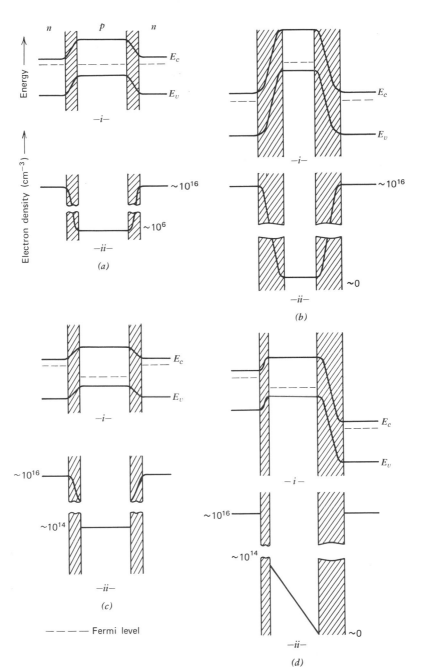

Figure 6.2 Energy-band diagrams (*i*) and electron-density sketches (*ii*) for the transistor sketched in Figure 6.1. (*a*) Equilibrium condition, (*b*) both junctions reverse-biased, (*c*) both junctions forward-biased, (*d*) one junction reverse-biased, and one junction forward-biased. The cross-hatched areas represent space-charge regions.

called *active bias*, are sketched in Figure 6.2*d*. As we see in the figure, the forward-biased junction injects electrons into the base because the barrier is reduced from its equilibrium value. The reverse-biased junction, on the other hand, sweeps any nearby electrons from the base into the *n*-region at the end. There is thus a flow of electrons (base minority carriers) from the forward-biased junction, which we call the *emitter* junction, to the reverse-biased junction, known as the *collector* junction. Electrons in the vicinity of the collector junction are swept rapidly through the space-charge region and into the *n*-type collector region.

As an introduction to transistors under active bias, and to clarify some of the qualitative discussion that we have presented, we consider briefly the amplifying or active-bias operation of the prototype transistor. In Figure 6.2*d* we have sketched as a straight line the electron density (minority-carrier density) in the transistor base. This result is justified by the analysis of the short-base diode that was carried out in Chapter 5 because the electrons in the base of a uniformly doped transistor are spatially distributed in the same way as they would be in the ideal short-base diode (Section 5.3). That is, the continuity equation that applies to this case has a negligible recombination term, and the doping density is constant. Hence, the solution for the electron density is linear in *x*. The boundary value at the base-emitter edge is exponentially dependent on the voltage V_{BE}; at the collector edge of the base, the electron density is negligible. The distribution of excess electrons n' is thus linear, as found in Equation 5.3.17 (for holes in that case). In terms of the geometry of Figure 6.1, the excess electron density $n' = n - n_{po}$ is

$$n' = n_{po}\left[e^{qV_{BE}/kT}\left(1 - \frac{x}{x_B}\right) - 1 \right] \qquad 0 \leqslant x \leqslant x_B \qquad (6.1.11)$$

The current of electrons through the base for V_{BE} greater than kT/q, as is usual for active bias, is easily found since it is purely diffusion flow. In terms of base doping N_a, since $J_n = qD_n\,(dn/dx)$, we find

$$J_n = \frac{-qD_n n_i^2 \exp(qV_{BE}/kT)}{N_a x_B} \qquad (6.1.12)$$

To derive this result from Equation 6.1.9, we note that Q_B for the prototype transistor (Equation 6.1.8) is

$$Q_B = qN_a x_B \qquad (6.1.13)$$

When this value of Q_B is used in Equation 6.1.10 and the negligible $\exp(qV_{BC}/kT)$ term in Equation 6.1.9 is dropped, the expression for current that we derived in Equation 6.1.12 is obtained. The significance of the integrated equations for transistor action (Equations 6.1.7 to 6.1.10) lies in their ability to deal with transistors that are not uniformly doped. The original development of this approach[2] to transistor analysis had as its purpose the optimization of various alternative transistor doping profiles. We shall return to consider further aspects of this point of view in Chapter 7.

Transistors for Integrated Circuits

Before deriving further equations for transistors under active bias, we devote a few paragraphs to integrated-circuit transistors. This permits us to contrast *IC* devices with the simple prototype structure of Figure 6.1 and allows us to develop theory and to discuss practical applications at the same time. The discussion here will be amplified in Section 6.5 after we have developed a perspective that will allow a more searching examination of the *IC* transistor.

Transistors for integrated circuits are made universally by the planar process described in Chapter 2. They therefore differ considerably in structure and in dopant configuration from the prototype device sketched in Figure 6.1. The prototype transistor is a fairly good model for a grown-junction transistor, a device of little commercial interest at present. We have used it in our discussion because it permits a clear focus on many phenomena that are important to the operation of all transistors.

A top view and a cross-sectional view of a junction–isolated integrated-circuit transistor are shown in Figures 6.3*a* and 6.3*b*, respectively. The structure is produced by a sequence of steps identical to the one outlined in Section 5.5 where the

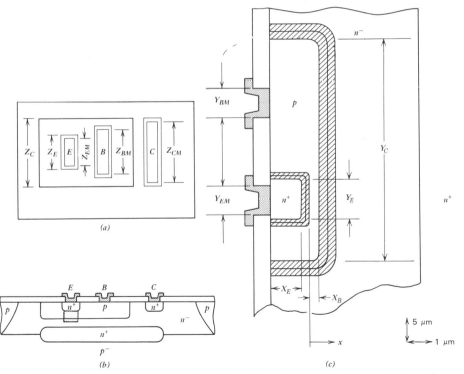

Figure 6.3 Top view (*a*) and cross sections (*b*) and (*c*) of a representative *npn IC* transistor. The region dominating transistor action is shaded in (*b*) and the area bounded by the base diffusion is rotated 90° and expanded in (*c*); *x* and *y* scales are indicated for (*c*) only.

fabrication of an array of junction-isolated diodes was described. To make transistors instead of diodes, a heavily doped n-region is diffused into the p-region. This additional diffusion step produces the emitter-base junction. As with the diode array, a lightly doped p-type wafer is used for the substrate. A heavily doped n-type buried layer added to reduce series resistance underlies the collector pn junction. More specific discussion about dopant densities, geometries, and trade-offs in the processing of transistors is continued in Section 6.5. Our focus for the present is on the basic device geometry.

One of the most critical parameters in transistor production, the width of the quasi-neutral base region (x_B in Figure 6.3c), is typically a few hundred nanometers ($\sim 10^{-5}$ cm). Most of the other dimensions in the transistor are substantially larger—10^{-4} cm and greater. Transistor action, as we have defined it, is quite critically dependent upon the proximity of the two interactive junctions. Because of this, under most transistor bias conditions, transistor action in the device is confined to the shaded region of the cross-sectional view in Figure 6.3b. In Figure 6.3c, the region that includes the base diffusion has been expanded to facilitate discussion. In this sketch, the emitter-base and base-collector space-charge regions are indicated by cross hatching. The cross section has been rotated by 90° with respect to Figures 6.3a and b to maintain an x-axis orientation consistent with Figure 6.1.

Figure 6.3c reveals that transistor action is essentially confined to an area defined by Y_E, the width of the emitter quasi-neutral region. This is nearly equal to the emitter-stripe width because Y_E is typically at least five or ten times as wide as the penetration of space charge into the emitter. Thus, although IC processing unavoidably makes the collector width Y_C much larger than the emitter width Y_E, the transistor is nonetheless well approximated for many purposes as a one-dimensional device. We can make use of this fact in specifying a single area A to convert current density J (Equation 6.1.9) to current I; for the transistor sketched in Figure 6.3, that area A is the product of Y_E and Z_E. As the theory of junction transistors is developed further, we shall refer to other dimensions on Figure 6.3, and examine the constraints imposed by the IC device structure on transistor equations.

In Section 6.2 we consider transistor action under *active bias* in more detail. As we shall see, the concepts involved are important for understanding both amplifying and switching uses of the transistor.

6.2 Active Bias

We have described the condition V_{BE} positive and V_{BC} zero or negative as active bias* for an npn transistor. This bias results in electron injection at one junction (the *emitter-base junction*) and electron collection at the other (*base-collector junction*). When V_{BC} is zero or negative and V_{BE} is substantially larger than kT/q, we

---

* Strictly, these polarities correspond to *forward-active bias*; if the polarities of V_{BE} and V_{BC} are interchanged, the transistor is under *reverse-active bias*.

see from Equation 6.1.9 that an electron current

$$J_n \approx -J_S \exp\left(\frac{qV_{BE}}{kT}\right) \tag{6.2.1}$$

flows from left to right across the collecting junction (J_2 in Figure 6.1). Using the standard convention that currents into a transistor are positive, J_n in Equation 6.2.1 is equal to $+J_C$, a positive collector-current density. Equation 6.2.1 shows that under active bias, collector current is exponentially related to emitter-base voltage. Experimentally, this relationship is found to hold over many decades of collector current (Figure 6.4).

Figure 6.4 shows experimental measurements of collector current I_C plotted on a logarithmic scale as a function of the base-emitter bias voltage V_{BE}. Over nearly the entire range of voltage, these measurements confirm an exponential dependence. Furthermore, after accounting for the ratio of natural to common logarithms, the slope of the J_C plot is just q/kT, as expected from Equation 6.2.1. (Alternatively, we see from Equation 6.2.1 that a decade change in $|J_S/J_n|$ occurs for $\Delta V_{BE} = (kT/q)\ln(10)$ or for $\Delta V_{BE} = 60$ mV at $T = 300\,K$.) The experimental precision of this dependence is sufficiently high to make possible the construction of accurate thermometers that take advantage of collector-current changes with temperature in transistors under fixed emitter bias. The intercept of an extrapolated

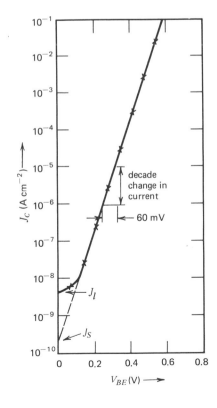

Figure 6.4 Semilogarithmic plot of collected current versus base-emitter voltage for an *IC npn* transistor under active bias at 300 K. Crosses designate data points. The extended straight line indicates J_S (Equation 6.2.1); J_l is a leakage component.

line drawn through the J_C measurements with the current axis at $V_{BE} = 0$ provides a value for J_S in Equation 6.2.1. From Equation 6.1.10 we find that the built-in base charge in the quasi-neutral region can be obtained once J_S is known.

$$Q_{B0} = \frac{q^2 n_i^2 \tilde{D}_n}{J_S} \qquad (6.2.2)$$

All of the other parameters in Equation 6.2.2 are well known.* We called Q_{B0} in Equation 6.2.2 the built-in base charge because, as shown in Equation 6.1.8, it represents the hole charge per unit area in the quasi-neutral base as the base-emitter bias tends to zero. This charge is built-in during processing of the transistor. As we see above, a value for Q_{B0} can be obtained through current-voltage measurements on the transistor. Because this was first described by H. K. Gummel,[3] the number of base-dopant atoms (per cm^2) in the quasi-neutral region

$$\int_0^{x_B} N_a(x)\,dx = \frac{Q_{B0}}{q} = \frac{q n_i^2 \tilde{D}_n}{J_S} \qquad (6.2.3)$$

is often called the *Gummel Number*.

Equation 6.2.3 emphasizes that J_S, the multiplying factor for transistor current at a given bias, is inversely proportional to the Gummel Number, that is, to the total base doping. The lower the built-in base charge Q_{B0}, the higher is the current at a given bias. Therefore, one might propose to design a prototype transistor with a low constant base doping. A major disadvantage of this design is that even a small forward bias might make the low-injection approximation invalid near the base-emitter junction in the transistor [i.e., $n_p(0)$ would approach N_a]. As we shall see in Chapter 7, high injection in a transistor causes a loss in performance. It can be avoided, and Q_{B0} can be kept small at the same time by making the base dopant profile taper from a maximum near the base-emitter junction. The dopant density is then large in the region where the injected minority density is large and small where the minority density becomes negligible. Fortunately, diffusion technology automatically provides transistors with graded-base doping and the advantage that we have described is obtained in *IC* transistors. Other advantages of graded-base transistors were first pointed out by Kroemer,[4] and are described in Chapter 7.

Control of Q_{B0} during transistor processing is the key step in the production of integrated circuits. For high-gain transistors, in which the Gummel Number is below 10^{12} cm^{-2}, extraordinary care is taken to control this parameter.

EXAMPLE   Gummel-Number Calculations

Assume that the data in Figure 6.4 were measured on a prototype *npn* transistor with a base width $x_B = 0.5$ μm. Find the Gummel Number (Equation 6.2.3) for the

---

* A possible exception is $\tilde{D}_n$, but diffusivity is not a very strong function of concentration (Figure 1.15) and thus can be approximated.

transistor and calculate the value of base-emitter voltage necessary to cause the electron density at the emitter edge of the base to be 1% of the base dopant density.

Solution

From Figure 6.4 we have $J_S = 2.4 \times 10^{-10}$ A cm^{-2}. From Equation 6.2.3, the Gummel Number GN is

$$GN = \frac{qn_i^2 D_n}{J_S} = 1.4 \times 10^{11} \tilde{D}_n$$

If we assume (Figure 1.15) that $\tilde{D}_n$ is 20 cm^{2}s^{-1}, then $GN = 2.8 \times 10^{12}$ dopant atoms cm^{-2}, and $N_a = \dfrac{GN}{x_B} = 5.6 \times 10^{16}$ cm^{-3}. Referring again to Figure 1.15, we see that if $N_a = 5.6 \times 10^{16}$, $\tilde{D}_n \approx 22$ instead of the value 20 which we had assumed. We can redo the calculation taking $\tilde{D}_n = 22$ which leads to $GN = 3.1 \times 10^{12}$ cm^{-2} and $N_a = 6.2 \times 10^{16}$ cm^{-3}, which is approximately consistent with the value of $\tilde{D}_n$ used in the calculation.

The required value for the Gummel Number is, therefore, 3.1×10^{12} dopant atoms cm^{-2}.

The second part of the question requires a value for V_{BE} such that

$$n'(x = 0) = \frac{n_i^2}{N_a} \exp \frac{qV_{BE}}{kT} = 0.01 \times N_a = 6.2 \times 10^{14} \text{ cm}^{-3}$$

The required value is $V_{BE} = 0.67$ V.

Current Gain

Our discussion of transistor action under active bias has thus far considered only current flowing between the collector and the emitter. This represents the output current in an active-biased transistor. Collector current is an exponential function of base-emitter voltage (Equation 6.2.1), because forward bias on the base-emitter junction causes electron injection into the base to vary exponentially. Under active bias, these electrons are collected efficiently by the field at the base-collector, space-charge region. The base-emitter terminals are thus the control electrodes for the collector current under active bias. The smaller the current that flows through these terminals for any given positive V_{BE}, the more effective is the transistor as an amplifier since the input power (the product of V_{BE} and the base-emitter current) is lower.

Several mechanisms give rise to base-emitter current in an active-biased transistor. Based on our previous considerations, the most straightforward of these is recombination of injected electrons with majority-carrier holes in the base. Another component of base-emitter current is due to recombination in the base-emitter space-charge region. A third component of current flows because forward

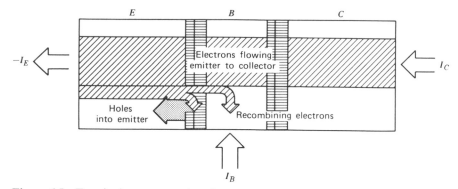

Figure 6.5 Terminal currents and major current components in an active-biased transistor. Not shown is the collector leakage current (J_l in Figure 6.4).

bias on the junction not only injects electrons into the base, but it also causes holes to be injected into the emitter. These various components are indicated pictorially in Figure 6.5. If a transistor is meant to amplify effectively, all of these current components should be kept much smaller than the collector current.

Most transistors are designed so that recombination of injected electrons in the base-emitter, space-charge region is of less consequence than the other components mentioned except under low current conditions. Accordingly, we shall consider recombination in the space-charge region separately in Chapter 7 when we discuss limitations on transistor performance.

Recombination within the base itself can be expressed readily using the theory developed in Section 5.3. There we found that recombination of excess minority carriers is directly proportional to their density (in this case n' the excess base electron density) so that the total base-region recombination current is

$$I_{rB} = qA_E \int_0^{x_B} \frac{\left[n - (n_i^2/N_a) \right] dx}{\tau_n} \tag{6.2.4}$$

where $A_E = Y_E \times Z_E$ in Figure 6.3 is the area of significant minority-carrier injection. To make I_{rB} as small as possible at a given bias, we see from Equation 6.2.4 that base lifetime τ_n should be maximized and base width x_B should be minimized.

Under active bias, the injected excess electron density n' is much greater than the equilibrium electron density (n_i^2/N_a) over most of the base. Also, lifetime is not strongly x dependent so that Equation 6.2.4 can be simplified to

$$I_{rB} = \frac{qA_E}{\tau_n} \int_0^{x_B} n \, dx \tag{6.2.5}$$

For the special case of a transistor with a uniformly doped base such as the prototype transistor sketched in Figure 6.1, n' depends linearly on x. Thus, the integration in Equation 6.2.5 is easily carried through to obtain

$$I_{rB} = \frac{qA_E n_i^2 x_B}{2N_a \tau_n} \left[\exp\left(\frac{qV_{BE}}{kT} \right) - 1 \right] \tag{6.2.6}$$

Although Equation 6.2.6 was derived for a special case, the proportionality of base recombination current to $[\exp(qV_{BE}/kT) - 1]$ is obtained in general.

The loss of carriers to recombination in the base region is measured by the base transport factor, which is usually given the symbol α_T and defined by

$$\alpha_T = \frac{|I_{nE}| - |I_{rB}|}{|I_{nE}|} = 1 - \left|\frac{I_{rB}}{I_{nE}}\right| \tag{6.2.7}$$

where I_{nE} is the electron current injected from the emitter. Using Equations 6.1.12 and 6.2.6 for a transistor with a uniformly doped base, we obtain

$$\alpha_T = 1 - \frac{x_B^2}{2D_n\tau_n} = 1 - \frac{x_B^2}{2L_n^2} \tag{6.2.8}$$

Although Equation 6.2.8 is not directly applicable to *IC* transistors, it is sometimes used for these devices. The equation does not apply because base doping in the *IC* transistor is not constant. Since a graded base, as obtained by diffusion technology, increases J_S, less minority-carrier injection is needed for a given output current. Thus, a graded base improves the transport factor and α_T as calculated by Equation 6.2.8 can be regarded as a "worst-case" parameter. If we take a typical diffusion length of 10 μm and consider x_B to be 0.3 μm, we find $\alpha_T = 0.9996$. Indeed, the loss of minority carriers to recombination in the quasi-neutral base is small in *IC* transistors.

The injection of base majority carriers (holes) into the emitter is the predominant cause for base current in most integrated-circuit transistors. Expressions for this current have already been derived in Section 5.3, since the base-emitter junction is just a forward-biased diode. For a given device, however, it is necessary to determine where recombination of the holes injected into the emitter takes place in order to write a correct expression for the hole current. Consider first the prototype transistor shown in Figure 6.1; it has an emitter contact spaced a distance x_E from the edge of the base-emitter space-charge region. If x_E is much greater than a diffusion length for holes, virtually all injected holes will recombine before reaching the ohmic contact. Hence, by the theory developed in Section 5.3, the excess holes are distributed in the emitter according to an exponentially decaying function (Equation 5.3.12) as sketched in Figure 6.6a. The hole current is proportional to the hole-density gradient at the edge of the emitter quasi-neutral region.

$$I_{pE} = \frac{-qA_En_i^2 D_{pE}}{N_{dE}L_{pE}}(e^{qV_{BE}/kT} - 1) \tag{6.2.9}$$

If the emitter contact is close to the base ($x_E \ll L_{pE}$), the hole concentration is a linear function of x (Figure 6.6b), and the hole current is

$$I_{pE}' = \frac{-qA_En_i^2 D_{pE}}{N_{dE}x_E}(e^{qV_{BE}/kT} - 1) \tag{6.2.10}$$

For many integrated-circuit transistors, the emitter contact is closer to the base than a diffusion length. In these devices, however, emitter doping is not constant

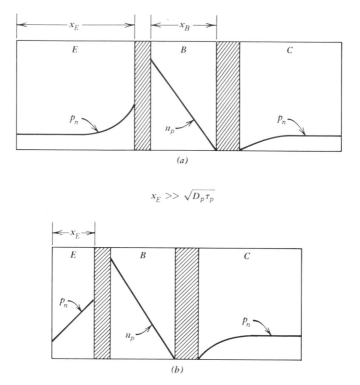

$$x_E \gg \sqrt{D_p \tau_p}$$

$$x_E \ll \sqrt{D_p \tau_p}$$

Figure 6.6 Minority-carrier distributions in the prototype transistor of Figure 6.1 under active bias. (a) $x_E \gg$ hole diffusion length in emitter. (b) $x_E \ll$ hole diffusion length.

so that injected holes in the emitter are influenced by a built-in electric field. Equations for this condition could be derived using a nearly analogous set of arguments to those employed in Section 6.1 for the transport of electrons in the base. The difference for this case is the presence of a nonzero majority-carrier (electron) current.

Emitter Injection: Nonuniform Doping. To discuss nonuniform doping, consider the *IC* transistor of Figure 6.3. The impurity profile causes a built-in electric field in the emitter at thermal equilibrium. Aside from the junction space-charge region, however, we noted in Section 4.1 that most nonuniformly doped regions can be treated as nearly neutral or quasi-neutral. This assumption means that $n_0(x) \approx N_d(x)$ and the built-in field at thermal equilibrium has the value obtained in Equation 4.1.13, which we repeat here for reference.

$$\mathscr{E}_0(x) = -\frac{kT}{q} \frac{1}{N_{dE}(x)} \frac{dN_{dE}(x)}{dx} \qquad (6.2.11)$$

If we consider the emitter-base junction under forward bias, we can expect that the field in the quasi-neutral emitter will be altered somewhat from the thermal-equilibrium value in Equation 6.2.11. For first-order analysis we express the field under bias as the sum of $\mathscr{E}_0$ plus an added term $\mathscr{E}_a$ that results from $V_{BE} \neq 0$. This is a reasonable assumption consistent with our consideration of low-level injection (i.e., $p_n' = n_n' \ll n_0$). With these considerations, the expressions for electron and hole currents in the emitter become

$$J_n = q\mu_n(n_{no} + p_n')(\mathscr{E}_0 + \mathscr{E}_a) + qD_n\left(\frac{dn_{no}}{dx} + \frac{dp_n'}{dx}\right)$$

and

$$J_p = q\mu_p(p_{no} + p_n')(\mathscr{E}_0 + \mathscr{E}_a) - qD_p\left(\frac{dp_{no}}{dx} + \frac{dp_n'}{dx}\right) \qquad (6.2.12)$$

Since the drift and diffusion components balance at equilibrium (when $\mathscr{E} = \mathscr{E}_0$), Equations (6.2.12) can be written more simply as

$$J_n = q\mu_n n_{no}\mathscr{E}_a + qD_n\frac{dp_n'}{dx} \qquad (6.2.13)$$

and

$$J_p = q\mu_p(p_{no} + p_n')\mathscr{E}_a + q\mu_p p_n'\mathscr{E}_0 - qD_p\frac{dp_n'}{dx} \qquad (6.2.14)$$

where we have neglected $p_n'(\mathscr{E}_a + \mathscr{E}_0)$ compared to $n_{no}\mathscr{E}_a$.

The first term in Equation 6.2.13 represents the ohmic flow of majority-carrier electrons. As with the ideal diode, we expect that only a small field $\mathscr{E}_a$ is necessary to provide this ohmic current. Typically $\mathscr{E}_a$ will be less than $\mathscr{E}_0$. Making this assumption in Equation 6.2.14 for the minority-carrier hole current, we conclude that the second term is appreciably larger than the first. Thus,

$$J_p = q\mu_p p_n'\mathscr{E}_0 - qD_p\frac{dp_n'}{dx} \qquad (6.2.15)$$

By using Equation 6.2.11 for the thermal-equilibrium field and combining terms, we derive

$$J_p = -\frac{qD_p}{N_d(x)}\frac{d}{dx}\left[p_n'(x)N_d(x)\right] \qquad (6.2.16)$$

an equation similar to Equation 6.1.4 for electron current in the base.

Now let us consider the important case of a short emitter region in which $x_E \ll L_{pE}$ so that negligible recombination of holes occurs in the bulk of the n-type region. Then the hole current is not a function of position, and Equation 6.2.16 can be integrated in the negative x direction from an arbitrary point x within the emitter to the ohmic contact where $p_n' = 0$.

$$J_p \int_x^{x_E} \frac{N_d(x')\,dx'}{D_p} = qp_n'(x)N_d(x) \qquad (6.2.17)$$

The hole current can be found by evaluating Equation 6.2.17 at the edge of the emitter-base, space-charge region $-x_n$, where the minority-carrier density is related to its thermal equilibrium value by Equation 5.3.8. The hole current in the emitter is then found to be

$$I_{pE}'' = -\frac{q\tilde{D}_p n_i^2 A_E(e^{qV_{BE}/kT} - 1)}{\displaystyle\int_{-x_n}^{x_E} N_d \, dx} \tag{6.2.18}$$

For a uniformly doped emitter, Equation 6.2.18 reduces to Equation 6.2.10.

If recombination in the quasi-neutral region is not negligible and some of the carriers recombine before traversing the emitter, Equation 6.2.16 is still valid, but it is not possible to remove J_p from the integral for an explicit solution as in Equation 6.2.18. The hole current, however, increases over the alue predicted by Equation 6.2.18. Further approximations may be used to extra ct a solution, but we will not consider them here.

For the two cases in which surface recombination dominat:s in determining I_{pE} (Equations 6.2.10 and 6.2.18), the proper value to use for A_E may not be straightforward. This is because surface recombination will almost certainly be far more effective at an ohmic contact than at an oxide surface. Hence, I_{pE} may be specified more accurately by using the contact area ($Y_{EM} \times Z_{EM}$ in Figure 6.3) instead of the junction area for A_E in either Equations 6.2.10 or 6.2.18. We comment further on this point in Section 6.5 when we consider diffused planar transistors in more detail.

The effectiveness of an emitter junction in injecting electrons into the base is measured by the emitter efficiency, usually denoted by the symbol γ.

$$\gamma = \frac{I_{nE}}{I_{nE} + I_{pE}} = \frac{1}{1 + I_{pE}/I_{nE}} \tag{6.2.19}$$

Since $(I_{nE} + I_{pE})$ is the total emitter current I_E, the electron current crossing the emitter-base junction I_{nE} is just γI_E. Let us make an approximation for γ for the IC transistor using the approach taken previously for α_T. We apply the simple theory for the prototype transistor (Figure 6.1) and take dimensions and average doping appropriate to the IC transistor. Then using Equations 6.1.12 and 6.2.10 in Equation 6.2.19, we have

$$\gamma = \frac{1}{1 + \dfrac{x_B N_{aB} D_{pE}}{x_E N_{dE} D_{nB}}} = \frac{1}{1 + \dfrac{GN_B \tilde{D}_{pE}}{GN_E \tilde{D}_{nB}}} \tag{6.2.20}$$

In the second form of the equation we have introduced the Gummel Numbers GN (Equation 6.2.3) both for the base and the emitter.

Equation 6.2.20 is a frequently quoted representation for the emitter efficiency in a BJT. A direct application of Equation 6.2.20 to IC transistors, however, can lead to error in predicting the emitter efficiency because two effects associated with heavy doping in the emitter were not considered in its derivation. The first effect, described in Section 1.1 (Figure 1.12 and Equation 1.1.33), is the narrowing of the bandgap and the corresponding increase in the intrinsic carrier density n_i when dopant concentrations in silicon exceed about 10^{18} cm^{-3}. Since the hole density injected into the emitter of an *npn* transistor is proportional to n_i^2 (Equation 6.2.10), bandgap narrowing results in increased minority-carrier injection and a corresponding reduction in the emitter efficiency. The second effect is the lifetime reduction that takes place when sufficient majority carriers are present to make Auger recombination (Equation 5.2.16) significant. The effective minority-carrier lifetime in the emitter becomes so short that recombination cannot be neglected, as is done in deriving Equation 6.2.10.

Because of these two effects, the derivation of Equation 6.2.10 for the reverse injection (of holes into the emitter) is not valid for a heavily doped emitter region. An exact analysis which includes the effects of heavy doping will not be carried out here. Instead, we will compensate for these effects by reducing the emitter Gummel Number in Equation 6.2.20 below the integrated dopant concentration. The reduction is sizable; if the dopant densities approach 10^{21} cm^{-3}, the effective emitter Gummel Number is only a few percent of the integrated dopant concentration in the emitter.

EXAMPLE   Emitter Efficiency of a *BJT*

Use Equation 6.2.20 to calculate the emitter efficiency of an *IC* transistor in which the distance from the contact to the edge of the charge-neutral region in the emitter is 0.8 μm and the base Gummel Number GN_B is 3×10^{12} cm^{-2}. Assume that the emitter dopant concentration is 6×10^{20} cm^{-3} at the surface and that it decreases roughly exponentially to 5×10^{16} cm^{-3} at $x = 0.8$ μm. Assume also that heavy-doping effects reduce the effective emitter Gummel Number (GN_E) to 2% of the integrated dopant density.

Solution

We need to know the emitter Gummel Number, and we shall need to estimate the ratio $\tilde{D}_{pE}/\tilde{D}_{nB}$. For the emitter, we assume an exponential variation in dopant density with a characteristic length equal to λ

$$N_{dE} = N_{dE0} \exp - \left(\frac{x}{\lambda} \right)$$

From the values at $x = 0$ and $x = 0.8$ μm, we solve for λ

$$\lambda^{-1} = \frac{\ln(1.2 \times 10^4)}{0.8 \times 10^{-4}} \quad \text{or} \quad \lambda = 85.2 \text{ nm}$$

The integrated dopant density in the emitter is therefore

$$\int_0^{x_E} N_{dE}(x) \, dx \approx N_{dE0}\lambda = 5.1 \times 10^{15} \text{ cm}^{-2}$$

The effective emitter Gummel Number is 2% of the integrated dopant density or $1.02 \times 10^{14} \, \text{cm}^{-2}$. For the emitter diffusion constant, we find the average emitter dopant density by dividing the effective emitter Gummel Number by the emitter depth, $N_{davg} = 1.28 \times 10^{18} \, \text{cm}^{-3}$. From Figure 1.15 we find $\tilde{D}_{pE} \approx 4.0 \, \text{cm}^2 \, \text{s}^{-1}$. The base diffusion constant $\tilde{D}_{nB}$ is approximately $22 \, \text{cm}^2 \, \text{s}^{-1}$ from the example considered in Section 6.2. Using these numbers in Equation 6.2.20, we find

$$\gamma = \frac{1}{1 + \dfrac{3 \times 10^{12} \times 4.0}{1.02 \times 10^{14} \times 22}} = 0.9947$$

The emitter efficiency is indeed high, but not quite as high as we found the base transport factor α_T (0.9996 in the example associated with Equation 6.2.8). This result is typical; the base-transport factor is closer to unity than is the emitter efficiency in *IC* transistors.

The magnitude of the ratio of the collector current I_C to the emitter current I_E under active bias is frequently given the symbol α_F. For our analysis, α_F is the product of γ and α_T.

$$\alpha_F = \gamma \alpha_T \qquad (6.2.21)$$

Since all currents into the transistor sum to zero by Kirchhoff's current law, we have

$$I_B + I_E + I_C = 0$$

$$I_B - \frac{I_C}{\alpha_F} + I_C = 0$$

or

$$I_C = \frac{\alpha_F I_B}{(1 - \alpha_F)} = \beta_F I_B \qquad (6.2.22)$$

where $\beta_F \equiv I_C/I_B$ is the current gain for the case in which input current flows between the base and emitter and output current flows into the collector. Because α_F is nearly unity, β_F is large (typically of order 100). For the case that we took as an example, $\alpha_F = 0.9947$ and $\beta_F = 188$. Small changes in α_F caused, for example, by process variations in fabricating the transistor are likewise magnified to large changes in $\beta_F [d\beta_F = d\alpha_F/(1 - \alpha_F)^2]$. This means that β_F is not a precisely controlled transistor parameter. Circuit designers can only be assured that β_F will be large; its value in any given process run can vary substantially.

As indicated in the example, the emitter efficiency γ is typically the factor that limits the size of the common-emitter current gain β_F in a *BJT*. Furthermore, as described above, both bandgap narrowing and Auger recombination place limitations on the improvement that can be obtained in γ by increasing the Gummel

Number in the emitter. It has been demonstrated, however, that more than an order-of-magnitude increase in β_F can be attained in *BJT*s if they are fabricated with a layer of polysilicon between the metal interconnection line and the doped emitter region. The exact physical mechanisms giving rise to this improvement have not been firmly established although it is fairly certain that the cause is associated with the establishment of a barrier to the transmission of minority carriers in the direction of the contacting electrode. This barrier impedes the flow of holes (in an *npn* transistor), thereby reducing their gradient at the base-emitter junction and increasing the emitter efficiency.

The theory that we have developed in this section applies to quiescent bias conditions, and the equations have been derived for total currents. For applications to amplifiers, equations that express the results of incremental excursions in bias around a quiescent bias point are needed. It is straightforward to obtain them, but we defer taking this step until Chapter 7 when we derive various equivalent circuits that are useful for design with transistors. We shall also consider frequency effects in transistors at that time. Our analysis thus far has considered only the dc (and low-frequency) case.

Before resuming our discussion of transistor action, we should point out that the analysis of hole injection into the emitter that we carried out for the case of nonuniform doping applies in general to *pn* junctions in integrated circuits. For example, the injected hole current given by Equation 6.2.18 is analogous to the result we would obtain for electron injection into the *p*-type region in the diffused *pn*-junction diode considered in Section 5.5 (Figure 5.18*b*).

6.3 Transistor Switching

Our discussion of transistor action in Section 6.1 has emphasized that injection of electrons (base-minority carriers) into the base region is necessary for current to flow between the collector and the emitter or vice versa. Thus, transistor switching can be understood in terms of storage, extraction, and transport of electrons in the base. Transistor models having widespread utility are derived from this approach to understanding transistor switching.

Let us consider first the switching transistor in the "cut-off state." This condition is assured by depleting the base region of minority carriers, that is, by making both V_{BE} and V_{BC} zero or negative. With these bias voltages applied, the base has at most its equilibrium population of electrons and, thus, only very small currents can flow between collector and emitter. When negative bias is simultaneously applied to both junctions, the base becomes depleted of even the few "built-in" minority carriers and the dc behavior of the transistor is very close to that of an open circuit. The minority-carrier populations in all three regions of a homogeneously doped transistor biased into cut-off are shown in Figure 6.7.

Although the cut-off transistor will not pass dc current, its behavior is not identical to that of an open circuit. As described in Section 4.3, a reverse bias on

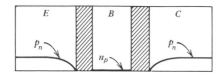

Figure 6.7 Minority-carrier densities in the prototype transistor of Figure 6.1 under cut-off conditions.

a *pn* junction exposes fixed donor and acceptor charges by extracting compensating free charges. For calculations, it is useful to make a graph that shows the variation in stored junction charge, which we denote Q_V, as a function of voltage. This is readily accomplished by using the theory of Chapter 4 for step or linearly graded junctions or else using the numerical data of Lawrence and Warner (reference 8 in Chapter 5). A graph of this type was developed in the example of Section 5.4 for an abrupt *pn*–junction diode. A similar graph with Q_V plotted for both an abrupt and a linearly graded junction is given in Figure 6.8. Derivation of the graph is considered in an example and in Problem 6.7. The inset in Figure 6.8 identifies Q_V the stored charge, and makes clear that only the excess above the junction space charge at thermal equilibrium is plotted.

To place a transistor in cut-off, the switching source must supply the junction with the incremental charge to be stored in the space-charge region; thus, the cut-off device has a capacitorlike behavior although the capacitance (dQ/dV) is not constant.

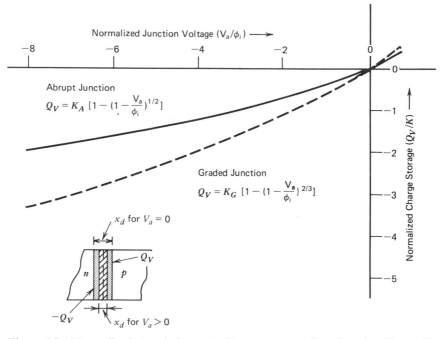

Figure 6.8 Normalized stored charge Q_V/K versus normalized junction bias V_A/ϕ_i in the space-charge regions of an abrupt junction and a linearly graded junction.

EXAMPLE   **Stored Charge and Transistor Switching**

An *npn* prototype transistor similar to that shown in Figure 6.1 has a base doping $N_a = 10^{16}$ cm^{-3}, and emitter and collector dopings $N_d = 10^{19}$ cm^{-3}. The transistor is biased in the cut-off mode with $V_C = 3$ V, $V_E = 0$ V, and $V_B = -3$ V. If the junction area is 10^{-5} cm^2, how much charge must be supplied to the base in order to bring the base voltage to $V = 0$ V?

Solution

From Equation 4.2.10 we find $\phi_i = 0.872$ V for both the base-emitter and base-collector junctions. Using Equation 4.3.1 we find K_A in Figure 6.8 is

$$K_A = \sqrt{2\epsilon_s q N_a \phi_i} = 53.8 \text{ nC cm}^{-2}.$$

Since the junction area is 10^{-5} cm^2, we have $K_A \times A = 0.54$ pC for the factor representing total charge at each junction. Initially, $V_a = -6$ V $= -6.88 \times \phi_i$ at the collector-base junction. From Figure 6.8, the charge stored at the collector $Q_{VC} = -1.8 \times K_A A = -0.972$ pC. When $V_B = 0$ V, $V_a = -3$ V $= -3.44 \times \phi_i$, and $Q_{VC} - -0.59$ pC. The charge supplied from the base is therefore 0.378 pC. At the base-emitter junction, the bias change is from -3 to 0 V; hence, the stored charge changes from -0.59 to 0 pC.

The total charge supplied from the base is the sum of these charges or 0.97 pC. If, for example, the source switching the base voltage could supply a maximum of 1 mA of current, it would require 0.97 ns to switch the voltages as described in this example. This type of calculation is frequently carried out by designers of transistor switching circuits.

A measure of the very small currents that flow in a cut-off transistor is available in the active-bias data plotted in Figure 6.4. We noted there that if the collector-base junction is reverse biased and V_{BE} is reduced nearly to zero, the collector current will begin to deviate from the exponential relationship predicted by Equation 6.2.1. Instead, J_C approaches a low value J_l that is independent of V_{BE}. The small size of J_l (nA cm^{-2}), indicated in Figure 6.4 is typical. At low base-emitter bias, the base-emitter junction injects such a small number of electrons that transistor action between the junctions is negligible. The collector-base junction behaves electrically like an isolated *pn*-junction diode under reverse bias, just as it would in cut-off, and J_l flows from collector to base.

As was discussed in Chapter 5, three separate mechanisms can be responsible for the small current in a *pn* junction under reverse bias: generation of holes and electrons in the space-charge region, generation of electrons in the *p*-type base, and generation of holes in the *n*-type collector. The first of these three components, generation in the space-charge region, dominates in silicon *pn* junctions, as we discovered in evaluating Equation 5.3.29.

The space-charge region generation component I_g is approximately proportional to the width of the space-charge layer as was shown in Section 5.3. It depends relatively weakly on collector-base bias (Equations 5.3.26–28), and is most
often written

$$I_l = J_l A_C = I_g = \frac{1}{2} \frac{q n_i x_d A_C}{\tau_0} \tag{6.3.1}$$

where x_d is the space-charge region thickness, A_C is the collector-base junction
area, and τ_0 is the effective lifetime in the region itself. Because of its small size I_l
is of no consequence to the operation of silicon transistors biased in the active
mode. It may be important for transistor switches, however. In terms of the *IC*
transistor structure in Figure 6.3, the area A_C is roughly $Y_C \times Z_C$ plus the vertical
portions of the collector-base junction that meet the surface. There is also a contribution to I_g from recombination centers at the surface (Section 5.3). With careful processing this contribution can be made very small.

Regions of Operation

Figure 6.9 is a useful aid for understanding transistor switching because it delin-
eates the regions of device operation in terms of the applied junction voltages. For
example, the cut-off region that we have been discussing consists of the third
quadrant in this map of a bias "space" having coordinates V_{BE} and V_{BC}. The fourth
quadrant (V_{BE} positive, V_{BC} negative) corresponds to the forward-active region that
was considered in Section 6.2.

In the second quadrant the polarities of both V_{BC} and V_{BE} are inverted from
the forward-active condition and the transistor is said to be biased in the reverse-

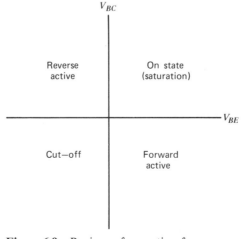

Figure 6.9   Regions of operation for an *npn*
transistor as defined by base-emitter and
base-collector biases.

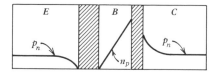

Figure 6.10 Minority-carrier densities in the prototype transistor of Figure 6.1 under bias in the reverse-active region.

active region. In this region an *npn* transistor injects electrons at the collector and collects them at the emitter. When operated in the reverse-active region, a one-to-one correspondence can be made to the parameters defined in this section for forward-active bias. These parameters are subscripted with an *R* to denote a reverse measurement. For example, output current is delivered to the emitter lead and the ratio of output current to input (base) current in a reverse-active condition will define $\beta_R = I_E/I_B$. The distribution of minority carriers in the prototype transistor of Figure 6.1 under reverse-active bias is sketched in Figure 6.10. Its similarity to Figure 6.6 should be noted.

For the prototype transistor of Figure 6.1, the distinction between forward-active bias and reverse-active bias is not critical because the device is symmetrical; the doping is the same in the emitter and collector, and the two junctions have equal areas.

In contrast to the prototype transistor, the integrated-circuit transistor (Figure 6.3) is asymmetric both in geometry and doping, and a sketch of minority-carrier distributions under reverse-active bias differs markedly from one made in the forward-active region. First, electrons injected from the normal collector travel against the built-in base field as they move toward the emitter. Second, injection from the collector results in losses at the base contact and at the passivating oxide that are not significant in the forward-active region. Third, the injection efficiency in the reverse-active mode is very much lower than in the forward-active region. These asymmetries reduce the gain in the reverse-active bias mode relative to forward-active bias and have important practical consequences that we discuss in Chapter 7.

The first quadrant in Figure 6.9 is defined by positive bias on both the base-emitter and base-collector junctions. This bias condition is called *saturation*.* A transistor switch in the closed or "on" position will be biased in this region. In saturation, both junctions inject electrons, and the minority carriers are distributed throughout the prototype transistor as shown in Figure 6.11. Note that in saturation the electron concentration is markedly increased throughout the entire base region. If the minority-carrier populations in Figure 6.11 are compared to the diagrams of Figures 6.6 and 6.10, we see that the saturated condition corresponds to the superposition of forward-active and reverse-active operation. The physical basis for this fact is that both junctions are injecting and collecting electrons *at the same time.* They inject because the built-in potential is reduced from its equilibrium value; they collect because the junction field is still of the proper

---

* The term "saturation" refers to a collector current that is determined by conditions in the circuit external to the transistor.

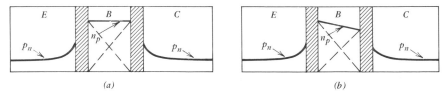

Figure 6.11   Minority-carrier densities in the prototype transistor of Figure 6.1 under bias in the saturation region, (*a*) with no current between collector and emitter, (*b*) with current flowing from collector to emitter.

polarity to sweep electrons from the base. This property of the saturated transistor being made up of a superposition of forward-active and reverse-active bias conditions does not depend on transistor geometry; it is valid for the *IC* transistor as well as for the prototype device.

We began this section with a general discussion of transistor switching and of the conditions in a cut-off transistor. This led us to a review of the regions of transistor operation, which are summarized in Figure 6.9. A transistor switch is biased alternately in regions 1 and 3 (saturation and cut-off) and moves through regions 2 and 4 only during switching transients. To move from saturation to cut-off, the charges stored in and near the base of a transistor must be altered. Much of the design of transistor switching circuits consists of assuring that these charging and discharging requirements are adequately met. Models that account for the physical mechanisms we have discussed and that are useful for transistor switching calculations are presented in Chapter 7. In the next section we introduce a transistor model that has useful dc applications and that quantifies some of the physical pictures we have developed.

6.4 Ebers-Moll Model

A simple and very useful model of the carrier injection and extraction phenomena occurring in bipolar transistors was developed in 1954 by J.J. Ebers and J.L. Moll.[5] Over three decades later, this model still provides the basic framework for much more complex computer-aided models of bipolar transistors (*BJTs*). The *Ebers-Moll Model* is based upon the understanding of the *BJT* in terms of interacting diode junctions, a viewpoint already presented in Section 6.1, and used to derive Equation 6.1.9. For the active-biased transistor (shown in Figure 6.5), we recognize that J_n in Equation 6.1.9 represents the electrons flowing between the emitter and the collector and linking these regions. We therefore call J_n (and its counterpart J_p for a *pnp* transistor) the *linking current*. The theory introduced in Section 6.1 treats only linking current, and therefore does not account for the components of base current shown in Figure 6.5.

To treat the base current, we consider separately the component flowing between the base and the emitter I_{BE}, and that flowing between the base and the collector

I_{BC}. Since the base-to-emitter contact forms a *pn* junction diode, we express currents through it by the ideal diode expression, denoting the saturation current as I_{0E}

$$I_{BE} = I_{0E}[\exp(qV_{BE}/kT) - 1]. \tag{6.4.1}$$

The total current in the emitter consists of the flow to the collector (the linking current) minus the base-emitter diode current. For the linking current we use Equation 6.1.9, assuming (for the present) a constant area A across the base and taking $I_S = J_S A$, where J_S is given by Equation 6.1.10.* Thus, the emitter current is

$$I_E = I_S[\exp(qV_{BC}/kT) - \exp(qV_{BE}/kT)] - I_{0E}[\exp(qV_{BE}/kT) - 1] \tag{6.4.2}$$

Similarly, for the collector current, we have

$$I_C = I_S[\exp(qV_{BE}/kT) - \exp(qV_{BC}/kT)] - I_{0C}[\exp(qV_{BC}/kT) - 1] \tag{6.4.3}$$

where the base-collector current is

$$I_{BC} = I_{0C}[\exp(qV_{BC}/kT) - 1]. \tag{6.4.4}$$

If we group the terms in Equations 6.4.2 and 6.4.3 according to their voltage dependences, we can write

$$I_E = -(I_S + I_{0E})[\exp(qV_{BE}/kT) - 1] + I_S[\exp(qV_{BC}/kT) - 1] \tag{6.4.5a}$$

and

$$I_C = -(I_S + I_{0C})[\exp(qV_{BC}/kT) - 1] + I_S[\exp(qV_{BE}/kT) - 1]. \tag{6.4.5b}$$

We now define

$$I_{ES} \equiv I_S + I_{0E}, \qquad I_{CS} \equiv I_S + I_{0C} \tag{6.4.6a}$$

and

$$\alpha_F \equiv \frac{I_S}{I_S + I_{0E}}, \qquad \alpha_R \equiv \frac{I_S}{I_S + I_{0C}}. \tag{6.4.6b}$$

In terms of these new variables, Equations 6.4.5 become

$$I_E = -I_{ES}[\exp(qV_{BE}/kT) - 1] + \alpha_R I_{CS}[\exp(qV_{BC}/kT) - 1] \tag{6.4.7a}$$

and

$$I_C = -I_{CS}[\exp(qV_{BC}/kT) - 1] + \alpha_F I_{ES}[\exp(qV_{BE}/kT) - 1] \tag{6.4.7b}$$

Equations 6.4.7 are the Ebers-Moll ($E-M$) equations for an *npn* transistor. In the corresponding equations for a *pnp* transistor, the current directions are changed to account for the polarity of the *pn* junctions. The diodes in a *pnp BJT* are under forward bias when V_{EB} and V_{CB} are positive (Problem 6.10).

---

* The base area A need not be constant (indeed it is not in an *IC* transistor) for the *Ebers-Moll* model to be applicable.

The $E-M$ equations predict directly the emitter and collector currents for the transistor; in conjunction with Kirchhoff's current law (the sum of all current into a node is zero), they also specify the base current. The $E-M$ model has four parameters (α_F, α_R, I_{ES}, and I_{CS}). From Equations 6.4.6, however, we see that only three parameters are independent; one can be obtained from the other three by the *reciprocity relationship*.

$$\alpha_F I_{ES} \equiv \alpha_R I_{CS} \equiv I_S \qquad (6.4.8)$$

The reciprocity relationship is discussed further at the conclusion of this section.

Equations 6.4.7 can be simplified by defining two new quantities, a diode current related to forward-active bias I_F, and one related to reverse-active bias I_R. These currents are expressed

$$I_F = I_{ES}[\exp(qV_{BE}/kT) - 1] \qquad (6.4.9a)$$

and

$$I_R = I_{CS}[\exp(qV_{BC}/kT) - 1]. \qquad (6.4.9b)$$

In terms of I_F and I_R,

$$I_E = -I_F + \alpha_R I_R \qquad (6.4.10a)$$

and

$$I_C = -I_R + \alpha_F I_F \qquad (6.4.10b)$$

An equivalent circuit that represents Equations 6.4.10 is shown in Figure 6.12. The circuit consists of diodes and current sources connected between the base and the emitter and between the base and the collector. The current sources are needed in the circuit to represent the current components that depend on voltages across a remote junction (I_R at the emitter and I_F at the collector). Applying Kirchhoff's current law to the circuit of Figure 6.12, we solve for the base current

$$I_B = -(I_E + I_C) = I_F(1 - \alpha_F) + I_R(1 - \alpha_R) \qquad (6.4.11)$$

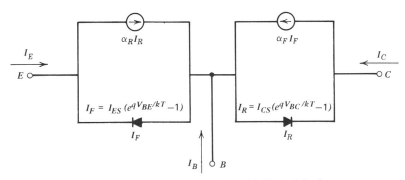

Figure 6.12 Equivalent circuit for the Ebers-Moll model of an *npn* transistor.

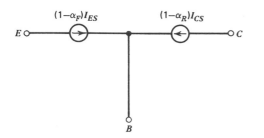

Figure 6.13 The Ebers-Moll representation for a transistor biased in the cut-off region.

Applications. To see how the $E-M$ model represents the transistor in the various regions of operation, consider first the cut-off region in which V_{BE} and V_{BC} are both negative. From Equations 6.4.9, 6.4.10 and 6.4.11, we obtain the equivalent circuit shown in Figure 6.13. The model reduces to two current sources that represent the reverse saturation currents of the two junctions.

In the forward-active region, the base-emitter junction is forward biased and the base-collector junction is reverse biased. We can rearrange Equations 6.4.10 to express collector current in terms of emitter current:

$$I_C = -\alpha_F I_E - I_R(1 - \alpha_F\alpha_R) \tag{6.4.12}$$

which, under active-bias conditions, becomes

$$I_C = -\alpha_F I_E + I_{CS}(1 - \alpha_F\alpha_R) \tag{6.4.13}$$

Similarly, reverse-active bias results in

$$I_E = -\alpha_R I_C + I_{ES}(1 - \alpha_F\alpha_R) \tag{6.4.14}$$

Inspection of Equations 6.4.13 and 6.4.14 shows that the parameters of the $E-M$ model, α_F, α_R, I_{ES}, and I_{CS}, can be obtained by measuring I_C versus I_E under forward-active bias or I_E versus I_C under reverse-active bias. If the forward-active measurements are used, the data should plot linearly with slope equal to $-\alpha_F$. The intercept of the plot with $I_E = 0$ corresponds to a measurement with open-circuited emitter, and the current flowing in this case is usually denoted I_{CO} (sometimes I_{CBO}) where

$$I_{CO} = I_C|_{I_E=0} = I_{CS}(1 - \alpha_F\alpha_R) \tag{6.4.15}$$

This current can be compared to the current I_{CEO} in the collector when the base is open circuited. From Equations 6.4.1 and 6.4.2,

$$I_{CEO} = I_C|_{I_B=0} = \frac{I_{CS}(1 - \alpha_F\alpha_R)}{(1 - \alpha_F)} = \frac{I_{CO}}{(1 - \alpha_F)} \tag{6.4.16}$$

The difference in magnitudes between I_{CO} and I_{CEO} can be traced to the boundary conditions enforced by the bias arrangements on the emitter-base junction. When I_{CO} flows, the emitter-base junction assumes a built-in reverse bias because of the extraction of some electrons from the emitter without their replacement from the external circuit. In this case the collector current is carried only by electrons generated in the base and by holes generated in the collector. In the second case, I_{CEO}, the base is open circuited instead of the emitter. Then, the base-emitter junction becomes forward biased, and the greater part of the collector current results from electrons carried across the base from the emitter. The leakage I_{CO} is therefore effectively multiplied by the transistor gain.

EXAMPLE   Ebers-Moll Equations

Calculate the reverse bias voltage that is present on the base-emitter junction of an *npn* *BJT* when the emitter is open-circuited and a reverse bias is placed on the base-collector junction. Assume $\alpha_F = 0.98$, $\alpha_R = 0.70$, $I_{CS} = 1 \times 10^{-13}$ A, $I_{ES} = 7.14 \times 10^{-14}$ A.

Solution

For this bias condition, the collector current is I_{CO} as given by Equation 6.4.15. Since the emitter current is zero, Equation 6.4.10a establishes that

$$I_F = \alpha_R I_R \approx -\alpha_R I_{CS} = I_{ES}\left(\exp\frac{qV_{BE}}{kT} - 1\right)$$

where we have used $I_R \approx -I_{CS}$.

Using the reciprocity relationship (Equation 6.4.8), we solve for V_{BE}

$$V_{BE} = \frac{kT}{q}\ln(1 - \alpha_F) = -0.10 \text{ V}$$

at $T = 300$ K.

The reverse bias on the base-emitter junction is seen to be dependent only on α_F for this case of an open-circuited emitter. This occurs because the bias is established by balancing the linking current with the current returned through the back-biased, base-emitter junction diode so that the emitter current is zero.

It is often very useful to obtain models in which the active elements (generators) are actuated by terminal currents. This is easily done for active bias with emitter current as the terminal variable using the equations we have developed. The equivalent circuit shown in Figure 6.14a is an emitter-actuated model that is consistent with Equation 6.4.13 if the leakage term at the collector is represented by I_{CO} (Equation 6.4.15). It is straightforward to derive expressions for the active-biased transistor when driven from the base and to show the validity of the circuit sketched in Figure 6.14b (Problem 6.11).

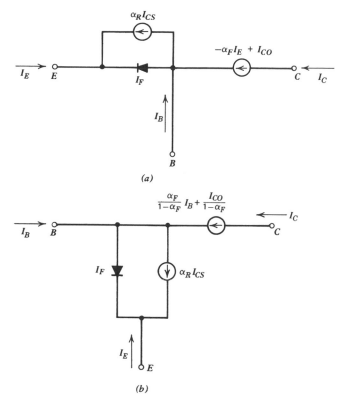

(a)

(b)

Figure 6.14   Equivalent circuits for *npn* transistors actuated
by terminal currents. (*a*) Emitter-current actuated circuit,
(*b*) base-current actuated circuit.

In saturation, the quantity of greatest interest is $V_{CE_{sat}}$, the voltage drop across
the "on-state" switch. In this case the $E-M$ model allows one to derive

$$V_{CE_{sat}} = \frac{kT}{q} \ln \left\{ \frac{\left[1 + \dfrac{I_C}{I_B}(1 - \alpha_R) \right]}{\alpha_R \left[1 - \dfrac{I_C}{I_B}\left(\dfrac{1 - \alpha_F}{\alpha_F} \right) \right]} \right\} \tag{6.4.17}$$

To obtain Equation 6.4.17, it is necessary to recognize that the exponential terms
in the expressions for the diode currents are large compared to unity when the
transistor is saturated (Problem 6.12). Since $V_{CE_{sat}}$ is small, a first approximation
that is often used for design purposes is to consider it negligible. If not considered
negligible, Equation 6.4.17 shows that $V_{CE_{sat}}$ changes only slowly with collector
current. Hence, an equivalent circuit consisting of a voltage source from collector
to emitter or, preferably of two sources connected from base to emitter and from
base to collector, respectively, would be appropriate. The $E-M$ equations do not
take account of any resistance in series with the junctions. The voltage drops

across series resistances, particularly in the collector regions of IC transistors, often exceed $V_{CE_{sat}}$ predicted by Equation 6.4.17; thus the "on-state" transistor switch is often modeled by a series voltage source and resistor denoted by $R_{C_{sat}}$.

In this section we have been discussing the static $E-M$ model, the model for quiescent conditions. Although this model can be modified to allow dynamic calculations (i.e., to solve the transient conditions), we shall find it advantageous to make use of another technique, known as charge-control modeling, for these calculations. The charge-control model is discussed in Chapter 7, where we shall also discuss a modification of the Ebers-Moll model that accounts for some important second-order effects.

Reciprocity

In deriving the Ebers–Moll equations we showed that there is a reciprocity relationship between the four model parameters (Equation 6.4.8). This relationship is a direct consequence of the incorporation of the expression for the linking current (Equation 6.1.9) in the Ebers–Moll model. In our derivation we proved that the reciprocity condition applies to a transistor with arbitrary base doping, constant base cross section, and negligible recombination since these conditions were imposed on Equation 6.1.9 in Section 6.1. The validity of reciprocity is, however, considerably wider in scope. It extends, as we shall argue, to IC transistors under their usual operating conditions. Even when not strictly valid, reciprocity and the Ebers–Moll model can usually be assumed without having theory stray too far from experiment. This is a very important point because the $E-M$ model is the cornerstone for virtually all large-signal, transistor-analysis techniques.

To explore this concept further, let us build upon the limited conditions for which reciprocity is proven: namely the case cited in Section 6.1. As a first complication, one might object to the condition $J_n = $ constant that was used to remove current from the integral (Equation 6.1.5). An IC transistor in the reverse-active mode should, for example, inject a great deal more charge at a given $V_{BC} = V_a$ than would the same transistor in the forward-active mode with $V_{BE} = V_a$. Thus, a greater component of base-recombination current would flow in the reverse-active mode than in the forward-active mode. Consider then the currents in an IC transistor under reverse-active bias. The magnitude of the linking current to the emitter (I_n) would be given by Equation 6.1.9 $|I_n| \approx I_S \exp(qV_a/kT)$ because the linking current is, in fact, not a function of x; thus the derivation of the equation is still valid. By the arguments of Section 6.3, the recombination component would be proportional to the exponential factor; we shall call it $I_{rB} = I_{RB} \exp(qV_a/kT)$. The total current crossing the collector will, by Equations 6.4.9 and 6.4.10, be $I_{CS} \exp(qV_a/kT)$ neglecting small voltage-independent components. Since we are, for the present, considering only extra current resulting from recombination, the total current crossing the collector junction is just the sum of the linking component and the recombination component:

$$I_{CS} \exp\left(\frac{qV_a}{kT}\right) = I_S \exp\left(\frac{qV_a}{kT}\right) + I_{RB} \exp\left(\frac{qV_a}{kT}\right) \qquad (6.4.18)$$

Because the magnitude of the ratio between emitter current and collector current is α_R and the magnitude of the current at the emitter is just $|I_n|$, we have

$$\alpha_R = \left|\frac{I_E}{I_C}\right| = \left|\frac{I_S \exp(qV_a/kT)}{(I_S + I_{RB}) \exp(qV_a/kT)}\right| \qquad (6.4.19)$$

which, using Equation 6.4.18, reduces to

$$\alpha_R = \frac{I_S}{I_{CS}} \qquad (6.4.20)$$

As we can easily see from the derivation, a component of junction current resulting from any parasitic (or unwanted) loss can be added to the linking component $|I_n|$ and we will still be able to obtain Equation 6.4.20 provided that the parasitic loss is proportional to $\exp(qV_a/kT)$ so that the voltage dependence can be factored and cancelled. Furthermore, if parasitic losses are considered for the IC transistor in the forward-active mode, we obtain $\alpha_F = I_S/I_{ES}$. This equation for α_F and Equation 6.4.20 for α_R are identical with Equation 6.4.8. Thus, reciprocity is established for the transistor with parasitic losses as well as for the simpler structure of Section 6.1. In particular, the asymmetry of the IC transistor does not invalidate reciprocity because the alphas for forward- and reverse-active operation vary inversely with the respective junction saturation currents. The constraint on this statement is that all junction currents must vary with voltage in the same way as the linking component. This constraint is violated if recombination in the space-charge region becomes significant [its voltage variation is $\exp(qV_a/2kT)$ as shown in Equation 5.3.24] or if high-level effects occur. We consider the latter in Chapter 7.

6.5 Devices: Planar Bipolar Amplifying and Switching Transistors

In discussing bipolar transistors for integrated circuits, it is helpful to divide them into two broad categories defined by their ultimate use: either amplification or switching. Transistors of either type are nearly always fabricated in an epitaxial layer of relatively high resistivity silicon. Typical values for resistivity of the epitaxial layer and for other design parameters of IC transistors are given in Table 6.1. The purpose of this epitaxial layer is to obtain a controlled, lightly doped region of collector material adjacent to the base junction. A buried layer of heavily doped silicon is added to maintain a high conductivity collector region below this.

The lightly doped region allows the collector-base junction to sustain relatively high voltages without breaking down (Section 4.4), while the higher doped buried-layer region reduces series resistance between the junction and the metallic collector contact. Inclusion of the buried layer reduces series ohmic resistance from the kilohm range to a few hundred ohms in typical devices. For some applications even this resistance is too much, and an extra processing step is added in which the heavily doped n^+ region under the collector contact is diffused until it reaches

Table 6.1   **Typical Design Parameters for *IC* Transistors**

	Amplifying (Junction-Isolated)	Switching (Junction-Isolated)	Switching (Oxide-Isolated)
Epitaxial Film			
Thickness	10 μm	3.0 μm	1.2 μm
Resistivity	1 Ω-cm	$0.3 - 0.8$ Ω-cm	$0.3 - 0.8$ Ω-cm
Buried Layer			
Sheet resistance		~ 20 $\Omega/\square$	~ 30 $\Omega/\square$
Up diffusion	2.5 μm	1.4 μm	0.3 μm
Emitter			
Diffusion depth in base	2.5 μm	0.8 μm	0.25 μm
Sheet resistance	5 $\Omega/\square$	12 $\Omega/\square$	30 $\Omega/\square$
Base			
Diffusion depth	3.25 μm	1.3 μm	0.5 μm
Sheet resistance	100 $\Omega/\square$	200 $\Omega/\square$	600 $\Omega/\square$
Substrate			
Resistivity		~ 10 Ω-cm	~ 5 Ω-cm
Orientation		(111)	(111)

the buried layer. This extended diffused region is called a *collector plug*, and its inclusion reduces series resistance to the order of 10 ohms.

The junction-isolation regions, base, emitter, and collector-contact regions are established by successive diffusions unless oxide-isolation between devices is to be used. Top and cross-sectional views of a typical junction-isolated *IC* transistor are shown in Figure 6.15. Plots of the impurity concentrations measured along a coordinate perpendicular to the surface and passing through the emitter, base, and collector are given in Figures 6.16 and 6.17. These figures show that the major differences in the design of switching and amplifying transistors lie in the thickness and resistivity of the epitaxial layer. Both quantities are greater in amplifying devices than in switching transistors; this results in increased breakdown voltage and reduces a parasitic effect (the Early effect) that will be described in Chapter 7. For switching devices, saturation ("on-state") resistance should be minimized, which necessitates epitaxial thicknesses of a few micrometers at resistivities in the order of some tenths of an Ω-cm.* If Schottky clamping is to be used on the switching transistor (as described in Chapter 3), the resistivity must be greater than 0.1 Ω-cm in order to obtain a good metal-semiconductor barrier. Schottky-clamped transistors differ from unclamped switching devices only in the extended finger of metal from the base contact to the undoped epitaxial region forming the collector as shown in Figure 6.18. The base doping level is ideally very low in order to

---

* When the BJT is saturated, the voltage drop between the collector and the emitter is equal to V_{CEsat} as given by Equation 6.4.17 in series with the drop across series resistance in the structure. Typically the largest series resistance is in the epitaxial layer.

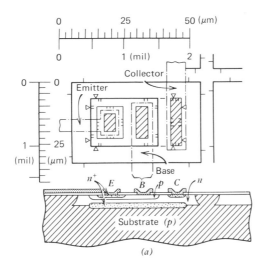

Collector

Emitter

0 ─ 0
          25
1 ─
(mil) (μm)

Base

n^+ E   B p C n

Substrate (p)

(a)

Drawings for layout of an integrated circuit often use the following code
for the various layers:

▽ ── Buried layer
└ ── Isolation diffusion
╨ ── Base diffusion
╫ ── Emitter diffusion
//////// Contact windows
─·─·─·─ Metal

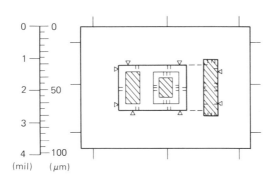

0 ─ 0
1 ─
2 ─ 50
3 ─
4 ─ 100
(mil)   (μm)

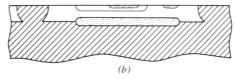

(b)

Figure 6.15 Top views (showing dimensions) and cross
sections of (a) a junction-isolated, digital IC transistor with
5 μm minimum dimensions, (b) an amplifying (35 V breakdown)
transistor.[8]

303

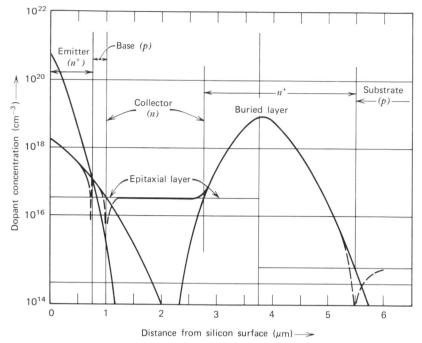

Figure 6.16   Cross section of the diffusion profile of the transistor in Figure 6.15a.[8]

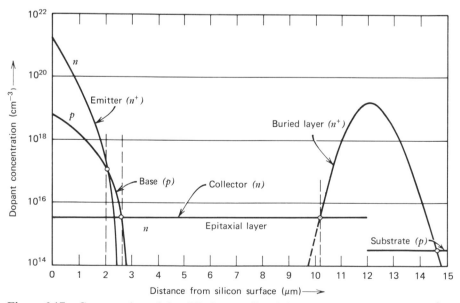

Figure 6.17   Cross section of the diffusion profile of the transistor in Figure 6.15b.[8]

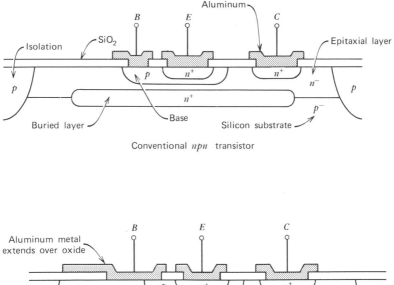

Conventional *npn* transistor

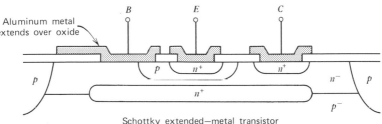

Schottky extended—metal transistor

Figure 6.18   Comparison of cross sections of (*a*) conventional *npn* transistor, (*b*) Schottky-clamped, extended-metal *npn* transistor.

maximize emitter injection efficiency γ, but it cannot be too low (not less than about 5×10^{16} cm^{-3}) without the possibility of a poor metal-semiconductor contact or even surface inversion (described in Chapter 8) at the metallic contact to the base. If the doping is too low, series resistance in the base also becomes problematic. The emitter is usually doped very heavily to increase emitter efficiency. When doped above about 10^{20} cm^{-3}, however, efficiency falls because of decreased hole lifetime in the emitter (which enhances hole injection from the base) and because of effects associated with the narrowing band gap of degenerately doped silicon (Section 1.1).[6]

Reducing the base width and decreasing the base conductivity are two means of increasing gain, as we have noted. As the base width is reduced to submicron dimensions, however, it becomes increasingly likely that the collector-base space-charge region may reach through to the emitter-base space-charge region, completely depleting the base region (Figure 6.19). This condition, called *punch-through* results in a highly conductive path from emitter to collector and can lead to damaging currents in a transistor. This is one mode of failure that is influenced by the collector-base bias. The other failure mode that is sensitive to V_{CB} is avalanching of the collector junction. Avalanche breakdown has been described in connection with diode effects in Section 4.4.

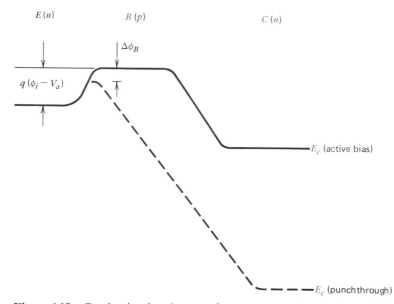

Figure 6.19   Conduction-band energy for an *npn* transistor under active bias with normal operating voltages (solid line). At high collector biases the entire base region may be depleted of mobile carriers, causing punchthrough (dashed curve) and a corresponding reduction in barrier height ($\Delta\phi_B$) for electron injection from the emitter.

Advanced Processes.   The size and parasitic capacitances associated with an *IC* transistor can be reduced below the values attained with the technologies discussed thus far by using more advanced processes. Figure 6.20 indicates the effects of some of these processes on the cross sections and surface areas of Schottky-clamped *BJT*s. A conventional, junction-isolated, Schottky-clamped transistor is shown first in Figure 6.20*a*.

To reduce the surface area of the transistor, a *washed emitter* can be used (Figure 6.20*b*). In a washed-emitter process, the thin oxide that forms over the emitter region during the emitter drive-in diffusion is etched away without a masking step (as would be used conventionally). The metal emitter contact covers the entire original emitter-diffusion pattern. A base-emitter short circuit is avoided (most of the time) because the emitter dopant diffuses laterally under the thicker oxide covering the base. The washed-emitter process thus allows the emitter to be as small as the permissible lithographic linewidth, and the overall size of the transistor can be reduced.

If oxide isolation is used (*cf* Section 2.6 and Figure 5.19), the base of the *BJT* can be in contact with the oxide (Figure 6.20*c*), substantially reducing both area and sidewall capacitance. In addition, oxide isolation eliminates the possibility of lateral transistor action with minority-carrier collection in the isolation-diffusion region as can happen when junction isolation is used. The size of the *BJT* can be

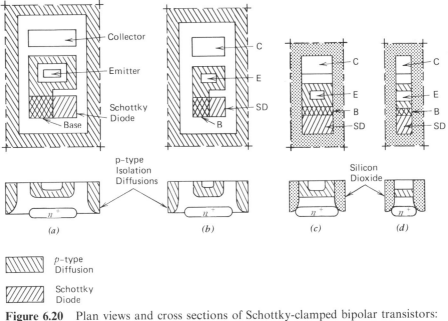

Figure 6.20   Plan views and cross sections of Schottky-clamped bipolar transistors: (*a*) conventional junction-isolated BJT, (*b*) washed-emitter BJT with junction isolation, (*c*) oxide-isolated transistor, (*d*) walled-emitter BJT. The p-type diffused regions and the Schottky diode are cross hatched. The contact regions are labeled.

reduced still further by allowing the emitter edge to abut the oxide isolation region in a design called a *walled emitter* (Figure 6.20*d*). Because the depth of the oxide isolation is limited to dimensions of the order of 1 μm, the base and emitter diffusions must likewise be scaled to smaller dimensions than with junction isolation. The surface dimensions are also typically reduced. Figure 6.21 shows the cross section and dopant profile for an advanced-process, oxide-isolated *BJT* for which the minimum surface dimension (the emitter width) is 1 μm.

General Considerations.   The symbol adopted to represent a bipolar transistor is so widely used that it is doubtless familiar to the reader. As sketched in Figures 6.22*a* and 6.22*b*, it consists of two angled lines that identify the collector and emitter leads; these are in contact with a straight line that stands for the base. An arrow in the direction of the *pn* junction of the emitter-base diode differentiates an *npn* transistor (Figure 6.22*a*) from a *pnp* transistor (6.22*b*). In reality, *npn* transistors in integrated circuits are automatically coupled to a parasitic *pnp* structure with the substrate acting as a collector (Figure 6.22*c*). For forward-active bias on the *npn* transistor, the substrate *pnp* is cut-off. In other modes of *npn* operation, the *pnp* can become active. It is important to be certain that the *pnp* transistor does not have any appreciable gain to avoid undue parasitic loss and transient misbehavior in these cases. This is another point where the buried layer is valuable.

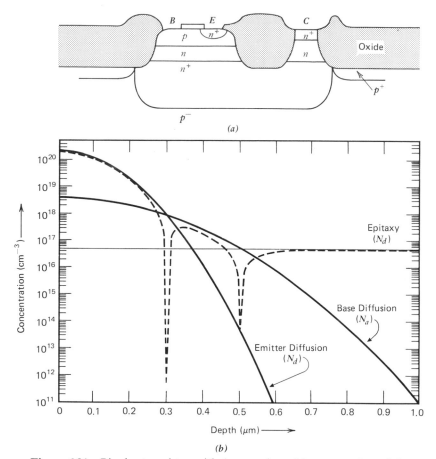

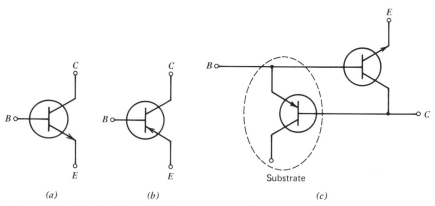

Figure 6.21 Bipolar transistor with 1 μm emitter: (*a*) cross section of the oxide-isolated device, (*b*) dopant profile (solid lines—individual dopant densities, dashed lines—net dopant densities).

Figure 6.22 Standard symbols for (*a*) *npn* transistor and (*b*) *pnp* transistor. (*c*) The *IC npn* transistor automatically has a parasitic *pnp* transistor (dashed circle) attached to it because of the fabrication process.

308

By increasing the base charge of the parasitic *pnp* transistor and by lowering lifetime in this region, it is not difficult to reduce α_F in the parasitic *pnp* transistor below 0.05 in typical cases. For junction-isolated bipolar transistors, the parasitic *pnp* transistor also includes laterally injected holes that travel through the *n*-epitaxial region to be collected at the reverse-biased, *p*-isolation-region diffusion. In practice, it is necessary to keep the outer edge of the base at least a few micrometers removed from the isolation junction to reduce this component to an acceptable value. This constraint is lessened if oxide isolation is used.

Lateral injection of electrons can also occur from the emitter into the base. We have not explicitly mentioned this injection loss in the discussion of transistor gain, but the dimensions of modern *IC* transistors have been so reduced that it can contribute a detectable loss. The transport factor for electrons injected laterally is considerably lower than that for electrons injected downward because the effective lateral base width is large and also because some of the electrons are in the vicinity of the ohmic base contact where they will be lost to recombination. Injection laterally is inhibited naturally by the fact that ϕ_i increases with doping (Equation 4.2.10) and electron injection at a given bias is smaller where ϕ_i is greater. It is also inhibited by the lack of any aiding field in the lateral direction across the base. The designer can keep this component small by keeping the emitter shallow and the lateral base dimension large.

Transistors for integrated circuits can be made in smaller surface areas than are needed for typical *IC* resistors and capacitors. It is thus advantageous to design circuits that make use of transistors whenever possible, and to incorporate other devices only when absolutely necessary. If bipolar transistors are used for the active devices, they will typically make up the overwhelming fraction of total devices in an integrated circuit. Transistors are very often used in place of simpler *pn*-junction diodes, as described in Chapter 5, even though they take up slightly more surface area. This is because an *IC* process that is needed to produce transistors will yield diodes with lower series resistance and shorter switching times when those diodes are made up of the transistor structure in a diode connection. There are a total of five ways to connect a transistor as a diode because only two of three possible regions need to be accessed. Of the various diode connections, shown in Figure 6.23, the lowest series resistance and fastest switching times are obtained in the connection ofFigure 6.23a, which is used most often.[7]

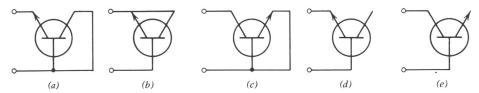

| (a) | (b) | (c) | (d) | (e) |

Figure 6.23 Five ways in which a transistor structure can be connected to obtain diode behavior.

Summary

An understanding of the operation of bipolar transistors can be obtained by focusing on the behavior of minority carriers at biased *pn* junctions. Because the space-charge region in a *pn* junction is a barrier to majority-carrier flow, it acts as a collector of minority carriers. Thus, when two *pn* junctions are located close together, bias on one of them can influence the minority-carrier populations in the vicinity of the second, and markedly alter its electronic behavior. This idea, which is called *bipolar transistor action*, has far ranging and significant consequences. Making use of transistor action, one can build effective switches and amplifying elements.

A useful formulation of the equations for transistor action follows from the fact that majority-carrier currents in the transistor *base* between the two junctions must be nearly zero because a barrier to majority-carrier flow is always encountered along this path. The current linking the two junctions consists, therefore, of minority carriers. The size of the linking current depends inversely on the total majority-carrier charge in the base. Currents are nonlinear functions of terminal voltages in transistors. The nonlinearity is very strong, and practical design of transistor circuits is frequently aided by considering approximating equations that are valid over selected regions of base-emitter and base-collector bias. Under active bias, the collecting junction is continuously maintained under reverse bias while the emitting junction is continuously forward biased. Current gain in the transistor in the active-bias mode is a function of the efficiency of the injection of minority carriers into the base region from where they can be collected, as well as of their loss to recombination in the base. These mechanisms depend strongly on processing procedures and on device geometry.

The *Ebers-Moll* model describes the operation of bipolar transistors over all ranges of bias. It is a very important first-order model and is the basis for more refined descriptions of the transistor. A reciprocity condition is part of the Ebers-Moll model. The reciprocity condition predicts that a transistor biased in the reverse-active mode will deliver the same output current as it will when biased in the forward-active mode provided that the respective terminal voltages are interchanged. This result applies to asymmetric as well as to symmetric transistors because loss mechanisms that cause a reduction in the current gain simultaneously cause an increase in the junction-saturation currents. Provided that the losses have the same voltage dependence as does the linking current, there is an exact inverse relationship and reciprocity is valid.

Transistors in integrated circuits for both switching and amplification are typically built in an epitaxial layer with buried layers beneath the collector regions. The chief difference between devices for these two applications is in the dimensions perpendicular to the surface. Switching transistors are usually built in thinner epitaxial regions, and more emphasis is placed on reducing the effects of series collector resistance and charge storage in these devices than in the design of amplifying transistors. The engineering trade-offs for the two classes of devices differ. In integrated circuits, conventional *npn* transistors are shunted by parasitic *pnp* transistors in which the substrate is the collector. Although the *pnp* transistor is cut off when the *npn* transistor is under active bias, it can become active during switching. Because they are smaller than other devices in bipolar integrated circuits, bipolar transistors are used by circuit designers whenever possible.

References

1. W. Shockley, *Bell Syst. Tech. J.*, **28**, 435 (1949).
2. J. L. Moll and I. M. Ross, *Proc. IRE*, **44**, 72 (1956).
3. H. K. Gummel, *Proc. IRE*, **49**, 834 (1961).
4. H. Kroemer, *Arch. Elek. Übertrag.*, **8**, May, August, November (1954).
5. J. J. Ebers and J. L. Moll, *Proc. IRE*, **42**, 1761 (1954).
6. J. W. Slotboom, *Solid-State Electronics* **20**, 279 (1977).
7. C. S. Meyer, D. K. Lynn and D. J. Hamilton, *Analysis and Design of Integrated Circuits*, McGraw-Hill., New York, 1968, pp. 248–258.
8. H. Camenzind, *Electronic Integrated Systems Design.* Copyright 1972 by Litton Educational Publishing, Inc. Reprinted by permission of Van Nostrand Reinhold Company.
9. P. E. Gray, D. DeWitt, A. R. Boothroyd, and J. F. Gibbons, *Physical Electronics and Circuit Models of Transistors*, Wiley, New York, 1964, p. 145.

Textbooks

A. G. Milnes *Semiconductor Devices and Integrated Electronics*, Van Nostrand Reinhold, New York, 1980.
B. G. Streetman, *Solid-State Electronic Devices*, Prentice-Hall Inc., Englewood Cliffs, N. J., 2nd Edition 1980.

Problems

6.1* Show by using Equation 6.1.2 that there will be a constant field in a region in which the doping density varies exponentially with distance. If a constant field of -4000 V cm^{-1} exists in the base of a transistor having a base width of 0.3 μm and the dopant density is 10^{17} at the emitter-base edge, what is the dopant density at the edge of the collector-base space-charge region?

6.2 Apply Equation 6.1.5 to an *npn* transistor under active bias. Consider that $I_C = -J_n A_E$ and prove, therefore, that

$$I_C = \frac{q A_E \tilde{D}_n N_a(x) n(x)}{\int_x^{x_B} N_a \, d\xi}.$$

6.3† Use the expression derived in Problem 6.2 to find the x dependence of the electron density in the base of the transistor of Problem 6.1. Sketch plots of $n(x)$ versus x in the base under active-bias conditions for two cases of *npn* transistors that are biased so that both pass the same collector current. Use a single set of axes for the two transistors: (i) the transistor described in Problem 6.1, and (ii) a transistor having a constant base doping of 10^{17} cm^{-3} but otherwise identical to that in Problem 6.1. Note that both plots have the same gradient at the edge of the collector space-charge region. Why is this the case?

6.4† Derive expressions for the total stored minority charge in the base in the two transistors described in Problem 6.3. Use this result to compare the values of β_F in the two devices under the assumption that base current results only from recombination in the base.

6.5* (a) Find Q_{BO}, the base-majority charge in the quasi-neutral region for the transistor data plotted in Figure 6.4.

   (b) Find the base width x_B if the average base doping is 10^{17} atoms cm^{-3}. (Assume that $\tilde{D}_n = 25$ cm^2 s^{-1}.)

6.6† Use approximating techniques to determine the net dopant atoms per unit area between the base-emitter junction plane and the base-collector junction plane in the switching transistor of Figure 6.16 and the amplifying transistor of Figure 6.17. Compare these values with the dopant atoms per unit area in the quasi-neutral base (Gummel Number in Equation 6.2.3) for the transistor of Problem 6.5 and explain any differences between the results.

6.7 Redo the problem considered in the example of Section 6.3 if both the base–emitter and base–collector junctions are linearly graded with the grading coefficient $a = 10^{22}$ cm^{-4} (where $N_d - N_a = ax$), and $\phi_i = 0.872$ V.

6.8* The accompanying table gives values of the net dopant density $(N_d - N_a)$ as a function of position relative to the junction plane between the emitter and base on a transistor with the doping profile of Figure 6.17.

Net dopant density $(N_d - N_a)$ (cm^{-3})	7.36×10^{17}	-1.25×10^{16}	-4.0×10^{16}	-1.9×10^{16}	-6.4×10^{15}	-1.5×10^{15}
Distance from base-emitter junction (μm)	-0.25	$+0.09$	$+0.23$	$+0.30$	$+0.41$	$+0.50$

Assume that the transistor has an active area of 2×10^{-6} cm^2. (a) Plot these values on a linear scale and argue that the profile can be roughly approximated by a one-sided step junction having $N_a \approx 2 \times 10^{16}$ cm^{-3}. (b) Make a plot similar to that of Figure 6.8 to show Q_{VE} versus total voltage for the junction over the bias range from breakdown to $+0.3$ V. Find values for ϕ_i and K_A as shown in Figure 6.8.

6.9 Use the result of Problem 6.1 and the data of Problem 6.8 to estimate the size of the field in the quasi-neutral base of the transistor having the doping profile sketched in Figure 6.17.

6.10 Draw an equivalent circuit similar to Figure 6.12 and write equations similar to Equations 6.4.9 and 6.4.10 to represent the Ebers–Moll model for a *pnp* transistor.

6.11 Show that Figure 6.14*b* represents the Ebers–Moll model for a transistor driven by base current.

6.12* Derive Equation 6.4.17 for the voltage drop across a saturated transistor as predicted by the Ebers–Moll model. Evaluate V_{CEsat} for $I_C/I_B = 10$, $\alpha_F = 0.985$, and $\alpha_R = 0.72$.

6.13 Use the Ebers–Moll equations for a *pnp* transistor (Problem 6.10) to find the ratio of the two currents I_{CEO} to I_{CBO} where I_{CEO} is the current flowing in the reverse-biased collector with the base open circuited, and I_{CBO} is the current flowing in the reverse-biased collector with the emitter open circuited. Explain the cause for the difference in the currents in terms of the physical behavior of the transistor in the two situations.

6.14† Consider an *npn* transistor that is biased in the active mode. At time $t = 0$, an intense visible light is focused on the collector-base space-charge region. The light produces G hole-electron pairs per unit time inside the space-charge zone. (Consider that qG is roughly of the same order as the dc base current I_B.)
   (a) If the emitter-base voltage and collector-base voltage are both held constant, what will be the values of base, emitter, and collector currents for $t > 0$?
   (b) Repeat part a if the base is driven by a current source so that the illumination does not alter I_B.

6.15 Near room temperature, the collector current I_C and the base current I_B are both normally positive for an *npn* transistor biased in the active mode. Assume that an *npn* transistor is under active bias and I_C is held constant while the temperature is increased. If one measures I_B, he will find its magnitude to decrease and ultimately to change sign. What physical effects account for this behavior? (Consider all currents at the base lead.)[9]

6.16 Consider an *npn* transistor in which both the base and emitter regions are nonuniformly doped and in which recombination of reverse-injected holes into the emitter takes place at the emitter contact. Show that the emitter efficiency γ can be written as $\gamma = (1 + Q_{BO}\tilde{D}_{pE}/Q_{EO}\tilde{D}_{nB})^{-1}$ where Q_{EO} is the emitter dopant in the quasi-neutral region of the emitter, Q_{BO} is the base dopant in the quasi-neutral region of the base, and $\tilde{D}_{pE}$ and $\tilde{D}_{nB}$ are the effective minority-carrier diffusion constants in these regions. Show the consistency of this result with that given for the prototype transistor when it is subjected to the same recombination constraints (Equation 6.2.20).

6.17* An *npn* transistor has a cross-sectional area of 10^{-5} cm^2 and a quasi-neutral base that is uniformly doped with $N_a = 4 \times 10^{17}$ cm^{-3}, $D_{nB} = 18$ cm^2 s^{-1}, and $x_B = 0.5$ μm.
   (a) If conditions in the emitter are similar to those described in Problem 6.16 and the emitter has a total dopant (Q_{EO}/q) of 8×10^9 atoms and $\tilde{D}_{pE}$ is 2 cm^2 s^{-1}, calculate γ.
   (b) Estimate α_T, the transport factor if the base lifetime is 10^{-6} s.
   (c) Calculate β_F for this transistor. What is the percentage error involved in taking $\beta_F \approx Q_{EO}\tilde{D}_{nB}/Q_{BO}\tilde{D}_{pE}$? This simplified form is sometimes used to estimate β_F.

6.18† Assume that the transistor of Problem 6.17 is placed in a radiation environment where the electron lifetime in the base decays according to the equation: $\tau_n = \tau_{n0} \exp -(t/t_d)$. In this equation t is measured in days and t_d is 10 days. Plot β_F as a function of time in days and determine the time interval until β_F drops to unity. What is the base lifetime when this occurs?

6.19† Consider the transistor structure shown in Figure P6.19.

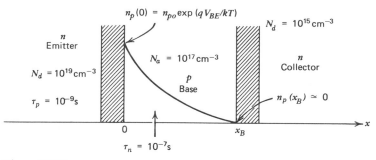

Figure P6.19

(a) Derive an expression for the electron distribution in the base as a function of x assuming that recombination takes place. (This will demand solving an equation similar to Equation 5.3.10.)

(b) The slope of the distribution at $x = 0$ is proportional to the injected electron current, and the slope at $x = x_B$ is proportional to the collected electron current. The difference represents current lost because of recombination in the base (supplied by the base lead). Find an expression for this base recombination current as a function of L_n and x_B/L_n $[L_n = (D_n \tau_n)^{1/2}]$.

(c) The dopant concentrations in the emitter, base, and collector are assumed to be uniform throughout each region with values of 10^{19}, 10^{17}, and 10^{15} cm^{-3}, respectively. The hole lifetime in the emitter is 10^{-9} s and the electron lifetime in the base is 10^{-7} s. The emitter and base widths are each 1 μm. Calculate the emitter efficiency and β_F. Use Figure 1.15 to obtain values for the diffusion constants. (Hint: For this problem, it is best to express solutions for the continuity equation as the sum of hyperbolic functions cosh (x/L_p) and sinh (x/L_p).)

6.20[†] By considering collector current as charge in transit ($I_C = qnv_dA$), use the known distribution of injected electrons in the base of an npn prototype transistor biased in the active mode to solve for the base transit time τ_B. That is, formulate the transit time as the integrated sum of incremental path lengths divided by velocity and then carry out the integration to prove that τ_B is given by $\tau_B = x_B^2/2D_n$ (i.e. take $\tau_B = \int_0^{x_B} dx/v$).

7

BIPOLAR TRANSISTORS II: LIMITATIONS AND MODELS

The discussion of bipolar transistors, thus far, has been framed in terms of fundamental operation. For example, the description of a transistor biased in the active mode has emphasized that carriers are injected with high efficiency from the emitter into the base. There, the injected carriers vastly increase the minority-carrier population and flow from the base-emitter contact toward the collector. Upon reaching the base-collector space-charge region, these added free carriers are swept across to the quasi-neutral collector region where they are majority carriers and flow as ohmic current through the collector lead.

In real transistors this simple description is remarkably accurate over a wide range of conditions. There is, however, more that needs to be considered than this basic ideal behavior. Real transistors are limited by processes that we have not specifically considered thus far. For example, a transistor as described in Chapter 6 would have a current-source output in the active mode; that is, output current would not depend in any way on the voltage applied between the base and collector terminals. In real transistors, however, both output currents and output voltages influence the device performance. Similarly, the theory that has thus far been presented would predict a current gain in the active mode that is insensitive to bias level. But this is only a rough approximation. A first objective of this chapter, therefore, is to investigate mechanisms that must be considered to obtain a more realistic picture of transistor operation.

Another major inadequacy of the theory developed thus far is that it does not deal with time-varying effects. Their consideration is therefore the second major topic to be developed in this chapter. A great simplification in considering time-varying effects will be obtained by modeling of transistors as charge-controlled devices. This viewpoint leads very naturally into the important topic of large-signal transient models for bipolar transistors. The amalgamation of the Ebers-Moll model with the charge-control model provides a powerful analysis technique. It will be straightforward to derive a small-signal ac equivalent circuit from the more general large-signal representation. This small-signal equivalent circuit, known as the *hybrid-pi* circuit, is relatively easy to characterize because it is composed of elements that relate directly to physical mechanisms in the transistor.

The production of *pnp* transistors within the confines of standard *IC* processing is described in a final section. Two basic types of *pnp* transistors, *substrate pnp* and *lateral pnp* transistors are typically fabricated. The substrate *pnp* transistor has only limited application because its collector is not isolated. The lateral *pnp* transistor has severely limited performance compared to that of *npn* transistors. Simple models for loss mechanisms in this device explain the limited performance. An important application of lateral *pnp* transistors—a dense form of logic circuitry known as *integrated-injection logic* (I^2L)—is described at the conclusion of the chapter.

7.1 Effects of Collector Bias Variation (Early Effect)

When we considered bipolar transistors under active bias in Chapter 6, the function of the voltage applied across the collector-base junction was merely to insure the efficient collection of base minority carriers and their delivery to the collector region. The magnitude of the bias only limited the range of the permissible collector voltage swing, provided that it was below the breakdown value. We have overlooked an important aspect of *pn*-junction electronics in taking this simplified view. As we noted in Chapter 4, the width of a reverse-biased *pn* junction is voltage dependent; in fact, this bias dependence made possible the operation of junction field-effect transistors. In the case of bipolar transistors, a changing collector-base bias causes variation in the space-charge layer width at the collector junction and, consequently, in the width of the quasi-neutral base region. These variations result in several effects that complicate the performance of the device as a linear amplifier. Base-width modulation as a consequence of variations in collector-base bias was first analyzed by James Early,[1] and the phenomenon is generally called the *Early effect.*

The dependence of collector current on collector-base bias can be formulated directly by using the integral equations for active bias (*npn* transistor) developed in Section 6.2. In particular, from Equations 6.2.1 and 6.1.10, we can write

$$I_C = \frac{q\tilde{D}_n n_i^2 A_E \exp(qV_{BE}/kT)}{\int_0^{x_B} p\, dx} \tag{7.1.1}$$

where the integration is performed over x_B, the width of the quasi-neutral base region and the other terms have been defined in Section 6.1.

Variation in the base width with voltage V_{CB} causes a variation in the collector current that can be written

$$\frac{\partial I_C}{\partial V_{CB}} = \frac{-q\tilde{D}_n n_i^2 A_E \exp(qV_{BE}/kT)p(x_B)}{\left[\int_0^{x_B} p\, dx\right]^2} \frac{\partial x_B}{\partial V_{CB}} \tag{7.1.2}$$

Several of the terms in Equation 7.1.2 can be combined to represent collector current itself, so that Equation 7.1.2, which represents the small-signal conductance at the collector-base junction, can be written

$$\frac{\partial I_C}{\partial V_{CB}} = -I_C p(x_B)\left[\frac{1}{\int_0^{x_B} p\, dx}\right]\left[\frac{\partial x_B}{\partial V_{CB}}\right]$$

$$= -\frac{I_C}{V_A} = \frac{I_C}{|V_A|} \tag{7.1.3}$$

Since the collector is reverse biased, the derivative in Equation 7.1.3, $\partial x_B/\partial V_{CB}$, is negative, and the Early effect results in an increase in I_C when V_{CB} is increased. This increase is evident if we examine *common-emitter* output characteristic curves such as the family shown in Figure 7.1a. These curves are actually plots of I_C versus V_{CE}, but V_{CE} has almost the same value as V_{CB} for bias in the active region ($V_{CE} \approx V_{CB} + 0.7$ V).

Equation 7.1.3 reveals that the Early effect varies linearly with collector current. The reciprocal of the factor multiplying the current has the dimension of a voltage and has been defined[2] as the *Early voltage.* The Early voltage is usually given the symbol V_A. From Equation 7.1.3, an equation for V_A for an *npn* transistor is

$$V_A = \frac{\int_0^{x_B} p\, dx}{p(x_B)\partial x_B/\partial V_{CB}} \tag{7.1.4}$$

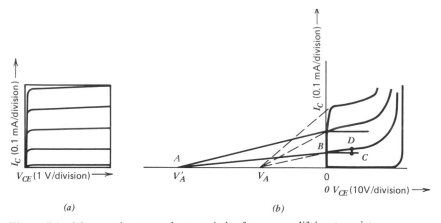

Figure 7.1 Measured output characteristics for an amplifying transistor: collector current versus V_{CE}. (*a*) vertical 0.1 mA/div., horizontal 1 V/div. (*b*) vertical 0.1 mA/div., horizontal 10 V/div.; tangents to the measured curves (at the edge of saturation) are extended to the voltage axis (dashed lines) to determine the Early Voltage V_A. Extension of lines drawn approximately tangential to the characteristic in the active region intersects the voltage axis at V_A' (solid lines).

Again, the derivative in Equation 7.1.4 is negative and, thus, the Early voltage is negative for an *npn* transistor. The analogous effect of emitter-base space-charge layer widening when the transistor is biased in the reverse-active mode can also be interpreted by using a different Early voltage, usually denoted as V_B.

Except for high-level effects, the three terms that define V_A depend only on the transistor manufacturing process and on collector-base voltage. In practice, the collector-base voltage dependence of V_A itself is usually treated as negligible and the Early voltage is approximated by its value at a single bias, often at $V_{CB} = 0$. By using this condition to specify V_A, we can expect that a series of tangents to curves of I_C versus V_{CB} (V_{CE} in practice), drawn at the edge of the forward-active region where V_{CB} is approximately zero, should intersect the V_{CE} axis at V_A. In Figure 7.1*b* dashed lines drawn from the point at which the transistor moves out of saturation do intersect at a common point, indicating the Early voltage. For circuit-design and analysis, however, interest is not in an Early voltage characterizing the edge of saturation, but rather in a parameter to use in the forward-active region. If tangents are drawn to the curves of I_C versus V_{CE} in the active region, they do not, in general, intersect each other on the voltage axis. It is usual, however, to approximate an intersection point appropriate to the range of bias of the transistor as at V_A', formed by the solid lines in Figure 7.1. This construction forms similar triangles AOB and BCD (if we neglect the small saturation-voltage offset of the *BJT* curves). We can infer Equation 7.1.3 directly from these similar triangles if the differential elements ∂I_C and ∂V_{CB} in Equation 7.1.3 are treated as incrementals ΔI_C (indicated by line CD) and ΔV_{CB} (approximately indicated by line BC) since line AO indicates V_A' and line OB indicates I_C.

A useful and informative alternative expression for V_A can be obtained by re-arranging some terms in Equation 7.1.4. First, we use Equation 6.1.8 to express the numerator in terms of the base majority-charge density Q_B in that portion of the base where transistor action is taking place:

$$\int_0^{x_B} p \, dx = \frac{Q_B}{q} \tag{7.1.5}$$

We then recognize that the denominator of Equation 7.1.4 represents the derivative of base charge Q_B with respect to V_{CB}:

$$qp(x_B) \frac{dx_B}{dV_{CB}} = \frac{dQ_B}{dV_{CB}} \tag{7.1.6}$$

The derivative on the right-hand side of Equation 7.1.6 can be related to C_{jc}, the small-signal capacitance per unit area at the collector-base junction.

$$\left| \frac{dQ_B}{dV_{CB}} \right| = C_{jc} \tag{7.1.7}$$

Therefore, the Early voltage is

$$|V_A| = \frac{Q_B}{C_{jc}} \tag{7.1.8}$$

To reduce the influence of collector-base voltage on collector current, V_A should be increased in magnitude. From Equation 7.1.8 we see that this can be accomplished in practice by increasing the ratio of base majority-charge per unit area to the capacitance per unit area at the collector-base junction. Physically, this reduces the movement of the base-collector boundary into the base region.

It is useful, for computer modeling purposes, to describe the widening of the base-collector space-charge layer through the Early voltage V_A, as we shall see in Section 7.6. It may, however, help conceptually to consider an alternative view of the Early effect that is physically more revealing for the prototype transistor introduced at the beginning of Chapter 6. For homogeneous doping and active-mode bias, the minority-carrier distribution in the base has a triangular shape (Figure 7.2). Increasing the collector-base voltage from V_{CB} to $(V_{CB} + \Delta V_{CB})$ moves the base-collector junction edge a distance Δx_B from x_{B0}. A second triangular distribution specifies the new minority-carrier profile, and the shaded area between the two distributions represents the decrease in stored base charge.

The collector current is thus increased in the ratio x_{B0}/x_{B1} by the change in V_{CB}. In addition, it becomes clear by looking directly at stored base charge that the Early effect reduces charge storage. This affects both the transient behavior of transistors and the dc base current since base recombination depends directly on stored base charge. We shall consider these effects in more detail in Section 7.5 when we discuss BJT models for circuit applications.

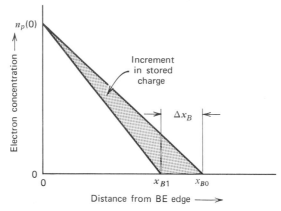

Figure 7.2   The influence of the Early effect on the minority–carrier distribution in the quasi-neutral base region of a prototype transistor. An increase in V_{CB} reduces the base width from x_{B0} to x_{B1}, increasing the gradient of n_p and decreasing the total minority charge in the base.

7.2 Effects at Low and High Emitter Bias

Some interesting behavior is revealed when a transistor is biased into the active region and the base and collector currents are measured as functions of the base-emitter voltage. Because of the physics of transistor action, such measurements may be more easily interpreted when plotted semilogarithmically with junction voltage on the linear scale. Typical data measured on an amplifying *IC npn* transistor are given in Figure 7.3. The excellent fit of the data for I_C and I_B to straight lines over the mid-range of currents indicates an exponential dependence on voltage, consistent with the theory that was developed in Chapter 6. Clearly, however, straight lines of constant slope do not adequately represent all of the measurements in Figure 7.3 either for I_C or I_B. At this point, it may seem surprising that collector current is "ideal" over a wider voltage range than is base current. The reason will be clear after some discussion. First, however, we consider the variation in I_B at very low voltages where the theory given thus far obviously disagrees with experiment.

Currents at Low Emitter Bias

At low base-emitter voltage there is a reduction in the slope of the straight-line variation of log I_B with V_{BE}. The experimental data indicate that the base current asymptotically approaches a curve that can be represented by

$$I_B = I_o \exp\left(\frac{qV_{BE}}{nkT}\right) \tag{7.2.1}$$

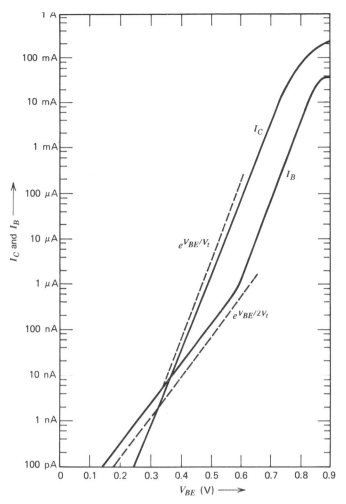

Figure 7.3 Typical behavior of collector and base currents as functions of base-emitter voltage for bias in the forward–active region.

as V_{BE} goes to zero. In the asymptotic form (Equation 7.2.1), n is a parameter that is generally found to be between one and two. Furthermore, the parameter I_o in Equation 7.2.1 is larger than is the corresponding multiplier for an exponential form that would fit the data at intermediate bias values.

The source for the added base-emitter current at low biases is clear from our discussion of current in *pn* junctions in Chapter 5; it is recombination in the base-emitter space-charge region. Space-charge-region recombination current in a diode structure was considered in Section 5.3, and Shockley-Hall-Read theory was used there to derive Equation 5.3.24. That equation for recombination current in the space-charge region has the same voltage dependence as does Equation 7.2.1 if n is taken equal to 2. Values for n between 1 and 2 can be justified by considering possible variation of different parameters affecting recombination within the space-charge region. As was shown in Equation 5.3.25, the importance of the recombination component relative to the injection components into the quasi-neutral regions increases as voltage is reduced. For the example used in the diode discussion that illustrated Equation 5.3.25, the recombination current dominated for junction voltages less than 0.48 V.

Recombination current in the space-charge region flows only in the base and emitter leads. Its dominance does not affect the collector current, which consists almost entirely of collected electrons that are injected at the base-emitter junction. Hence, the collector current continues to be represented by Equation 7.1.1 as V_{BE} is decreased until injection becomes so low that the collector current is dominated by generation in the space-charge region as expressed by Equation 6.3.1. Thus, at low biases the collector current is a smaller fraction of emitter current than it is in the intermediate bias range. This behavior is brought more clearly into focus in a plot of $|I_C/I_B|$, that is β_F as defined in Equation 6.2.22. Accordingly, the data of Figure 7.3 has been used to make the plot shown in Figure 7.4. The decrease in β_F as base-emitter bias decreases represents a clear limitation to the application of transistors for the amplification of low voltages.

There is great importance and enormous commercial significance to maintaining large current gains at low bias levels and, therefore, to reducing space-charge recombination current. Such applications for transistors and integrated circuits as hearing-aid amplifiers and *pacemaker* circuits for the medical treatment of heart

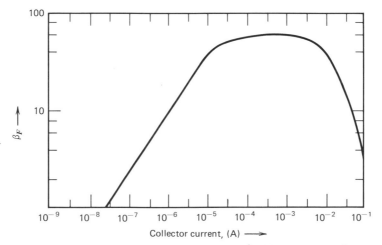

Figure 7.4 Current gain (β_F) as a function of collector current for the transistor of Figure 7.3.

patients depend on achieving adequate performance at the lowest possible current levels. For circuit applications such as these, processing efforts are usually concentrated on maintaining as high a lifetime as possible within the base-emitter space-charge region.

High-Level Injection

The transistor theory considered thus far has relied on the low-level injection approximation, that is, that majority-carrier populations under junction bias are essentially unperturbed from their values at thermal equilibrium. As transistors are biased more heavily in the active mode, violations of the low-level injection assumption should first become apparent either at locations where there is a high density of injected minority carriers or else where there is a low density of dopant atoms. The first situation can be expected in the base region at the emitter edge, while the second might occur in an *IC* transistor near the base-collector junction. Both possibilities lead to observable effects in practical *IC* transistors. We shall consider them separately, concentrating first on high-level injection near the emitter junction.

High-Level Injection at the Base-Emitter Junction. The basic equation for collector current under active bias (Equation 7.1.1) was derived by considering a special case of the equation for transistor action (Equation 6.1.7). The exponential dependence of collector current on voltage predicted by Equation 7.1.1 is seen in Figure 7.3 to be valid over nearly eight decades of current until it deviates under high bias on the base-emitter junction.

One cause for the deviation is high-level injection into the base. If the injection of electrons into the base is sufficient to cause a significant increase in the population of base majority carriers, then the full dependence of current on voltage is not stated explicitly in Equation 7.1.1. In this case, the integrated majority charge in the denominator becomes bias dependent. (An associated, but second-order effect is a change in $\tilde{D}_n$.) Because of quasi-neutrality in the base, the expression for the integral of $p(x)$ should be written

$$\int_0^{x_B} p(x)\, dx = \int_0^{x_B} [N_a(x) + n'(x)]\, dx \qquad (7.2.2)$$

where $n'(x)$ is the injected electron density and $N_a(x)$ represents the base doping profile. To proceed further, one would need to specify the x-dependences in the integral and, therefore, to consider a specific dopant profile.

To continue a qualitative discussion, however, we can focus on the boundary value for the injected-electron density $n'(0)$. Since this is the maximum density, it dominates in the integral expression of Equation 7.2.2. The boundary value can be determined by making use of Equation 5.3.20 together with the quasi-neutrality assumption. Equation 5.3.20, which states that the *pn* product at the emitter-base boundary is related to the applied bias by the expression $p(0)n(0) = n_i^2 \exp(qV_{BE}/kT)$, remains valid* provided that the current levels do not become of the same order as the balanced equilibrium flow tendencies (estimated at $\sim 10^5$ A cm^{-2} in Section 5.3). Currents of this size do not flow in practice without causing irreparable thermal damage to the junction.

The minority-carrier boundary value $n(0)$, which under active bias is almost the same as the injected electron density at the boundary $n'(0)$, is therefore

$$n(0) = \frac{N_a(0)}{2}\left[\left(1 + \frac{4n_i^2 \exp(qV_{BE}/kT)}{N_a^2(0)}\right)^{1/2} - 1\right] \qquad (7.2.3)$$

At low or moderate injection levels, Equation 7.2.3 reduces to Equation 5.3.7 as expected. Under high-injection conditions, however, the second term in the square root becomes dominant and $n(0)$ approaches an asymptote that varies as $\exp(qV_{BE}/2kT)$. Under this condition, there is a partial cancellation of the exponential increase predicted by the numerator of Equation 7.1.1. Collector current tends likewise to become proportional to $\exp(qV_{BE}/2kT)$.

High-Injection Effects at the Collector (Kirk Effect).[†] The introductory comments about low-level injection pointed out the weakness of this assumption in locations where the material is lightly doped. Inspection of a typical doping profile for an amplifying *IC* transistor (Figure 6.17) shows that minimal doping occurs in the collector, close to the base region where the epitaxial material has not been affected by any diffusions. This region serves several useful purposes. It allows for a significant increase in the breakdown voltage of the collector-base junction, it reduces collector capacitance, and it reduces the widening of the base-collector space-charge layer into the base (Early effect). In this section, we shall see that several complicated and undesirable consequences also arise from this low collector doping.

Under amplifying conditions, the base-collector junction is reverse biased so that any minority carriers that impinge upon its edges are quickly swept across to the opposite region. We have used this physical picture to justify our assumption of essentially zero density for minority carriers at the edges of the reverse-biased collector-base space-charge region. This assumption is clearly an approximation, however; it must be violated if any carriers are to flow across the space-charge region. Since free carriers are limited in their velocities, a current density J requires a density of minority carriers at least equal to J/qv_l, where v_l is the scattering-limited velocity (Equation 1.2.12). This charge density will be neutralized by majority carriers in the base region, but it will add algebraically to the fixed background charge associated with the uncompensated dopant densities in the space-charge region. The presence of this charge component associated with collector current will begin to modify the theory that has been developed thus far when the density

---

* The validity of Equation 5.3.20 under high-level-injection conditions is also necessary to permit the application of the basic equation for transistor action (Equation 6.1.7) since the former was used in the derivation of the latter.

of the injected, current-carrying charge becomes comparable to the background density of dopant ions. As the current level in a transistor is increased, this mechanism can significantly alter important transistor properties.

Qualitatively, the effect may be described as follows: the free carriers entering the base-collector space-charge region modify the background charge in that region and thus affect the electric field. For a constant collector-base voltage, the integral of the field across the space-charge region is constant. Hence, any space-charge modification must be accompanied by a change in the width of the region. We shall see shortly that the width of the space-charge region tends to decrease and, therefore, the neutral base tends to widen. This mechanism is often called the *Kirk effect* after the author of the first paper to analyze it.[3]

In Figure 7.5a we have sketched a portion of the doping profile for an *IC* amplifying transistor from Figure 6.17. We consider a case for which this device

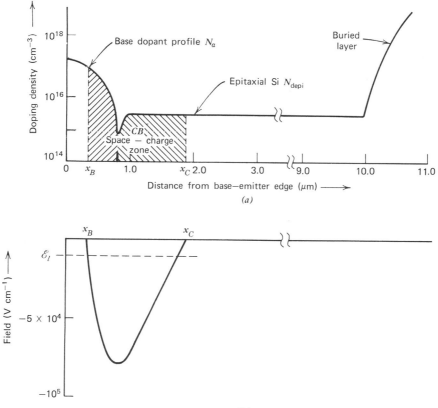

Figure 7.5   Space charge and field at the collector-base junction of the amplifying transistor shown in cross section in Figure 6.17. (*a*) The space-charge region for 10 V collector-base reverse bias, (*b*) the electric field in the collector-base space-charge region. The approximate field $\mathscr{E}_l$ at which free electrons reach limiting velocity is marked on the plot.

is biased in the active mode with 10 V applied to the collector-base junction. A plot of the electric field with little or no current flowing is sketched in Figure 7.5b. Noted on the plot is $\mathscr{E}_l$ at which free electrons reach their limiting drift velocity v_l (Figure 1.17). Poisson's equation for the collector-base space-charge region of the *npn* transistor takes the form

$$\frac{d\mathscr{E}}{dx} = \frac{1}{\epsilon_s}\left[qN(x) - \frac{J_C}{v(x)}\right] \qquad (7.2.4)$$

where $N(x)$ is the net dopant density $(N_d - N_a)$ in the space-charge region (negative acceptor ions in the diffused region near the base and positive donor ions in the high-resistivity epitaxial layer and in the heavily diffused buried layer). If we consider constant collector-base bias V_{CB}, we obtain a second equation involving the field $\mathscr{E}$

$$V_{CB} + \phi_i = \int_{x_B}^{x_C} - \mathscr{E}\, dx \qquad (7.2.5)$$

where ϕ_i is the built-in junction voltage (Equation 4.2.10). Equation 7.2.5 implicitly specifies the width of the space-charge layer, that is $(x_C - x_B)$. Referring again to Equation 7.2.4, we note that the second term on the right-hand side is negligible at low currents, but will dominate at high currents. The critical current J_1 dividing the low- and high-current regions may be found by equating the two terms

$$J_1 = qN(x)v(x) \qquad (7.2.6)$$

The critical current will be reached first where $N(x)$ is a minimum; that is, in the epitaxial region, where for most bias levels $v(x)$ is at its maximum value v_l and N is N_{epi}. Thus, $J_1 = qN_{epi}v_l$. An analysis of this effect is of practical importance because in many applications the collector current density equals and exceeds the critical current J_1. For example, if $A_E \approx 10^{-7}$ cm^2 and $N_{epi} = 5 \times 10^{15}$ cm^{-3}, the critical collector current is 0.8 mA.

To carry out the mathematical treatment, we first multiply Equation 7.2.4 by x and then integrate between x_B and x_C, the boundaries of the collector-base space-charge layer.

$$\int_{x_B}^{x_C} x \frac{d\mathscr{E}}{dx}\, dx = \frac{1}{\epsilon_s}\int_{x_B}^{x_C} x\left[qN(x) - \frac{J_C}{v(x)}\right]dx. \qquad (7.2.7)$$

The left-hand side of Equation 7.2.7 can be integrated by parts and used with Equation 7.2.5 to yield

$$\int_{x_B}^{x_C} x\, d\mathscr{E} = -\int_{x_B}^{x_C} \mathscr{E}\, dx = V_{CB} + \phi_i \qquad (7.2.8)$$

where $\mathscr{E}(x_C)$ and $\mathscr{E}(x_B)$ have been taken to be zero. Thus, the collector voltage can be expressed as

$$V_{CB} = \frac{1}{\epsilon_s}\int_{x_B}^{x_C} x\left[qN(x) - \frac{J_C}{v(x)}\right]dx - \phi_i \qquad (7.2.9)$$

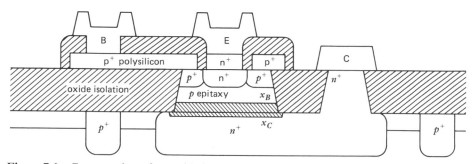

Figure 7.6 Cross section of an oxide-isolated *npn* transistor made using a *p*-type epitaxial layer and polysilicon connections to the base and emitter. The collector-base space-charge layer extending from x_B to x_C is shown by the (left-slanting) cross-hatching.

Application of Equation 7.2.9 to the transistor represented in Figure 7.5 is not straightforward because of the shape of the doping profile. Mathematical complications can be avoided and a worthwhile physical picture of the Kirk effect can still be obtained by considering first, the less common structure shown in Figure 7.6. For this "epitaxial-base" structure, the epitaxial region is a high-resistivity *p*-type layer instead of the more commonly used *n*-type material. For a transistor made in a *p*-type epitaxial layer, the base-collector space-charge region would extend outward from the n^+ buried layer, and the point x_C would be relatively insensitive to bias changes because of the high density of donors in the buried layer. We therefore consider x_C as constant and solve Equation 7.2.9 to find the variation with current density of $x_{CB} = x_C - x_B$, the width of the collector-base space-charge region. For this case of acceptor doping, if we take the charge $qN(x)$ in Equation 7.2.9 to be a constant equal to $-qN_{epi}$ up to the buried layer, and assume $v = v_l$, then

$$V_{CB} = \frac{1}{2\epsilon_s}\left[qN_{epi} + \frac{J_C}{v_l}\right]x_{CB}^2 - \phi_i$$

$$= \frac{qN_{epi}}{2\epsilon_s}\left[1 + \frac{J_C}{J_1}\right]x_{CB}^2 - \phi_i \qquad (7.2.10)$$

which can solved for x_{CB}.

At zero current the collector space-charge region has width x_{CO} where

$$x_{CO} = \left[\frac{2\epsilon_s(V_{CB} + \phi_i)}{qN_{epi}}\right]^{1/2} \qquad (7.2.11)$$

The variation with current of the collector-base space-charge region x_{CB} can then be expressed as

$$x_{CB} = \frac{x_{CO}}{(1 + J_C/J_1)^{1/2}} \qquad (7.2.12)$$

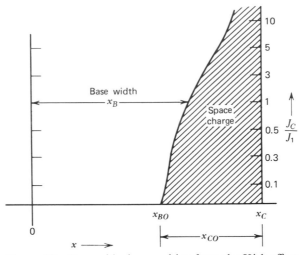

Figure 7.7   Base widening resulting from the Kirk effect in a transistor having the cross section sketched in Figure 7.6. The variation shown has been calculated using Equation 7.2.12.

This variation is shown in Figure 7.7, where we see that the Kirk effect results in a widening of the charge-neutral base region as currents approach and exceed the critical value J_1. This base-layer widening decreases the transistor current gain at high currents and also degrades the frequency response of the device. We consider the latter effect in the next section, but first let us return to the case of the Kirk effect in *npn* transistors with an *n*-type epitaxial layer, the most widely used structure.

   The complication in the case of an *n*-type epitaxial layer arises because the dopant concentration varies considerably through the collector-base space-charge region, going from $-qN_a(x_B)$, in the tail of the *p*-type base diffusion, to $+qN_{epi}$, the fixed-donor density in the epitaxial region and finally to $+qN_d(x)$ at the outer edge of the out-diffusing buried layer. In this case, as the current level is increased above $J_1 = qN_{epi}v_l$, the sign of the space charge in the epitaxial layer changes from plus to minus. This sign reversal causes the electric field in the epitaxial layer to vary with position in the opposite sense from the currentless case when current densities exceed J_1. Electrical flux lines that end on the negative acceptor ions in the base when currents are very low become terminated instead on the current-carrying free electrons when the Kirk effect occurs. The entire space-charge region is thereby pushed toward the heavily doped collector region, moving from an initial position at the *pn* junction formed by the base acceptor diffusion to a final position at the edge of the buried layer at high currents. Under this high current condition, the electric field increases linearly with distance, and the space-charge region is compressed against the highly doped collector edge (Problem 7.10).

Computer calculations of the Kirk effect for this case have been carried out[4] and several of the results obtained are shown in Figure 7.8.

The doping profile of the transistor is shown in Figure 7.8a, while the field and electron densities at various current levels are indicated in Figures 7.8b and 7.8c, respectively. At low currents, the field in the collector-base region has the roughly triangular shape sketched in Figure 7.5b. (Note that the *negative* of the field is plotted in Figure 7.8b.) The critical current density J_1 for the structure of Figure 7.8 is of the order of 500 A cm^{-2}. In the neighborhood of this current density, the high-field region is seen in Figure 7.8b to migrate toward the highly doped buried layer. At still higher currents, the field becomes compressed against the boundary between the epitaxial region and the buried layer. The plot of electron concentration

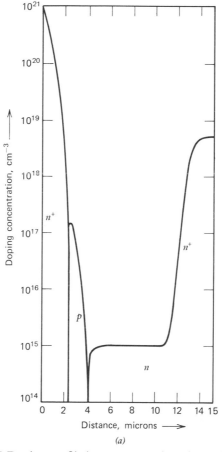

(a)

Figure 7.8   (*a*) Doping profile in an *npn* transistor for which a computer analysis of the Kirk effect has been carried out.[4] (*b*) Electric field as a function of distance for various collector currents in the transistor of (*a*). (*c*) Base minority-carrier (electron) density as a function of distance for various collector currents in the transistor of (*a*).

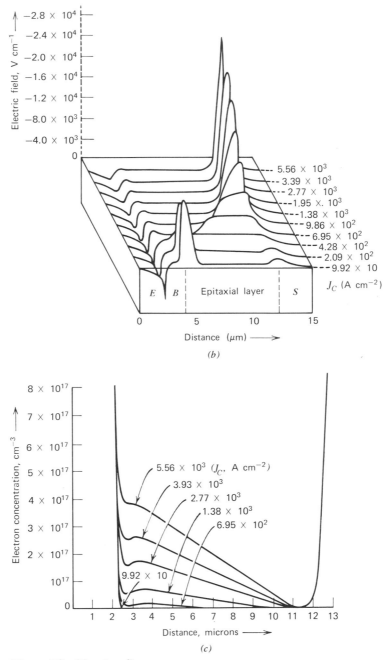

Figure 7.8   (*Continued*)

in Figure 7.8*c* has a steep gradient near the emitter to overcome the built-in opposing field in this region. The extension of the base region with increasing current is evident on the same figure. Note that at higher currents, the plots of electron density have nearly constant gradients over most of the epitaxial region.

Base Resistance

Because current gain β_F in amplifying transistors is typically so high, one may be tempted to assume that base current is negligible and therefore that resistance in the base region is of little consequence to transistor operation. This simplified viewpoint fails to consider that small voltage differences in the base region are magnified by the exponential diode factor, and thus can cause much larger differences in current densities across the emitter-base *pn* junction.

Consider the cross section of the transistor shown in Figure 7.9. As V_{BE} is increased from zero, injection of electrons from the emitter region will be the greatest at that part of the junction where the base doping N_a is the smallest. Because the base is diffused, this will be across the bottom plane of the emitter diffusion. Majority–carrier base current will flow into this region to supply carriers for recombination and to be injected into the emitter. A typical base thickness, however, is less than one micrometer and therefore a sizable resistance will generally be present between the base electrode and the active area of the transistor. Since the emitter-base current is spread over the active region of the emitter (Figure 7.9), the base current decreases continuously as one moves toward the center line of the emitter. Therefore, we cannot calculate a resistance value that will account in a straightforward way for the ohmic drop in the base region. More important, the potential dropped along the base region decreases the base-emitter bias along the same path. The density of injected-electron current therefore decreases from its highest value in the active region nearest the base electrode to a minimum at the center of the emitter. This *current crowding* toward the perimeter of the emitter increases with bias and causes localized heating at current levels that might be

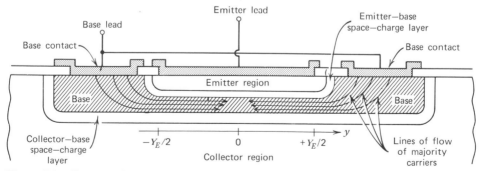

Figure 7.9 Cross section of a transistor under active bias. The base current is supplied from two side base contacts and flows toward the center of the emitter causing the base-emitter voltage to vary with position.

tolerable if the current were distributed uniformly. The high-injection effects that were considered earlier in this section also occur at lower currents because of the uneven current density across the active region. In order to reduce base resistance, power transistors are made with large interleafing comb-like base and emitter contacts.

A plot of collector current versus base-emitter voltage for a transistor in which base resistance is significant is shown in Figure 7.10. Because base resistance acts to reduce the junction bias, we might try to fit the measured data from Figure 7.10 to the expression

$$I_C = I_S \exp\left[\frac{q(V_{BE} - I_B R_B)}{kT}\right]$$
(7.2.13)

To do so, we recognize from our discussion of current crowding that the resistance R_B in Equation 7.2.13 (called the *base spreading resistance*) must be variable. Using the measurements of I_C, I_B, and V_{BE}, we find that to correspond to Equation 7.2.13, R_B must decrease with increasing current (Figure 7.11). The initial reduction in base resistance apparent in Figure 7.11 corresponds to the decreasing path length from the base electrode to the active region of the transistor as current increases.

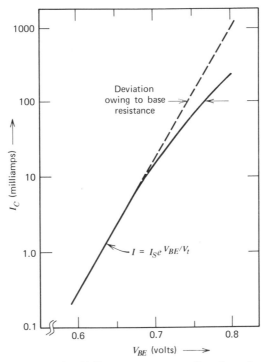

Figure 7.10   Collector current as a function of base-emitter voltage showing the deviation from ideal behavior at high currents.

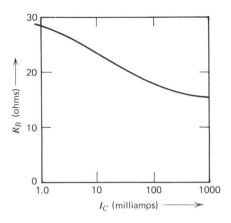

Figure 7.11 Base resistance in the transistor of Figure 7.10. Values of R_B obtained using measured data in Equation 7.2.13.

If only base spreading resistance were important, an asymptotic value of R_B would be approached when all of the current had crowded toward the perimeter of the emitter and all resistance was associated with inactive regions of the base. However, the onset of the other high-level effects, previously mentioned, may cause R_B to decrease to a lower value.

It is possible to analyze exactly the current-crowding effects that we have described by making use of a spatially distributed form of the diode equation. The mathematics become relatively tedious, however, and may obscure the relevant physical mechanisms. We shall, therefore, consider an approximate analysis for which the transistor is divided into sections. Each section is assumed to have the same current gain as the original device and to follow the ideal transistor equation, that is, to have negligible base resistance. Each section is characterized by its proportional share of the overall saturation current (I_S in Equation 7.2.13) and is separated from the adjacent section by a resistance corresponding to part of the physical resistance along the base region. By increasing the number of pieces into which the transistor is sliced, we can increase the accuracy of the analysis and ultimately approach the exact distributed case. It is only necessary, however, to use the four-section structure in Figure 7.12 to illustrate both the physical mechanism of current crowding and the method of solution. The calculation of base resistance from the network sketched in Figure 7.12 can be accomplished by assuming a current in the last leg of the ladder and then applying network laws to calculate voltage and current at the input terminal. Base resistance is then obtained by using Equation 7.2.13 at each value of input current I_B and voltage V_{BE}. Further consideration of base resistance is found in the problems.

In this section we have considered several bias-dependent physical effects that were not included in the basic theory of junction transistors. The effects described provide limits on the useful range of bias for an amplifying bipolar transistor. Analysis of these effects shows that the equation for transistor action (Equation 6.1.7) adequately describes high-level injection at the base-emitter junction provided that the total majority charge in the base is properly specified. The basic

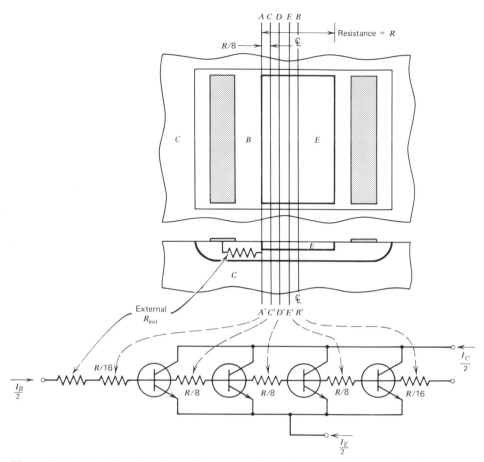

Figure 7.12 Modeling the effect of base-spreading resistance. The base region for an IC transistor is divided into eight sections (four on each half) and total built-in resistance between the base contacts is divided between eight transistors.[9] Network analysis allows us to determine the effective resistance seen at the base-emitter junction as bias is varied (Problem 7.12).

theory of Chapter 6 must be extended, however, in order to treat low-level injection, the Kirk effect, or base resistance. Further consequences of these effects are described in the discussion of transistor speed and models that follows.

For purposes of clarity, the discussion of high-level injection has treated effects one-by-one. In practice, of course, they occur simultaneously. Some interactions between effects, such as a reduction in base resistance because of high-level injection at the base-emitter junction may be surmised; others are more subtle. Computer methods have been developed that treat high-level injection from first principles and thus give an accurate overall analysis. The computer analysis[5] shows that the Kirk effect, sometimes called *base push-out*, plays a dominating role under high-level injection conditions.

The list of limiting effects that has been treated is not all-inclusive. The mechanisms described are, however, concerned with internal transistor operation. They are thus distinguished from performance limits imposed by parasitic elements that are introduced in the *IC* fabrication process. Limiting effects resulting from these elements can usually be treated by using circuit analysis as illustrated in the discussion of lateral *pnp* transistors in Section 7.7.

7.3 Base Transit Time

The transit time of minority carriers across the quasi-neutral base of a transistor under active bias is a significant parameter. The delay associated with this transit time was the dominant limitation on the transient behavior of the first transistors. The search for methods of reducing base transit time was responsible for many advances in semiconductor processing technology. In the present design of bipolar transistors for integrated circuits, the base transit time has been so greatly reduced that it is just one among several important time limitations. Its discussion, however, can elucidate considerably the interrelationships between transistor design and performance.

Let us first consider the magnitude of the excess minority charge in the quasi-neutral base of a transistor under active bias. This charge, which we designate Q_{nB}, can be expressed*

$$Q_{nB} - \int_0^{x_B} q A_E n'(x)\, dx \tag{7.3.1}$$

The injected charge Q_{nB} carries the collector current of Equation 7.1.1 that flows because of transistor action between the emitter and collector. Accordingly, a characteristic *base transit time* τ_B for the transfer of minority carriers across the base is given by the quotient of Q_{nB} divided by I_C:

$$\tau_B = \frac{Q_{nB}}{I_C} \tag{7.3.2}$$

Calculation of the base transit time in the prototype transistor sketched in Figure 6.1 is particularly simple. Because $n'(x)$ is linear (Equation 6.1.11), $Q_{nB} = \frac{1}{2}qn'(0)x_B A_E$ and I_C is given by Equation 6.1.12. Hence, for this case

$$\tau_B = \frac{x_B^2}{2\tilde{D}_n} \tag{7.3.3}$$

Equation 7.3.3 gives a measure of the time required for diffusion transport across a region. Some appreciation of the low speed of diffusion transport relative to drift of carriers in an electric field can be gained by calculating τ_B for a lightly doped base ($N_a \approx 10^{16}$ cm^{-3}) having a width x_B of 1 μm. Application of Equation 7.3.3

---

* A similar definition for excess hole charge in an *n*–type region in a diode was made in Equation 5.4.1.

gives $\tau_B \simeq 144$ ps. If a field were present in the base, the carriers would experience a transit time of $x_B/\mu_n \mathscr{E}$ or $x_B^2/\mu_n V$ for a constant field. The drift transit time would be smaller than the diffusion time for any voltage drop greater than about 50 mV.

For the arbitrarily doped transistor, an expression for τ_B can be written using Equations 7.1.1 and 7.3.1 in Equation 7.3.2.

$$\tau_B = \frac{\int_0^{x_B} p \, dx \int_0^{x_B} n' \, dx}{\tilde{D}_n n_i^2 \exp(q V_{BE}/kT)} \tag{7.3.4}$$

Equation 7.3.4 is particularly useful when considering the effects of various base dopant profiles on τ_B.

Application of Equation 7.3.4 to the prototype transistor under high-level injection conditions will also prove enlightening. For the prototype transistor under active bias, the carrier densities vary linearly with position, and the integrals in Equation 7.3.4 are therefore particularly simple to evaluate. A plot of the minority-carrier density in the base is triangular and, because of charge neutrality, the majority carriers take on a trapezoidal distribution, decreasing away from the edge of the base-emitter space-charge region. Sketches of the densities are shown in Figure 7.13. Under these conditions, the integrals in Equation 7.3.4 can be evaluated, and the base transit time can be written:

$$\tau_B = \frac{[\frac{1}{2}n'(0)x_B][\frac{1}{2}(n'(0) + 2N_a)x_B)}{\tilde{D}_n[n'(0)(n'(0) + N_a)]}$$
$$= \frac{x_B^2}{4\tilde{D}_n}\left[1 + \frac{N_a}{n'(0) + N_a}\right] \tag{7.3.5}$$

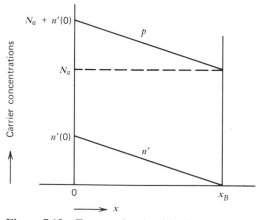

Figure 7.13 Free-carrier densities in the base of a homogeneously doped transistor. Charge neutrality causes the hole density to increase above the dopant concentration N_a.

Equation 7.3.5 shows that under low or moderate–injection conditions, for which $n'(0)$ is much less than N_a, τ_B approaches $x_B^2/2\tilde{D}_n$. High bias levels actually reduce the base transit time, with τ_B approaching $x_B^2/4\tilde{D}_n$ asymptotically. This factor of two reduction in base transit time for homogeneously doped transistors under high-level injection was first predicted by W. M. Webster[6] and is frequently called the *Webster effect*. The physical source for the reduction in τ_B is the field associated with the higher density of majority carriers that exists throughout the base under high-level injection conditions. To balance a diffusion tendency of these majority carriers resulting from their nonuniform distribution, an electric field must exist directed between the emitter and collector. This electric field balances the diffusion tendency and prevents majority-carrier flow toward the collector. At the same time, the field aids the flow of minority carriers and acts to reduce their transit time through the base. In the case of uniform base doping, the transport of minority carriers changes from a pure diffusion process at low biases to drift-aided diffusion at higher biases.

Although *IC* transistors are typically not made with uniformly doped base regions, we have seen in Section 7.2 that high-level injection conditions extend the base into the uniformly doped epitaxial region (Kirk effect). For this reason, high-level injection analyses often take τ_B to be $x_B^2/4\tilde{D}_n$.[3,4]

An alternative form for τ_B can be derived from Equations 6.1.5 and 7.3.4. If the density of electrons at the collector boundary $n(x_B)$ is taken to be zero, then Equation 6.1.5 shows that the minority-carrier density as a function of x is

$$n(x) = -\frac{J_n}{q\tilde{D}_n p(x)} \int_x^{x_B} p(\xi)\,d\xi \tag{7.3.6}$$

Under the usual active-bias approximation that $n \approx n'$, Equation 7.3.6 can be combined with Equation 7.3.4 to give

$$\tau_B = \frac{1}{\tilde{D}_n} \int_0^{x_B} \frac{1}{p(x)} \left[\int_x^{x_B} p(\xi)\,d\xi \right] dx \tag{7.3.7}$$

If the distance variable x in Equation 7.3.7 is normalized to x_B by introducing $y = x/x_B$, then Equation 7.3.7 becomes

$$\tau_B = \frac{x_B^2}{\tilde{D}_n} \left\{ \int_0^1 \frac{1}{p(y)} \left[\int_y^1 p(\xi)\,d\xi \right] dy \right\}$$
$$= \frac{x_B^2}{v\tilde{D}_n} \tag{7.3.8}$$

where the factor v represents the reciprocal of the nested definite integral in the first form of Equation 7.3.8. Equation 7.3.8 shows that the effect of an inhomogeneous base doping on τ_B can be incorporated into the value of v.

For constant base doping $p(x) = N_a$, it is easy to use Equation 7.3.8 to find that $v = 2$. When graded profiles are considered, τ_B can be reduced from $x_B^2/2\tilde{D}_n$ by about an order of magnitude. Limits on reducing τ_B are imposed by the need to maintain acceptable injection efficiency [which constrains $N_a(0)$ to be appreciably lower than the emitter doping] and the requirement of an extrinsic p-type doped material at $x = x_B$ [which forces $N_a(x_B)$ to be greater than $N_{d\,epi}$].

7.4 Charge-Control Model

A framework for bipolar-transistor equations that is especially useful for time-dependent analysis is called the *charge-control model*.[7] In this model the controlled variable is not current or voltage; instead, equations are framed in terms of controlled charges within regions of the device.

A typical charge-control relationship for a transistor under active bias was derived in the previous section in Equation 7.3.2. This equation, $I_C = Q_{nB}/\tau_B$ relates the minority charge stored in the quasi-neutral base Q_{nB} to the current I_C carried by transistor action between the emitter and the collector. The charge and current are linearly related with τ_B, the transit time in the quasi-neutral base, as the proportionality factor. Because Equation 7.3.2 represents only minority-carrier transport across the base, however, it is just one portion of the charge-control model for a transistor.

Control in an amplifying *npn* transistor is exercised by the bias on the base-emitter junction. This bias affects not only Q_{nB}, but other charge components as well. The major additional components to be considered are the charges represented by holes injected into the emitter, which we designate as Q_{pE}, and the charges stored on the base-emitter and base-collector depletion capacitances, which are given the symbols Q_{VE} and Q_{VC}, respectively. Figure 7.14 shows these components for a prototype transistor. We first discuss the two injection components, Q_{nB} and Q_{pE} that are responsible for steady-state base current. The other charge

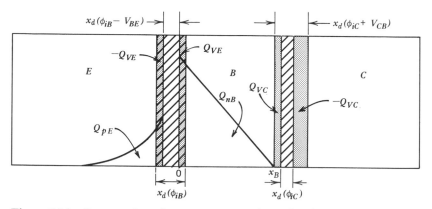

Figure 7.14 Cross section of a prototype transistor showing the locations of the charge components used in charge-control modeling. The charges Q_V represent storage at the edges of the space-charge regions. The cross-hatching indicates the two junction space-charge regions at thermal equilibrium.

components shown in Figure 7.14 influence the time-varying behavior of the *BJT* and will be considered later. Since both Q_{nB} and Q_{pE} increase when the base-emitter voltage is increased, their sum $(Q_{nB} + Q_{pE})$ is called Q_F (because Q_F is increased when the *BJT* is under *forward*-active bias). The sign of Q_F is the sign of the controlling (base majority-carrier) charge, positive for an *npn* transistor and negative for a *pnp* transistor. The steady-state collector current can be written in terms of Q_F if a characteristic time τ_F is introduced. The equation for current (in analogy to Equation 7.3.2) is

$$I_C = \frac{Q_F}{\tau_F} \tag{7.4.1}$$

Note that all that is formally required to make Equation 7.4.1 accurate is that I_C must be linearly related to Q_F (with τ_F taken to be constant).

Because Q_F represents the sum of the magnitudes of the excess minority charge in the quasi-neutral emitter and base regions, its voltage dependence is generally that of a diode, and we may write

$$Q_F = Q_{FO} \left[\exp\left(\frac{qV_{BE}}{KT}\right) - 1 \right] \tag{7.4.2}$$

where Q_{FO} is a function of the dopant profiles and device geometry. Since Q_F is at least somewhat greater than Q_{nB}, τ_F must be greater than the base transit time τ_B (Equation 7.3.2). Further comments about τ_F are deferred until the charge-control model is developed more fully.

The steady-state current flowing in the base lead is proportional to the rate at which Q_{nB} recombines in the quasi-neutral base plus the rate at which holes are injected into the emitter to replenish Q_{pE}. These two rates are proportional to the diode factor $[\exp(qV_{BE}/kT) - 1]$, and therefore (by Equation 7.4.2) proportional to Q_F. It is therefore possible to write a charge-control expression for the input (base) current of the transistor

$$I_B = \frac{Q_F}{\tau_{BF}} \tag{7.4.3}$$

A considerable amount of physical analysis, involving emitter efficiency and the recombination processes for excess carriers in both the emitter and base, would be required to express τ_{BF} or τ_F in terms of more fundamental parameters. This analysis is not of special value because both τ_{BF} and τ_F can be obtained in practice from measurements, and also because the charge-control model aims at representing the device for design, not at exploring the physical electronics of transistor models.

Equations 7.4.1 and 7.4.3 can be used to show that the steady-state current gain is simply the ratio of the two characteristic times:

$$\frac{I_C}{I_B} = \beta_F = \frac{\tau_{BF}}{\tau_F} \tag{7.4.4}$$

For example, Equations 7.4.1, 7.4.3 and 7.4.4 can be applied to the prototype transistor of Figure 6.1. If the emitter efficiency is very high in the prototype device, then $Q_F \approx Q_{nB}$ with $Q_{nB} = \frac{1}{2}qn'(0)x_B A_E$. Since only base recombination is significant for this case, $\tau_{BF} = \tau_n$ and $\tau_F = x_B^2/2\tilde{D}_n$ as derived in Equation 7.3.3. Therefore, from Equation 7.4.4 β_F is $2L_n^2/x_B^2$ where $L_n = \sqrt{\tilde{D}_n \tau_n}$. This result for dc current gain can be compared with earlier analysis of the same problem in Section 6.2. There, α_F was derived as the base transport factor to be $[1 - (x_B^2/2L_n^2)]$ in Equation 6.2.8. If we use this result for α_F in the equation $\beta_F = \alpha_F/(1 - \alpha_F)$, we find an identical value for β_F to that obtained from the charge-control analysis.

A full charge-control model for the bipolar transistor is derived by adding terms to represent the currents that flow because of time variations in stored charge. Clearly, if Q_F increases with time, there will be a component of base current equal to dQ_F/dt. Likewise, changes in the charges stored at the base-emitter and base-collector junctions (Q_{VE} and Q_{VC}) result in added base current. An overall expression for the base current is, therefore,

$$i_B = \frac{Q_F}{\tau_{BF}} + \frac{dQ_F}{dt} + \frac{dQ_{VE}}{dt} + \frac{dQ_{VC}}{dt} \tag{7.4.5}$$

The first three components of current in Equation 7.4.5 flow from the base to the emitter; the last flows from the base to the collector. Combining Equation 7.4.1 and 7.4.5 and using Kirchhoff's current law, one obtains a set of charge-control equations for the transistor under active bias:

$$i_C = \frac{Q_F}{\tau_F} - \frac{dQ_{VC}}{dt}$$

$$i_B = \frac{Q_F}{\tau_{BF}} + \frac{dQ_F}{dt} + \frac{dQ_{VE}}{dt} + \frac{dQ_{VC}}{dt} \tag{7.4.6}$$

$$i_E = -Q_F\left(\frac{1}{\tau_F} + \frac{1}{\tau_{BF}}\right) - \frac{dQ_F}{dt} - \frac{dQ_{VE}}{dt}$$

We have thus derived a set of *linear* equations relating currents and charges in a bipolar transistor. These equations are in contrast to the nonlinear expressions that relate currents and voltages in the transistor.

The circuit diagram of Figure 7.15 represents the various terms in Equation 7.4.6. The diode from the base to the emitter passes the steady-state current and has a saturation current $I_{ES} = Q_{FO}[(1/\tau_F) + (1/\tau_{BF})]$ as indicated by Equation 7.4.2. The elements storing the charges Q_F, Q_{VE}, and Q_{VC} are shown as capacitors with a line across them to indicate that they store charge (like capacitors), but are voltage dependent. Now that a basic set of charge-control equations has been written for a *BJT*, we are better able to gain a perspective on the limitations of this viewpoint.

The basic premise underlying charge-control analysis is the existence of a constant proportionality between quantity of charge and current. Another way of stating this is that the characteristic times in the charge-control equations must

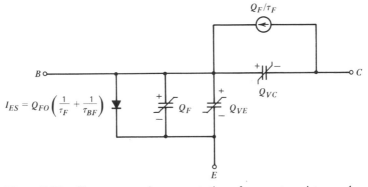

Figure 7.15   Charge-control representation of an *npn* transistor under active bias with junction-charge storage and injected-base charge taken into account.

not themselves be functions of charge or of bias voltages. Under this premise, the characteristic times may be derived for dc conditions and then be used to express dynamic conditions. Strictly this premise is not correct; the characteristic times are functions of the charge. For example, to affect current at the collector, minority carriers must disperse through the base region after being injected at the edge of the base-emitter junction. Thus, during transient conditions, collector current does not have the same ratio to base charge as it does in the steady state.

If a dynamic problem is analyzed with the charge-control model, the solutions for charge are constrained to be a time sequence of differing "steady-state" solutions. Hence, these solutions are sometimes called *quasi-static approximations*. For the greater part of the transient, however, the charge-control solution is usually a fair representation of the more exact result. The time scale over which there is significant error in the charge-control solution is of the order of the transit time in the base, that is, the order of τ_F. For most applications the time scale of interest is appreciably greater than τ_F, and solution by means of the charge-control model is perfectly adequate. A specific example in the next section will make this more clear.

Applications of the Charge-Control Model

Before discussing the bipolar charge-control model further, it is worthwhile to illustrate the use of Equations 7.4.6 in an application. A simple circuit appropriate for this purpose is shown in Figure 7.16*a*, and the circuit plus charge-control model is sketched in Figure 7.16*b*. What is sought is the collector-current response of an *npn* transistor under active bias when driven by a current source at the base. For hand calculations, several simplifications are appropriate. First, the current dQ_{VE}/dt can be considered negligible since the base-emitter voltage varies only slightly under forward bias. This simplification is usually made for active-bias operation. Because we have chosen a simple circuit in which the collector voltage is constant, it is also possible to treat the current dQ_{VC}/dt as negligible.

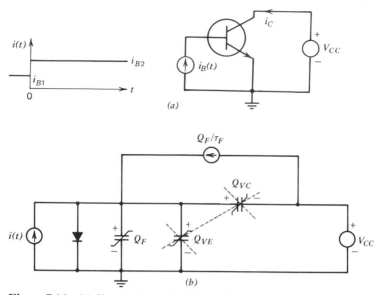

Figure 7.16 (*a*) Simple circuit for illustration of charge-control model. (*b*) Equivalent-circuit model. The cancelled elements carry negligible currents.

With these approximations the equation for the base current has only one unknown, the controlled charge Q_F, and takes the form

$$i_B = \frac{Q_F}{\tau_{BF}} + \frac{dQ_F}{dt} \tag{7.4.7}$$

Since $i_B(t)$ is specified by

$$i_B = i_{B1}(t < 0)$$
$$= i_{B2}(t > 0)$$

the solution for Q_F will contain terms associated with both the homogeneous and particular forms of Equation 7.4.7. When the boundary values $Q_F(t = 0) = i_{B1}\tau_{BF}$ and $Q_F(t \to \infty) = i_{B2}\tau_{BF}$ are used, the solution becomes

$$Q_F = \tau_{BF}[i_{B2} + (i_{B1} - i_{B2})\exp(-t/\tau_{BF})] \tag{7.4.8}$$

Under the approximations that we have made, the collector current is given by Q_F/τ_F so that its time dependence becomes (using Equation 7.4.4)

$$i_C = \beta_F[i_{B2} + (i_{B1} - i_{B2})\exp(-t/\tau_{BF})] \tag{7.4.9}$$

The collector current thus changes from its initial to its final value following an exponential function with a characteristic time constant equal to τ_{BF} (Figure 7.17).

As mentioned in the previous section, the transient solution given in Equation 7.4.9 is in error for small values of t. At zero time, for example, Equation 7.4.9 predicts an abrupt change in the slope of the collector current equal to

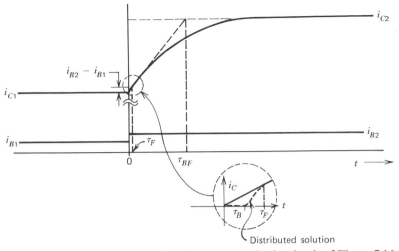

Figure 7.17   Time variation of collector current in the circuit of Figure 7.16 as calculated from the charge-control model.

$(i_{B2} - i_{B1})/\tau_F$ whereas collector current will not change until the extra electrons injected at the emitter side of the base reach the collector. A more complete analysis that takes account of the distributed nature of base charging predicts that i_C does not change at all for times close to $t = 0$; initially, the increments in the base and emitter currents are equal. The collector current first begins to change when t reaches the base transit time τ_B. The behavior of i_C then rapidly approaches the transient solution predicted by the charge-control model and essentially matches that result for times larger than $t = \tau_F$. (See inset in Figure 7.17.)

The foregoing example may seem artificial because of the number of simplifications that have been imposed, first in the choice of circuit and second in the approximations that have been made. These physical simplifications, however, have avoided mathematical complications that might tend to obscure the use of the model.

One such complication would arise if, for example, the collector in Figure 7.16a were not connected to an ac ground, but rather was connected to the source through a load resistor R_L. Since V_{CB} would be variable in this case, the current required to charge Q_{VC} would not be negligible. The solution for i_C can be obtained readily if we define an effective capacitance C_{jC} equal to the average of dQ_{VC}/dV_{CB} over the collector voltage interval. The solution for i_C is then found to be equal to Equation 7.4.9 provided that the time constant τ_{BF} in Equation 7.4.9 is replaced by [Problem 7.19]:

$$\tau_{BF}' = \tau_{BF}\left(1 + \frac{R_L C_{jC}}{\tau_F}\right) \qquad (7.4.10)$$

To derive Equation 7.4.10, it is necessary to note that the presence of the collector capacitance affects the base current much more than it does the collector current.

Hence, we can still approximate $i_C = Q_F/\tau_F$, but i_B must include the transient charge storage at the collector junction dQ_{VC}/dt.

This result shows that a collector resistor lengthens the time of the current transient by the binomial factor in Equation 7.4.10.

Large-Signal Model. The most extensive use of charge-control equations is for the solution of large-signal transient problems, and their utility is especially apparent when one is dealing with switching between regions of operation, typically between cut-off and saturation. Writing charge-control equations for large-signal switching is straightforward because all regions of operation have in common the injection and extraction of charge at the two junctions. These injection and extraction processes can be represented by adding together two sets of charge-control equations; one set that characterizes forward-active operation (Equations 7.4.6) and a companion set having the same form to represent operation under reverse-active bias. The total controlled charge thus becomes a superposition of charges representing forward-active bias Q_F and reverse-active bias Q_R.

The full set of equations for an *npn* transistor is*

$$i_E = -\frac{dQ_F}{dt} - Q_F\left(\frac{1}{\tau_F} + \frac{1}{\tau_{BF}}\right) + \frac{Q_R}{\tau_R} - \frac{dQ_{VE}}{dt}$$

$$i_C = \frac{Q_F}{\tau_F} - \frac{dQ_R}{dt} - Q_R\left(\frac{1}{\tau_R} + \frac{1}{\tau_{BR}}\right) - \frac{dQ_{VC}}{dt} \tag{7.4.11}$$

$$i_B = \frac{dQ_F}{dt} + \frac{Q_F}{\tau_{BF}} + \frac{dQ_R}{dt} + \frac{Q_R}{\tau_{BR}} + \frac{dQ_{VE}}{dt} + \frac{dQ_{VC}}{dt}$$

A circuit model corresponding to these equations is shown in Figure 7.18. Its form clearly indicates the superposed forward-active and reverse-active charge-control models. The dc components in the circuit of Figure 7.18 have a one-to-one correspondence to the components of the Ebers-Moll large-signal representation sketched in Figure 6.12. The Ebers-Moll model is also built up from superposed forward-active and reverse-active equivalent circuits.

In Equations 7.4.11, parameters associated with the reverse-active region of bias are defined analogously to those representing forward bias. For example, the controlled charge Q_R is related to V_{BC} by

$$Q_R = Q_{RO}\left[\exp\left(\frac{qV_{BC}}{kT}\right) - 1\right] \tag{7.4.12}$$

and this charge represents storage in both the base and collector quasi-neutral regions.

The density of dopant atoms in the collector region of an *IC* transistor is normally of the same order of magnitude or lower than the density in the base

---

* Charge-control equations for a *pnp* transistor are given in Problem 7.17.

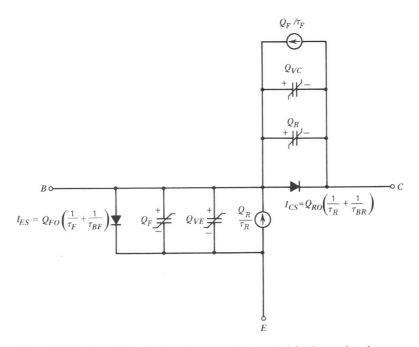

Figure 7.18 Complete bipolar charge-control model for large-signal applications.

near the collector junction. Hence, when a transistor is under reverse-active bias, the efficiency of electron injection into the base is typically quite low (approximately 60 to 85%). Thus, a relatively large amount of charge is stored in the collector. In fact, because the base is so narrow, this charge is typically the dominating component of Q_R. For example, Figure 7.19 is a sketch of the components of Q_F and Q_R for two cases of saturated transistors. Figure 7.19a shows the case of a homogeneously doped transistor with a lightly doped collector region and Figure 7.19b shows a typical IC transistor. The large amount of collector charge Q_C in either transistor when it is saturated must be removed to bring it out of saturation. This can cause considerable delay. One means of speeding its removal, the incorporation of a large number of recombination centers, usually by doping the collector region with gold, was extensively employed in early designs of fast switching integrated circuits. Superior switching results can be obtained by preventing the transistor from entering the saturated mode of operation. An effective method to accomplish this was described in Section 3.6 where Schottky clamping was discussed. An IC realization of a Schottky-clamped switching transistor was shown in Figure 6.18. Figure 7.20 shows the comparative switching times of a gold-doped switching transistor and a Schottky-clamped transistor having the same dimensions. The impressive reduction in the time delay before the output voltage rises makes clear why Schottky-clamped transistors have become so important for fast digital logic circuits.

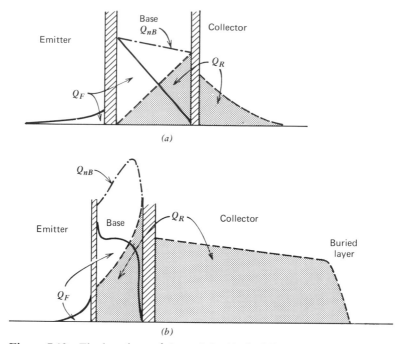

Figure 7.19   The locations of Q_F and Q_R (dashed lines) for saturated conditions: (a) in a homogeneously doped transistor with a lightly doped collector region and (b) in an epitaxial diffused *IC* transistor. The dot-dash lines represent the total base charge Q_{nB}

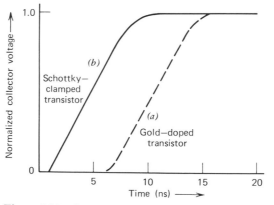

Figure 7.20   Comparison of turn-off times of (a) gold-doped and (b) Schottky-clamped transistors. The gold-doped device is delayed 7 *ns* for recombination to take place before it begins to change state.[11]

Saturation Transient.[†] As a final application of charge-control modeling, we consider the solution for the transient behavior of a bipolar transistor within the saturated region of operation, that is, a transistor that has not been Schottky clamped. For this case all terms in Equations 7.4.11 need to be retained.

The circuit is shown in Figure 7.21. We assume that $V_S \gg V_o$, the "turn-on voltage" of the base-emitter diode that was introduced in Section 3.6. We also take V_C to be much greater than the drop across the transistor in saturation (Equation 6.4.13) and assume that the base drive is strong enough to saturate the transistor fully; that is, that $\beta_F V_S/R_S \gg V_C/R_L$. If the switch is closed at $t = 0$, the transistor will enter the active-bias region and current will build in the collector circuit according to Equation 7.4.9 with the modification of the time constant given in Equation 7.4.10. The final value of current $\beta_F i_2$ will not be reached because the transistor will enter saturation when V_{BC} becomes roughly 0.5 V. The collector current will then be limited to V_C/R_L as shown in Figure 7.22a. Before the transistor saturates, the stored charge Q_F will increase with a wave shape similar to that of i_C (Figure 7.22b). Although the collector current stabilizes after the transistor saturates, the charges Q_F and Q_R are still changing with time, and the transistor is not in a steady-state "on" condition until these quantities are constant. Before saturation is reached, the collector-base junction is back biased, and Q_R is essentially zero. After saturation Q_R increases, as does Q_F.

To analyze the saturated behavior, a set of independent simultaneous equations for Q_F and Q_R must be solved. The set consists of equations for i_B and i_C.

$$i_B = \frac{Q_F}{\tau_{BF}} + \frac{dQ_F}{dt} + \frac{Q_R}{\tau_{BR}} + \frac{dQ_R}{dt}$$

$$i_C = \frac{Q_F}{\tau_F} - Q_R\left(\frac{1}{\tau_R} + \frac{1}{\tau_{BR}}\right) - \frac{dQ_R}{dt} \qquad (7.4.13)$$

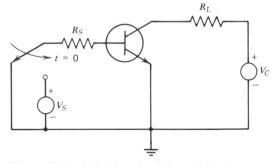

Figure 7.21 Switching circuit in which a transistor undergoes transient behavior from cut-off to saturation. The source voltage V_S is taken to be much greater than the "on voltage" V_o for the base-emitter diode. The base drive is taken to be sufficient to cause transistor saturation ($\beta_F V_S/R_S \gg V_C/R_L$).

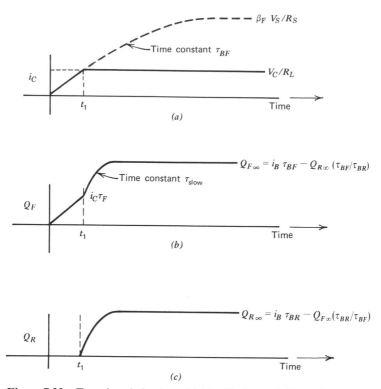

Figure 7.22 Transient behavior of (a) i_C, (b) Q_F, and (c) Q_R for the switching circuit in Figure 7.21.

In writing these equations, the terms representing charging currents for Q_{VC} and Q_{VE} are dropped as negligible because V_{BC} and V_{BE} are roughly constant in saturation.

Equations 7.4.13 represent simultaneous differential equations for which the natural frequencies are solutions of s that satisfy the equation:

$$\left(s + \frac{1}{\tau_{BF}}\right)\left(s + \frac{1}{\tau_R} + \frac{1}{\tau_{BR}}\right) + \left(s + \frac{1}{\tau_{BR}}\right)\frac{1}{\tau_F} = 0 \qquad (7.4.14)$$

The roots of this quadratic are approximately given by

$$|s_1| = \frac{1}{\tau_{FAST}} = \left(\frac{1}{\tau_F} + \frac{1}{\tau_R} + \frac{1}{\tau_{BR}} + \frac{1}{\tau_{BF}}\right)$$

and

$$|s_2| = \frac{1}{\tau_{SLOW}}$$

$$= \tau_{FAST}\left(\frac{1}{\tau_F \tau_{BR}} + \frac{1}{\tau_R \tau_{BF}} + \frac{1}{\tau_{BF} \tau_{BR}}\right) \qquad (7.4.15)$$

These roots represent normal modes of the transient solution for Q_F and Q_R in saturation. As the subscripted names imply, the root s_1 is responsible for a transient solution that is completed long before the steady state is reached, while the root s_2 leads to the transient behavior of Q_F and Q_R that dominates while the transistor remains saturated. Hence, a good approximation to the transient behavior of Q_F and Q_R in saturation (denoted by Q_{FS} and Q_{RS}) is

$$Q_{FS} = (Q_{F\infty} - Q_{F1})\left[1 - \exp\frac{-(t - t_1)}{\tau_{\text{SLOW}}}\right] + Q_{F1}$$

$$Q_{RS} = (Q_{R\infty} - Q_{R1})\left[1 - \exp\frac{-(t - t_1)}{\tau_{\text{SLOW}}}\right] + Q_{R1} \qquad (7.4.16)$$

where t_1 is the time at which saturation begins, $Q_{F\infty}$ and $Q_{R\infty}$ are the final values, and Q_{R1} and Q_{F1} are the values of the two controlled charges at $t = t_1$.

The total saturation charge $Q_{ST} = (Q_{FS} + Q_{RS})$ is sometimes combined in terms of the forward charge $Q_F = I_C \tau_F$ at the edge of saturation and an added portion that enters the channel after the transistor saturates.

$$Q_{ST} = I_C \tau_F + Q_S\left[1 - \exp\frac{-(t - t_1)}{\tau_{\text{SLOW}}}\right] \qquad (7.4.17)$$

where

$$Q_S = Q_{F\infty} + Q_{R\infty} - I_C \tau_F$$

For the prototype transistor in which essentially all charge is stored in the base, these definitions aid in understanding the physical mechanisms. They can be simply represented as shown in Figure 7.23. The important feature, emphasized by Figure 7.23, is that transistor switching is not completed when the collector current assumes a steady value (which occurs at the time of saturation), but that the steady state is only reached when Q_R and Q_F have reached their final values.

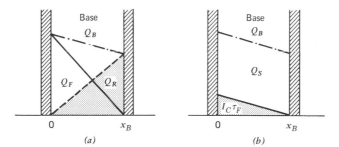

(a)   (b)

Figure 7.23 Two representations for total charge at steady state in a saturated, homogeneously doped transistor with high emitter and collector doping. (a) The two charge-control variables Q_F and Q_R are preserved. (b) The total charge is broken up into $I_C\tau_F$, which is present at the onset of saturation, and Q_S, which is added during the saturation transient.

We have just considered the "turn-on" transient. To calculate the "turn-off" transient a similar two-part analysis would need to be carried out. We would first find solutions in the saturated region while Q_R and Q_F are reduced to the values applying at the edge of active-mode bias and then solve for the transient through the active region to transistor cut-off.

It will not be of much benefit to carry these calculations any further; they have already served to indicate the analytical means afforded by charge-control analysis. We see that substantial mathematical labor is involved in obtaining hand-calculated solutions to the charge-control equations. We also have seen that their format is a useful representation, not only because they provide a set of linear simultaneous equations, but also because it is straightforward to identify the physical basis for each term and to simplify the equations in many special cases. Fortunately, the charge-control equations are a particularly simple set to implement for solution by computer routines when detailed solutions are desired. Further applications of the model to switching circuits are extensively described in reference 8.

7.5 Small-Signal Transistor Model

When transistors are biased in the active region and used for amplification, it is often worthwhile to approximate their behavior under conditions of small voltage variations at the base-emitter junction. If these variations are smaller than the thermal voltage $V_t = (kT/q)$, it is possible to represent the transistor by a linear equivalent circuit. This representation can be of great aid in the design of amplifying circuits. It is called the small-signal transistor model. It can be derived readily by making use of several of the charge-control relationships that were discussed in Section 7.4.

When a transistor is biased in the active mode, collector current is related to base-emitter voltage by Equation 6.2.1, which is repeated here for convenient reference.

$$I_C = I_S \exp\left(\frac{qV_{BE}}{kT}\right) = I_S \exp\left(\frac{V_{BE}}{V_t}\right) \tag{7.5.1}$$

Hence, if V_{BE} varies incrementally, I_C will also vary according to

$$\frac{\partial I_C}{\partial V_{BE}} = \frac{I_S}{V_t} \exp\left(\frac{V_{BE}}{V_t}\right) = \frac{I_C}{V_t} \equiv g_m \tag{7.5.2}$$

This derivative is recognized as the transconductance defined in Equation 4.5.13 and given the usual symbol g_m. Notice that g_m is directly proportional to the bias current in the transistor. The variation of base current with base-emitter voltage can be found most directly by making use of the charge-control expressions for I_C and I_B (Equations 7.4.1 and 7.4.3).

$$\frac{\partial I_B}{\partial V_{BE}} = \frac{\partial(Q_F/\tau_{BF})}{\partial V_{BE}} = \frac{\partial(I_C\tau_F/\tau_{BF})}{\partial V_{BE}} = \frac{\tau_F g_m}{\tau_{BF}} = \delta g_m \tag{7.5.3}$$

where the ratio of τ_F to τ_{BF} is expressed as a *defect factor* δ. Referring to Equation 7.4.4, we see that δ is equal to β_F^{-1}.

The base minority charge Q_F varies with base-emitter voltage according to

$$\frac{\partial Q_F}{\partial V_{BE}} = \frac{\partial(I_C\tau_F)}{\partial V_{BE}} = g_m\tau_F \equiv C_D \tag{7.5.4}$$

where the symbol C_D (often called the *diffusion capacitance*) represents the capacitance associated with incremental changes in the injected minority-carrier charge.

If the incremental voltages and currents in Equations 7.5.2, 7.5.3, and 7.5.4 are identified with ac signals, then the system of equations can be represented by the equivalent circuit shown in Figure 7.24. The small-signal incremental currents and voltages are denoted by lowercase symbols. The base-emitter input circuit is a parallel RC network having a time constant τ_F/δ that is just the base time constant τ_{BF}.

The collector-emitter output circuit consists of a current source activated by the input voltage. The output current for a given v_{BE} is proportional to g_m, and therefore dependent on the dc bias I_C as shown in Equation 7.5.1. The equivalent circuit of Figure 7.24 emphasizes the fact that, to first order, the input is decoupled from the output and the output is insensitive to collector-base voltage variations. From the analysis of Section 7.1, we know that the voltage across the collector-base junction does influence collector current, chiefly as a result of the Early effect. The variation of I_C with V_{CB} was shown in Section 7.1 to be the ratio of the collector current I_C to the Early voltage V_A. In terms of the small-signal parameters,

$$\left|\frac{\partial I_C}{\partial V_{CB}}\right| \equiv \frac{I_C}{|V_A|} = \frac{g_mV_t}{|V_A|} = \eta g_m \tag{7.5.5}$$

where a new parameter $\eta \equiv V_t/|V_A|$ has been introduced to represent the ratio of the change in I_C when V_{CB} is varied to the change in I_C when V_{BE} is varied.

Variation in I_C with V_{CB} must also result in a change in the controlled charge Q_F with V_{CB}. This can be calculated from the charge-control relationship of Equation 7.4.1.

$$\left|\frac{\partial Q_F}{\partial V_{CB}}\right| = \left|\frac{\partial(I_C\tau_F)}{\partial V_{CB}}\right| = \tau_F\eta g_m = \eta C_D \tag{7.5.6}$$

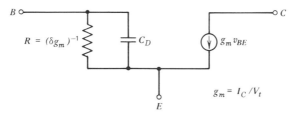

Figure 7.24 Equivalent circuit representing small-signal active bias for a bipolar junction transistor. Only first-order effects have been considered.

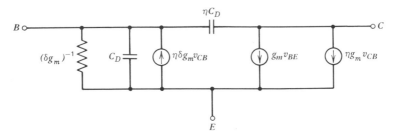

Figure 7.25 Small-signal equivalent circuit for a bipolar junction transistor with Early-effect elements.

Any change in base minority charge results in a change in base current, as well as in collector current. Thus, variation in V_{CB} results in a change in I_B that is given by

$$\left|\frac{\partial I_B}{\partial V_{CB}}\right| = \left|\frac{\partial(Q_F/\tau_{BF})}{\partial V_{CB}}\right| = \frac{\eta g_m \tau_F}{\tau_{BF}} = \eta \delta g_m \qquad (7.5.7)$$

The variations calculated in Equations 7.5.5, 7.5.6, and 7.5.7 can be incorporated into the linear equivalent circuit of Figure 7.24 by adding three elements as shown in Figure 7.25. The variation in I_C calculated in Equation 7.5.5 represents a change in the current flowing from collector to emitter in response to a change in collector-base voltage. It is thus modeled as a current generator activated by collector-base voltage. The current generator is directed from collector to emitter because an increase in V_{CB} increases I_C as described in Section 7.1. The change in injected charge storage in response to a change in collector-base voltage calculated in Equation 7.5.6 is modeled as a capacitor from collector to base. The variation in current calculated in Equation 7.5.7 flows between the base and the emitter and is caused by a changing collector-base voltage. It is thus modeled as a current generator directed from the emitter to the base. This direction is consistent with a reduced base current as a consequence of a reduction in Q_F.

The equivalent circuit sketched in Figure 7.25 can be simplified by employing two procedures common in circuit analysis. First, the generators involving v_{CB} can be drawn between the collector and base nodes by using the equivalence that is illustrated in Figure 7.26. In this way some generators will be activated by the voltage that appears across their terminals and can be replaced by passive

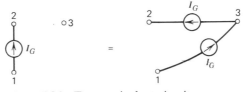

Figure 7.26 Two equivalent circuits representing the same current flowing between nodes.

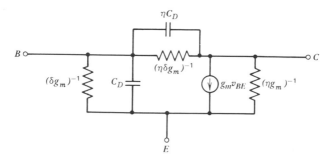

Figure 7.27 Simplified small-signal equivalent circuit including Early-effect elements. In general, $\delta \ll 1$ and $\eta \ll 1$.

elements. The second procedure involves re-expressing the activating voltages for several generators by making use of the identity: $v_{CE} \equiv v_{CB} + v_{BE}$. When these operations are performed and the admittance of elements in parallel is summed, the circuit can be simplified to the form shown in Figure 7.27. This form of the small-signal equivalent circuit is widely referred to as the *hybrid-pi circuit* for the transistor. The term *hybrid* is used because the generator is a voltage-activated current source, and it therefore relates quantities of differing dimensions. The reference to *pi* denotes the general geometric shape of the circuit in the form of a Greek letter Π.

For accurate representation of the transistor small-signal behavior in some situations, two additional effects remain to be considered. The first of these is base resistance, discussed in Section 7.2. There we saw that current-crowding effects cause an overall dc base resistance R_B (as defined implicitly in Equation 7.2.13) that is a function of collector current. The dependence of R_B on I_C for a typical *npn* transistor is shown in Figure 7.11. To incorporate this result into a small-signal equivalent circuit, the interdependences of the small-signal variations in voltage and current must be considered. To do this properly, it is necessary to consider the functional relationships between V_{BE}, I_B, I_C, and R_B. Total differentiation of V_{BE} with respect to I_C, for example, will result in three terms:

$$\frac{dV_{BE}}{dI_C} = \frac{\partial V_{BE}}{\partial I_C}\bigg|_{I_B,R_B} + \frac{\partial V_{BE}}{\partial I_B}\bigg|_{I_C,R_B}\left(\frac{dI_B}{dI_C}\right) + \frac{\partial V_{BE}}{\partial R_B}\bigg|_{I_C,I_B}\left(\frac{dR_B}{dI_C}\right) \tag{7.5.8}$$

From Equation 7.2.13 we have

$$V_{BE} = I_B R_B + V_t \ln\left(\frac{I_C}{I_S}\right) \tag{7.5.9}$$

Therefore, using Equation 7.5.8, we obtain

$$\frac{dV_{BE}}{dI_C} = \frac{V_t}{I_C} + R_B \frac{dI_B}{dI_C} + I_B \frac{dR_B}{dI_C} = \frac{1}{g_m} + \delta R_B + I_B \frac{dR_B}{dI_C} \tag{7.5.10}$$

where we have used the previously defined symbols g_m and δ while retaining the term involving the derivative of R_B. In practice this derivative can be obtained from a plot similar to that in Figure 7.11.

To obtain an equivalent-circuit representation from Equation 7.5.10, we may solve for the base input resistance R_I where

$$R_I = \frac{dV_{BE}}{dI_B} = \frac{dV_{BE}}{dI_C} \cdot \frac{\partial I_C}{\partial I_B} = \frac{dV_{BE}}{dI_C} \cdot \frac{1}{\delta} \tag{7.5.11}$$

Using Equation 7.5.10 in 7.5.11, we find

$$R_I = \frac{1}{\delta g_m} + \left(R_B + \frac{I_B}{\delta} \frac{dR_B}{dI_C} \right) \tag{7.5.12}$$

Thus, in order to take account of base resistance, a resistor of value

$$r_b = R_B + I_C \frac{dR_B}{dI_C} \tag{7.5.13}$$

must be added in series with the base-emitter resistance $(\delta g_m)^{-1}$ that was previously determined. This has been done in the low–frequency circuit sketched in Figure 7.28. In Figure 7.28 we have simplified the equivalent circuit from that shown in Figure 7.27 by omitting capacitors (valid at low frequency) and large resistors in order to focus our attention on the effects of base resistance. As is apparent from this circuit, when base resistance is taken into account, the current generator in the output circuit is no longer actuated by the applied base-emitter voltage, but is rather a function of an internal node-pair voltage. It is left as a problem to show that the current gain in the presence of base resistance is also properly modeled when the circuit of Figure 7.28 is used.

Our consideration of base resistance has been limited to dc and low-frequency effects. Base resistance can lead to more complex behavior at higher frequencies because the base resistance and junction capacitances behave like distributed transmission lines. These effects can be modeled with fair accuracy by using an equivalent shunt RC network in place of r_b. A full discussion of this topic is provided in reference 9.

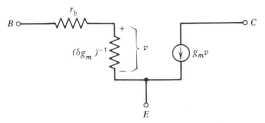

Figure 7.28 Low frequency, small-signal equivalent circuit including base resistance.

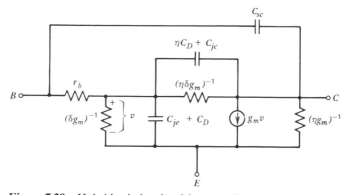

Figure 7.29 Hybrid–pi circuit with space-charge capacitances and base resistance.

The last complication that we shall add to the hybrid-pi circuit is the capacitance associated with the junction space-charge regions. This capacitance is in parallel with the base-emitter and base-collector capacitors shown in Figure 7.27. The added capacitance is usually denoted by C_{je} and C_{jc} and is calculated from the junction-capacitance equations already derived in Chapter 4. For increased accuracy, it is sometimes necessary to divide the total collector junction capacitance into a portion across the base impedance element (referred to as C_{sc}) and a portion C_{jc} that is returned to the intrinsic transistor base node, that is, the node connected to δg_m. This division implies that some parts of the collector capacitance will not be charged through the base resistance. The overall equivalent circuit including these effects is shown in Figure 7.29.

The circuit taken in its entirety may appear formidable. Fortunately, it is seldom necessary to deal directly with the overall hybrid-pi circuit in hand calculations. Either several elements in the circuit have negligible effect under given conditions and the circuit can therefore be simplified, or else the calculations are carried out by a computer.

Equivalence Between Models

Three models representing bipolar transistors have now been discussed: the Ebers-Moll model in Section 6.4, the charge-control model in Section 7.4, and the hybrid-pi model in Section 7.5. The first two of these models are valid for all ranges of bias, whereas the third is a small-signal equivalent circuit for use only in the region of active bias. Since the models have regions of overlapping validity, a series of relationships between their parameters can be expected. For example, in Problem 7.20 it is found that the Ebers-Moll parameters α_F and α_R are related to the charge-control parameters τ_F and τ_{BF} by

$$\alpha_F = \frac{\tau_{BF}}{\tau_F + \tau_{BF}}$$

and

$$\alpha_R = \frac{\tau_{BR}}{\tau_R + \tau_{BR}} \tag{7.5.14}$$

respectively. In the dc Ebers-Moll representation, the six parameters of the basic charge-control model (Q_{FO}, Q_{RO}, τ_F, τ_{BF}, τ_R, and τ_{BR} are reduced to only four parameters α_F, α_R, I_{ES}, and I_{CS}).

Relationships such as Equation 7.5.14 are useful to obtain model parameters by measurements and to establish properties of the models. For example, the reciprocity condition of the Ebers-Moll model $\alpha_F I_{ES} = \alpha_R I_{CS}$ (Equation 6.4.8) can be used with the charge-control model to derive

$$\frac{Q_{FO}}{\tau_F} = \frac{Q_{RO}}{\tau_R} \tag{7.5.15}$$

A useful relationship that allows the experimental determination of the charge-control parameter τ_F can be derived by amalgamating the small-signal equivalent circuit with the charge-control model under active bias (Figure 7.15). Consider that only dc bias is applied to the collector. Hence, the ac small-signal equivalent circuit has the collector shorted to ground. Under this condition, it is usually a good approximation to neglect the Early-effect elements in the circuit of Figure 7.29 and also to consider base resistance as insignificant. The circuit of Figure 7.29 can then be simplified to the form shown in Figure 7.30. The current gain i_C/i_B in this circuit is

$$\begin{aligned}
\frac{i_C}{i_B} &= \frac{(1/\delta)(1 - j\omega C_{jc}/g_m)}{1 + j\omega[(C_{je} + C_{jc})/g_m\delta + \tau_F/\delta]} \\
&\approx \left(\frac{\tau_{BF}}{\tau_F}\right)\left[1 + j\omega\left(\tau_{BF} + \frac{(C_{je} + C_{jc})\tau_{BF}}{g_m\tau_F}\right)\right]^{-1}
\end{aligned} \tag{7.5.16}$$

where we have used Equation 7.4.4 and omitted the frequency-dependent term in the numerator. This numerator term is only of consequence at frequencies appreciably above those at which the imaginary term in the denominator is dominant.

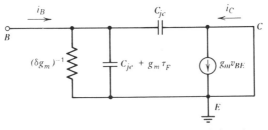

Figure 7.30 Equivalent circuit for obtaining the interrelationship between f_T and τ_F.

As the frequency is increased, the current gain decreases and, from Equation 7.5.16, the magnitude of the current gain will be unity at a frequency f_T which is approximately given by

$$f_T = \frac{1}{2\pi \left(1 + \dfrac{C_{je} + C_{jc}}{g_m \tau_F}\right) \tau_F}$$ (7.5.17)

Solving this equation for τ_F, we have

$$\tau_F = \frac{1}{2\pi f_T} - \frac{(C_{je} + C_{jc})}{g_m}$$ (7.5.18)

Thus, measurements of short-circuit gain (i.e., current gain with the collector in an ac short-circuit connection) as a function of frequency provide a means to obtain the charge-control parameter τ_F. The parameter f_T can be obtained by extrapolating a plot of gain versus frequency to unity gain. A value for τ_F is then obtained by using f_T in Equation 7.5.18. The low frequency gain β_F can then be used to calculate $\tau_{BF} = \beta_F \tau_F$ (Equation 7.4.4).

   If these measurements are repeated for the transistor under reverse-active bias, the parameters β_R and τ_{BR} can be obtained in a similar manner. Measurements of leakage current under active bias permit the calculation of I_{CS} and I_{ES} by using Equation 6.4.11 and the corresponding relationship for reverse-active bias. Thus, all parameters for the basic transistor models can be extracted from a series of measurements when the equivalences between parameters of the various transistor models are considered.

7.6 Bipolar Transistor Model For Computer Simulation[†]

For computer simulation of transistors, precision takes precedence over conceptual or computational simplicity. To maximize the usefulness of computer programs, models for transistors should be accurate for both large- and small-signal applications and they should also be readily characterized by parameters that are relatively easy to obtain and to verify. These requirements have been met most successfully thus far by simulations that are based on the Ebers-Moll equations, which were introduced in Section 6.4.

   The starting point for our discussion is the so-called "transport version" of the Ebers-Moll equations. This consists of Equations 6.4.2 and 6.4.3, which specify transistor currents in terms of the linking current between the emitter and the collector and additional base-emitter and base-collector diode components.

   In Equation 6.1.9 we derived an expression for the linking-current density J_n, which we write here in terms of total current I_n.

$$I_n = I_S \left[\exp\left(\frac{V_{BC}}{V_t}\right) - \exp\left(\frac{V_{BE}}{V_t}\right) \right]$$ (7.6.1)

This equation and the forms derived in Equations 6.4.7a and 6.4.7b allow the Ebers-Moll equations to be written in the following form:

$$I_C = -I_n - \frac{I_S}{\beta_R}\left[\exp\left(\frac{V_{BC}}{V_t}\right) - 1\right]$$

$$I_E = I_n - \frac{I_S}{\beta_F}\left[\exp\left(\frac{V_{BE}}{V_t}\right) - 1\right]$$

$$I_B = \frac{I_S}{\beta_F}\left[\exp\left(\frac{V_{BE}}{V_t}\right) - 1\right] + \frac{I_S}{\beta_R}\left[\exp\left(\frac{V_{BC}}{V_t}\right) - 1\right] \tag{7.6.2}$$

In this formulation, the three parameters I_S, β_F, and β_R suffice to characterize the basic Ebers-Moll relationships. Additional terms must be added to the equations

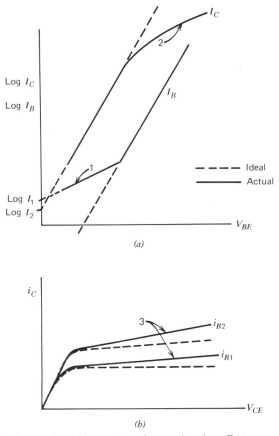

(a)

(b)

Figure 7.31 The results of second-order effects on bipolar transistor characteristics in the active mode. The numbers on the figures refer to the effects enumerated in the text. The base current extrapolated to zero base-emitter voltage is I_1 in Equation 7.6.3.

in this set to represent effects not included in the Ebers-Moll model, for example, the phenomena described earlier in this chapter. Gummel and Poon[10] have shown relatively straightforward methods by which Equations 7.6.2 can be modified to incorporate three important second-order effects: (1) recombination in the emitter-base space-charge region at low emitter-base bias, (2) the current-gain decrease experienced under high-current conditions, and (3) effects of space-charge-layer widening (Early effect) on the linking current between the emitter and the collector. The consequences of these second-order effects lead to the deviations from ideal performance that are revealed in the sketches in Figure 7.31a and 7.31b. Inclusion of these effects leads to the *Gummel-Poon model*, which is useful for computer simulation.

Recombination in the Space-Charge Regions. As we saw in Chapter 5, recombination in the space-charge region leads to modified diode relationships for the junction currents. These can be modeled by adding four parameters to the Ebers-Moll model to define base current in terms of a superposition of ideal and nonideal-diode components.

$$I_B = \frac{I_S}{\beta_F}\left[\exp\left(\frac{V_{BE}}{V_t}\right) - 1\right] + I_1\left[\exp\left(\frac{V_{BE}}{n_e V_t}\right) - 1\right]$$
$$+ \frac{I_S}{\beta_R}\left[\exp\left(\frac{V_{BC}}{V_t}\right) - 1\right] + I_2\left[\exp\left(\frac{V_{BC}}{n_c V_t}\right) - 1\right] \quad (7.6.3)$$

The new parameters I_1, I_2, n_e, and n_c are found in practice by measurements made at low base-emitter biases. For example, I_1 is obtained from the intercept of a plot of $\log I_B$ versus V_{BE} extrapolated to $V_{BE} = 0$ (Figure 7.31).

Early Effect and High-Level Operation. Both the high-current effect (2) and the Early effect (3) can be incorporated by modifying the value of I_S, the multiplier for the linking current between the emitter and the collector. In Section 6.1 it was shown that J_S depends inversely on the base majority-charge density Q_B. If we rewrite Equations 6.1.8 and 6.1.10 to represent the total base charge Q_{BT} and total saturation current I_S, we have

$$I_S = J_S A_E = \frac{q^2 A_E^2 n_i^2 \tilde{D}_n}{Q_{BT}} \quad (7.6.4)$$

where

$$Q_{BT} = q A_E \int_0^{x_B} p(x)\, dx \quad (7.6.5)$$

In the Gummel-Poon model, Q_{BT} is represented by components having a bias dependence that can be calculated readily. First, there is the "built-in" base charge Q_{BO} where

$$Q_{BO} = q A_E \int_0^{x_B} N_a(x)\, dx \quad (7.6.6)$$

In addition to this term, there are emitter and collector charge-storage contributions (Q_{VE} and Q_{VC}) plus the charge associated with forward and reverse injection of base-minority carriers. These are all summed to represent Q_{BT} by the equation:

$$Q_{BT} = Q_{BO} + C_{je}V_{BE} + C_{jc}V_{BC}\frac{A_E}{A_C} + \frac{Q_{BO}}{Q_{BT}}\tau_F I_S\left[\exp\left(\frac{V_{BE}}{V_t}\right) - 1\right]$$
$$+ \frac{Q_{BO}}{Q_{BT}}\tau_R I_S\left[\exp\left(\frac{V_{BC}}{V_t}\right) - 1\right] \tag{7.6.7}$$

By defining several parameters, Equation 7.6.7 can be put into a more manageable format.

$$q_b \equiv \frac{Q_{BT}}{Q_{BO}}; \qquad I_{KF} \equiv \frac{Q_{BO}}{\tau_F};$$

$$I_{KR} \equiv \frac{Q_{BO}}{\tau_R}; \qquad |V_A| \equiv \frac{Q_{BO}}{C_{jc}}\frac{A_C}{A_E}$$

$$|V_B| \equiv \frac{Q_{BO}}{C_{je}} \tag{7.6.8}$$

The key variable, total base charge Q_{BT} is normalized in Equation 7.6.8 to Q_{BO}, and its dimensionless counterpart is designated as q_b. The two-charge control time constants τ_F and τ_R together with Q_{BO} define "knee currents" I_{KF} and I_{KR} having a significance that will shortly become apparent. The definition of the Early voltage V_A and the equivalent Early voltage for reverse operation V_B is the same as was derived in Equation 6.1.8.

In terms of the normalized parameters, Equation 7.6.7 can be written in the following form:

$$q_b = q_1 + \frac{q_2}{q_b} \tag{7.6.9}$$

where q_1 and q_2 are auxiliary variables as defined by

$$q_1 = 1 + \frac{V_{BC}}{|V_A|} + \frac{V_{BE}}{|V_B|}$$
$$q_2 = \frac{I_S}{I_{KF}}\left[\exp\left(\frac{V_{BE}}{V_t}\right) - 1\right] + \frac{I_S}{I_{KR}}\left[\exp\left(\frac{V_{BC}}{V_t}\right) - 1\right] \tag{7.6.10}$$

These new variables provide a convenient indication of the significance of the second-order effects. If the Early effect is negligible, q_1 will approach unity. If high-level injection effects are not important, q_2 will be small.

Thus, base-width modulation effects have been modeled through the introduction of the two Early voltages while high-level bias effects are specified through the knee currents I_{KF} and I_{KR}. The Gummel-Poon model thus requires the specification of three variables I_S, β_F, and β_R for the basic Ebers-Moll model and then adds four more, I_1, I_2, n_e, and n_c to model space-charge-region recombination effects. The Ebers-Moll parameters that are specified should be valid in the mid-bias range, where high-level effects are not present.

Finally, base-width and majority-charge modulation are modeled by specifying a variable q_b that depends on the values of four additional variables I_{KF}, I_{KR}, V_A, and V_B. The overall model is thus specified by 11 parameters plus the temperature (to enable a calculation of V_t). The collected equations making up the model for an *npn* transistor are:

$$I_B = \frac{I_S}{\beta_F}\left[\exp\left(\frac{V_{BE}}{V_t}\right) - 1\right] + I_1\left[\exp\left(\frac{V_{BE}}{n_e V_t}\right) - 1\right]$$

$$+ \frac{I_S}{\beta_R}\left[\exp\left(\frac{V_{BC}}{V_t}\right) - 1\right] + I_2\left[\exp\left(\frac{V_{BC}}{n_c V_t}\right) - 1\right]$$

$$I_C = \frac{I_S[\exp(V_{BE}/V_t) - \exp(V_{BC}/V_t)]}{q_b} - \frac{I_S}{\beta_R}\left[\exp\left(\frac{V_{BC}}{V_t}\right) - 1\right]$$

$$- I_2\left[\exp\left(\frac{V_{BC}}{n_c V_t}\right) - 1\right] \tag{7.6.11}$$

$$q_b = \frac{q_1}{2} + \frac{\sqrt{q_1^2 + 4q_2}}{2}$$

$$q_1 = 1 + \frac{V_{BE}}{|V_B|} + \frac{V_{BC}}{|V_A|}$$

$$q_2 = \frac{I_S}{I_{KF}}\left[\exp\left(\frac{V_{BE}}{V_t}\right) - 1\right] + \frac{I_S}{I_{KR}}\left[\exp\left(\frac{V_{BC}}{V_t}\right) - 1\right]$$

To illustrate the validity of this equation set we might consider low-level, active-mode operation for which $q_2 \simeq 0$ (because the collector current in Figure 7.32 is

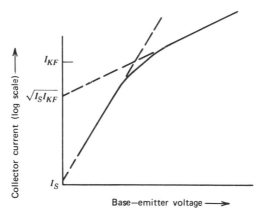

Figure 7.32 Logarithm of collector current versus V_{BE} in the active mode to illustrate the Gummel-Poon model for high-level effects. The asymptotes for low and high bias intersect at the "knee" current $I_C = I_{KF}$.

much less than the knee current I_{KF}). For this case

$$I_C \simeq \frac{I_S \exp\left(V_{BE}/V_t\right)}{1 + V_{BC}/|V_A|} \simeq I_S \exp\left(\frac{V_{BE}}{V_t}\right)\left(1 - \frac{V_{BC}}{|V_A|}\right) \tag{7.6.12}$$

and

$$\frac{\partial I_C}{\partial V_{CB}} = \frac{I_C}{|V_A|} \tag{7.6.13}$$

as was derived from fundamental considerations in Equation 7.1.3.

For operation at high-current levels, for which the Early effect is of far lesser consequence than are high-level injection effects, we have $q_2 > q_1$. Under this condition, the normalized base charge q_b has a high-bias asymptotic behavior of the form

$$q_b = \sqrt{\frac{I_S}{I_{KF}}} \exp\left(\frac{V_{BE}}{2V_t}\right) \tag{7.6.14}$$

Thus, collector current will vary as (*cf* Figure 7.32)

$$I_C = \sqrt{I_S I_{KF}} \exp\left(\frac{V_{BE}}{2V_t}\right) \tag{7.6.15}$$

It is left as a problem to show that the intercept of the asymptotic behavior of the Gummel-Poon model at low bias with the high-bias asymptote (Equation 7.6.15) occurs at the "knee" current $I_C = I_{KF}$. The physical basis for the form represented by Equation 7.6.15 is that sufficiently high injection into the base region leads to a bias dependence for the base majority-carrier concentration, as was described in the discussion of Equation 7.2.3.

7.7 Devices: *pnp* Transistors

There is a major advantage in circuit yield and economy if an integrated-circuit process is directed toward the production of only one type of transistor. Basically because of the higher mobility of electrons than of holes, *IC* processes are centered on the production of high quality *npn* transistors. If a *pnp* transistor is needed in a circuit, it is usually made without complicating the *IC* process. Two types of *pnp* transistors have been designed that meet this constraint: *substrate* and *lateral pnp transistors*.

Substrate *pnp* Transistors

One reason for the excellent performance of *npn* transistors is that the region in which transistor action takes place in these devices has been designed to be away from the surface and uniform over the broad area of the junction that is parallel to the surface plane. A *pnp* transistor with these features can be obtained using a

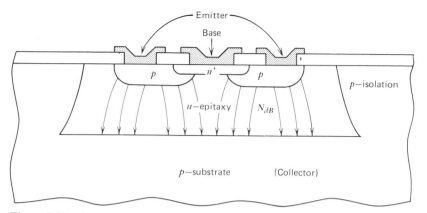

Figure 7.33 Cross section of a substrate *pnp* transistor. Flow lines for the linking current are sketched on the figure.

standard planar process by making the *pnp* emitter from the *p*-type diffusion that is normally used for the *npn* base region. The structure of this *pnp* transistor is shown in cross section in Figure 7.33.

The epitaxial region serves as the *pnp* base and the grown *np* junction at the interface between the epitaxial layer and the substrate is the collector junction. Because the collector region is the substrate of the integrated circuit it is not isolated from other *pnp* transistors formed in the same way. For this reason, these so-called *substrate pnp transistors* can only be used in an integrated circuit when the collector junction is an ac ground as in an emitter-follower circuit. This is a valuable component for many integrated circuits, but it obviously cannot be used in every case for which a *pnp* transistor is desirable. Although substrate *pnp* transistors do not have the built-in base field resulting from a graded base that is found in double-diffused *npn* transistors, they can be made with values of β up to about 100 at 1 mA. Depending on the process used, vertical *pnps* have been designed to operate in the range 1 μA to roughly 10 mA and to have values of f_T as high as 10 MHz.

Lateral *pnp* Transistors

There is a convenient way to make a *pnp* transistor with an isolated collector using standard processing. This is to employ the standard *npn* base *p*-type diffusion for both the emitter and the collector by spacing two *p*-regions close together as shown in the cross-sectional view in Figure 7.34. The device formed by this construction is known as a *lateral pnp transistor* because transistor action takes place laterally—that is, parallel to the surface between the emitter and collector regions. This design sacrifices the advantages normally gained by removing transistor action from the surface region. As a result, the performance of lateral *pnp* transistors is markedly inferior to that of standard *npn* transistors. Nevertheless, lateral *pnp* transistors are frequently used in both analog and digital integrated circuits.

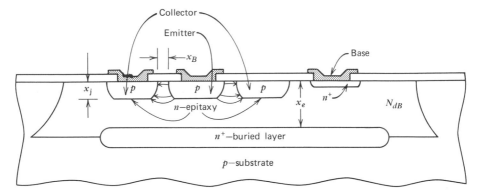

Figure 7.34 Cross section of a typical lateral *pnp* transistor for *IC* applications. The diffused collector region completely surrounds the emitter. Flow lines for the linking current are sketched on the figure.

The two collector regions for the lateral *pnp* transistor shown in Figure 7.34 are joined; typically, the collector completely surrounds the emitter region to improve current gain in the transistor. A buried layer is also included, as shown in Figure 7.34. The buried layer improves transistor gain and frequency response in two ways: (1) by reducing base resistance and (2) by suppressing the collection of holes at the junction between the epitaxial layer and the substrate. One way of understanding the second improvement is to note that the built-in field that results from the doping gradient in the buried layer repels any holes that are incident from the epitaxial region. Alternatively, one can consider that the inclusion of a buried layer causes an increase in the base doping Q_B of the parasitic substrate *pnp* transistor. As seen in Equation 6.1.10, this reduces the loss of holes to the parasitic device and, thereby, improves the gain of the lateral *pnp*.

Collector Current. The current linking the emitter and the collector in a lateral *pnp* transistor follows a two-dimensional path as can be seen from the flow lines sketched in Figure 7.34. Because the epitaxial region is uniformly doped, the boundary value for the injected hole density in the base at a given emitter-base bias will be uniform along the edge of the emitter-base junction regions. The density gradient that causes the injected holes to diffuse toward the collector will be maximum near the surface where the spacing between the junctions is minimum. If one moves away from the surface along the emitter-base junction, the gradient in the hole density decreases slowly as the distance between the diffused *p*-regions increases. Thus, the base width for the lateral *pnp* transistor is only approximated by the spacing between the junctions at the surface (x_B in Figure 7.34), and the hole current is nonuniform along the emitter-base junction. The configuration of the flow lines for the linking current is also influenced by the thickness of the epitaxial layer and by the geometry of the buried layer.

An analysis of lateral *pnp* transistors[12] has considered these effects in detail and has given an empirical means of relating the linking current I_p to the flow in a one-dimensional transistor of base width x_B and emitting area $P_E x_j$ where P_E is the perimeter of the emitter and x_j is the depth of the diffused junction. The actual current is written

$$I_p = F \frac{q P_E x_j D_p n_i^2}{N_{dB} x_B} \exp\left(\frac{q V_{EB}}{kT}\right) \tag{7.7.1}$$

where F can be shown to be a function of the two dimensionless ratios: x_e/x_j and x_B/x_j.[12] As seen in Figure 7.34, x_j is the depth of the emitter and collector diffusions, x_e is the thickness of the epitaxial region up to the edge of the buried layer, and x_B is the separation of the emitter and collector at the oxide-silicon interface. Curves showing $F(x_e/x_j, x_B/x_j)$ from reference 12 are given in Figure 7.35. A typical value for both of these arguments might be 2, in which case F from Figure 7.35 would be roughly 1.8. Thus, the total current is nearly doubled from that predicted by one-dimensional analysis.

The spacing x_B is determined by the separation of the emitter and collector windows in the photolithographic mask minus the lateral diffusion under the oxide toward one another by the acceptors forming the emitter and the collector. The

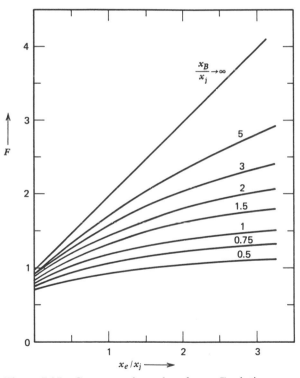

Figure 7.35 Geometry-dependent factor F relating actual collector current to that obtained from a one-dimensional model.[12]

photolithographic spacing is typically limited to about 3 μm, and the sideways diffusion is somewhat less than the diffusion in the vertical direction. Thus, x_B is generally of the order of 2 μm although by diffusing deeper, one can attain smaller spacings. By reducing x_B, one improves gain and frequency response, but suffers in reproducibility and reliability.

The epitaxial layer is typically lightly doped (between 10^{15} and 10^{16} donors cm^{-3}). Hence, only moderate forward bias at the emitter-base junction is required to bring the base into a high-level injection condition (Problem 7.9). Chou[12] has shown that high-level injection results in a fall-off of the gain of lateral *pnp* transistors because of three effects: (1) a lessened dependence of $p'(0)$—the excess hole density at the emitter-base junction—on emitter-base voltage, which can be accounted for by writing the equation analogous to Equation 7.2.3 for holes, (2) an effective variation in the diffusion constant D_p because of the *Webster effect* (Section 7.3), and (3) voltage drops in series with the applied emitter-base bias because of resistances in the base and the emitter.

Base Current. When the *npn* transistor was considered, it was only necessary to take account of three components of base current for an accurate device model. The three components were caused by injection of minority-carrier holes into the emitter (the dominant source for base current under most conditions of bias for the *npn* transistor), recombination of injected electrons in the base, and recombination of injected electrons in the emitter-base depletion region (important at low emitter currents. Currents analogous to these three components, in which the roles of electrons and holes are interchanged, are also important in lateral *pnp* transistors. Because of the presence of the oxide-silicon surface and the lateral geometry of the device, however, two other significant causes of base current are present in the lateral *pnp* transistor. These other components are caused by extra recombination at the oxide-silicon interface and in the neighborhood of the buried layer. An adequate representation of base current in the lateral *pnp* transistor can be obtained when these five current components are considered. The five components of base current enumerated above are indicated schematically in Figure 7.36.

The three contributions to I_B that are analogous to the significant components of base current in *npn* transistors can be expressed by equations similar to those already presented. An exception is the formulation of an equation analogous to Equation 6.2.4 for base recombination. The two-dimensional flow pattern in the lateral *pnp* transistor complicates a proper definition for the volume of the base. Chou[12] has shown that base recombination can be written

$$I_{B2} = \frac{qn_i^2[\exp(qV_{EB}/kT) - 1]}{N_{dB}\tau_p} V_B^* \qquad (7.7.2)$$

where τ_p is the lifetime of holes in the epitaxial region and V_B^* represents the volume defined in Figure 7.36 by a surface that bisects the base width x_B.

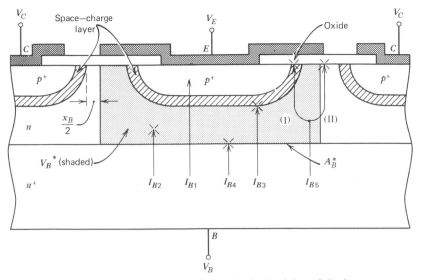

Figure 7.36 Schematic illustration of the physical origins of the base current in a lateral *pnp* transistor.[12] I_{B1} represents electron injection into the emitter, I_{B2} represents base recombination, I_{B3} represents space-charge region recombination away from the surface, I_{B4} accounts for recombination at the buried layer and collection by the substrate, and I_{B5} represents surface recombination of holes.

It has proven convenient to account for the flow of holes to the buried layer (I_{B4}) by assigning a recombination velocity s_{nn+} to the interface between the un-doped epitaxial region and the buried layer. As introduced in Equation 5.2.22, the recombination velocity is multiplied by the incident excess carrier density to express the total recombination rate at a surface. Assuming a relatively long lifetime in the epitaxial layer, one can use the boundary value for excess holes at the base-emitter junction to write

$$I_{B4} = q s_{nn+} A_B^* \frac{n_i^2}{N_{dB}} \left[\exp\left(\frac{qV_{EB}}{kT}\right) - 1 \right] \tag{7.7.3}$$

where A_B^* is the area at the lower surface of V_B^* (Figure 7.36). In the lateral *pnp* transistor, there is really no plane having the recombination velocity s_{nn+}. Rather, the recombination velocity is an effective parameter that can be used to account for all the hole current into the buried layer. The incident holes may recombine at the interface, recombine within the buried layer, or else be collected across the junction between the buried layer and the substrate. Recombination at the oxide-silicon interface is also treated by defining a recombination velocity s_{os} and expressing current by equations similar to Equation 7.7.3.

Once all the expressions for base current are written, the current itself can be calculated if appropriate values for s_{nn+}, s_{os}, and the hole lifetimes in the bulk and

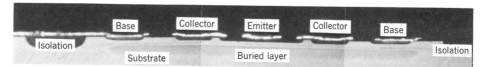

Figure 7.37   Stained angle-lap section of a lateral *pnp* transistor. (*Courtesy Signetics Corporation.*)

space-charge regions can be obtained. In general, special test structures are needed to determine these parameters.[12] Experiments on lateral *pnp* transistors have shown that for typically low values of s_{os} ($s_{os} \sim 1$–5 cm s^{-1}), the recombination current at the oxide-silicon surface I_{B5} is generally negligible. All of the other components have significance in one or another range of useful emitter-base bias. There appears to be a fairly wide range for s_{nn^+}, the recombination velocity at the buried layer. Values between 10 and 2000 cm s^{-1} have been reported. This variation apparently arises from differing dopant densities, buried-layer geometries and processing schedules. When s_{nn^+} becomes roughly 100 or greater, the vertical hole current generally becomes important.

The current gain β in lateral *pnp* transistors is substantially lower than that exhibited by *npn* transistors. Values of 20 or less are common although, with care, values of β up to about 100 have been achieved. In the low microampere range, β is typically lower than one. It increases with current until I_C is roughly 100 μA (for an emitter area $\sim 10^{-7}$ cm^2). At higher currents, β decreases strongly as collector current increases. The increasing β at low biases corresponds to the lessening importance of recombination in the emitter-base space-charge region. The decrease in β at higher currents accompanies the onset of high-level injection effects.

Figure 7.37 is a composite photograph (enlarged ~ 1500 times) of a cross section through a lateral *pnp* transistor. It was made by using angle-lapping and staining techniques that delineate diffused regions in integrated circuits. Such photographs are of great value in assessing *IC* process control and in locating errors.

Integrated-Injection Logic.   The lateral *pnp* transistor is an important component in a family of densely packed digital integrated circuits. This circuit family, usually called *integrated-injection logic* or I^2L has been described as "superintegrated" because it merges together *pnp* and *npn* transistors. The collector of one device functions simultaneously as the base of the other. Because this is the case, much of the surface area that is normally required for contacts, interconnections, and isolation diffusions in digital circuits can be saved.

An understanding of the ideas underlying I^2L can be gained by considering the basic I^2L gate shown in cross section in Figure 7.38a. A circuit diagram that corresponds to the gate is shown in Figure 7.38b. Note that the emitter of the lateral *pnp* transistor in the circuit is adjacent to the *p*-base of what appears in the cross-sectional view to be a conventional *npn* transistor. In I^2L, the *npn* transistor operates in the inverted mode, that is, with the uppermost n^+ region (normal

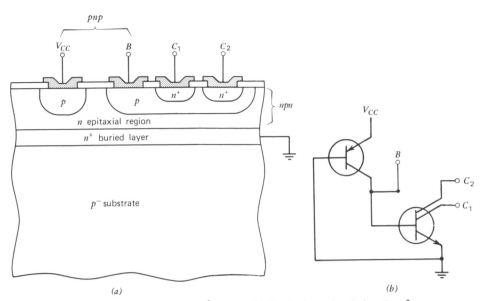

Figure 7.38 (a) Cross section of an I^2L gate. (b) Equivalent circuit for the I^2L gate.

emitter) functioning as a collector. When the *npn* transistor is connected in this way, many digital circuits can be crowded into a small surface area on the chip.

When bias is applied to the gate shown in Figure 7.38, the holes delivered by the lateral *pnp* transistor to the base of the *npn* device cause the latter to saturate unless base current is drawn through electrode *B*. The gate is thus bistable, with the *npn* transistor either biased in cut-off or else in saturation depending upon the presence or absence of current through electrode *B*. The gates shown in Figure 7.38 can be interconnected to obtain flip-flops or to carry out binary logic functions. A full analysis of the I^2L cell can be carried out by enumerating all of the physical origins for currents in the structure. The procedure is similar to that employed in our analysis of the lateral *pnp* transistor.

Summary

The basic theory of transistor action introduced in Chapter 6 must be augmented by considering several important physical effects to obtain acceptable accuracy for many design situations. For transistors biased in the active mode, one such effect is the variation in collector current as a result of variation in collector-base bias. The phenomenon, generally called the *Early effect*, can be conveniently described by the introduction of an *Early voltage*, which indicates the output current variation at a given bias point. Other effects result in limits to the application of transistors at different ranges of bias. The operation of bipolar transistors at low current levels is generally limited by recombination within the base-emitter space-charge region. This recombination reduces the injection of minority carriers into the base and results in a decrease in current gain (β_F) as the quiescent bias in the devices is reduced to very low levels.

When transistors are biased at high current levels, several important effects can take place. One of these is a reduction in emitter efficiency when the base minority-carrier density begins to approach the dopant density at the edge of the base-emitter space-charge region. Another effect at high current levels is the modification of the space-charge configuration within the collector-base depletion region. When this high–current condition is reached, the boundaries of the quasi-neutral base region are affected; in general, widening of the base region takes place. To analyze this so-called *Kirk effect*, it is necessary to consider in detail the doping profile of the transistor and to obtain a consistent set of solutions of Poisson's equation and the equations expressing mobile space charge in terms of bias currents. Base spreading resistance is another source for the degradation of transistor performance at higher current levels. The effect of base spreading resistance is to reduce the bias of those parts of the base-emitter junction remote from the base contact because of the ohmic voltage drop associated with the delivery of base majority carriers to the active base region. In a given device all the high-current effects may occur simultaneously at a given bias level. Understanding them in detail is more important in the modification of device design than it is in making use of the transistors in circuits. For circuit design, an empirical model of high-level performance is usually all that is necessary.

An equation for the transport of injected carriers across the quasi-neutral base of a transistor (the *base transit time*) can be formulated in the general case of arbitrary base doping. An understanding of this equation and of the steps in its derivation is useful in developing the charge-control representation of the transistor. Consideration of the effect of high-level operation on the base transit time leads to an understanding of the *Webster effect* in which minority-carrier transit time across the base is reduced by the field associated with excess majority carriers introduced into the base.

The charge-control model of the transistor comprises a set of linear differential equations that provide an extremely useful description of the transistor for circuit-design purposes. The model is useful when considering the device from its terminals outward; that is, the charge-control model does not give accurate information about any of the distributed effects within the device. When considered on an incremental-voltage basis, the charge-control model can be used to derive a small-signal equivalent circuit (*hybrid-pi model*) that is especially helpful for the design of amplifying circuits.

Interrelationships between the parameters of the charge-control model, the Ebers-Moll model, and the hybrid-pi model are useful both in gaining an understanding about the various models and in selecting experimental means to obtain model parameters. A model that successfully represents most of the significant physical effects in bipolar transistors and that is suitable for computer analysis of circuit behavior has been presented (*Gummel-Poon* model). In this model, several parameters are added to the basic Ebers-Moll transistor representation. Because of the processing requirements for integrated circuits, lateral *pnp* transistors—that is, transistors in which the separation between the emitter and collector is parallel to the surface—are often employed. The performance of lateral *pnp* transistors is frequently dominated by effects different from those that are significant for *npn* transistors. Lateral *pnp* transistors also play an important role in the operation of some *integrated-injection logic* circuits (I^2L).

References

1. J. M. Early, *Proc. IRE*, **40**, 1401 (1952).
2. F. A. Lindholm, and D. J. Hamilton, *Proc. IEEE*, **59**, 1377 (1971).
3. C. T. Kirk, *IRE Trans. Electron Devices*, ED-9, 164 (1962).
4. H. C. Poon, H. K. Gummel, and D. L. Scharfetter, *IEEE Trans. Electron Devices*, ED-16, 455, (1969): Reprinted by permission.
5. O. Manck, H. H. Heimeier, and W. L. Engl, *IEEE Trans. Electron Devices*, ED-21, 403 (1974).
6. W. M. Webster, *Proc. IRE*, **42**, 914 (1954).
7. R. Beaufoy, and J. J. Sparkes, *Automat. Teleph. Elect. J.*, **13**, Reprint 112 (1957).
8. P. E. Gray, and C. L. Searle, *Electronic Principles: Physics, Models and Circuits*, Wiley, New York, 1969.
9. P. E. Gray, D. DeWitt, A. R. Boothroyd, and J. F. Gibbons, *Physical Electronics and Models*, SEEC Volume II, Wiley, New York, 1964.
10. H. K. Gummel, and H. C. Poon, *Bell Syst. Tech. J.*, **49**, 827 (1970).
11. R. N. Noyce, et al, *Electronics*, July 21, 1969, p. 74.
12. S. Chou, *Solid-State Electron.*, **14**, 811 (1971).
13. A. S. Grove, *Physics and Technology of Semiconductor Devices*, Wiley, New York, 1967, pp. 229, 240.
14. J. Logan, *Bell System Tech. J.*, **50**, 1105 (April 1971).
15. A. Bar–Lev, *Semiconductor and Electronic Devices*, Prentice–Hall International, Englewood Cliffs, N.J., 1984.
16. I. Getreu, *Modeling the Bipolar Transistor*, Tektronix, Inc., Beaverton, OR 97077, 1976.

Problems

7.1 Show that the relationship for the Early voltage V_A that is derived in Equation 7.1.3 properly specifies $\partial I_C/\partial V_{CB}$ for the prototype transistor by considering collector current to be carried purely by diffusion between the emitter and the collector.

7.2 Calculate the value of V_A at $V_{CB} = 0$ for a prototype transistor in which the base is doped with 10^{17} atoms cm^{-3} of boron and the collector is doped with 10^{16} atoms cm^{-3} of phosphorus if the neutral base width is 2.5 μm. Consider the junction to be a step between the two concentrations. What is the indicated slope of I_C owing to the Early effect?

7.3* Compare the Early voltages for the two transistors that are considered in Problem 6.3. Assume that $\partial x_B/\partial V_{CB}$ is approximately equal for both devices.

7.4[†] Consider an *npn* transistor in which the base doping varies *linearly* across the quasi-neutral region, going from 10^{17} cm^{-3} at the emitter side to 10^{16} at the collector side. The base width is 1 μm and both emitter and collector have $N_d = 10^{19}$ cm^{-3}.
  (a) Sketch the minority-carrier densities (i) at thermal equilibrium, and (ii) under low-level, active-bias conditions.
  (b) Sketch the "built-in" electric field in the base.
  (c) Derive an expression for the field and give its maximum value.
  (d) Determine the approximate ratio between the two Early voltages (i) V_A that is encountered in forward-active bias, and (ii) V_B that characterizes reverse-active bias.

7.5 Using Equation 7.1.4, discuss qualitatively the dependence of V_A on collector-base bias for (a) the prototype transistor, and (b) an *IC* amplifying transistor.

7.6 One criterion of the onset of current crowding is a drop in the transverse base voltage exceeding kT/q. Estimate the corresponding collector-current level for a transistor that has a β_F of 50, the impurity distribution shown in Figure P7.6, and a stripe geometry with $Z_E = 0.1$ cm and $Y_E = 2 \times 10^{-3}$ cm (see Figure 6.3). It can be shown that, for the stripe geometry, the base spreading resistance is

$$R_B = \frac{\bar{\rho}_B Y_E}{6 x_B Z_E}$$

The average resistivity of the base region $\bar{\rho}_B$ is given by

$$\bar{\rho}_B \simeq \frac{A_E x_B}{\mu_p Q_{BO}}$$

where μ_p is the mobility that corresponds to $N_a = Q_{BO}/q A_E x_B$.[13]

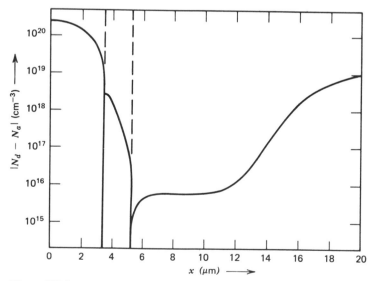

Figure P7.6

7.7* Consider a *pn* junction having the properties described in Problem 5.12. If the junction is to be used as the emitter-base of a *pnp* transistor, what is the maximum value of β_F that can be attained?

7.8† Reconsider Problems 6.17 and 6.18 in which β_F was to be found in a transistor and degradation due to radiation was considered. In the present case, assume that the transistor is to be used at a voltage such that $J_t/J_r = 10$ in Equation 5.3.25.
   (a) Compute β_F for this case.
   (b) If lifetime in the space-charge region varies in the same way as base minority lifetime resulting from the radiation damage, calculate the information required in Problem 6.18.

7.9* Derive an expression and make a plot of β_F/β_{FO} versus collector current for an *npn* transistor in which high-level injection effects at the emitter edge of the base (as described in Section 7.2) degrade current gain. The current gain β_{FO} is the value obtained in the middle bias range. Consider a prototype transistor with $A_E = 5 \times 10^{-5}$ cm^2, $N_a = 5 \times 10^{16}$ cm^{-3}, $D_n = 20$ cm^2 s^{-1}, and $x_B = 5$ μm. Consider that the base current is dominated by reverse injection into the emitter. Under this condition, base current will remain proportional to exp (V_{BE}/V_t). Thus, beta will fall off as $n(0)$ begins to depart from its middle-range dependence on V_{BE}.

7.10† Consider the extreme case of the Kirk effect (described in Section 7.2) in which the collector space-charge region is moved to the edge of the n^+ buried layer. Take the case that the negative space charge is completely due to electrons in transit and the positive space charge is provided by a very high donor concentration that starts abruptly at the edge of the buried layer. Calculate the field and space-charge-layer width as a function of current under the assumption that the electron velocity v is given by the following two cases: (a) $v = \mu\mathscr{E}$ and (b) $v = v_l$. These are two cases of space-charge-limited currents in solids.

7.11 (a) Show that R_B in Equation 7.2.13 can be obtained from a plot of I_C versus measured V_{BE} such as in Figure 7.10. In particular, if I_{CA} is the *actual* collector current and I_{CI} is the collector current that would flow in the absence of base resistance (the *ideal current*), show that

$$R_B = \frac{V_t \beta_F}{I_{CA}} \ln\left(\frac{I_{CI}}{I_{CA}}\right)$$

   (b) Use the following data for the transistor of Figure 7.10 to obtain a plot of R_B versus I_{CA}. $I_S = 3 \times 10^{-14}$ A, $V_t = 0.0252$ V, $\beta_F = 100$ (assumed constant).

$V_{BE}(V)$	$I_{CA}(mA)$
0.70	11.25
0.72	22.4
0.75	56.2
0.80	200

7.12† If the total resistance R in the transistor shown in Figure 7.12 is 150 Ω (0.75 square at 200 Ω/square), and the external resistor is 20 Ω, use the network in the inset of the figure to investigate the crowding of current due to base resistance. Assume that each of the transistor segments has $\frac{1}{8}$ of the I_S given in Problem 7.11, but that each has $\beta_F = 100$.

(a) Assume currents of 1 and 10 mA in the innermost transistor segment. Work through the network to calculate the total current flowing in the overall transistor.

(b) Use the results of Problem 7.11 to calculate R_B and the total applied bias between the external base and emitter leads.

(c) Assuming that base resistance is the only important high-current effect, what valu^{...} ꞈoes R_B approach at the highest current levels?

(d) Show that R_B approaches the sum of the external resistance plus $11R/128$ at very low currents.

(You will need to maintain about four significant figures for V_{BE} to derive accurate results; a calculator is highly desirable.)

7.13   Prove the statements made in the paragraph containing Equation 7.3.3.

7.14[†]   Use Equation 7.3.8 to calculate τ_B for a transistor having a constant (built-in) base field resulting from an exponential variation in base doping. Specifically, take the case that the built-in voltage drop between $x = 0$ at the emitter side to $x = x_B$ at the collector side of the base is κV_t where V_t is the thermal voltage kT/q.

(a) Show that $v = \kappa^2/(\kappa - 1 + e^{-\kappa})$ and that v behaves properly as $\kappa \to 0$.

(b) Calculate τ_B for $\kappa = 20$, $x_B = 0.5\ \mu m$, $D_n = 20\ cm^2\ s^{-1}$, and explain why this value of κ is about as large as can be realized practically.

7.15[†]   Consider the influence of the Kirk-effect results that are sketched in Figure 7.8 on τ_B. Use Equation 7.3.8 to make a semiquantitative plot of τ_B versus collector current. The indicated fall-off in high-frequency performance is an important consequence of the Kirk effect.[3]

7.16   Find an expression for τ_{BF} (as introduced in Equation 7.4.3) in terms of transistor geometry and minority-carrier lifetime in the case of a prototype transistor for which the emitter efficiency is given by Equation 6.2.20. Formulate the conditions required in order that τ_{BF} will become nearly equal to τ_n, the electron lifetime in the transistor base.

7.17   Show that the set of charge-control equations for a *pnp* transistor is of the form:

$$i_C = -\frac{Q_F}{\tau_F} + \frac{dQ_R}{dt} + Q_R\left(\frac{1}{\tau_R} + \frac{1}{\tau_{BR}}\right) + \frac{dQ_{VC}}{dt}$$

$$i_E = \frac{dQ_F}{dt} + Q_F\left(\frac{1}{\tau_F} + \frac{1}{\tau_{BF}}\right) - \frac{Q_R}{\tau_R} + \frac{dQ_{VE}}{dt}$$

$$i_B = -\frac{dQ_F}{dt} - \frac{Q_F}{\tau_{BF}} - \frac{dQ_R}{dt} - \frac{Q_R}{\tau_{BR}} - \frac{dQ_{VE}}{dt} - \frac{dQ_{VC}}{dt}$$

7.18   Consider the physical nature of the space charges represented by Q_{VE} and Q_{VC} to argue that the signs for the derivatives of these quantities are correct in Equation 7.4.11, and in the equation set of the preceding problem (for a *pnp* transistor). *Hint: Consider the sign of the charges supplying the base current and the sign of the Q_V terms Reference to Figure 7.14 may be helpful.*

7.19   Carry through the steps necessary to show the validity of Equation 7.4.10.

7.20   Analyze the circuit shown in Figure 7.18 (which corresponds to Equation 7.4.11) under dc conditions. Show that if one defines

$$\alpha_F = \frac{\tau_{BF}}{\tau_F + \tau_{BF}}$$

$$\alpha_R = \frac{\tau_{BR}}{\tau_R + \tau_{BR}}$$

and

$$I_{ES} = Q_{FO}\left(\frac{1}{\tau_F} + \frac{1}{\tau_{BF}}\right)$$

$$I_{CS} = Q_{RO}\left(\frac{1}{\tau_R} + \frac{1}{\tau_{BR}}\right)$$

Equations 7.4.11 will reduce to the Ebers-Moll equations (Equations 6.4.10).

7.21[†] Show that τ_{SLOW} in Equation 7.4.15 can be expressed by

$$\tau_{SLOW} = \frac{(\beta_F + 1)\tau_{BR} + (\beta_R + 1)\tau_{BF}}{1 + \beta_F + \beta_R}$$

This form for τ_{SLOW} is frequently used in practice.

7.22 (a) Show that Q_{FO} as defined in Equation 7.4.2 is linearly related to $n_{po}(0)$ and write its value for a prototype transistor. *Hint: Consider Equations 7.1.1 and 7.4.1*

(b) Show that V_{CE} for a saturated transistor is given by

$$V_{CE} = \frac{kT}{q} \ln\left[\frac{Q_{RO}(Q_F + Q_{FO})}{Q_{FO}(Q_R + Q_{RO})}\right]$$

7.23* A transistor has the following charge-control parameters:

$$\tau_F = 12 \text{ ns}, \qquad \beta_F = 100, \qquad \tau_R = 36 \text{ ns}, \qquad \beta_R = 10$$

(a) Evaluate the forward stored charge Q_F if the collector current $I_C = 2$ mA and the transistor operates just on the boundary of the saturation region with $V_{CB} = 0$.

(b) Determine the base-charge components Q_F and Q_R if the base current is now made $I_B = 0.5$ mA with I_C remaining at 2 mA.

(c) Compare the charge stored for cases a and b.

7.24 For a calculated variation in charge with voltage to be modeled by a capacitance, not only must the variation be activated by the terminal voltage associated with the charge, but also the sign of the charge variation must be consistent with that of a capacitor. Show that this condition is fulfilled when the variation in Q_F (as calculated in Equation 7.5.6) is modeled in the capacitor ηC_D of Figure 7.27.

7.25 Show by considering the signs of v_{BE} and the generator that the equivalent circuit shown in Figure 7.27 is valid either for *npn* or for *pnp* transistors.

7.26 Discuss the limitation of the small-signal equivalent circuit to variations in base-emitter voltage that are less than V_t. What steps in the derivation of the circuit require this limitation?

7.27 Carry out the steps necessary to reduce the equivalent circuit of Figure 7.25 to the form sketched in Figure 7.27. (Make use of the fact that $\delta \ll 1$ and $\eta \ll 1$).

7.28 Verify Equation 7.5.15.

7.29 An increment of emitter current dI_E is applied to a transistor under active bias with quiescent current I_E and base-emitter voltage V_{BE}. Assume that the simplified hybrid-pi circuit of Figure 7.30 (with C_{jc} negligibly small) is applicable. Show that a time

$$\tau_E \simeq \tau_F + (C_{je}V_t/I_C)$$

is required in order to bring the base-emitter voltage to a new steady-state value. This is one of the delays that affects the measured f_T in a transistor.

7.30* (a) Obtain the hybrid-pi representations for the two transistors described in Problems 6.1 and 6.3. Take $I_C = 2$ mA and $\phi_i + V_{CB} = 10$V for both transistors. In the uniform base transistor, take $\tau_n = 100$ ns.
       (b) Comment on the relative performance of the two transistors when they are used as small-signal amplifiers.

7.31[†] Consider an *npn* transistor that is biased in the active mode and illuminated in the collector-base space-charge region. The radiation produces hole-electron pairs at a rate r pairs per unit time. Consider r to be a sinusoidal function of time.
       (a) Briefly indicate the flow of the generated carriers.
       (b) Indicate how the effects of radiation might be incorporated in the low-frequency hybrid-pi circuit (Figure 7.28) (take the Early effect to be negligible).
       (c) Use the circuit of part b to compute i_C produced by the radiation if v_{BE} is zero (base-emitter ac shorted).
       (d) Repeat part c if the base-emitter is ac open-circuited (i.e., $i_B = 0$).

7.32 Show that a symmetrical equivalent circuit diagram like that in Figure P7.32 represents the "transport version" of the Ebers-Moll equations (Equations 6.4.2 and 6.4.3), provided that we define

$$I_{AA} = I_{ES}'\left[\exp\left(\frac{qV_{BE}}{kT}\right) - 1\right]$$

$$I_{BB} = I_{CS}'\left[\exp\left(\frac{qV_{BC}}{kT}\right) - 1\right]$$

where I_{ES}' and I_{CS}' differ from I_{ES} and I_{CS} in the original Ebers-Moll equations. Advantages for this representation over that of the more conventional Ebers-Moll equivalent circuit (Figure 6.12) are discussed by J. Logan.[14]

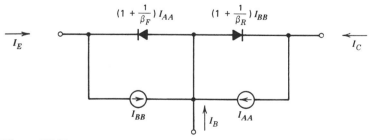

Figure P7.32

7.33[†] Derive Equations 7.6.2.

7.34[†] Derive the solution for q_b in Equation 7.6.9.

7.35[†] Using the Gummel-Poon equations derived and discussed in Section 7.6, consider active-bias and low-level conditions to show that a decrease in $\beta_F(I_C/I_B)$ is predicted as current decreases. Find the variation in β_F as a function of I_C at low levels. Show that the parameter n_e can be obtained from the behavior noted and plot a reasonable sketch for β_F in a transistor having $\beta_F = 100$ at currents above 0.9 mA with a dropoff asymptote that intersects the midrange β_F at $I_C = 0.5$ mA. How might the curve of β_F be used to obtain a value for I_1?

7.36[†] (a) Show that the intercept of the two asymptotic forms expressing collector current as a function of base-emitter bias in the Gummel-Poon model occurs at the "knee" current $I_C = I_{KF}$ (Figure 7.32).

(b) Show that the Gummel-Poon model predicts that, in the high-current region, β_F becomes proportional to I_C^{-1}.

8

PROPERTIES OF THE
METAL-OXIDE-SILICON SYSTEM

In our discussion of the electronics and technology of materials in Chapters 1 and 2, we noted that silicon had become overwhelmingly important as a semiconductor material because of some special properties. The chief reason for its prominence among possibly competitive materials for semiconductors devices is the ability to produce by compatible technologies both a semiconductor (single-crystal silicon) and an insulator (amorphous silicon dioxide) that have superb electrical and mechanical properties. This ability has made planar technology possible and, in turn, the reliable production of large-scale integrated circuits. Thus, properties of the oxide-silicon system are fundamental to the performance of integrated-circuit devices. Knowledge of these properties and of their control has been responsible for many advances in device design and performance. Despite years of work in the area, research on the oxide-silicon system is still ongoing, and new applications are continually being found.

A useful starting point for the consideration of the oxide-silicon system is the construction of an energy-band diagram. The utility of the band diagram is enhanced considerably, however, by adding to the figure a third material, a metal overlay above the oxide. The metal provides an electrode at which the voltage can be fixed, and the resultant three-component, metal-oxide-silicon (MOS) system is useful in understanding several important integrated–circuit structures: most notably the metal-oxide-silicon field-effect transistor (MOSFET), sometimes called the insulated-gate field-effect transistor (IGFET).*

The metal-oxide-semiconductor field-effect transistor is a device of such major importance that the succeeding two chapters are devoted to it. The advent of dense, large-scale integrated circuits has made the MOSFET more important than the bipolar transistor among the devices used in integrated circuits. The discussion of the MOS system in this chapter will help greatly to focus on the physical electronics that underlies the operation of the MOSFET.

After we have incorporated the effects of oxide charge into MOS theory, we can deduce the electronic behavior of oxide-silicon systems without a metal overlay. Since most *pn* junctions made by the planar process intersect the silicon surface at an oxide-silicon interface, conditions at this interface influence the properties of *pn* junctions. Therefore, the topics of this chapter are necessary additions to our discussion of the properties of *pn*-junction devices.

A direct application of MOS electronics has been the fabrication of precise capacitors for integrated circuits. Arrays of MOS capacitors can be used to make charge-coupled devices (*CCDs*). The *CCD*, invented as a result of detailed studies of the oxide-silicon system, has been applied to integrated circuits for optical imaging and signal processing.

Acceptor-doped silicon is considered throughout the chapter for uniformity in the presentation of material. The oxide-silicon system with donor doping is the subject of several problems, and results for this system are included in a summary of important equations. In most cases, the modifications to be made when the doping type is changed are clear.

Throughout the discussion that follows, we describe the characteristics of the silicon dioxide–silicon system. However, as MOS technology develops and the oxides required for high-performance *IC*s become thinner, insulating materials other than pure silicon dioxide are being considered. For example, combinations of silicon dioxide and silicon nitride may be useful, although at the expense of a less perfect insulator–silicon interface. Most of our remarks about the oxide–silicon system can be applied to these more complex systems.

8.1 The MOS Structure

To derive an energy-band diagram for the metal-oxide-silicon system, we apply the basic principles that we have used previously in studying systems of metals and semiconductors, and of *p*-type and *n*-type silicon. The starting point is recognition that systems at thermal equilibrium are characterized by a constant Fermi

---

* Strictly speaking, IGFET denotes a broader class of devices than those made from metal-oxide-silicon structures, although in most cases MOSFET and IGFET can be used interchangeably.

energy. In this case of a three-component system, the Fermi level is constant throughout all three materials: the metal, the oxide, and the silicon.

Thermal-Equilibrium Energy-Band Diagram

For the present, we shall idealize the MOS system by considering that the interfaces between the materials consist of planes and are free of charges. The Fermi levels in the various materials are equalized by the transfer of negative charge from the materials with the higher Fermi levels (smaller work functions) across the interfaces to the materials with lower Fermi levels (greater work functions). As discussed in Chapter 3 the vacuum level is a continuous function of position and, hence, knowledge of the electron affinities of the insulator and semiconductor, and of the work functions for the semiconductor and metal permit the construction of a unique energy-band diagram. In Figure 8.1, aluminum (work function = 4.1 eV), silicon dioxide (electron affinity ~0.95 eV), and uniformly doped p-type silicon (electron affinity 4.05 eV) having a work function of 4.9 eV are considered.* The vacuum level is designated E_0, and the various energies when the materials are separated are indicated on the figure.

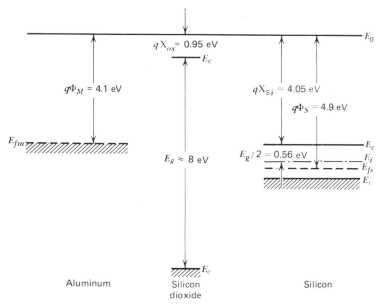

Figure 8.1   Energy levels in three separated components that form an MOS system: aluminum, thermally grown silicon dioxide, and p-type silicon containing $N_a \approx 1.1 \times 10^{15}$ cm^{-3}. (Although recent measurements indicate E_g in SiO$_2$ to be ~9 eV, many experimental data appear to be consistent with the 8 eV value that we will use.)

---

* There is considerable variation in tabulated values for work functions and electron affinities. The values given here provide accurate correspondence to theory in a number of MOS experiments.

When the materials form a system at equilibrium, negative charge has been transferred from the aluminum into the silicon because the work function of the metal is 0.8 eV less than the work function of the silicon. The insulator, which is incapable of transferring charge (since it ideally possesses zero mobile charge), sustains a voltage drop because of the charge stored on either side of it. That charge, in turn, consists of a thin sheet of positive charge (a plane in the ideal case of a perfect conductor) at the surface of the metal and of negatively charged acceptors extending into the semiconductor from its surface. The voltage corresponding to this energy difference therefore divides across the oxide and the space-charge region at the surface of the silicon. It may be puzzling to consider charge transfer in the MOS system with reference to an oxide that is an almost perfect insulating material. In fact, if the system being considered were fabricated without any path for charge flow between the metal and the silicon other than the oxide, the materials could exist in a condition of nonequilibrium (i.e., with Fermi levels unequal) for long periods. Nearly every MOS system of interest does, however, have some alternative path for the transfer of charge that is much more transmissive to charge flow than is the oxide. Hence, we can assume that there is thermal equilibrium between the metal and the semiconductor. Under these assumptions the band diagram for the MOS system formed with the materials of Figure 8.1 is sketched in Figure 8.2. The vacuum level, which plays no significant role in most device analysis, has been omitted from Figure 8.2. In terms of charge flow, the diagram of Figure 8.2 would result from the transfer of holes in the p-type silicon to an ohmic contact (not shown) where the holes are freely converted into electrons. These electrons are then supplied from the aluminum of the MOS system. The charge imbalance in the aluminum because of the transfer of these electrons leaves a sheet of positive charge at the metal surface near the oxide (as close as

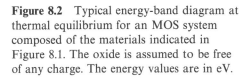

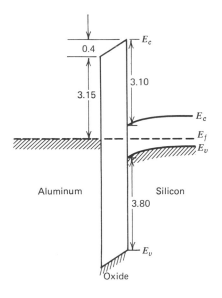

Figure 8.2 Typical energy-band diagram at thermal equilibrium for an MOS system composed of the materials indicated in Figure 8.1. The oxide is assumed to be free of any charge. The energy values are in eV.

it can get to the equal quantity of negative charge stored near the silicon surface). The drop of 0.4 eV across the oxide shown in Figure 8.2 is typical. The exact value of the energy difference between the metal and the surface of the semiconductor depends upon specific oxide properties as we shall see shortly in an example.

The band diagram of Figure 8.2, in which the p-type silicon is depleted of holes at its surface, can be compared to the diagram for gold and n-type silicon sketched in Figure 3.5. The space charge at the surface of the silicon in both cases consists of ionized dopant atoms, and the energy bands in the silicon are curved in the region of the space charge with a spatial curvature that can be calculated by solving Poisson's equation. The presence of the oxide in the case of Figure 8.2 acts to reduce the surface field by separating the surface charges, but there is otherwise no important difference in the band diagram within the silicon itself.

In terms of electrical characteristics, there are, of course, major differences in the two situations. In the MOS system, electrons cannot pass freely in either direction across the oxide. This distinction shows itself on the band diagram by the presence of abrupt steps in the energy of the allowed states for electrons in the MOS system. Referring to Figure 8.2, we note that electrons in the metal that exist at the Fermi level are 3.15 eV lower in energy than the "conduction-band edge" in the silicon dioxide.* Because of this separation of the allowed energies for free electrons, it is proper to characterize the metal-oxide interface by a 3.15 eV barrier for electron emission into the oxide. By the same reasoning, there exists a barrier of 3.10 eV at the oxide-silicon interface for electrons in the conduction band of the silicon and a barrier of 4.20 eV for electrons in the valence band of the silicon. The validity of these barrier heights has been confirmed directly by measurements of the photon energies required to emit electrons from the metal and semiconductor into silicon dioxide.

EXAMPLE   MOS Energy-Band Diagram

What is the thickness of the silicon-dioxide layer for the MOS energy-band diagram sketched in Figure 8.2?

Solution

Figure 8.2 is the energy-band diagram of the MOS system at thermal equilibrium for the materials shown in Figure 8.1. There is a voltage difference between the metal and the silicon brought about by the differing work functions equal to Φ_{MS} which is (4.9–4.1) or 0.8 V. Since Figure 8.2 indicates that 0.4 V is dropped across the oxide, the voltage drop at the silicon surface is also 0.4 V.

---

* As mentioned previously, there are theoretical objections to describing an amorphous material like silicon dioxide in terms of energy bands. Nonetheless, within the context of our present discussion, the concept is useful.

Because there is no charge in the SiO_2, the oxide field $\mathscr{E}_{ox}$ is constant and the voltage across the oxide V_{ox} is simply $\mathscr{E}_{ox} \times x_{ox}$, where x_{ox} is the oxide thickness. Therefore, x_{ox} can be found once $\mathscr{E}_{ox}$ is known.

Because the oxide-silicon interface has been assumed to be charge-free, the electric displacement D, perpendicular to the interface, is continuous and the field in the oxide is therefore related to the field at the surface of the silicon $\mathscr{E}_{s0}$ by the equation

$$\mathscr{E}_{ox} = \frac{\epsilon_s}{\epsilon_{ox}} \mathscr{E}_{s0}$$

There is a depletion layer at the surface of the silicon with a constant charge density qN_a that extends a distance x_d away from the Si–SiO_2 interface. In this region, the field and voltage dependences on x are the same as in the Schottky diode that was considered in Section 3.2. By similar analysis to that used to obtain Equations 3.2.2 and 3.2.3, we can write expressions for the surface field $\mathscr{E}_{s0}$ in the silicon and for the depletion-layer width x_d.

$$\mathscr{E}_{s0} = \frac{qN_a x_d}{\epsilon_s}$$

and

$$x_d = \left[\frac{2\phi_s \epsilon_s}{qN_a} \right]^{1/2}$$

The acceptor density N_a in the silicon can be obtained by using Equation 1.1.27 with $p = N_a$ and $(E_i - E_f)$ as given in Figure 8.1. From Figure 8.1, we have

$$(E_i - E_f) = (4.9 - 4.05 - 0.56) = 0.29 \text{ eV}$$

and, from Equation 1.1.27,

$$p = N_a = n_i \exp[(E_i - E_f)/kT]$$

which is 1.1×10^{15} cm^{-3}.

Using the value $\phi_s = 0.4$ V, we calculate $x_d = 685$ nm, and $\mathscr{E}_{s0} = 1.17 \times 10^4$ V cm^{-1}. Therefore,

$$\mathscr{E}_{ox} = 3.505 \times 10^4 \text{ V cm}^{-1} \qquad x_{ox} = \frac{V_{ox}}{\mathscr{E}_{ox}} = 114 \text{ nm}.$$

The Effect of Bias Voltage

We have seen that for an energy-band diagram of the idealized MOS system at thermal equilibrium, the metal and the semiconductor form two plates of a charged capacitor. The capacitor is charged to a voltage that corresponds to the difference between the metal and semiconductor work functions. The application of a bias voltage between the metal and the silicon causes the system to depart from thermal equilibrium and to change the amount of charge that is stored on the capacitor. For the particular case considered in Figure 8.2, a negative voltage applied to the metal with respect to the silicon opposes the built-in voltage on the capacitor. It therefore tends to reduce the charge stored on its plates from the equilibrium value.

In a particular case of interest the applied voltage is set to a value that exactly compensates the difference in the work functions of the metal and the semiconductor. The stored charge on the MOS capacitor is then reduced to zero and the fields in the oxide and the semiconductor vanish. In this situation, the energy bands in the silicon are level or flat in the surface region as well as in the bulk Figure 8.3). Because of the effect of the applied voltage on the band diagram, the voltage applied to achieve flat bands in the silicon is called the *flat-band voltage* and usually designated V_{FB}. The flat-band voltage varies with the dopant density in the silicon as well as with the specific metal used for the MOS system. Notice that the MOS system is not at thermal equilibrium under the flat-band condition; therefore, the Fermi energy is different in the metal than in the silicon (Figure 8.3). The voltage applied to the ideal MOS system to bring it to the flat-band condition equals the difference in the work functions of the metal and the silicon.

$$V_{FB}{}^0 = \Phi_M - \Phi_S \equiv \Phi_{MS} \tag{8.1.1}$$

(The result is written $V_{FB}{}^0$ in Equation 7.1.1 because the true flat-band voltage V_{FB} is a function of additional parameters.)

Continuing with our consideration of the system of Figure 8.2, if the silicon is held at ground and the voltage applied to the metal is kept negative but increased in magnitude above $V_{FB}{}^0$, the MOS capacitor will begin to store positive charge at the silicon surface. This positive charge is made up of an increase in the hole population at the surface. The surface, therefore, has a greater density of holes than N_a, the acceptor density. This condition is called *surface accumulation*, and the

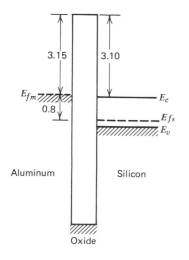

Figure 8.3 Energy-band diagram of the MOS system of Figure 8.2 under flat-band conditions. A voltage equal to V_{FB} has been applied between the metal and the silicon to achieve this condition, which does not correspond to thermal equilibrium.

region at the surface containing the increased hole population is known as the *accumulation layer.*

The surface accumulation layer is a space-charge layer composed of free carriers. Hence, the solution of Poisson's equation within the accumulation layer is the same as the solution for a Schottky ohmic contact, which was described in Section 3.4. From consideration of Equation 3.4.2 it was shown that half of the space charge of free carriers exists within $\sqrt{2}$ times the Debye length L_D of the surface, and the total extent of the accumulation layer is therefore a few times the value of L_D at the surface. To grasp the order-of-magnitude of these figures, consider that a *p*-type silicon wafer with $N_a = 10^{15}$ cm^{-3} is accumulated so that the surface density is 10 times the bulk density. The Debye length at the surface for this case is about 40 nm (from Equation 3.4.3), roughly the same thickness now used for many MOSFET gate oxides. Sketches of the energy-band diagram and of the charge configurations for the case of MOS surface accumulation are given in Figures 8.4*a* and 8.4*b*, respectively.

We have already seen that the MOS system of Figure 8.2 with zero voltage applied between the metal and the silicon stores negative charge at the silicon surface and positive charge on the metal. This is consistent with a built-in positive voltage between the metal and the silicon. If this built-in voltage is aided by applying a positive voltage between the metal and the silicon, the silicon will become further depleted as more acceptors become exposed at its surface. Correspondingly, the positive charge on the metal increases. Because of the behavior of the silicon surface charge, this condition is called *surface depletion.* The energy-band diagram and the charge configuration for depletion bias in the MOS system of Figure 8.2 are sketched in Figure 8.5. Under depletion bias, the energy-band diagram behaves very similarly to the back-biased Schottky-barrier diode. Figure 8.5*a* can be compared to Figure 3.6*b* to demonstrate this similarity (with due allowance for the fact that the silicon is *n*-type in Figure 3.6 and *p*-type in Figure 8.5).

If the voltage applied to the metal in the MOS system is increased further, however, behavior unlike that of a metal-semiconductor diode occurs. In the MOS

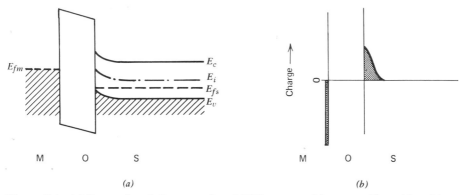

Figure 8.4   (*a*) Energy-band diagram of an MOS system with *p*-type silicon biased into accumulation, (*b*) charge in the same MOS system.

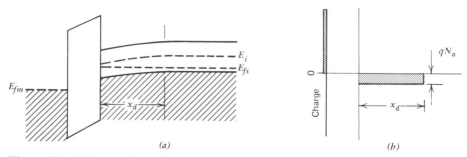

Figure 8.5 (a) Energy-band diagram of an MOS system with p-type silicon biased into depletion, (b) charge in the same MOS system.

case, as the metal voltage is increased, the field at the surface of the silicon increases and the energy bands bend considerably away from their levels in the bulk of the silicon. In the surface region, the majority carriers have been depleted and generation of carriers will therefore exceed recombination (Equation 5.2.9). The generated hole-electron pairs are separated by the field, the holes being swept into the bulk and the electrons moving to the oxide-silicon interface where they are held because of the energy barrier between the conduction band in the silicon and that in the oxide. If the voltage at the metal is changed slowly enough, this generation process can bring the populations of free carriers at the silicon surface into localized equilibrium with the bulk of the silicon. In that case a constant Fermi level can be drawn from the bulk to the oxide interface, and Fermi-Dirac statistics can be used to calculate the concentrations of carriers in the silicon. In fact, all of the energy-band diagrams drawn in this section have assumed this to be true. We shall have occasion later to consider alternative behavior. If the Fermi level remains constant in the silicon while the bands bend with applied voltage, at sufficiently high voltages, the intrinsic Fermi level E_i at the silicon surface will cross the Fermi level. The conduction-band edge at the oxide-silicon interface will then be closer to the Fermi level than the valence-band edge is to the Fermi level. The applied voltage has therefore created an *inversion layer*, so-called because the surface contains more electrons than holes, even though the material has been doped with acceptor impurities. The applied voltage between the metal and the silicon has induced a *pn* junction near the surface. A sketch of the energy-band diagram for the MOS system of Figure 8.2 when biased into inversion is shown in Figure 8.6a.

When E_i is just slightly below E_f at the surface, there is a low electron density (order n_i) in the inversion layer, and the MOS system is said to be biased in the *weak inversion* region. On the other hand, when $(E_c - E_f)$ at the surface is less than $(E_f - E_v)$ in the bulk, the electron density in the inversion layer is greater than the hole density in the bulk, and the system is in the *strong inversion* region. The dividing point between the two cases is conveniently taken when the electron density at the surface equals the acceptor concentration. While this division between weak and strong inversion is somewhat arbitrary, it is a convenient distinction for many purposes.

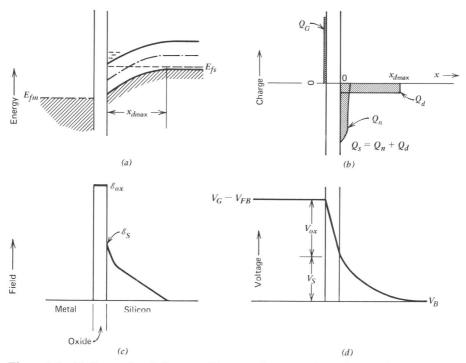

Figure 8.6 (a) Energy-band diagram, (b) space-charge configuration, (c) field, and (d) potential distribution for an MOS system with p-type silicon biased into inversion.

In the depletion and inversion regions of bias, the applied voltage induces negative charge in the semiconductor by repelling holes from the surface to create the depletion region and inducing electrons to form the inversion layer. Figure 8.6b shows the charge storage in the MOS system of Figure 8.2 when it is biased into inversion. The free-electron charge density in the inversion layer, which we call Q_n (C cm^{-2}), is close to the surface because of the attractive force of the surface field. The acceptor charge is distributed throughout the depletion region, which extends away from the surface toward the bulk of the silicon. The depletion-charge density is called Q_d, as in Chapters 3 and 4. The sum $Q_n + Q_d$ is Q_s, the silicon-charge density.

Once the silicon is biased into strong inversion, the free-electron population at the surface is nearly an exponential function of the potential at the surface so that there is little change in surface potential with increasing metal voltage. Thus, the total potential drop across the depletion region and the depletion-layer width in the silicon are both relatively constant when the surface is inverted. The maximum width of the depletion layer is usually denoted $x_{d\max}$ as in Figure 8.6. The field and potential variations in the inverted MOS structure are shown in Figures 8.6c and 8.6d, respectively. The field at the interface between the oxide and the silicon is discontinuous, falling from $\mathscr{E}_{ox}$ to $\mathscr{E}_s$ because of the change in the permittivity of the two materials. The total voltage across the structure ($V_G - V_B - V_{FB}$) is made up of a drop across the oxide and a drop across the space-charge region in the silicon. Equations for these two voltages are derived in Section 8.3.

This qualitative discussion of surface charge, voltage, and fields has served to introduce the important effect of the voltage applied to the overlying metal in determining the properties of the silicon surface. Although the system is basically only a capacitor, the various forms that the surface charge in the silicon can assume cause very significant differences in the electrical properties of the silicon surface. For example, the surface can be highly conducting and electrically connected to the bulk when it is accumulated; it can be highly insulating when it is depleted of free carriers, or else it can be highly conducting, but disconnected from the bulk, when it is inverted. These three conditions can be controlled by the bias applied between the metal and the bulk of the silicon. Because of this control, the metal layer is usually called the *gate*, and the voltage on the metal is denoted V_G.

Silicon Gate. For many device applications, it is desirable to construct the gate from heavily doped silicon rather than from a metal (Problem 8.2). The silicon may be conveniently deposited over the gate oxide by chemical vapor deposition (*CVD*) as discussed in Chapter 2. Because the silicon layer is stable at high temperatures, dopant atoms can be diffused into the substrate after the gate has been deposited, and the whole system can be passivated with an overcoating oxide. The advantages of this technology become more apparent in the discussion of MOS integrated circuits in Chapters 9 and 10. Because the silicon is deposited over amorphous silicon dioxide, it is a polycrystalline film typically consisting of sub micrometer-sized crystallites. Although the gate is silicon, its electrical function in the MOS system is similar to that of the metal gate, and silicon-gated structures are usually described as MOS systems.

8.2 Capacitance of the MOS System

As with the Schottky barrier and the *pn* junction, analysis of the behavior of the small-signal capacitance measured between the two output electrodes provides valuable insight into the electrical behavior of the MOS system. In the case of the MOS system, this analysis has been central to research that has resulted in the present well-developed state of understanding of the oxide-silicon system and its technology. The qualitative discussion of the previous section provides a valuable framework on which to build an understanding of the behavior of the small-signal capacitance as the applied gate voltage is changed.

Consider first that the MOS system is biased with a steady voltage that causes the silicon surface to be accumulated. For a *p*-type silicon sample such as that of Figure 8.2, this corresponds to a negative applied voltage and would result in a charge configuration like that sketched in Figure 8.4. The excess holes at the surface are pulled very close to the oxide. If a small ac voltage v_g is superposed on the dc bias V_G, it causes small variations in the charges stored on the metal gate and at the silicon surface. If the system is now connected to an instrument that measures the small-signal capacitance associated with these variations, the capacitance measured will be close to that of the oxide itself because the spatial extent of the modulated charge in the silicon is small compared to the oxide thickness. The more the surface is accumulated, the thinner will be the accumulation layer (effectively, the Debye length at the surface will be reduced by the added carriers); hence, the capacitance will asymptotically approach the capacitance associated with the pure oxide. Thus, the capacitance per unit area C in accumulation approaches

$$C_{ox} = \frac{\epsilon_{ox}}{x_{ox}}$$ (8.2.1)

where x_{ox} is the oxide thickness. When the gate voltage is changed in the direction of its flat-band value, the surface accumulation decreases to zero and the capacitance decreases as the Debye length at the surface increases. To obtain an exact formulation for capacitance in this bias range, it is necessary to solve Poisson's equation under the condition that free electrons, free holes, and dopant atoms all contribute to the total space charge at the surface. This analysis was first carried out by Kingston and Neustadter[1], and their results have been widely applied to obtain various electrical properties of MOS systems. A particular result that is readily calculated from the Kingston-Neustadter theory is the capacitance per unit area when $V_G = V_{FB}$. If this quantity is called C_{FB}, it can be expressed[2] as

$$C_{FB} = \frac{1}{\dfrac{x_{ox}}{\epsilon_{ox}} + \left[\dfrac{kT}{q^2\epsilon_s N_a}\right]^{1/2}}$$

$$= \frac{1}{1/C_{ox} + L_D/\epsilon_s}$$ (8.2.2)

where L_D, the extrinsic Debye length has been defined in Equation 4.2.14.

When the gate voltage becomes more positive than the flat-band voltage, holes are repelled from the surface of the silicon and the system is in depletion. Under this condition, relatively straightforward electrostatic analysis shows that the overall capacitance C corresponds to the capacitance obtained by a series connection of the oxide capacitance and the capacitance C_s across the surface depletion region (Problem 8.3):

$$C = \frac{1}{1/C_{ox} + 1/C_s} = \frac{1}{1/C_{ox} + x_d/\epsilon_s}$$ (8.2.3)

where x_d is the width of the surface depletion layer, which depends upon gate bias as well as the doping and oxide properties. From Equation 8.2.3, we see that the capacitance of the system decreases as the depletion region widens (Figure 8.7).

When the gate bias is increased sufficiently to invert the surface, a new feature must be considered to describe the MOS capacitance behavior. We recall that the

inversion layer at the MOS surface results from the generation of minority carriers. Hence, the population of the inversion layer can change only as fast as
carriers can be generated within the depletion region near the surface. This limitation causes the measured capacitance to be a function of the frequency of the ac
signal used to measure the small-signal capacitance of the system.

The simplest case arises when both the dc gate-bias voltage and the small-
signal measuring voltage are changed very slowly so that the silicon can always
approach equilibrium. In this case, the signal frequency is low enough so that the
inversion-layer population can "follow" it. The capacitance of the MOS system is
just that associated with charge storage on either side of the oxide; its value is
therefore approximately C_{ox}. Under these conditions a plot of measured capacitance versus gate bias follows the dashed curve marked "low frequency" in Figure
8.7: going from C_{ox} in the accumulation region of bias through a decreasing region
as the surface traverses the depletion region and moving back up to C_{ox} when the
surface becomes inverted.

The results of Problem 8.6 show that a characteristic time to form an inversion
layer at the surface of an MOS system biased to inversion is of the order of
$(2N_a\tau_0/n_i)$ where τ_0 is the minority-carrier lifetime at the surface.* For typical
values of lifetime (1 μs) and dopant concentrations (10^{15} cm^{-3}), this time is roughly
0.2 s. Therefore, the small-signal measuring voltage must be changed very slowly
to observe the low-frequency C-V curve (dashed curve of Figure 8.7).

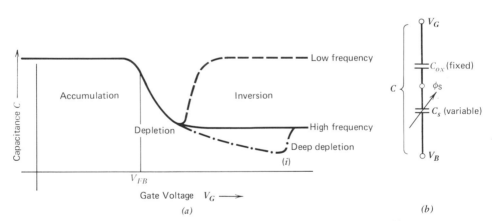

Figure 8.7 (a) Small-signal capacitance of an MOS system with p-type silicon.
Low-frequency behavior: both the bias voltage and the ac measuring signal vary
slowly (less than ~10 Hz). High-frequency behavior: bias voltage varies slowly, but
the ac measuring signal varies rapidly (typical circuits use 1 MHz); deep-depletion
behavior: both the gate bias voltage and the ac measuring signal vary rapidly. The
behavior at point (i) is described in the text. (b) The equivalent circuit for the
overall capacitance C is the series connection of C_{ox} (fixed) and C_s (variable) with
voltages V_G, ϕ_s, and V_B at the gate, interface, and substrate, respectively.

---

* Generation from surface states is considered negligible (discussed further in Section 8.5.) Note
that τ_0 can vary considerably in practice.

The presence of some means, such as surface illumination, that can stimulate the surface generation rate will increase the range of the low-frequency behavior. If the inversion layer is able to make ohmic contact to a region with which it can exchange electrons, the low-frequency curve can be observed at frequencies extending into the MHz range. This occurs because the electrons in the inversion layer can then be supplied and withdrawn rapidly through the ohmic connection.

When the ac measuring signal is changed rapidly while the dc bias voltage is varied slowly, the inversion layer cannot respond to the measuring signal. The number of charges in the silicon space-charge layer is modulated instead by the movement of holes at the far edge of the depletion region. The capacitance then corresponds to the series combination of the oxide capacitance and the depletion-region capacitance, as was true under depletion bias. Since the depletion region reaches a maximum width x_{dmax} when the system goes into strong inversion, the measured capacitance approaches a value corresponding to the series connection of the oxide capacitance and the capacitance associated with the maximum depletion-region width. It remains constant at this value as the bias voltage is increased further. This high-frequency C-V curve is shown by the solid line in Figure 8.7.

A final capacitance behavior is sketched in Figure 8.7. This is the curve shown dot-dash and marked *deep depletion*. It corresponds to the experimental situation in which both the gate bias voltage V_G and the small-signal measuring voltage vary at a faster rate than can be accommodated by generation in the surface depletion region. Since the inversion layer cannot form, the depletion region becomes wider than x_{dmax} and the name *deep depletion* is properly descriptive. The capacitance in this mode is given by Equation 8.2.3 as was true in the normal depletion mode; here, however, x_d exceeds the inversion value x_{dmax} and C does not reach a minimum. One means of generating a "deep depletion" capacitance curve is to sweep the gate bias V_G with a low-frequency triangular wave on which is superposed the sinusoidal ac measuring frequency. The generation rate of carriers increases as the depletion layer is widened, however, and the deep-depletion curve is frequently observed to relax to the high-frequency curve at higher biases. The relaxation is indicated schematically at point (*i*) on the deep-depletion curve of Figure 8.7.

8.3 MOS Electronics

The previous two sections have given a qualitative introduction to the charge induced at the silicon surface by the voltage applied between the overlying gate and the substrate. The particular application of this theory to the MOSFET in the next two chapters demands a more quantitative background, especially for conditions under inversion bias. To provide this background, an analysis based on the depletion approximation that helped to simplify the theories of Schottky barriers and *pn* junctions will be carried out. The analysis avoids an exact representation of the free-carrier densities in the inversion layer because solutions of the exact case cannot be written in closed form. The complexity of the exact forms is not needed to develop a model of the MOS system that adequately fits our purposes. The complete solutions have been published[1,3] and are available for reference.

Thermal-Equilibrium Analysis

For the first level of analysis, it is advantageous to consider the silicon surface region to be in thermal equilibrium with the bulk. Although this assumption will be removed shortly, the results obtained for thermal equilibrium will be enlightening in the more general case. They will also be readily generalized for the nonequilibrium case. We define potential in the silicon as in Equation 4.1.2:

$$\phi(x) = \frac{1}{q}\left[E_f - E_i(x)\right] \tag{8.3.1}$$

Since we consider thermal equilibrium, E_f is constant, but E_i can vary with position. The potential ϕ_p in the neutral bulk of the silicon of Figure 8.8 is thus negative because the material is p-type and E_f is less than E_i. At the surface, the potential ϕ_s is expressed as

$$\phi(0) = \phi_s = \frac{1}{q}\left[E_f - E_i(0)\right] \tag{8.3.2}$$

The carrier densities are related to $\phi(x)$ by Equations 1.1.26 and 1.1.27

$$p = n_i \exp\left(-\frac{q\phi}{kT}\right)$$

$$n = n_i \exp\left(\frac{q\phi}{kT}\right) \tag{8.3.3}$$

From these equations and the definitions of ϕ_p and ϕ_s, we can express the surface free-carrier densities n_s and p_s in terms of the potential drop $(\phi_s - \phi_p)$ across the depletion region at the silicon surface.

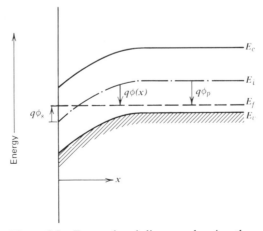

Figure 8.8 Energy-band diagram showing the potential as defined in Equation 8.3.1 in the vicinity of the silicon surface in an MOS system. The sketch corresponds to a positive value of surface potential (ϕ_s).

$$p_s = N_a \exp\left[\frac{q(\phi_p - \phi_s)}{kT}\right]$$

$$n_s = \frac{n_i^2}{N_a} \exp\left[\frac{q(\phi_s - \phi_p)}{kT}\right] \tag{8.3.4}$$

These equations can be used to give convenient "benchmarks" for the surface potential at various free-carrier densities. Table 8.1 lists these ϕ_s values for p-type silicon together with corresponding gate-bias ranges and descriptions of the surface-charge conditions at each value of ϕ_s.

The strong inversion condition, the last entry in Table 8.1, holds the greatest importance for device applications of the MOS system. Once strong inversion is reached, the surface potential ϕ_s remains relatively constant at $-\phi_p$ because n_s is such a sensitive function of ϕ_s (cf. Equation 8.3.4). For example, if n_s is increased from N_a to $10 \times N_a$, ϕ_s is only increased by 58 mV at room temperature (Problem 8.10). For uniform doping, we can relate the surface potential at the onset of strong inversion to the maximum width of the depletion layer through the use of the depletion approximation as was done for the Schottky barrier in Chapter 3 (Equation 3.2.3)

$$\phi_s = -\phi_p = \frac{q}{2\epsilon_s} N_a x_{dmax}^2 + \phi_p \tag{8.3.5}$$

Equation 8.3.5 can be solved for the maximum width of the depletion layer.

$$x_{dmax} = \sqrt{\frac{4\epsilon_s |\phi_p|}{qN_a}} \tag{8.3.6}$$

Table 8.1   MOS Surface-Charge Conditions for p-type Silicon

$(V_G - V_{FB})$	ϕ_s	Surface Charge Condition	Surface Carrier Density
Negative	Negative $\|\phi_s\| > \|\phi_p\|$	Accumulation	$p_s > N_a$
0	Negative $\phi_s = \phi_p$	Neutral (Flat-band)	$p_s = N_a$
Positive (small)	Negative $\|\phi_s\| < \|\phi_p\|$	Depletion	$n_i < p_s < N_a$
Positive (larger)	0	Intrinsic	$p_s = n_s = n_i$
Positive (larger)	Positive $\|\phi_s\| < \|\phi_p\|$	Weak inversion	$n_i < n_s < N_a$
Positive (larger)	Positive $\phi_s = -\phi_p$	Onset of strong inversion	$n_s = N_a$
Positive (larger)	Positive $\|\phi_s\| > \|\phi_p\|$	Strong inversion	$n_s > N_a$

The total charge stored in the depletion layer (per unit area) is designated as Q_d where

$$Q_d = -qN_a x_{dmax} = -\sqrt{4\epsilon_s q N_a |\phi_p|} \qquad (8.3.7)$$

EXAMPLE   Potential near the Oxide-Silicon Interface

Derive an expression for the potential distribution in an ideal MOS capacitor in the depletion condition in terms of the surface potential ϕ_s and the depletion width at the surface x_d taking the zero for potential in the silicon bulk. The silicon is doped p–type and $x = 0$ at the oxide-silicon interface.

Solution

Using the depletion approximation and Poisson's equation, the space charge and field gradient are constant and negative in the silicon for $(0 \leq x \leq x_d)$. Hence, if the surface field is $\mathscr{E}_s$, the field as a function of position away from the surface is

$$\mathscr{E}(x) = \mathscr{E}_s(1 - x/x_d) \qquad 0 < x < x_d$$

and

$$\begin{aligned} \phi(x) &= \phi_s - \int_0^x \mathscr{E}\, dx \\ &= \phi_s - \mathscr{E}_s x + \mathscr{E}_s x^2/2x_d \\ &= \tfrac{1}{2}\mathscr{E}_s x_d - \mathscr{E}_s x + \mathscr{E}_s x^2/2x_d \\ &= \frac{\mathscr{E}_s}{2x_d}(x_d - x)^2 \end{aligned}$$

and

$$\phi(x) = \phi_s\left(1 - \frac{x}{x_d}\right)^2 \qquad 0 < x < x_d$$

is the required expression for potential as a function of x.

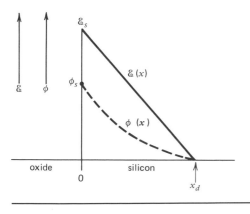

The surface potential ϕ_s corresponds to the area under the curve of $\mathscr{E}$ versus x. Therefore $\phi_s = \tfrac{1}{2}\mathscr{E}_s x_d$.

Nonequilibrium Analysis

Once the MOS system has been biased into inversion, a *pn* junction exists between the surface and the bulk of the silicon. If there is a nearby *n*-type diffused region that contacts the inverted surface as shown in Figure 8.9, it is possible to apply a bias to the *pn* junction. Application of bias in this manner corresponds to a nonequilibrium condition within the silicon, and some current will flow between the inversion layer at the surface and the bulk. However, in practical applications of MOS systems, the junction will be reverse biased, and the currents will be small.

An energy-band diagram for the case of bias applied to an inverted surface is characterized by two quasi-Fermi levels (Equations 1.1.28 and 1.1.29) one for the *p*-region and one for the *n*-region. Just as for the *pn* junction under reverse bias (Chapter 4), the two quasi-Fermi levels are separated by the applied reverse bias. This situation is sketched in Figure 8.10 for a reverse bias $(V_C - V_B)$ applied between the inversion layer (or *channel*) and the substrate (or *bulk*).

An applied reverse bias between the induced surface *n*-region and the bulk increases the charge Q_d in the depletion layer. Since the negative charge induced by $V_G - V_B$ is shared between the depletion and inversion layers, an increase of the charge in the depletion layer means that there is less charge available to form the inversion layer for a given gate voltage. Looked at another way, more gate voltage must be applied to induce the same number of electrons in the inversion layer when there is a reverse bias. With reverse bias present, the surface potential at the onset of strong inversion becomes $\phi_s = -\phi_p + (V_C - V_B)$ rather than

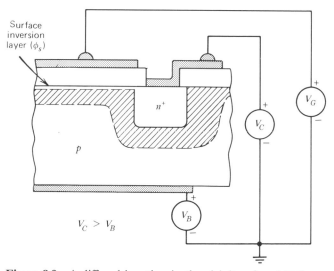

Figure 8.9 A diffused junction in the vicinity of an MOS capacitor can be used to bias the induced junction between the bulk of the silicon and an inversion layer formed at the oxide-silicon interface. The cross-hatching indicates the extent of the space-charge region in the depleted silicon.

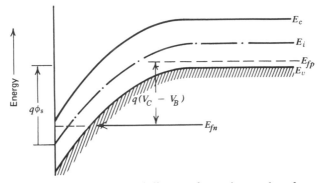

Figure 8.10 Energy-band diagram for an inverted surface on p-type silicon with a voltage ($V_C - V_B$) applied between the inversion layer and the substrate.

$\phi_s = -\phi_p$. The applied voltage extends the range of gate voltages under which the surface region is in depletion (consequently, x_{dmax} is larger). The bias at the surface prevents an inversion layer from forming as readily as when no bias is present by draining away electrons that could form an inversion layer until the surface potential ϕ_s reaches $|\phi_p| + V_C - V_B$. At this surface potential a channel is formed connected ohmically to the diffused electrode.

In the thermal-equilibrium analysis at the beginning of this section, we noted that ϕ_s does not change appreciably after the surface becomes inverted because the density of free electrons increases exponentially with increasing ϕ_s in inversion. Described in circuit terms, the reverse bias "clamps" the surface potential at a value $|\phi_p| + V_C - V_B$ instead of the unbiased inversion potential $|\phi_p|$. Consequently, the change of the surface potential between flat-band and strong inversion is $2|\phi_p| + V_C - V_B$ rather than $2|\phi_p|$. The corresponding maximum depletion-region width x_{dmax} and the depletion-layer charge Q_d (per unit area) become

$$x_{dmax} = \sqrt{\frac{2\epsilon_s(2|\phi_p| + V_C - V_B)}{qN_a}} \tag{8.3.8}$$

and

$$Q_d = -\sqrt{2\epsilon_s qN_a(2|\phi_p| + V_C - V_B)} \tag{8.3.9}$$

These two quantities are plotted in Figure 8.11 as functions of the dopant concentration for several values of bias between the inversion layer and the bulk.

The previous discussion about charge in the MOS system showed that flat band ($V_G - V_B = V_{FB}$) corresponds to the condition of charge neutrality in the silicon. Therefore, ($V_G - V_B$) − V_{FB} is the effective voltage tending to charge the MOS capacitor. In this context, the flat-band voltage V_{FB} for the MOS system is analogous to ϕ_i, the built-in voltage for the pn junction; that is, both act like offsets of the zero level in equations relating stored charge to an applied bias. To express the charge in terms of applied voltage, we carry out the following analysis.

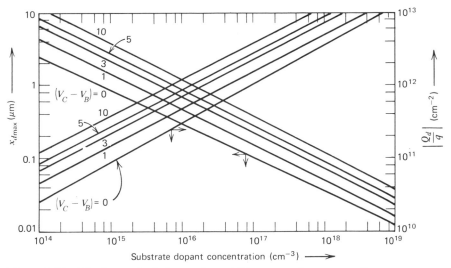

Figure 8.11 Maximum depletion-region width $x_{d\text{max}}$ and corresponding area density of charges in the depletion region Q_d/q as functions of the substrate dopant concentration with the applied channel-to-substrate bias $(V_C - V_B)$ as a parameter. The curves are solutions to Equations 8.3.8 and 8.3.9.

The charging voltage $[(V_G - V_B) - V_{FB}]$ is the sum of a drop V_{ox} across the oxide and a drop in the silicon $(\phi_s - \phi_p)$ (cf. Figure 8.6d).

$$V_G - V_B - V_{FB} = V_{ox} + \phi_s - \phi_p \tag{8.3.10}$$

The field in the insulating oxide is constant in the absence of any oxide charge. In terms of the applied voltage and oxide thickness, this field $\mathscr{E}_{ox}$ is

$$\mathscr{E}_{ox} = V_{ox}/x_{ox} = [(V_G - V_B - V_{FB}) - (\phi_s - \phi_p)]/x_{ox} \tag{8.3.11}$$

Just inside the silicon and adjacent to the oxide (before any silicon charge is encountered), the normal displacement D will be constant and the field $\mathscr{E}_{s0}$ will be

$$\mathscr{E}_{s0} = \frac{\epsilon_{ox}\mathscr{E}_{ox}}{\epsilon_s} \tag{8.3.12}$$

If Equation 8.3.12 is combined with Equation 8.3.11 and the definition for oxide capacitance (per unit area) $C_{ox} = \epsilon_{ox}/x_{ox}$ (Equation 8.2.1) is used, we find

$$\epsilon_s\mathscr{E}_{s0} = C_{ox}[(V_G - V_{FB}) - (\phi_s - \phi_p)] \tag{8.3.13}$$

Gauss' law states that the charge contained in a volume equals the permittivity times the electric field emanating from the volume. Applying Gauss' law to a volume extending from just inside the silicon at the oxide-silicon interface to the field-free bulk region, we can write

$$-\epsilon_s\mathscr{E}_{s0} = Q_s = Q_n + Q_d \tag{8.3.14}$$

where Q_s, the total charge induced in the semiconductor, is composed of the mobile electron charge Q_n and the depletion-region charge Q_d (all per unit area). These quantities are shown in Figure 8.6b. Using Equation 8.3.14 in Equation 8.3.13, we can express the mobile charge Q_n as

$$Q_n = -C_{ox}[(V_G - V_{FB} - V_B) - (\phi_s - \phi_p)] - Q_d \tag{8.3.15}$$

We insert the values of ϕ_s and Q_d into Equation 8.3.15 in order to relate the mobile charge Q_n to the applied voltages. In strong inversion with an applied reverse bias between the channel and the bulk, $\phi_s = -\phi_p + (V_C - V_B)$, and Equation 8.3.15 becomes

$$Q_n = -C_{ox}(V_G - V_{FB} - V_C - 2|\phi_p|) + \sqrt{2\epsilon_s q N_a(2|\phi_p| + V_C - V_B)} \tag{8.3.16}$$

Equation 8.3.16 reduces to the simpler expression

$$Q_n = -C_{ox}(V_G - V_{FB} - V_B - 2|\phi_p|) + \sqrt{4\epsilon_s q N_a|\phi_p|} \tag{8.3.17}$$

with no applied bias between the channel and the bulk. Note that the first terms on the right-hand side of either Equation 8.3.16 or 8.3.17 are negative while the second terms in both equations are positive. The positive terms are, however, smaller in magnitude (since $|Q_s| > |Q_d|$), so that the difference Q_n is negative, as must be the case for the inversion layer on a p-type substrate.

From these equations, we can directly express the gate voltage necessary to induce a conducting channel at the surface of the semiconductor. This voltage, known as the threshold voltage V_T, is defined as the gate voltage that results in $Q_n = 0$. From Equation 8.3.16, an expression for V_T can be written as

$$V_T = V_{FB} + V_C + 2|\phi_p| + \frac{1}{C_{ox}}\sqrt{2\epsilon_s q N_a(2|\phi_p| + V_C - V_B)} \tag{8.3.18}$$

Taken one by one, the terms in Equation 8.3.18 are readily seen to have the proper qualitative behavior. First, V_T contains V_{FB} because a gate voltage equal to V_{FB} is necessary to bring the silicon to a charge-neutral condition. (In the system of Figure 8.2, V_{FB} is negative, tending to reduce the size of V_T.) Second, increasing the channel voltage V_C increases the gate voltage necessary to induce a given charge near the silicon surface. Third, $2|\phi_p|$ volts must be applied to cause the silicon energy bands to be bent to an inverted condition. Finally, the square-root term accounts for the uniform distribution of space charge in the depletion region. This term is inversely proportional to the oxide capacitance. It increases with $V_C - V_B$ to reflect the redistribution of semiconductor charge Q_s from the inversion layer (where it contributes to Q_n) into the depletion layer (where it forms part of Q_d).

The charge in the inversion layer can be expressed very simply in terms of the difference between the applied gate voltage and the threshold voltage. From Equations 8.3.16 and 8.3.18

$$Q_n = -C_{ox}(V_G - V_T) \tag{8.3.19}$$

Equation 8.3.19 should be used with caution for V_G approximately equal to V_T, because it is derived on the assumption that there are no electrons at the surface until the onset of strong inversion when the surface potential $\phi_s = -\phi_p + V_C - V_B$. This is, of course, an approximation that can be removed by considering the exact solutions of Poisson's equation for the MOS system.[1,3]

The more frequently used equations that we have derived for the MOS system have been collected in Table 8.2 which is included at the end of the chapter for ready reference. Results for n-type substrates as well as p-type substrates are included.

8.4 Oxide and Interface Charge

The theory presented thus far has neglected consideration of an important characteristic of the oxide-silicon system. This is the influence of charge within the oxide and at its surfaces. The presence of charge in the oxide and at the oxide-silicon interface is unavoidable in practical systems. To appreciate the importance of oxide charge, it is worthwhile to estimate the order of magnitude of the charge densities that have been considered in discussing the MOS system. Just above the threshold voltage where the system enters the inversion region, for example, the surface density of electrons Q_n/q will be of the same order of magnitude as the density of dopant atoms (per unit area). For homogeneously distributed dopant atoms, the area density is $N_a^{2/3}$, or 10^{10} cm^{-2} if N_a is 10^{15} cm^{-3}. When this density is compared to the area density of silicon atoms $(5 \times 10^{22})^{2/3}$ or 1.35×10^{15} cm^{-2}, it becomes clear that a surface-charge density only about 10^{-5} times as large as the atom density can cause the MOS system to depart from the ideal analysis. Fortunately, when they are formed carefully, interfaces between thermally grown, amorphous silicon dioxide and single-crystal silicon can contain charge densities of the order of 10^{10} cm^{-2} or lower. This low charge density is unique and is one of the outstanding attributes of the oxide-silicon system. We shall consider specific sources for oxide charge after we have calculated its influence on MOS properties.

Theoretical Analysis. Consider that a density of charge Q_{ox} is located at the plane $x = x_1$ within the oxide as shown in Figure 8.12a. The charges at x_1 will induce equal and opposite charges that are divided, in general, between the silicon and the metal gate. The closer is x_1 to x_{ox} the oxide-silicon interface, the greater will be the fraction of induced charge within the silicon. Because this induced charge changes the charge stored in the silicon at thermal equilibrium, it correspondingly alters the flat-band voltage from the value predicted in the ideal MOS analysis (Equation 8.1.1).

The size of the shift in the flat-band voltage is readily found by using Gauss' law to obtain the value of gate voltage that causes all of the oxide charge Q_{ox} to be mirrored in the gate electrode—so that none is induced in the silicon. This condition is illustrated in Figure 8.12b. Referring to the figure and employing Gauss' law, we note that the field is constant between the metal (at $x = 0$) and Q_{ox} (at x_1) and zero between x_1 and the silicon surface (at $x = x_{ox}$). Its value $\mathscr{E}_{ox}$

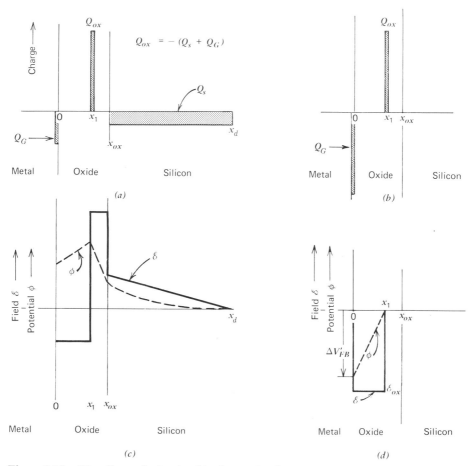

Figure 8.12   The effects of a fixed oxide-charge density Q_{ox} on the MOS system. (a) Charge configuration at zero bias: $Q_{ox} = Q_s + Q_G$; (b) charge at flat band: $Q_{ox} = Q_G$; (c) field (solid line) and potential (dashed line) at zero bias, (d) field (solid line) and potential (dashed line) at flat band. The silicon bulk is taken as the reference for potential in the diagrams for (c) and (d).

between the gate and x_1 is

$$\mathscr{E}_{ox} = -\frac{Q_{ox}}{\epsilon_{ox}} \qquad 0 < x < x_1 \qquad (8.4.1)$$

The gate voltage resulting from the presence of Q_{ox} is the negative integral of $\mathscr{E}_{ox}$. Because it contributes to the flat-band voltage, we call it $\Delta V_{FB}{}'$ (i.e., the change in $V_{FB}{}^0$ (Equation 8.1.1) from the ideal MOS case because of a planar sheet of oxide charge). The field and potential variations for the zero-bias case (Figure 8.12a) are shown in Figure 8.12c; those for flat band (Figure 8.12b) are shown in Figure 8.12d.

An expression for $\Delta V_{FB}{}'$ is

$$\Delta V_{FB}{}' = x_1 \mathscr{E}_{ox} = -\frac{x_1 Q_{ox}}{\epsilon_{ox}} \tag{8.4.2}$$

Equation 8.4.2 can be rewritten by using Equation 8.2.1 to express $\Delta V_{FB}{}'$ in terms of C_{ox} the oxide capacitance per unit area.

$$\Delta V_{FB}{}' = -\frac{Q_{ox} x_1}{C_{ox} x_{ox}} \tag{8.4.3}$$

The maximum value for $\Delta V_{FB}{}'$ occurs expectedly when Q_{ox} is situated at the oxide-silicon interface ($x_1 = x_{ox}$), because the charge induced by Q_{ox} is then contained entirely in the silicon. In contrast, when Q_{ox} is adjacent to the gate, it has no effect on $\Delta V_{FB}{}'$.

The results given for the sheet of charge at $x = x_1$ can be generalized to account for the shift in flat-band voltage by an arbitrary distribution of charge $\rho(x)$ by superposing and integrating the increments that result from charges distributed throughout the oxide. The overall result is

$$\Delta V_{FB} = -\frac{1}{C_{ox}} \int_0^{x_{ox}} \frac{x}{x_{ox}} \rho(x)\,dx \tag{8.4.4}$$

Fixed charge at the oxide-silicon interface is frequently treated separately from charge that is incorporated in the oxide itself, even though surface charge can be accounted for in the formulation of Equation 8.4.4. The fixed interface charge density is designated $Q_f{}^*$, and its contribution to the flat-band voltage is

$$\Delta V_{FB}{}' = -\frac{Q_f}{C_{ox}} \tag{8.4.5}$$

An expression for V_{FB} that includes the effects of differing work functions in the gate and in the silicon, as well as the influence of fixed oxide charge, can be written by combining Equations 8.1.1, 8.4.4 and 8.4.5.

$$V_{FB} = \Phi_{MS} - \frac{Q_f}{C_{ox}} - \frac{1}{C_{ox}} \int_0^{x_{ox}} \frac{x}{x_{ox}} \rho(x)\,dx \tag{8.4.6}$$

From Equation 8.4.6, we see that the effect of oxide charge is to shift the flat-band voltage from its value in the ideal case. If the oxide charge is stable, the shift in flat band produces a corresponding shift in the threshold voltage V_T (Equation 8.3.18). Experimentally, this threshold shift would cause the capacitance versus gate-voltage curves to be translated along the V_G axis. A typical result for a high-frequency capacitance curve is sketched in Figure 8.13 (dashed curve).

In some cases, oxides and oxide-silicon interfaces can contain unstable charges that can be influenced by the applied voltage. In this case, the threshold voltage is itself dependent on the gate voltage. The capacitance versus voltage curve will then be distorted as in the dotted curve in Figure 8.13. To understand this behavior as well as the influence of fixed charge, we discuss the physical sources of oxide charge.

---

* The fixed interface-charge density is also often called Q_{ss}.

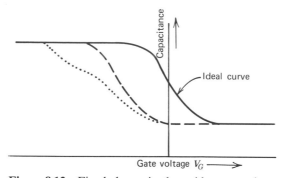

Figure 8.13   Fixed charge in the oxide causes the capacitance–voltage curve to translate along the V_G axis without distortion (dashed curve); charge that is influenced by the gate voltage causes distortion (dotted curve).

Origins of Oxide Charge

It is useful to consider separately four distinct types of charge in the oxide-silicon system. These are shown in Figure 8.14a along with the generally accepted names and symbols for the four charge types: Q_f the *fixed interface charge density*, Q_{ot} the *oxide trapped-charge density*, Q_{it} the *interface trapped-charge density*, and Q_m the *mobile charge density*.

We have already discussed the fixed interface-charge density Q_f (with corresponding numerical density $N_f = Q_f/q$). This charge is positive and, as shown in

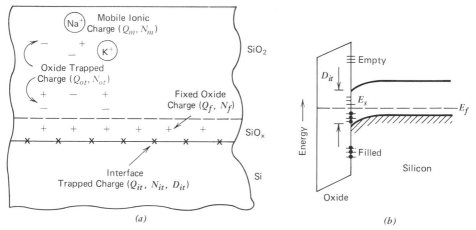

(a) (b)

Figure 8.14   (a) Four categories of oxide charge in the MOS system. The symbols for the charge densities Q (C cm^{-2}), and state densities N (states cm^{-2}) or D (states cm^{-2} eV^{-1}) have been standardized.[4] (b) Energy levels at the oxide-silicon interface. The interface trapping levels are distributed with density D_{it} (states cm^{-2} eV^{-1}) within the forbidden-gap energies.

Figure 8.14a, located within a very thin (one or two nm) layer of nonstoichiometric silicon oxide (labeled SiO$_x$). The oxide trapped-charge density (Q_{ot}) is both positive and negative (typically preponderantly negative), and located in traps distributed throughout the oxide layer. Only a small amount of oxide trapped charge is usually introduced by processing (typically a negligible quantity). This charge is fixed except under unusual conditions (discussed in Chapter 10). We defer, for the present, discussion of the mobile ionic charge and consider the fourth component shown in Figure 8.14a, the interface trapped-charge density Q_{it} residing in trapping levels N_{it}.

The trapping levels N_{it} (traps cm^{-2}) are directly at the oxide-silicon interface and, like the traps considered in Chapter 5, have energy levels within the forbidden-gap region (Figure 8.14b). They are distributed with density D_{it} (traps cm^{-2} eV^{-1}) through the forbidden-gap energies. A discussion of various sources for extra allowed energy levels at a silicon surface was given in Section 3.5 where we considered metal-semiconductor contacts. That section, which would be profitably reviewed at this time, pointed out that even clean surfaces will have extra allowed energy levels different from those in the bulk of a crystal. The inevitable presence of impurities incorporated during wafer processing is a source for still more allowed levels. Electrons in these extra levels and ions associated with them both contribute to interface charge.

To relate the behavior of these traps to the distorted $C-V_G$ curve shown in Figure 8.13 (dotted curve), consider an oxide-silicon interface characterized by interface-trapping levels at a given energy E_s shown in Figure 8.14b. If the gate voltage causes the Fermi level at the surface to cross E_s, the charge state of these levels will change. This introduces a voltage-dependent term Q_{it}/C_{ox} into Equation 8.4.6, making both the flat-band and threshold voltages (Equation 8.3.18) vary with V_G, and leading to the distorted $C-V_G$ variation. The presence of interface-trap densities that approach typical inversion-layer-charge densities (order of 10^{10} cm^{-2} or higher) is generally unacceptable for reliable device design. Using modern MOS technology these trap densities can be reduced to tolerable limits, although higher densities seemed unavoidable for many years and were a major cause for the lagging development of MOS devices through the 1960s and early 1970s. The density of interface-trapping states is typically reduced by annealing the oxidized silicon wafer in hydrogen or forming gas (a mixture of hydrogen and nitrogen).

Limited densities of fixed charge at the interface [where some is always found (Q_f)] and within the oxide [where it is less likely (Q_{ot})] can be tolerated unless the fixed charge densities are so high that the threshold voltages needed for circuit operation become impractical. The fixed charge always present at the interface (Q_f) is thought to arise from uncompleted silicon-to-silicon bonds. The density of atoms at the surface of a silicon crystal depends on the crystal orientation, and thus Q_f is sensitive to the orientation of the silicon wafer. Because more bonds are unfilled in the transition from (111)-oriented silicon to silicon dioxide than in a similar transition for (100)-oriented silicon, Q_f is generally larger when (111)-oriented silicon is used to make MOS devices. For this reason essentially all commercial MOS processing is done on (100)-oriented silicon. The density of fixed interface charge is also dependent, however, on the high-temperature processing that the silicon undergoes, especially in the last processing steps. Some of the unfilled bonds can generally be completed (and Q_f reduced) by annealing at high temperature.

The mobile charge Q_m shown in Figure 8.14a results from alkali-metal ions (mainly sodium and potassium) that are readily absorbed in silicon dioxide. Sodium is especially widely distributed in many metals and chemicals, and easily transmitted to the oxide by human contact. Since the ions carry positive charge, we see from Equation 8.4.4 that the distribution of the charges, as well as their total concentration, influences ΔV_{FB}. The alkali ions have sufficient mobility to drift in the oxide when relatively low voltages are applied. Their mobility increases with temperature and the problem of flat-band instability is likewise magnified as temperature is raised. Since the metal ions are positively charged, negative gate voltage causes the ions to migrate to the metal-oxide interface where they do not affect flat-band voltage. Positive voltage, however, can move the ions to the oxide-silicon interface where their effect is maximal. Consequently, the characteristics of an MOS structure with mobile ions in the oxide are unstable. For threshold voltage stability of about 0.1 V, less than 2×10^{10} cm^{-2} mobile ions can be tolerated in the oxide. This source of oxide charge was another vexing problem in the development of practical MOS systems. It is avoided by careful processing and by the introduction of impurities that immobilize the alkali ions. Chlorine and its compounds, particularly HCl have been introduced into the thermal oxidation process for silicon and these have led to the production of stable MOS systems. Another technique is to add phosphorus to the oxide.

Charge can also be introduced into the MOS system by irradiating it. The radiation may consist of energetic electrons incident on the device during its fabrication or of high-energy particles or photons encountered during operation (as in a space environment). Both charge in the oxide Q_{ot} and in interface trapping states D_{it} can be altered by irradiation.

As an example, high-energy photons (those with energies greater than the approximately 8 or 9 eV bandgap of silicon oxide) can generate electron-hole pairs in the oxide, just as light with energy greater than the bandgap of silicon can create electron-hole pairs in the silicon. After electron-hole pairs are generated in the oxide, however, their behavior differs from that of carriers that are generated in a semiconductor. Since the oxide is thin and contains so few free carriers, the probability of recombination is small. Instead, most of the electrons are swept out of the oxide by any field that may be across it. There are typically a large number of hole traps within the oxide, however, and many holes are therefore immobilized and increase the positive charge in the oxide. This shifts the flat-band voltage as predicted by Equation 8.4.4.

As can be seen in Figure 8.2, electrons can be excited into the oxide by photons of appreciably lower energy than the silicon-oxide band gap. These electrons may be photoemitted from the metal if the photon energy exceeds the energy barrier (3.15 eV in the case of aluminum) or 3.1 eV if the emission is from the conduction band of the silicon. In practice, appreciable emission from the silicon only occurs when photon energies are sufficient to excite electrons from the silicon valence band (which requires photons having greater than ~ 4.2 eV of energy). If an oxide has been charged by trapped holes, that charge can be reduced by causing the photoemission of electrons from either the metal or the silicon. Some of the photoemitted electrons will recombine with the trapped holes, thereby reducing the positive oxide charge.

As described in Section 4.4, energetic electrons are created in silicon when an avalanching field is present. Thus, avalanche within the silicon provides another means for electrons to gain sufficient energy to surmount the barrier at the oxide interface. This mechanism is especially significant for device applications because it is under electrical control. In Chapter 10, we consider device effects more fully. For the present discussion, avalanche near the silicon surface can simply be regarded as an alternative to photoemission for providing a supply of energetic electrons that can enter the oxide.

We conclude this section describing oxide charge by remarking that each type of charge Q_f, Q_{ot}, Q_{it}, and Q_m will, in general, affect the flat-band (Equation 8.4.6) and threshold voltages (Equation 8.3.18). The only charge density mentioned explicitly in these equations is Q_f; the other charge densities are kept low and, if present, represented in the integral expression for charge in the oxide in the equation for V_{FB}.

8.5 Surface Effects on *pn* Junctions[†]

Several important device effects can occur when a *pn* junction exists in the vicinity of an oxide that is covered by an overlying gate (as in the structure sketched in Figure 8.9). We have already seen in the discussion of Figure 8.9 (in Section 8.3) that the bias applied to the junction can modify the charge Q_n stored in the channel and the charge Q_d stored in the depleted regions of an inverted MOS system. This is an effect of the junction on the MOS system. There is also a strong need for the *IC* designer to understand the effect of the MOS system on the properties of the junction. Most of the *pn* junctions in devices made by the planar process intersect an oxide-silicon interface. The only exceptions are those formed with buried layers. Thus, the properties of the oxide-silicon system can exert a significant influence on the circuit performance of *pn*-junction devices such as bipolar junction transistors.

Figure 8.15 shows the cross section of a planar n^+p integrated-circuit diode. If this diode were, for example, situated above a second *pn* junction, it might represent the emitter and base regions of a bipolar transistor. In Section 5.3, it was shown that the generation of electron-hole pairs in the depletion region of a *pn* junction is generally the dominant source for the reverse leakage current of the junction. Likewise, at low forward bias, recombination in the space-charge region is the major current component. Generation and recombination in the space-charge region not only cause departures from the ideal diode law but, more seriously, as discussed in Chapter 6, these processes degrade transistor performance by adding

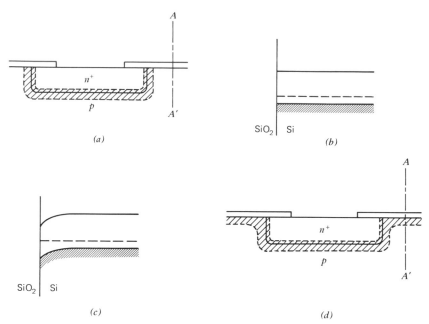

Figure 8.15 (*a*) Planar junction diode. (*b*) Ideal analysis considers flat band along section AA'. (*c*) Real oxide-to-*p*-silicon interface is usually depleted because of positive oxide charge. (*d*) Surface depletion region is appended to junction depletion region.

components of base current that are not delivered to the collector. Some special features of these processes in surface space-charge regions deserve our attention.

To be more specific, consider section *A-A'* through the diode in Figure 8.15*a*. The ideal *pn*-junction analysis considered the oxide-silicon surface as being free of charge, characterized by flat bands as sketched in Figure 8.15*b*. The discussion of the MOS system, however, has made clear that the flat-band condition does not correspond to thermal equilibrium. It must usually be imposed by applying a gate-to-substrate bias equal to V_{FB}. If there is no electrode over section *A-A'*, then the most significant influence on the surface is that of oxide charges. These are almost always positive, causing either depletion or even inversion of *p*-type silicon (Figure 8.15*c*). Conversely, positive charge in the oxide tends to cause accumulation in *n*-type silicon. (The consequence of these effects on the reliable production of bipolar transistors is considered in Problem 8.13.) Returning to consideration of the *p*-region in Figure 8.15, we see that surface depletion in the vicinity of section *A-A'* will enlarge the overall junction depletion region because it connects to the depletion region of the diffused junction. If the surface is inverted, there is an extension of the *n*-region along the oxide surface as well as an enlargement of the depletion region.

One effect of the enlarged depletion region is to provide more volume for the generation of current under reverse bias. Of greater significance, however, is the fact that the oxide-silicon surface will generally contain generation-recombination sites in greater density than in the bulk. Furthermore, the activity of these sites is dependent on the surface potential at the oxide-silicon surface as was shown in the Shockley-Hall-Read (*SHR*) theory for generation and recombination (Section 5.2).

Gated-Diode Structure. To consider the effect of the surface potential on the *pn* junction, we introduce the concept of a *gated diode*.[5] This is the structure shown in Figure 8.16*a* in which a metal gate overlays both the *p* and *n* regions of a diode. To focus attention on the physical mechanisms in the gated diode, we consider that the bulk of the silicon is at ground potential ($V_B = 0$), and we take the *n*-region to be biased positively to a voltage V_R (reverse bias on the junction).

At gate voltages V_G that are negative with respect to flat band, the surface of the *p* type region is accumulated, and the depletion region at the surface is small. The depletion region near the surface in the *n*-type region is hardly changed by the gate voltage because it is heavily doped. Thus, for negative V_G, the leakage current of the diode is approximately determined by generation of holes and electrons in the fabricated or "metallurgical" junction region. From Equation 5.3.26, this current, which we denote by I_M, is given by

$$I_M = \frac{qn_i}{2\tau_0} x_i A_M \qquad (8.5.1)$$

where τ_0 is the lifetime, A_M is the area of the metallurgical junction, and x_i is the active region for generation as described in Section 5.3. In practice, x_i can be taken to be equal to the bulk depletion-layer width x_{db} and, hence, to vary with V_R in the same way as does x_{db}. The current is therefore nearly insensitive to V_G for this bias condition.

If V_G is made more positive than the flat-band voltage, the surface of the *p*-region under the gate electrode will be depleted as shown in Figure 8.16*b*. Two components of current will now flow in addition to I_M as given in Equation 8.5.1. First,

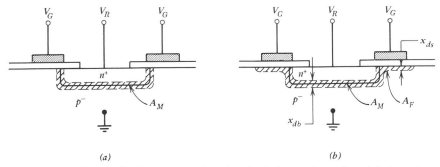

(a)   *(b)*

Figure 8.16 Gated-diode structure showing depletion regions when (*a*) the gate voltage is V_{FB} for the *p*-region and (*b*) when the gate causes a depletion region of width x_{ds} at the silicon surface.

there will be a current because of the generation of carriers in the depletion region induced by the gate. Again from Equation 5.3.26 this component, which we call I_F (arising from the field-induced junction), will be given by

$$I_F = \frac{qn_i}{2\tau_0} x_{ds} A_F \tag{8.5.2}$$

where x_{ds}, the width of the depletion region at the surface, is a function of the applied gate voltage V_G; and A_F, the area of the surface depletion region, is determined by the coverage of the gate electrode over the p-region. The second component of current that arises when the p-region becomes depleted results from the activity of surface generation sites. This component, which we call I_S, is best described in terms of a parameter called the surface recombination velocity s that we introduced in Section 5.2. There, it was shown that the surface recombination velocity is directly proportional to N_{st} the density of generation-recombination sites at the surface (Equation 5.2.22).* If we consider that the sites have energies near E_i, then (applying Equation 5.2.19 for p_s, $n_s \ll n_i$) we calculate I_S as q times the generation rate

$$I_S = \frac{qn_i s_o A_F}{2} \tag{8.5.3}$$

where s_o has been defined in Equation 5.2.2, $s_o = N_{st} v_{th} \sigma$, v_{th} is the thermal velocity, and σ is the capture cross section associated with the generation-recombination site. The value of s_o is directly proportional to the number of surface generation-recombination centers—and it is therefore strongly influenced by device processing and annealing procedures.

If the gate voltage is increased until the surface of the p-type silicon becomes inverted, I_F in Equation 8.5.2 rises to a maximum when x_{ds} increases to x_{dmax}. Once inversion occurs, however, the surface electron density n_s becomes much greater than the intrinsic density n_i, and the surface recombination velocity decreases markedly from s_o following the prediction of Equation 5.2.22. For typical values (Problem 8.16) I_F is smaller than I_S when the surface is depleted; after inversion I_S, the surface generation component, is negligible and the reverse-bias current becomes the sum of I_M and I_F. Typical measured behavior for reverse leakage current in a gated diode is sketched in Figure 8.17. When the pn junction is under forward bias at low voltages, the surface space-charge region also affects the recombination current, but the effect is typically small.

In calculating the behavior of the surface-charge layer, the dependence of the threshold voltage on the reverse bias across the junction must be considered. Quantitatively, the presence of a reverse bias on the junction can be incorporated into the threshold voltage calculation by using Equation 8.3.18 and letting the channel voltage V_C be equal to the diode reverse-bias voltage V_R while holding the bulk voltage V_B at zero.

---

* The states N_{st} are a subset of the interface trapping-state density N_{it} introduced in Section 8.4. The N_{st} states are characterized by energies near the intrinsic Fermi level and by nearly equal rates of interchange with electrons and holes.

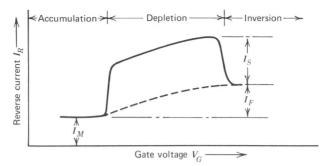

Figure 8.17 Reverse current in the gated diode as a function of V_G, showing the marked increase in leakage when the surface is depleted. The currents I_S and I_F are formulated in Equations 8.5.2 and 8.5.3.

In the absence of a gated-diode structure, the surface condition is determined by the oxide charge. In many cases of IC processing for bipolar transistors, the surface-charge condition creates a wide depletion region over the p-region surfaces. In these cases, pn junctions turn out to be leaky. Often an annealing step can be carried out to reduce s_o and the oxide charge content. This step usually improves the junction characteristics significantly.

8.6 MOS Capacitors and Charge-Coupled Devices (CCDs)

Probably the most straightforward device use of the oxide-silicon system is to make high-quality, precisely controlled capacitors. One particular example of this application is shown in Figure 8.18a, which is an enlarged view of an integrated circuit whose function is the conversion of analog signals to digital representation.[6] The A/D conversion is accomplished by a sequential comparison of the signal with fractions of a reference voltage. The reference is divided by applying the comparison through an array of capacitors having capacitances that are successively reduced by factors of two.

The capacitors in the A/D conversion circuit are MOS devices visible as square structures in Figure 8.18a. For accurate conversion of analog signals, precise control over the ratios of the capacitance values in the circuit is necessary. This control is successfully achieved with MOS capacitors having the structure shown in Figure 8.18b. The floating metal strips in Figure 8.18b serve only to assure reliable precision during etch steps for capacitor fabrication and do not have any circuit function.[6]

Charge-Coupled Devices. Charge-coupled devices (CCDs)[7] consist almost entirely of closely spaced arrays of MOS capacitors. The capacitors are built close enough to one another so that the free charge stored in the inversion layer associated with one MOS capacitor (the channel) can be transferred to the channel

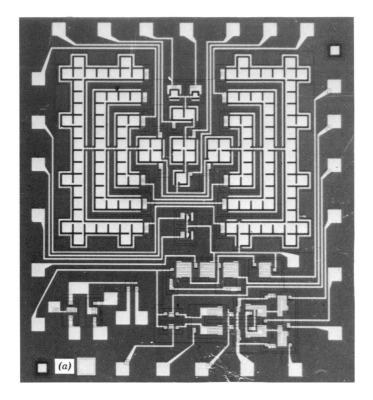

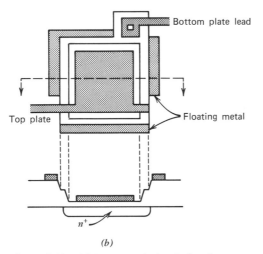

(b)

Figure 8.18 (a) Integrated circuit for the conversion of analog signals to a digital representation.[6] The circuit makes use of many precision MOS capacitors having the structure shown in (b).

region of the adjacent device. The exchange of charge is under the control of the voltages applied to the gates of the MOS capacitors.

In CCDs, the electrical signal is represented by the channel charge. The basic CCD operation depends upon the fact that the channel does not form instantaneously under a gate to which the threshold voltage has been applied; a relatively long time period must elapse before sufficient generation has occurred to charge the channel (Problem 8.6). This time delay needed to establish thermal equilibrium between the silicon surface and the bulk is responsible for the different capacitance versus gate voltage variations that were sketched in Figure 8.7. If the gate of the MOS structure is biased into the strong inversion region, the capacitance can range from C_{ox} when the semiconductor surface is in equilibrium with the bulk to a much lower value when all of the semiconductor space charge is contained in the depletion layer at its surface (deep depletion). At thermal equilibrium, the Fermi level in the silicon is constant and the band diagram is sketched in Figure 8.8. Under nonequilibrium conditions, the Fermi level in the silicon is not constant and a band diagram similar to that sketched in Figure 8.10 applies.

The significant idea that is exploited in CCDs is that if the gate bias is greater than the threshold voltage, the channel can be charged to any surface electron density between Q_n equals zero, which corresponds to the deep-depletion mode, and a value of Q_n equal to the inversion density at thermal equilibrium. The channel charging can be accomplished either by transferring Q_n from an adjacent gate or, in imaging applications, by localized stimulation of the generation rate by incident radiation. There is thus a lower limit for the speed of CCD operation. The signal must be transferred into and out of a stage at a rate that is fast compared to the background generation rate at a depleted silicon surface. No stage in a CCD line can be held for a long time in a deep-depletion condition, and noise will be added continuously because of thermal generation over the period during which the signal traverses a CCD circuit.

These constraints impose very severe requirements on the quality of the oxide-silicon system. As we saw from the discussion of surface generation rates, the requirements imposed can be met if the density of generation-recombination sites at the surface is kept very low. Manufacturers of CCD circuits routinely hold these state densities to values in the low 10^9 cm^{-2} range, which results in surface generation currents that are in the range of 1 to 10 nA cm^{-2}. A small amount of charge is involved in many CCD circuits. Because CCD gates are typically a few $(\mu m)^2$ in area, values of detected charge range from roughly 10 electrons to about 10^7 electrons. By way of comparison, the thermal generation current of 1 nA cm^{-2} corresponds to roughly 60 electrons s^{-1} under a 1 $(\mu m)^2$ gate.

To discuss the basic operation of a CCD, we assume that charge is transferred to and held on MOS stages in times that are short compared to the time that would be necessary for thermal generation of an inversion layer. Consider the MOS system shown in Figure 8.19 in which three capacitors are situated side by side on a silicon surface. Assume that the middle gate has a higher voltage V_2 applied to it than do either of the side gates, and that V_2 is greater than the MOS threshold

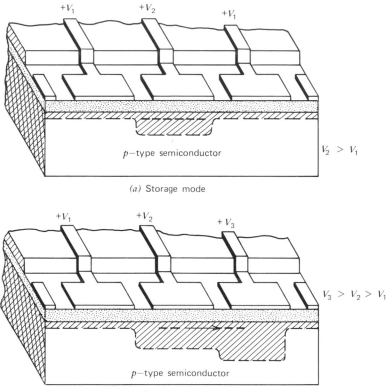

(a) Storage mode

(b) Transfer mode

Figure 8.19 Basic transfer mechanism in a *CCD*. (*a*) In the storage mode, charge is held under the center gate, which has a channel beneath it. (*b*) Application of $V_3 > V_2$ on the right-hand gate causes transfer of charge to the right.

voltage. If electrons have been introduced into the surface region, they will therefore reside in the channel under the middle gate. The application of a more positive voltage V_3 to the right-hand gate will cause the electrons to be transferred to the channel beneath it (Figure 8.19*b*). The voltage on the middle gate can now be decreased to V_1, and then the voltage at the right-hand gate can be reduced to V_2. The net result is a shift of the channel charge one stage to the right.

In digital applications for CCDs, ones and zeros are usually represented by the presence or absence of channel charge, and long rows of capacitors are laid out side-by-side. An example of a nine-stage CCD is shown in Figure 8.20 where channel charge exists at stages 1 and 7 in the storage mode of Figure 8.20*a* while no charge is stored on stage 4. To initiate a transfer of "information" to the right, the gate voltages are changed as shown in Figure 8.20*b*, and the signals are seen in Figure 8.20*c* to exist on gates 2, 5, and 8. Thus, signals are clocked through a CCD at a rate set by the timing of the gate voltages. The CCD structure is useful

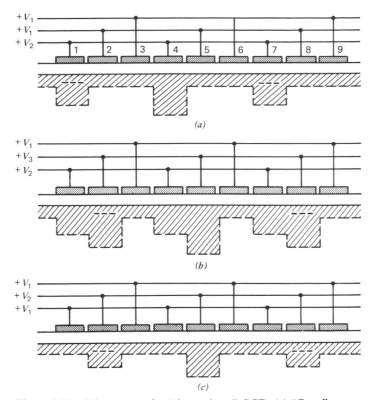

Figure 8.20 Nine gates of a "three-phase" CCD. (*a*) "Ones" are stored on gates 1 and 7; a "zero" on gate 4. (*b*) The binary information is transferred. (*c*) The information has been shifted one stage along the CCD.

as a memory for digital information which is to be stored for a precise time; that is, the time it takes to traverse all stages in a CCD array. Analog (i.e., continuously variable) signals can also be stored in CCD arrays because the amount of charge in the channel can be varied continuously. Thus, CCDs are useful for analog delay lines. By accessing the signal after differing delay periods, various useful signal processing tricks such as convolution, which involves multiplication of a time-varying signal by a delayed value of itself, can be implemented easily with CCDs.

Another analog use for a CCD array is as an imaging device. For this application, a whole field of CCD stages is biased into deep depletion and exposed to a focused image for a time interval. The generation of carriers under the CCD gate is enhanced during the exposure time according to the brightness of the image. The channel under each gate or "picture element" (*pixel*) therefore becomes charged to a level that represents the brightness at its location. The analog information can then be clocked out to sense amplifiers that are built on the edges of the CCD image array. This design is used for integrated–circuit television cameras.

EXAMPLE   **Charge-Coupled Devices**

It is required to design a charge-coupled device as an image sensor with square gates 5 μm on a side functioning as picture elements (*pixels*). The detectable charge threshold is 2500 electrons per pixel, and the charge on each pixel is accessed and reset to zero every 10 ms. The inversion-layer charge density for the CCD at thermal equilibrium would be 10^{13} electrons cm^{-2}.

If the thermal (unilluminated) generation of electrons is described by exponential time behavior (as derived in problem 8.6), determine the required minority-carrier lifetime τ_0 in 12 Ω-cm p-type silicon such that less than 5% of the detectable threshold charge would be delivered by thermal generation.

Solution

At thermal equilibrium, on each gate there are $10^{13} \times (5 \times 10^{-4})^2 = 2.5 \times 10^6$ electrons. The detectable charge threshold is 2500 electrons and the permissible number of thermally generated electrons is 2500 times $0.05 = 125$ on each gate.

Hence $2.5 \times 10^6 \ (1 - e^{-t/\tau_a}) = 125$ which implies $t/\tau_a = 5 \times 10^{-5}$. For a generation time $t = 10^{-2}$ s, the minimum surface-generation lifetime τ_a that will meet the requirements is

$$\tau_a = \frac{10^{-2}}{5 \times 10^{-5}} = 2 \times 10^2 \ \text{s}.$$

From the analysis in problem 8.6,

$$\tau_0 = \frac{n_i}{2N_a} \times \tau_a$$

From Figure 1.14, $N_a = 10^{15}$ cm^{-3} for 12 Ω-cm Si; therefore, the minimum acceptable lifetime is

$$\tau_0 = \frac{1.45 \times 10^{10}}{2 \times 10^{15}} \times 2 \times 10^2 = 1.45 \times 10^{-3} \ \text{s} = 1.45 \ \text{ms}$$

The results of this problem emphasize that the characteristic time τ_a for surface generation in a CCD is much longer (200 s) than the minority-carrier lifetime τ_o (1.45 ms).

A key factor in the production of a CCD is the spacing of adjacent gates. These must be close enough to one another to permit the fringing fields to cause the charge to be transferred when it is desired. Although systems have been built utilizing only metal gates, there are advantages to employing silicon gates singly or in two or three levels, especially to implement more elaborate clocking schemes than we have described in this brief introduction to CCDs.

For many imaging applications, light collection is carried out in photodiodes fabricated in a two-dimensional array. Each diode is accessed under the control

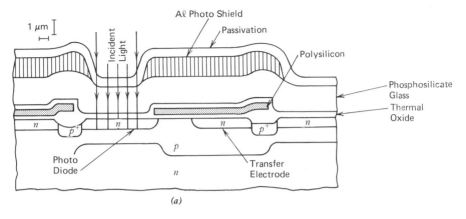

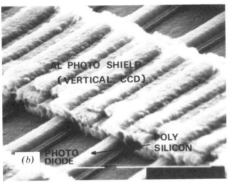

Figure 8.21 (*a*) Unit-cell cross section of an interline CCD image sensor. (*b*) Scanning-electron microphotograph of the surface layers of the image sensor. (*Courtesy NEC Corporation*)[10]

of an overlying gate and the signals are read out to amplifiers on the periphery of the CCD imager. A cross section for this type of an array, called an interline imaging CCD, is shown in Figure 8.21*a*. A scanning-electron microphotograph of the surface of this CCD is shown in Figure 8.21*b*. As seen in the cross section, each photodiode functions as a pixel responding to light transmitted to the silicon surface through transparent layers over the diode. The photogenerated carriers are collected by the surface *np* junction and, under control of the polysilicon gate, transferred to the *n*–region. Except for the photodiode, the silicon surface is shielded from incident light by an aluminum film deposited on a layer of phosphosilicate glass (PSG).

A portion of an ultra-high density, frame-transfer, imaging CCD containing 588 lines with 604 pixels on each line is shown in Figure 8.22. The overall chip is $5.46 \times 7.00 \text{ mm}^2$ and the pixels are $7 \times 11 \text{ } \mu\text{m}^2$. This CCD makes use of triple layers of polysilicon and an *n*-type substrate with an implanted *p*-well. The peripheral circuits to operate the CCD are implemented in CMOS technology (described in Chapter 9).

Figure 8.22 (*a*) A high-density frame-transfer CCD. The sensor array is on the left side of the chip and the light-shielded storage area is on the right. This CCD has 588 lines of 604 pixels each. The chip area is 38.22 mm². (*b*) Peripheral CMOS circuitry for the CCD. (*Courtesy Philips Research Laboratories*)[11]

From these examples, we can see that the CCD is a very useful device, especially for imaging applications. It has also proven useful for signal processing and digital memories. These uses have evolved from a detailed understanding of the technology and electronics of the MOS system.

Summary

The electronic behavior of the oxide-silicon system is gainfully studied by considering the effects of voltages applied to a metal-oxide-silicon (MOS) structure. An important parameter of the MOS structure is the *flat-band voltage* V_{FB}. The significance of V_{FB} becomes clear when an energy-band diagram is constructed for the MOS system. If V_{FB} is applied between the metal and the substrate, there will be no field or charge at the surface of the silicon. The flat-band voltage is determined by several parameters. For ideal (charge-free) oxides and interfaces, it is a function only of the work-function difference between the metal and the semiconductor. Practical structures, however, contain charges both in the oxide and at the oxide-silicon interface, and the flat-band voltage is also a function of these charges. If the oxide charges are affected by the applied voltage, unstable MOS characteristics result. Generally, variable oxide-charge densities must be kept below about 10^{10} cm^{-2} to produce acceptable MOS devices. Alkali metals, particularly sodium, are especially troublesome oxide impurities because they can be moved by the field applied to an MOS oxide.

In many cases there are practical advantages to making the "metal" electrode in an MOS device of doped silicon. The silicon is usually formed over amorphous silicon dioxide by chemical vapor deposition. It is composed of many small crystallites and is therefore known as *polycrystalline silicon.*

A voltage applied across an MOS capacitor between the metal or silicon *gate* and the substrate can accumulate or deplete the silicon surface of bulk majority carriers. It can also bias the silicon to inversion, in which case a *pn* junction is induced near the silicon surface. The charge that is controlled from the gate is then distributed, partly as fixed ions in the depletion layer and partly as free carriers in the inversion layer. The three conditions, *accumulation, depletion* and *inversion* can be sensed by measuring the MOS capacitance. The measured MOS capacitance is frequency dependent because of the time required to establish an inversion layer at the surface.

When a surface is inverted, the inversion layer is often called a *channel.* If the channel exists adjacent to a *pn* junction, it is possible to bias the induced *pn* junction by means of the diffused *pn* junction. This method of biasing can alter the distribution of gate-induced charge between the depletion layer, where the charge is fixed, and the inversion layer, where it is free.

The space-charge layer at the silicon surface can affect the characteristics of *pn* junctions. The activity of surface states depends upon the surface potential, as can be demonstrated through experiments with *gated diodes.* An important direct application of the MOS system is for the production of *IC* capacitors. Arrays of closely spaced MOS capacitors make up *charged-coupled devices* (CCDs). Important applications of CCDs include imagers and signal-processing circuits.

Table 8.2   **Formulas for the Oxide-Silicon System**

p-type substrate (*n*-channel)	*n*-type substrate (*p*-channel)

Flat-band voltage   (*Equation 8.4.6*)

$$V_{FB} = \Phi_{MS} - \frac{Q_f}{C_{ox}} - \frac{1}{C_{ox}} \int_0^{x_{ox}} \frac{x}{x_{ox}} \rho(x)\,dx$$

Bulk potential   (*Equation 4.2.9*)

$\phi_p = -\dfrac{kT}{q} \ln\left(\dfrac{N_a}{n_i}\right)$	$\phi_n = \dfrac{kT}{q} \ln\left(\dfrac{N_d}{n_i}\right)$

Surface potential for strong inversion   (*Table 3.1*)

Thermal equilibrium $\phi_s = |\phi_p|$

$\phi_s - \phi_p = 2|\phi_p|$	$\phi_s = -|\phi_n|$
	$\phi_s - \phi_n = -2|\phi_n|$

With bias $(V_C - V_B) = V_{CB}$

$\phi_s = |\phi_p| + V_{CB}$	$\phi_s = -|\phi_n| - |V_{CB}|$

Maximum depletion width, x_{dmax}   (*Equation 8.3.6*)

Thermal equilibrium

$\sqrt{\dfrac{4\epsilon_s|\phi_p|}{qN_a}}$	$\sqrt{\dfrac{4\epsilon_s|\phi_n|}{qN_d}}$

With bias V_{CB} (*Equation 8.3.8*)

$\sqrt{\dfrac{2\epsilon_s(2|\phi_p| + V_{CB})}{qN_a}}$	$\sqrt{\dfrac{2\epsilon_s(2|\phi_n| + |V_{CB}|)}{qN_d}}$

Work-function difference, Φ_{MS}

$\Phi_M - (X + E_g/2q + |\phi_p|)$	$\Phi_M - (X + E_g/2q - |\phi_n|)$

Threshold voltage V_T (arbitrary reference)   (*Equation 8.3.18*)

$V_{FB} + V_C + 2|\phi_p|$	$V_{FB} + V_C - 2|\phi_n|$
$+ \dfrac{1}{C_{ox}} \sqrt{2\epsilon_s qN_a(2|\phi_p| + V_C - V_B)}$	$- \dfrac{1}{C_{ox}} \sqrt{2\epsilon_s qN_d(2|\phi_n| + V_B - V_C)}$

REFERENCES

1. R. H. Kingston and S. F. Neustadter, *J. Appl. Phys.* **26**, 718 (1955).
2. S. M. Sze, *Physics of Semiconductor Devices*, *2nd Edition*, Wiley-Interscience, New York, 1981, p. 372.
3. C. E. Young, *J. Appl. Phys.* **32**, 329 (1961).
4. B. E. Deal, *J. Electrochem. Soc.* **127**, 979 (1980).
5. A. S. Grove and D. J. Fitzgerald, *Solid-State Electron.* **9**, 783 (1966).
6. J. L. McCreary and P. R. Gray, *IEEE J. Solid-State Circuits*, **SC-10**, 371 (1975). Reprinted by permission.

7. W. S. Boyle and G. E. Smith, *Bell Sys. Tech. J.* **49**, 587 (1970).
8. C. N. Berglund, *IEEE Trans. Electron Devices*, **ED-13**, 701 (1966).
9. C. Jund and R. Poirer, *Solid-State Electron.* **9**, 315 (1966).
10. N. Teranishi, A. Kohno, Y. Isihara, E. Oda, and K. Arai, *IEEE Trans. Electr. Devices*, **ED-31**, 1829 (Dec. 1984).
11. A. J. P. Theuwissen, C. H. L. Weijtens, L. J. M. Esser, J. N. G. Cox, H. T. A. R. Duyvelar and W. C. Keur, *Tech. Digest IEEE Int. Electr. Devices Mtg.*, 40 (Dec. 1984).

Problems

8.1* Sketch the energy-band diagrams (i) at thermal equilibrium and (ii) at flat band for ideal MOS systems made with aluminum gates (a) to 1 Ω-cm *n*-type silicon, and (b) to 1 Ω-cm *p*-type silicon.

8.2 Repeat the sketches required in Problem 8.1 for an MOS system with a polycrystalline silicon gate. Assume that the silicon gate has a band structure similar to single-crystal silicon but that (i) the gate over *n*-type silicon is doped with acceptors until it is just at the edge of degeneracy and (ii) the gate over *p*-type silicon is doped with donors until it is just at the edge of degeneracy. (These conditions correspond to usual silicon-gate technology for reasons to be described in Chapter 9).

8.3 Prove that the small-signal capacitance of an MOS capacitor *C*, biased into depletion, is given by Equation 8.2.3. That is, show that *C* is equal to the capacitance of a series connection of two capacitors: (1) a capacitor made with one plate in the bulk of the silicon and the other plate at the oxide-silicon interface, and (2) a capacitor that has its plates separated by the oxide. (*Hint. Use Gauss' law to express the charge* $\Delta Q = \epsilon_{ox}\Delta\mathscr{E}_{ox}$. *Then, show that the voltage across the capacitor is* $\Delta V = \Delta\mathscr{E}_{ox}x_{ox} + \Delta\mathscr{E}_{ox}\epsilon_{ox}x_d/\epsilon_s$ *and evaluate* $C = \Delta Q/\Delta V$.)

8.4 Take $V_{FB} = -0.5$ V and use Equation 8.2.3 to show the behavior of the overall capacitance *C* for an MOS system in the depletion region. Sketch a plot of C/C_{ox} versus V_G. Consider that the silicon oxide is 100 nm thick and the silicon is *p*-type with 1 Ω-cm resistivity. Locate the flat-band capacitance C_{FB} using Equation 8.2.2.

8.5 The value of ϕ_s (the surface potential) is frequently needed for experimental studies of MOS systems.

(a) By using the results of Problem 8.3, show that when the gate voltage V_G is changed on an MOS capacitor biased in the depletion region, it is possible to find the corresponding change in ϕ_s by using the measured capacitance of the MOS system. The change in ϕ_s can be calculated from the relationship

$$\phi_s(V_{G2}) - \phi_s(V_{G1}) = \int_{V_{G1}}^{V_{G2}} \left(1 - \frac{C}{C_{ox}}\right)dV_G$$

This technique is known as *Berglund's method* after its originator.[8] It can be used conveniently if V_{G1} is taken to be V_{FB} at which point *C* is given by Equation 8.2.2.

(b) If V_{G1} is taken as V_{FB} sketch a low-frequency MOS capacitance curve for *p*-type silicon (normalized to C_{ox}) and indicate (by shading) an area on the curve equal to $\Delta\phi_s$.

8.6† Consider that an MOS system on *p*-type silicon is biased to deep depletion by the sudden deposition of a total charge Q_G on the gate at $t = 0$. Carrier generation in the space-charge region at the silicon surface results in a charging current for the channel charge Q_n as described in the discussion of Equation 5.3.26. This allows one to write

$$\frac{dQ_n}{dt} = -\frac{qn_i(x_d - x_{df})}{2\tau_0}$$

where x_d is the (time dependent) depletion-region-width at the surface and τ_0 is the electron lifetime as given in Equation 5.2.14. The quantity x_{df} is the space-charge region width at thermal equilibrium; that is, when $x_d = x_{df}$, channel charging by generation goes to zero.

(a) Show that a differential equation for Q_n is

$$Q_n + \left(\frac{2\tau_0 N_a}{n_i}\right)\left(\frac{dQ_n}{dt}\right) = -[Q_G - qN_a x_{df}]$$

(b) Solve this equation subject to $Q_n(t = 0) = 0$ and thus show that the characteristic time to form the surface inversion layer is of the order of $2N_a\tau_0/n_i$. (Reference 9).

8.7 Sketch capacitance-voltage curves of the MOS structures shown in Figures P8.7a, P8.7b, and P8.7c. The capacitance is the small-signal value normalized to that of the oxide and measured at 100 kHz. In all cases, the gate dc bias is varied slowly. Show (by using dotted curves) what effect an increase in positive Q_f would have on the C-V_G curves. Label each region on the curves (accumulation, depletion, and inversion). Assume that the substrate resistivity is of the order of 10 Ω-cm in each case and make your sketches qualitatively correct.

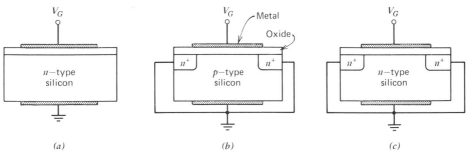

(a)   (b)   (c)

Figure P8.7

8.8 Sketch the curves as described in a through d below for an MOS capacitor on an n-type substrate that has been biased to inversion. Consider that $V_{FB} = -2$ V mainly because of the presence of fixed oxide charge Q_f. The sketches should show (a) the band diagram, (b) all charge in the system, (c) the electric field, and (d) the potential. (Use the silicon bulk as the reference for potential.)

8.9 Construct a table similar to Table 8.1 to represent the surface-charge conditions for n-type silicon.

8.10 Using the formulas in Section 8.3, prove that for $n_s = 10\,N_a$, ϕ_s is only 58 mV greater than $-\phi_p$.

8.11† Consider the dependence on $(V_C - V_B)$ of the expressions for Q_n (Equation 8.3.16) and V_T (Equation 8.3.18) in order to sketch a qualitative family of (low frequency) curves for C/C_{ox} versus V_G as $(V_C - V_B)$ is varied. This dependence was studied by Grove and Fitzgerald.[5]

8.12* Find the threshold voltage (a) in 1 Ω-cm p-type silicon and (b) in 1 Ω-cm n-type silicon. The MOS systems for each case are characterized by: (i) aluminum gate for which $q\Phi_M = 4.1$ eV, (ii) 100 nm silicon dioxide, (iii) the oxide is free of charge except for a surface density $(Q_f/q) = 5 \times 10^{10}$ cm^{-2}. The channel is not biased except from the gate $(V_C = V_B = 0)$.

8.13† Consider the effects of oxide charge on the surfaces of n and p regions as described in Section 8.4. Apply these results to the high-resistivity collector region in a double-diffused bipolar transistor. In particular, use sketches and develop arguments that show why these effects make it harder to manufacture reproducible and stable double-diffused pnp bipolar transistors than to produce npn bipolar transistors.

8.14† In practical MOS systems, measurements of capacitance versus voltage sometimes show hysteresis effects; that is, the C-V_G curves look like the sketch in Figure P8.14.* The sketch refers to measurements made when V_G is swept with a very low frequency triangular wave (~ 1 Hz) and the ac measurement frequency is of the order of 1 kHz or higher. The sense of the hysteresis on such a curve can be observed experimentally to be either counterclockwise, as shown in the sketch, or else clockwise. The hysteresis sense allows one to differentiate between the two most common causes of nonideal behavior. (a) Show this by considering the following nonideal effects: (i) field-aided movement of positive ions in the insulator and (ii) trapping of free carriers from the channel in traps at the oxide-silicon interface Q_{it}. (b) Using qualitative reasoning, prepare a table with sketches of the expected C-V_G plots for n- and p-type substrates; on each sketch indicate the sense of the hysteresis (i.e., clockwise or counterclockwise).

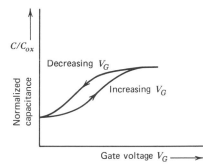

Figure P8.14

8.15* Compare the maximum capacitance that can be achieved in an area $100 \times 100\ \mu$m^2 by using either an MOS capacitance or a reverse-biased pn-junction diode. Assume an oxide breakdown strength of 8×10^6 V cm^{-1}, a 5 V operating voltage, and a safety factor of two (i.e., design the MOS oxide for 10 V). The pn junction is built by diffusing boron into n-type silicon doped to 10^{16} cm^{-3}.

8.16†* Calculate the area density of surface states that would lead the surface generation rate I_S (Equation 8.5.3) of a fully depleted surface to equal twice the generation rate in the surface depletion region I_F (Equation 8.5.2). Consider the states to be characterized by a capture cross section of 10^{-15} cm^2 and the thermal velocity to be 10^7 cm s^{-1}. Assume that the surface depletion region is 1 μm in width and that the time constant τ_0 is 1 μs.

* A laboratory test for oxide "quality" is to cycle the C-V_G measurements at an elevated temperature and check the total amount of voltage hysteresis present—a few tens of millivolts at 125°C is frequently the limit allowed.

9

MOS FIELD-EFFECT TRANSISTORS I:
Basic Theories and Models

The electronic properties of the metal-oxide-silicon system make possible a different type of transistor from the bipolar junction device that was discussed in Chapters 6 and 7. This transistor is usually called a *Metal-Oxide-Silicon Field-Effect Transistor* (MOSFET) or, less frequently, an *Insulated-Gate Field-Effect Transistor* (IGFET). The simplicity of the MOSFET and the high component density possible when it is employed in integrated circuits have made MOSFETs of great commercial importance, especially in digital circuits.

The concept of the MOSFET was actually developed well before the invention of the bipolar transistor. In the early 1930s patents were issued for devices that resembled the modern silicon MOSFET, but which were made from combinations of materials not including silicon. Poor control of the insulator-semiconductor interfaces in use at that time and a lack of a full understanding of insulator-semiconductor systems made practical use of these inventions impossible. It was not until after the advent of the silicon planar process and the technology to produce a well-behaved oxide-silicon interface that the MOSFET became practical.

The basic structure of an *n*–channel MOSFET is shown in Figure 9.1. It consists of the MOS structure that was discussed extensively in Chapter 8 with a surface inversion layer or *channel* extending between two diffused junctions. These diffused junctions are electrically disconnected unless there is an *n*–type inversion

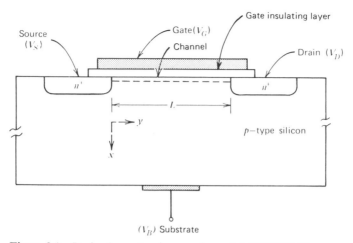

Figure 9.1 Basic elements of an n–channel, MOSFET. The source to drain spacing is designated L and the device width (in the z direction) is W.

layer at the surface to provide a conducting channel between them. When the surface is inverted and a voltage is applied between the junctions, electrons can enter the channel at one junction, called the *source*, and leave at the other, called the *drain*. The electrons flow primarily in regions of the silicon where they are majority carriers, a fundamental difference from bipolar transistors, in which current through the base is carried by injected minority carriers.

For a MOSFET structure, the difference in gate and silicon work functions (Φ_{MS}) and the presence of charges in the oxide can lead to an inverted surface or *channel* between the junctions when no gate voltage is applied. If this is the case the MOSFET is called a *depletion-mode* device because the gate is then usually employed to reduce the conductance of the built-in channel. In most cases, however, gate voltage must be applied to induce a channel; MOSFETs of this type are called *enhancement-mode* devices. In most cases in this chapter we discuss enhancement-mode MOSFETs, which are used more frequently in *ICs* than are depletion-mode MOSFETs.

The use of the terms *source* and *drain* is similar for p–channel MOSFETs; holes (i.e., carriers in the channel region) are supplied at the source and removed at the drain. Hence, conventional current flows from the source to the drain in p–channel MOSFETs and in the opposite direction in n–channel MOSFETs. The theory in this chapter is developed for n–channel MOSFETs, but the corresponding behavior in p–channel MOSFETs is analogous. Equations for both p– and n–channel transistors are collected in Table 9.2 at the end of this chapter, and several problems deal with the properties of p–channel devices.

Comparing the MOSFET shown in Figure 9.1 to the JFET shown in Figure 4.16, we see that each device has source and drain contacts through which output currents flow and each has a gate that controls the flow of electrons in a connecting

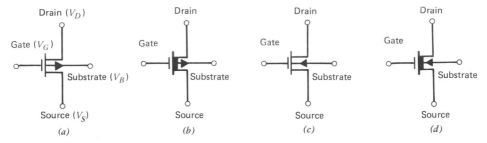

Figure 9.2 Electrical symbols for MOSFETs: (*a*) *p*–channel enhancement, (*b*) *p*–channel depletion, (*c*) *n*–channel enhancement, (*d*) *n*–channel depletion devices.

channel. In the JFET, this channel is defined by *pn*–junction depletion regions that vary in extent when the reverse bias on the junction changes. In the MOSFET, the density of free charge in the channel is controlled by a field that extends from the gate electrode to the silicon through the insulating layer. The properties of the MOS system that were described in Chapter 8 are the framework for the electronics of the MOSFET.

The symbols that have been adopted for MOSFETs are sketched in Figure 9.2*a* and *b* for *p*–channel enhancement and depletion devices, and in Figure 9.2*c* and *d* for *n*–channel enhancement and depletion devices. Note that *n*– and *p*–channel MOSFETs are differentiated by an arrow that represents the *pn* junction between the substrate and induced channel at the surface. In many circuit diagrams the connection to the substrate is not indicated on the symbol because it is evident to the designer. The symbol for depletion-mode MOSFETs has a heavy line between the source and drain, indicating the presence of a channel when no voltage is applied between the source and the gate. Unless an asymmetric geometry is used, the MOSFET is a bilateral device at its output terminals, and the source and drain electrodes are drawn identically.

9.1 Basic Theory

The *n*–type source and drain diffusions in the MOSFET shown in Figure 9.1 are separated by a lateral distance known as the *channel length* (L). The channel length extends along the *y*–axis in the figure, while the direction into the silicon (perpendicular to the oxide) is conventionally the *x*–direction. We shall see that there are advantages to making L small; in typical MOSFETs, channel lengths are of the order of a few micrometers, with minimum lengths determined mainly by limitations in lithography. The *channel width* (W) is perpendicular to the figure (in the *z*–direction). Channel widths are selected to meet circuit-design requirements. The circuit designer specifies a value for W to achieve a desired conductance for a given bias on the MOSFET. The thickness of the oxide (x_{ox}) separating the gate from the channel is typically about 40 nm although it can range from roughly one fourth to four times this value.

The MOSFET is a four-terminal device with connections to the source, drain, gate, silicon substrate (or bulk). The voltages at these terminals are shown in Figure 9.1 with appropriate subscripts. In this chapter, the MOSFET analysis is carried out without assigning any of the terminal voltages to be a reference. All voltages are therefore expressed in the equations we derive to retain maximum flexibility.

Charge-Control Analysis

As in the analysis of the JFET, we consider the MOSFET first at low drain-source voltages (V_{DS} positive but small) so that the channel charge does not vary strongly with position between the source and drain electrodes. Under this condition the surface space-charge region is uniform along the channel (Figure 9.3a).

If both V_S and V_B are at zero volts and V_G is increased from zero, the source and drain are initially isolated by a reverse-biased pn junction. When V_G becomes larger than V_{FB}, a depletion region forms along the channel. The depletion region

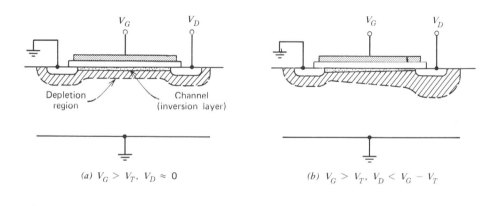

(a) $V_G > V_T$, $V_D \approx 0$ (b) $V_G > V_T$, $V_D < V_G - V_T$

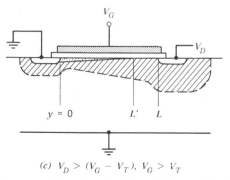

(c) $V_D > (V_G - V_T)$, $V_G > V_T$

Figure 9.3 MOSFET cross sections showing bias effects on the depletion regions: (a) drain voltage small, depletion region nearly uniform along the channel, (b) drain voltage large enough to cause significant variation in depletion–region thickness, (c) drain voltage exceeds the saturation value; channel extends only to $L' < L$.

then widens to terminate the increasing density of electric flux lines emanating from the gate. As V_G continues to increase, the mid-gap energy E_i at the surface is eventually pulled below the Fermi level, causing the surface potential ϕ_s to change from negative to positive values (Equation 8.3.2), and surface inversion commences. When this occurs, electrons in the surface inversion layer can flow and carry current between the source and the drain. For surface potentials less than the strong inversion or threshold value $|\phi_p|$, however, only small, so-called *subthreshold currents* are able to flow between the drain and the source. These subthreshold currents are not large enough to be useful for most applications, but they are sufficiently large to cause troublesome leakage, which can complicate the design of some circuits. Subthreshold currents will be considered in greater detail in Chapter 10.

At higher gate voltages, the surface potential of the semiconductor reaches its strong inversion value $|\phi_p|$.* Once strong inversion occurs, higher gate voltages cause little change in Q_d, the depletion-layer charge, and additional gate voltage primarily increases the electron concentration in the inversion layer Q_n. The electron concentration in the inversion layer then varies approximately linearly with applied gate voltage in excess of the threshold voltage, as does I_D the current flowing from drain to source. The current also increases linearly with drain-source voltage as long as this voltage difference remains small.

Lumped Analysis. It is straightforward to derive equations to represent MOSFET behavior under these conditions by employing a charge-control analysis as was done for the bipolar transistor in Chapter 7. The drain current I_D is related to the total charge in the channel Q_N by the transit time along the channel T_{tr} through the equation

$$I_D = -\frac{Q_N}{T_{tr}} \tag{9.1.1}$$

Because current in the channel is primarily a drift flow, T_{tr} is just the channel length L divided by the drift velocity $v_d = -\mu_n \mathscr{E}_y = \mu_n(V_D - V_S)/L = \mu_n V_{DS}/L$; hence

$$T_{tr} = \frac{L^2}{\mu_n V_{DS}} \tag{9.1.2}$$

The mobility μ_n in Equation 9.1.2 is typically about one half of the bulk value because of additional free-carrier scattering at the surface. The total channel charge Q_N in Equation 9.1.1 can be written most simply by using Equation 8.3.19 together with the channel dimensions

$$Q_N = Q_n WL = -C_{ox}(V_G - V_T)WL \tag{9.1.3}$$

where Q_n is the inversion charge per unit area, and the threshold voltage V_T is given by Equation 8.3.18 with $V_C = V_S$. Hence, the current for low drain bias is

---

* Referenced to the substrate, the surface potential at the onset of strong inversion is $\phi_s = \phi_p - (-\phi_p) = 2|\phi_p|$.

approximately

$$I_{D0} = \mu_n C_{ox} \frac{W}{L} (V_G - V_T) V_{DS} \qquad (9.1.4)$$

Equation 9.1.4* agrees with intuition; for low drain bias the drain current varies linearly with V_{DS}, and the conductance between the source and the drain $[\mu_n(W/L)C_{ox}(V_G - V_T)]$ is proportional to the effective gate bias $(V_G - V_T)$.

If V_D is now increased until it is no longer negligible compared to V_G, the analysis just given becomes inaccurate because the drain voltage V_D affects the channel bias near the drain and acts to reduce Q_N as described in Section 8.3. The space-charge region and channel charge for this situation are sketched in Figure 9.3b.

Distributed Analysis. Since the channel voltage varies continuously between the source and the drain, we require a differential equation to represent the channel charge exactly. The equation is derived by following the procedure used in Chapter 4 for analysis of the JFET. The incremental voltage drop along the channel is first found as a function of the channel current. Then, integration along a path extending from the source to the drain leads to an equation for I_D in terms of the applied voltages.

We assume the *gradual-channel* approximation to be valid, as we did in the JFET analysis in Chapter 4. This approximation states that fields in the direction of current flow are much smaller than are fields in the direction perpendicular to the silicon surface (mathematically that $|\partial\phi/\partial y| \ll |\partial\phi/\partial x|$). This assumption validates the use of a one-dimensional MOS analysis (as carried out in Chapter 8) to find the carrier concentrations and the depth of the depletion region under the channel.

For the present analysis, we also assume that the channel is appreciably longer than the depletion region at the drain (i.e., $L'/L \approx 1$). In very short channel devices, this may not be true, and the channel charge can be influenced strongly by the drain voltage as well as by the gate voltage. This short-channel effect is considered further in Chapter 10. Finally, we assume that since drain current consists of electron flow through n–type regions, it is carried exclusively by the drift process.

Figure 9.4 shows an n–channel MOSFET that is biased at moderate drain voltage. An incremental length dy along the channel sustains a voltage drop dV_C that can be expressed (Problem 9.2) as the product of the drain current I_D and the incremental resistance dR:

$$dV_C = I_D dR = -\frac{I_D dy}{W \mu_n Q_n(y)} \qquad (9.1.5)$$

where Q_n (<0) is a function of the channel voltage $V_C(y)$, as well as the gate voltage V_G. Separating variables and integrating Equation 9.1.5 from the source to the

---

* The subscript 0 on I_{D0} in Equation 9.1.4 is used to distinguish this zero-order formulation for MOSFET drain current. Other expressions for I_D and V_D will be subscripted in ascending order as they are introduced.

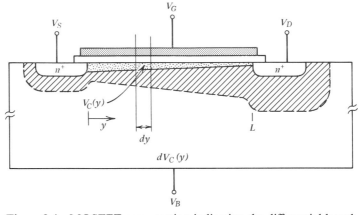

Figure 9.4 MOSFET cross section indicating the differential length dy along the channel. An ohmic voltage drop $dV_C = I_D\, dR$ is sustained across dy. The channel is W units wide.

drain, we obtain

$$I_D = \frac{-\mu_n W}{L} \int_{V_S}^{V_D} Q_n(V_C)\, dV_C \qquad (9.1.6)$$

The dependence of Q_n on channel voltage needed for the integral in Equation 9.1.6 was considered in Equation 8.3.16 and is repeated here for reference.

$$Q_n = -C_{ox}(V_G - V_{FB} - 2|\phi_p| - V_C) + \sqrt{2\epsilon_s q N_a(2|\phi_p| + V_C - V_B)} \qquad (9.1.7)$$

As seen in Equation 9.1.7, channel charge depends on V_C through a linear term and also through a term embedded in the square root which represents depletion-layer charge Q_d (cf Equation 8.3.9). Equation 9.1.7 can be considerably simplified if Q_d is represented only by its value at the source and its variation along the channel is neglected. This approximation, which linearizes the equation, leads to a useful description of MOSFET behavior. Neglecting the variation of Q_d with V_C, Equation 9.1.7 can be written

$$Q_{n1}(V_C) = -C_{ox}[V_G - V_T - (V_C - V_S)] \qquad (9.1.8)$$

where, as in Equation 9.1.3, V_T refers to the threshold voltage *at the source*. Using Equation 9.1.8 in Equation 9.1.6, we obtain

$$I_{D1} = \mu_n C_{ox} \frac{W}{L}\left[(V_G - V_T)V_{DS} - \frac{V_{DS}^2}{2}\right] \qquad (9.1.9)$$

where $V_{DS} = V_D - V_S$. We refer to Equation 9.1.9 as the *charge-control* equation for drain current. The name "charge control" is applied here because Equation 9.1.9 can be obtained by considering the averaged charge-control expression for Q_n. Equation 9.1.9 is clearly an approximation arising from the assumed form for the channel charge (Equation 9.1.8), but we shall see that it is still useful and frequently employed.

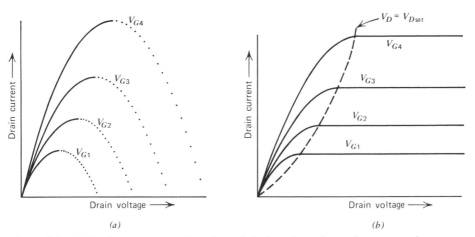

Figure 9.5 (a) Drain current as a function of drain voltage for various gate voltages as predicted by Equation 9.1.9. The dotted portions of the curve are unreasonable on physical grounds. (b) Overall $I_D - V_D$ curves as predicted by Equations 9.1.9 and 9.1.11. The dashed curve represents values of V_{Dsat} from Equation 9.1.10. The gate voltage increases from bottom to top in both curve families, and V_S has been taken to be zero.

To study the MOSFET behavior predicted by Equation 9.1.9, we have sketched curves of I_{D1} versus V_{DS} for varying values of $V_G\ (> V_T)$ in Figure 9.5a. The curves have maximum slope in the vicinity of the origin where the conductance is greatest, and Equation 9.1.4 applies. For higher V_D, the channel charge and hence the conductance diminish, and the curves consist of a series of downward-facing parabolas having maxima at $V_{DS} = (V_G - V_T)$. In Figure 9.5a, the curves are drawn as dotted lines for $V_{DS} > (V_G - V_T)$ because they do not apply in this range of drain bias. The negative incremental conductance $(\partial I_D/\partial V_D)$, indicated in this region, is an intuitive clue to their being physically unreasonable. The reason that our derivation breaks down can be seen by considering Equation 9.1.8 when the voltage along the channel $(V_D - V_S)$ approaches and then exceeds $(V_G - V_T)$. Under these conditions, Equation 9.1.8 (applied at the drain where $V_C = V_D$) predicts that Q_{n1} goes to zero and thereafter changes sign. This sequence is a physical impossibility because Q_{n1} represents an inversion-charge density that must always be negative for the n–channel MOSFET being analyzed.

Although the theory breaks down when $V_{DS} > (V_G - V_T)$, it is quite possible to deduce the behavior of I_D versus V_{DS} in this bias range. To do this, we recognize that electrons in the channel do not "see" a barrier as they approach the drain. On the contrary, they approach a high-field region with a very low electron density in which they are accelerated toward the drain (usually reaching limiting drift velocities) when $V_{DS} > (V_G - V_T)$. The drain current I_D is thus determined by the rate at which these electrons arrive at the edge of the high-field region. In the first-order analysis, this rate is insensitive to V_{DS}, and the drain current thus becomes constant or *saturates* for $V_{DS} > (V_G - V_T)$. Hence, the maxima in the curves of Figure 9.5a occur at a drain voltage known as the *saturation voltage* V_{Dsat} at which $Q_{n1} \to 0$ at the drain end of the channel.

$$V_{Dsat\ 1} = (V_G - V_T) + V_S$$

$$= V_G - V_{FB} - 2|\phi_p| - \frac{1}{C_{ox}}\sqrt{2q\epsilon_s N_a[2|\phi_p| + (V_S - V_B)]} \qquad (9.1.10)$$

The constant current for $V_{DS} > V_{Dsat}$ is sketched in Figure 9.5b. In this range of bias, the end of the channel (i.e., the point at which $Q_n \to 0$) is no longer at $y = L$, but rather occurs at the point $y = L'$ where the channel voltage $V_C(y) = V_{Dsat}$. The current when $V_D > V_{Dsat\ 1}$ (denoted $I_{Dsat\ 1}$) can therefore be calculated by using Equation 9.1.10 in Equation 9.1.9

$$I_{Dsat\ 1} = \frac{\mu_n W C_{ox}}{2L}(V_G - V_T)^2 \qquad (9.1.11)$$

In Equation 9.1.11, V_T refers to the threshold voltage at the source—and depends, therefore, on V_S. Since both Equations 9.1.11 and 9.1.9 are valid (within the assumptions made) at $V_D = V_{Dsat\ 1}$, a plot of $I_{Dsat\ 1}$ versus $V_{Dsat\ 1}$ on the I_D versus V_D axes is an upward-facing parabola (dashed curve in Figure 9.5b). The overall I_D versus V_D curves (excluding breakdown effects) are thus plotted in Figure 9.5b by using Equation 9.1.9 for $V_D < V_{Dsat\ 1}$ and Equation 9.1.11 for $V_D > V_{Dsat\ 1}$.

The channel charge and depletion regions for $V_D > V_{Dsat\ 1}$ are shown in Figure 9.3c. This figure emphasizes that L' decreases as V_D increases. This reduction in L' causes $I_{Dsat\ 1}$ to increase slightly with increasing V_D. The dependence on V_D arising from this effect can be incorporated by rewriting Equation 9.1.11 with L' in place of L and then expressing L' as a function of V_D. This modification will be considered at a later point in the discussion.

Water Analogy. Physically, the operation of the MOSFET can be understood in terms of a water analogy. The free carriers correspond to water droplets. The source and drain are deep reservoirs whose relative elevation difference is analogous to the source-to-drain voltage difference. The channel region is like a canal with a depth that depends on the local value of the gate-to-channel voltage as sketched in Figure 9.6.

If the drain and source are held at the same potential, the water surface is level through the source, canal, and drain in our analogy (Figure 9.6a). When a drain-to-source voltage is imposed, the surface of the drain reservoir is lowered, causing a flow of water along the canal from the source to the drain. The flow speeds up as the elevation difference (analogous to V_{DS}) increases. Since the flow is continuous, the water velocity increases as the depth of water in the canal decreases toward the drain reservoir. At first, the flow through the canal depends both on its dimensions (as controlled by the gate) and on the elevation difference between the source and drain (Figure 9.6b). At the condition analogous to saturation, the flow becomes entirely limited by the flow capacity of the canal. After saturation, if the drain reservoir is lowered further, its surface becomes abruptly disconnected from the water surface at the drain end of the canal. In this condition, the flow into the drain resembles the free fall of water over a waterfall (Figure 9.6c). The rate of flow is equal to the delivery rate to the lip of the waterfall, independent of the total drop over the cataract [which is analogous to $(V_D - V_{Dsat})$].

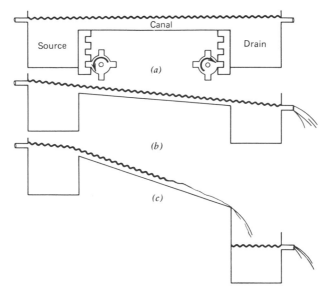

Figure 9.6 Water analogy for a MOSFET. (*a*) When the source and drain are level, there is no flow ($V_{DS} = 0$). The water depth in the canal can be varied by the gear and track (V_{GS}). (*b*) When the drain is lower than the source, water flows along the canal. (*c*) The flow is limited by the channel capacity; lowering the drain further only increases the height of the waterfall at its edge.

Variable Depletion-Charge Analysis. It is not necessary to neglect the depletion-charge variation along the channel and approximate $Q_n(V_C)$ by Equation 9.1.8, in order to derive an explicit equation for I_D. If Equation 9.1.7 is inserted directly into Equation 9.1.6, the resulting integration has many terms, but is easily carried out (Problem 9.2). The result can be written

$$I_{D2} = \mu_n \frac{W}{L} \left\{ C_{ox} \left(V_G - V_{FB} - 2|\phi_p| - \frac{1}{2} V_D - \frac{1}{2} V_S \right) V_{DS} \right.$$
$$\left. - \frac{2}{3} \sqrt{2E_s q N_a} \left[(2|\phi_p| + V_D - V_B)^{3/2} - (2|\phi_p| + V_S - V_B)^{3/2} \right] \right\} \quad (9.1.12)$$

As noted previously, the analysis leading to Equation 9.1.12 is valid as long as inversion charge exists along the entire length of the channel. When V_D is increased sufficiently to cause $Q_n(L)$ to decrease to zero, the drain current saturates. To find the corresponding drain voltage V_{Dsat}, therefore, we let $V_C = V_D$ in Equation 9.1.7 and solve for V_{Dsat} by letting $Q_n(L)$ equal zero.

$$Q_n(L) = 0 = - C_{ox}(V_G - V_{FB} - 2|\phi_p| - V_{Dsat}) + \sqrt{2\epsilon_s q N_a (2|\phi_p| + V_{Dsat} - V_B)}$$
$$(9.1.13)$$

From Equation 9.1.13, the saturation drain voltage is

$$V_{Dsat\ 2} = V_G - V_{FB} - 2|\phi_p| - \frac{\epsilon_s q N_a}{C_{ox}^2}\left[\sqrt{1 + \frac{2C_{ox}^2}{\epsilon_s q N_a}(V_G - V_{FB} - V_B)} - 1\right] \quad (9.1.14)$$

As seen in Equation 9.1.14, the saturation drain voltage is independent of the source voltage because it is defined by a zero free-charge condition at the drain end of the channel.

To compare these results from the variable depletion-charge analysis with the equations derived earlier using charge-control analysis, we take corresponding voltages $V_S = V_B = 0$. Under this condition Equation 9.1.12 reduces to

$$I_{D2} = \frac{\mu_n W}{L}\left\{C_{ox}\left(V_G - V_{FB} - 2|\phi_p| - \frac{1}{2}V_D\right)V_D - \frac{2}{3}\sqrt{2\epsilon_s q N_a}\right.$$
$$\left. \times\ [(2|\phi_p| + V_D)^{3/2} - (2|\phi_p|)^{3/2}]\right\} \quad (9.1.15)$$

which can be compared to Equation 9.1.9 when the expression for V_T (Equation 8.3.18) is used in the equation. When that is done, Equation 9.1.9 takes the form

$$I_{D1} = \mu_n\frac{W}{L}\left[C_{ox}\left(V_G - V_{FB} - 2|\phi_p| - \frac{1}{2}V_D\right)V_D - 2V_D\sqrt{\epsilon_s q N_a|\phi_p|}\right] \quad (9.1.16)$$

Equation 9.1.16, the charge-control expression, can be thought of as a first-order representation for drain current because account has not been taken of the variation of depletion charge along the channel. Equation 9.1.15 treats the depletion-charge variation and is, therefore, a second-order representation. We shall see that although both representations are useful, often there are further effects to be considered.

EXAMPLE   Charge-Control and Variable Depletion-Charge Analyses

Compare the output I_D versus V_D characteristics predicted by the charge-control equations with those obtained using the variable depletion-charge equations for the MOSFET. Consider a device for which $\mu_n W/L = 1.2 \times 10^4$ cm^2 V^{-1} s^{-1}, $C_{ox} = 3.98 \times 10^{-8}$ F cm^{-2} (87 nm oxide), $N_a = 2 \times 10^{16}$ cm^{-3}, and $V_{FB} = -0.5$ V.

Solution

The comparison can be made by using Equation 9.1.16 to represent the charge-control theory and Equation 9.1.15 for the variable depletion-charge analysis. As can be seen from the plot based on this comparison (Figure 9.7), the simpler charge-control equations (Equations 9.1.16 and 9.1.9) predict larger values for I_{Dsat} and V_{Dsat} for each value of gate voltage. The general behavior of the two sets of curves is, however, similar, and the currents predicted by both theories become

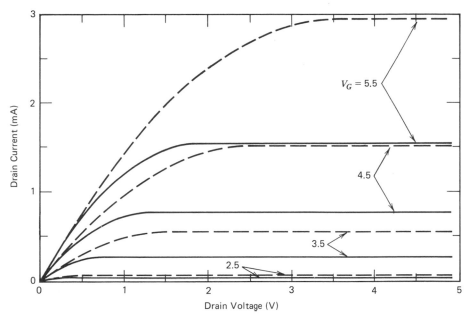

Figure 9.7   Theoretical predictions of I_D versus V_D using Equations 9.1.16 (dashed curves) and 9.1.15 (solid curves) for MOSFET having parameters given in the text. (I_D is taken to be equal to I_{Dsat} for $V_D > V_{Dsat}$.)

the same as the drain voltage approaches zero. This agreement is expected because both theories reduce to Equation 9.1.4 as the drain voltage approaches the source voltage (Problem 9.4). The two theories begin to deviate significantly when I_{D1} becomes roughly 20% of $I_{Dsat\,1}$ for the MOSFET having the characteristics of Figure 9.7.

Measurements of the I_D versus V_D characteristics of MOSFETs with channels longer than roughly 15 μm are in reasonable agreement with the predictions of the variable depletion-charge analysis.[2] Extensive analysis yielding satisfactory design information has, however, been carried out on similar devices using the charge-control analysis with empirically adjusted parameters.

Equations for Circuit Analysis.   As shown in Figure 9.7, the gross behavior of the MOSFET is properly modeled by both analyses. Once sufficient gate voltage has been applied to form a surface channel, curves of drain current versus drain voltage exhibit two distinct regions of behavior: (1) a low-voltage region in which the channel extends all the way from the source to the drain, and (2) a current-saturated region for $V_D > V_{Dsat}$ in which the current flows across a high-field space-charge region near the drain.

There are many cases of circuit design for which an approximation to device behavior is adequate. For these cases, the increased accuracy of the variable depletion-charge theory over the charge-control theory is not necessarily an advantage when computational complexity is considered. Often, for example, the designer is interested only in MOSFET operation in the saturated region; he may require only a value for I_{Dsat}. To obtain I_{Dsat} from the variable depletion-charge analysis, it is first necessary to calculate $V_{Dsat\,2}$ from Equation 9.1.14 and then to use this voltage to calculate $I_{Dsat\,2}$ from Equation 9.1.12 or 9.1.15. This lengthy process is greatly simplified in the charge-control theory in which $I_{Dsat\,1}$ is calculated directly from Equation 9.1.11.

From Figure 9.7, we see that $I_{Dsat\,1}$ is always larger than $I_{Dsat\,2}$. It has been found empirically, however, that by adjusting the prefactor in Equation 9.1.11, one can use this equation to obtain the variable depletion-charge result $I_{Dsat\,2}$ with good precision. Expressed mathematically, the equation to obtain the saturation current is

$$I_{Dsat\,3} = k' \frac{W}{2L} (V_G - V_T)^2 = \frac{k}{2} (V_G - V_T)^2 \qquad (9.1.17)$$

where k' (or k) is obtained in practical cases by measuring $I_{Dsat\,3}$ at a given gate bias. Correspondingly, if the MOSFET is not saturated, analogously to Equation 9.1.9, we write

$$I_{D3} = k \left[(V_G - V_T)V_{DS} - \frac{V_{DS}^2}{2} \right] \qquad (9.1.18)$$

By comparing Equation 9.1.17 to Equations 9.1.11 and 9.1.9, it is clear that k' would equal $\mu_n C_{ox}$ if charge-control theory were to be used. Since MOS processing procedures do not allow for the precise control of μ_n, the apparent arbitrary choice of k' is not a practical problem.

Most hand calculations for circuit design with MOSFETs make use of Equations 9.1.17 and 9.1.18. More complicated equations are typically implemented only in computer analyses.

EXAMPLE   **Approximate Equation for I_{Dsat}**

(a) Determine the extent to which $I_{Dsat\,3}$ (as defined in Equation 9.1.17) agrees with $I_{Dsat\,2}$ as predicted by the variable depletion charge theory (Equation 9.1.15) by considering the MOSFET analyzed for Figure 9.7. Specifically, choose a value for $k/2$ in Equation 9.1.17 to match the currents predicted by Equation 9.1.15 at the maximum value of V_G. (b) Compare the predictions of these two equations at the other V_G values used in Figure 9.7.

Solution

The characteristics plotted in Figure 9.7 are calculated for $V_S = 0$. Using $V_S = 0$, it is straightforward to calculate $V_{Dsat\,2}$ from Equation 9.1.14 with a programmable calculator. With this value, we obtain $I_{D\text{sat }2}$ from Equation 9.1.15. The tabulated results are

V_G (V)	$V_{D\text{sat }2}$ (V)	$I_{D\text{sat }2}$ (mA)
5.5	1.94	1.55
4.5	1.34	0.77
3.5	0.77	0.27
2.5	0.25	0.03

To obtain $k/2$ we must calculate V_T. By applying Equation 8.3.18, we find $V_T = 1.98$ V. We then use Equation 9.1.17 to write

$$\frac{k}{2} = \frac{I_{D\text{sat }3}}{(V_G - V_T)^2} = \frac{1.55 \times 10^{-3}}{(5.5 - 1.98)^2} = 1.25 \times 10^{-4}\,\text{AV}^{-2}$$

Using this value of $k/2$ in Equation 9.1.17 at the other gate biases, we find

V_G (V)	$I_{D\text{sat }3}$ (mA)
5.5	1.55
4.5	0.79
3.5	0.29
2.5	0.03

The agreement between $I_{D\text{sat }3}$ and $I_{D\text{sat }2}$, shown in this example, is quite adequate for circuit design, and Equation 9.1.17 is widely employed. The generality of this result is checked further using different MOSFET parameters in Problem 9.6. There is also a nearly constant ratio between $V_{D\text{sat }2}$ from the distributed theory and $V_{D\text{sat }3}$, which is simply $(V_G - V_T)$. This relationship is considered further in Problem 9.7.

9.2 MOSFET Parameters

A method frequently employed to obtain the dc parameters that describe a MOSFET is to tie together the gate and the drain of the transistor, and then to measure drain current as a function of applied drain voltage. The experimental arrangement is shown in Figure 9.8 (inset). Because V_D is set equal to V_G in this circuit, we can see from Equation 9.1.10 that the transistor is in the saturated region of operation. Hence, we can apply Equation 9.1.17, which indicates that a plot of $\sqrt{I_D}$ versus V_G should be linear. Typical measured data obtained in this way are shown in Figure 9.8. The utility of this analysis technique is apparent. The threshold voltage V_T is readily obtained from the intercept with the voltage axis and the slope of the plotted data can be used to determine the value of $k/2$. The dotted section of the curve in Figure 9.8 results from currents that flow below the threshold voltage. We discuss these *subthreshold currents* more completely in Section 10.1.

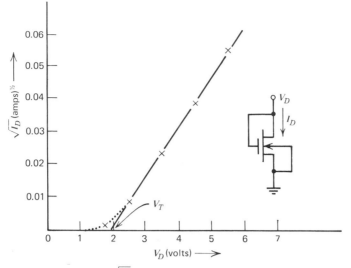

Figure 9.8 Plot of $\sqrt{I_D}$ versus V_D for n–channel MOSFET in saturation. Inset shows the circuit arrangement. The threshold voltage is indicated by the intercept of the straight line with the voltage axis.

It is also possible to define a "threshold voltage" as the gate voltage at which a specific small drain current flows; a common definition is the "threshold voltage" at which 1 μA of drain current flows per μm of drain width for $V_{DS} = 1$ V. This definition for threshold voltage is useful to characterize a process, but the determination of V_T by the intercept method, as illustrated in Figure 9.8, is preferable for modeling the device.

The I_D versus V_{DS} characteristic taken as $V_{DS} \to 0$ can be used to determine $[\mu_n C_{ox}(W/L)]$ and, therefore, the effective channel mobility. To do this most conveniently, we note from Equations 9.1.9 or 9.1.12 that the slope at the origin $(\partial I_D/\partial V_{DS}$, the zero-bias conductance) is given by

$$\left. \frac{\partial I_D}{\partial V_{DS}} \right|_{V_{DS} \to 0} = \mu_n C_{ox} \frac{W}{L}(V_G - V_T) \tag{9.2.1}$$

By repeating this procedure at several values of gate voltage, a reliable value for the prefactor can be found. Note that it is not proper to use the slope of $\sqrt{I_D}$ versus V_G in saturation to find $\mu_n C_{ox}(W/2L)$ as is sometimes erroneously done. That slope gives the value of $k/2$ (from Equation 9.1.17), which is different because of the inaccuracy of the charge-control model.

Body Effect. A circuit similar to that shown in the inset of Figure 9.8, except with a reverse bias (V_{SB}) applied between the source and the substrate (or *bulk*), can be used to measure the variation in the threshold voltage resulting from the source-substrate bias (often called the *body effect*). This circuit is shown in the inset of Figure 9.9.

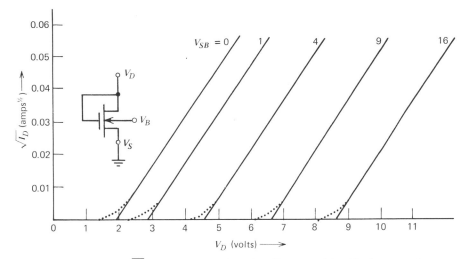

Figure 9.9 Plots of $\sqrt{I_D}$ versus V_D showing the effects of a bias V_{SB} between the source and the bulk. The observed shift in threshold voltage is predicted by Equation 9.2.2.

A reverse bias applied between the source and the substrate reduces the free-charge density in the channel. Hence, with source-substrate reverse bias, for either enhancement or depletion-mode MOSFETs, the threshold voltage becomes *more positive* for n–channel transistors, and *more negative* for p–channel transistors. As seen from the plots of $\sqrt{I_D}$ versus V_{GS} or V_{DS} in Figure 9.9, the curves are accordingly translated to the right along the voltage axis for this n–channel MOSFET.

The change in threshold voltage ΔV_T for the case of constant substrate doping can be calculated from Equation 8.3.18 assuming an abrupt *pn* junction.

$$\Delta V_T = \frac{\sqrt{2\epsilon_s q N_a}}{C_{ox}} \left(\sqrt{2|\phi_p| + |V_{SB}|} - \sqrt{2|\phi_p|} \right)$$
$$= \gamma \left(\sqrt{2|\phi_p| + |V_{SB}|} - \sqrt{2|\phi_p|} \right) \tag{9.2.2}$$

where we have defined a *body-effect* parameter γ. Note from Equation 9.2.2 that the units of γ are $V^{1/2}$.

$$\gamma = \frac{\sqrt{2\epsilon_s q N_a}}{C_{ox}} \tag{9.2.3}$$

In practical circuit analysis, γ is used to calculate the variation in the threshold voltage when a source-substrate bias is present.

EXAMPLE   **Parameters for a depletion-mode MOSFET**

Assume that a depletion-mode n–channel MOSFET can be described by Equations 9.1.17 and 9.1.18 if a negative value is used for the threshold voltage. The substrate doping is 1.63×10^{15} cm^{-3}, and the body-effect parameter $\gamma = 0.5$ V$^{1/2}$. The

MOSFET is connected in the circuit shown, and a current equal to 30 μA is measured with the supply voltage $V_{SS} = 0$. When V_{SS} is raised to 1 V, the current is reduced to 23.1 μA.

Calculate the threshold voltage $V_T(0)$ when the source voltage $V_S = 0$ V, and the prefactor $k'W/2L = k/2$ in Equation 9.1.17.

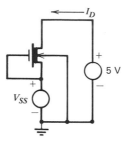

Solution

Since $V_{DS} \geq 4$ V for both measurements and $V_{GS} = 0$ V in the circuit, we assume that $V_{DS} > (V_{GS} - V_T)$ so that Equation 9.1.17 can be applied to both measurements. We will check the calculated value of V_T later to see that this assumption is true. Because the source voltage changes, we must consider the body effect using Equation 9.2.2, and we therefore require a value for $|\phi_p|$. From the given dopant concentration in the substrate, we calculate $|\phi_p| = 0.3$ V (Equation 4.2.9b).

At $V_{SS} = 0$ V, the source is at 0 volts and

$$30 = \frac{k}{2}[0 - V_T(0)]^2$$

At $V_{SS} = 1$ V, the source is at 1 volt and, from Equation 9.2.2

$$\Delta V_T = 0.5[\sqrt{0.6 + 1} - \sqrt{0.6}] = 0.245 \text{ V}$$

Since the body effect makes the threshold-voltage more positive for this n–channel MOSFET, $V_T(1) = V_T(0) + 0.245$. Using the measured current at $V_{SS} = 1$ V, we have

$$23.1 = \frac{k}{2}[0 - (V_T(0) + 0.245)]^2$$

Hence,

$$\left[\frac{I_D(0)}{I_D(1)}\right]^{1/2} = 1.14 = \frac{-V_T(0)}{-V_T(0) - 0.245}$$

From this equation, we calculate $V_T(0) = -2$ V. Using either measured current value, we have $k = 15$ μA V^{-2}. The calculated value of the threshold voltage is consistent with our assumption that the MOSFET is in saturation since $V_{DS} > (0 - V_T)$ for either of the measured conditions.

In practice, the effect on V_T of a bias between the source and the substrate predicted by Equation 9.2.2 is observed as long as the channel length of the MOSFET is considerably greater than the depletion width of the reverse-biased, source-substrate np junction. If this is not the case, the one-dimensional analysis used to solve Poisson's equation (which led to Equation 8.3.18) becomes seriously in error. A two-dimensional theory for the space-charge configuration is then needed to obtain an accurate theoretical expression for ΔV_T. The curves in Figure 9.9 apply to a MOSFET with the characteristics of Figure 9.7 (8 μm channel length), and the agreement between Equation 9.2.1 and the observed shift in threshold voltage is good.

Transconductance. The small-signal amplification of a MOSFET is most usefully characterized by its transconductance because the output (drain) current typically varies in response to a changing input (gate) voltage. From Equation 9.1.12, we find

$$g_m \equiv \frac{\partial I_D}{\partial V_G} = \mu_n C_{ox} \frac{W}{L} V_{DS} \qquad V_D < V_{Dsat} \qquad (9.2.4)$$

which increases linearly with drain voltage but is independent of gate voltage. When $V_D > V_{Dsat}$, the transconductance (as obtained from the variable depletion-charge analysis) is given by Equation 9.2.4 with $V_D = V_{Dsat\,2}$ (from Equation 9.1.14).

$$g_{msat} = \mu_n C_{ox} \frac{W}{L} \left\{ V_G - V_{FB} - 2|\phi_p| - V_S \right.$$
$$\left. - \frac{\epsilon_s q N_a}{C_{ox}^2} \left[\sqrt{1 + \frac{2C_{ox}^2}{\epsilon_s q N_a}(V_G - V_{FB} - V_B)} - 1 \right] \right\} \qquad (9.2.5)$$

Thus, g_{msat} is independent of V_D, but nearly linearly dependent on V_G. If we use the more convenient representation for I_{Dsat} that is given in Equation 9.1.17, we obtain the simpler expression

$$g_{msat} = (k'W/L)(V_G - V_T) \qquad (9.2.6)$$

which clearly exhibits the proportionality between transconductance and the effective gate bias $(V_G - V_T)$. The transconductance is also proportional to the ratio of channel width to length, the mobility, and the oxide capacitance per unit area because an increase in any of these terms increases the output current per unit change in gate-to-source voltage.

Speed of Response. There are two intrinsic limits on the speed of response of a MOSFET. First, as in all current amplifiers, a basic limit is set by the time for charge transport along the channel: that is, the transit-time limitation (described for the bipolar transistor in Section 7.3). Second, a limit is imposed by the charging of capacitances inherent in the device structure. In practical applications, a third limit on speed can be set by parasitic capacitances that are unavoidable in a given integrated circuit, but not inherent in the device itself. Limitations of this third type are properly considered by adding components external to the transient model for the intrinsic device; we shall comment on them briefly later.

Analysis of the speed limitations in the MOSFET itself depends upon whether the device is biased into saturation or not. We shall consider only operation in the saturation region. This restriction leads to considerable simplification, and is really all that is warranted for most applications.

When the MOSFET is operated in the current-saturated region, an approximate solution for the field along the channel $\mathscr{E}_y(y)$ can be derived fairly easily. This is accomplished by writing the differential equation for channel voltage (Equation 9.1.5) with the use of Equation 9.1.8 to express $Q_n(y)$. The equation then takes the form

$$dV_C(y) \simeq \frac{I_D dy}{\mu_n C_{ox} W [V_G - V_T - V_C(y)]} \tag{9.2.7}$$

In Equation 9.2.7, as in Equation 9.1.8, V_T refers to the threshold voltage at the source and the variation of depletion-layer charge Q_d with y is not represented. Equation 9.2.7 can be used to find a solution for the field $\mathscr{E}_y(y) = -\partial V_C / \partial y$ (Problem 9.20).

$$\mathscr{E}_y(y) = -\frac{(V_G - V_T)}{2L} \frac{1}{\sqrt{1 - (y/L)}} \tag{9.2.8}$$

The transit time along the channel T_{tr} is then found directly by using Equation 9.2.8 in the expression

$$T_{tr} = \int_0^L \frac{1}{v_y} dy = -\int_0^L \frac{1}{\mu_n \mathscr{E}_y} dy \tag{9.2.9}$$

which leads to

$$T_{tr} = \frac{4}{3} \frac{L^2}{\mu_n (V_G - V_T)} \tag{9.2.10}$$

If we consider an n–channel MOSFET having $L = 3$ μm, $\mu_n = 660$ cm^2 V^{-1} s^{-1}, and $(V_G - V_T) = 5$ V, Equation 9.2.10 predicts a transit time of 3.6×10^{-11} s. This time is at least an order of magnitude shorter than the fastest switching times that are obtained in MOSFET circuits. Accordingly, we conclude that *the speed of response of the MOSFET is governed not by the channel transit time but rather by the time needed to charge the capacitances associated with the device and the elements to which it is connected in a circuit.* For this reason, the calculation of MOSFET circuit transients is carried out assuming that currents in the MOSFET are governed by the static equations for the device.

Channel-Length Modulation. The MOSFET theory developed thus far has treated the channel length L as being constant. In fact, because the space-charge region at the drain junction varies with the drain voltage, L is a function of V_{DS}.

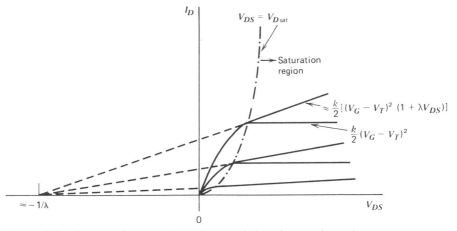

Figure 9.10 Output (I_D versus V_{DS}) characteristics of an n–channel MOSFET showing the effects of channel-length modulation.

Channel length is reduced as V_{DS} increases, resulting in an increase in the drain current over that predicted by the simple theory described thus far. The effect of channel shortening is greater in shorter channel MOSFETs. Although an exact theory of channel shortening is complicated, for most design calculations it can be modeled adequately[3] as being linearly proportional to V_{DS} with a proportionality constant λ. Hence, for example, if channel-length modulation is important, Equation 9.1.17 becomes

$$I_{D\text{sat}} = \frac{k}{2}(V_G - V_T)^2(1 + \lambda V_{DS}) \qquad (9.2.11)$$

The *channel-length-modulation parameter* λ is typically in the range 0.1 to 0.01 V^{-1}. By comparing Figure 9.10 with Figure 7.1, it is apparent that $1/\lambda$ is a voltage that is similar to the Early voltage in bipolar transistors in its effect on the output characteristics of the MOSFET. Figure 9.10 also indicates that an empirical value for $1/\lambda$ can be obtained by finding an approximate intercept with the voltage axis of tangents to measured curves of I_D versus V_{DS}. To see that this is the case, consider that the intercept of the extended tangents to the curves for I_D occurs at V_1; then $\lambda(V_{DS} - V_1) = I_D/I_{D\text{sat}} = 1$ at $V_{DS} = V_{D\text{sat}}$. Hence, $V_1 = V_{D\text{sat}} - 1/\lambda \approx -1/\lambda$ since λ is usually very small.

Small-Signal Circuit Model. Based upon the understanding of the MOSFET behavior that has been developed, the construction of a small-signal circuit model for the device is straightforward. A suitable model, consisting of elements that represent the basic device equations together with elements representing the inherent capacitances and resistances within the transistor structure is shown in Figure 9.11. Of the four capacitors connected to the gate in the figure, only two

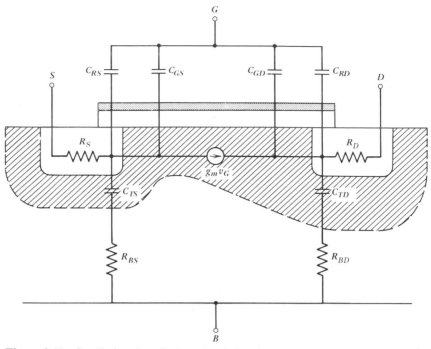

Figure 9.11   Small-signal equivalent circuit for the MOSFET.

(C_{GS} and C_{GD}) are inherent to the device. These capacitors represent the flux linkages to the channel charge which give rise to the basic operation of the MOSFET. The speed limitations associated with charging C_{GS} and C_{GD} are fundamentally related to the transit time of charge along the channel (Problem 9.22). The values of these two capacitors are bias-dependent. If V_{DS} is small, they are each equal to $C_{ox}WL/2$; when the MOSFET is saturated, C_{GS} becomes $\frac{2}{3}C_{ox}WL$ and C_{GD} approaches zero (Problem 9.22), representing the fact that few electric flux lines link the gate to the drain.

Capacitors C_{RS} and C_{RD}, between the gate and the source and between the gate and the drain, are parasitic elements that result from misalignment and overlap of the gate with respect to the source and drain diffusions. The two capacitors connected between the substrate and the source and between the substrate and the drain (C_{TS} and C_{TD}) are depletion-region capacitances at the reverse-biased *pn* junctions in these regions. The resistors R_S and R_D, which are typically of the order 10 to 100 Ω, represent ohmic and contact resistance between the external electrodes and the MOSFET channel. Resistors R_{BS} and R_{BD} account for ohmic resistance between the edges of the depletion regions and the contact to the substrate. The transconductance g_m for the current generator has been specified in Equations 9.2.4 and 9.2.6. Because MOSFETs are far more frequently employed in digital circuits than in analog applications, the small-signal circuit is less often employed than are the large-signal equations for the device.

9.3 MOSFET Design

The first large-scale development of MOS integrated circuits employed p–channel devices made with aluminum gates. Semiconductor manufacturers chose to make p–channel MOSFETs first because the oxide-charge densities (which are generally positive) were sizable and quite variable in early MOS technologies that had evolved directly from bipolar *IC* processes. Positive oxide charge tends to accumulate an n–type surface but to deplete or invert a p–type surface. Hence, typical oxide-charge densities on a p–substrate (needed for n–channel MOS) often result either in an inverted surface when $V_{GS} = 0$ (*depletion*–mode n–channel MOSFET) or else an n–channel MOSFET with an uncontrollably small threshold voltage. In addition, inversion in the *field region* surrounding each n–channel transistor can lead to undesirable conducting paths between adjacent MOSFETs.

Threshold Voltage and its Control

Control of the threshold voltage of MOSFETs is a major concern in the production of MOS integrated circuits. As described previously, the designer must not only consider the threshold voltage of the MOSFETs themselves, but also introduce *channel stops* which increase the threshold voltage in the region below circuit interconnection lines.

For the successful design of any MOS *IC*, the first level of device threshold-voltage control is to be able to produce either enhancement or depletion-mode MOSFETs reliably, according to circuit requirements. For most applications, enhancement-mode devices are desired. Hence, for p–channel MOSFETs V_T must be less than zero, whereas for n–channel devices V_T must be greater than zero. The requirement for p–channel devices is easily met, but that for n–channel MOSFETs can cause difficulty. Consideration of the equations for the threshold voltage makes clear why this is true. With $V_S = V_B = 0$, the threshold-voltage equations can be written (Table 8.2)

$$V_{Tn} = V_{FB} + 2|\phi_p| + \frac{|Q_{d_i}|}{C_{ox}} \tag{9.3.1}$$

for the n–channel MOSFET and

$$V_{Tp} = V_{FB} - 2|\phi_n| - \frac{|Q_d|}{C_{ox}} \tag{9.3.2}$$

for the p–channel MOSFET. In both equations Q_d represents the depletion-charge density that exists when the surface becomes inverted.

The flat-band voltage is typically negative both for p– and n–channel MOSFETs (Problem 9.1). Since the remaining terms in Equation 9.3.2 are also negative, V_{Tp} is assuredly negative and therefore the production of enhancement-mode p–channel MOSFETs poses no difficulties. For the n–channel device, however, the sum of the last two terms in Equation 9.3.1 must be larger than $|V_{FB}|$ to produce enhancement-mode MOSFETs. This requirement sets a lower limit on the allowable dopant concentration for n–channel devices. For 100 nm-thick oxides and densities of $|Q_f/q|$ near 10^{11} cm^{-2}, the level of doping for marginal operation is in the low 10^{15} cm^{-3} range. To design with a safety factor (e.g., to obtain threshold voltages of nearly a volt), the dopant density should be of order 10^{16} cm^{-3}. Such a high dopant density is undesirable because it leads to high substrate capacitance and relatively low junction-breakdown voltages. Compounding this problem is the fact that boron, the commonly used p–type dopant, tends to segregate into the oxide as the oxide grows, thus reducing the doping N_a at the surface to a lower value than that in the substrate. An alternative means of changing the threshold voltage in the desired direction is to decrease C_{ox} by increasing the gate-oxide thickness. This is undesirable, however, because it reduces the gain of the MOSFET (k' in Equation 9.1.17).

It is also possible to alter the threshold voltage by applying a substrate bias so that V_T is shifted by the effect illustrated in Figure 9.9. This technique (sometimes called *body biasing*) is employed in a number of commercial n–channel MOSFET circuits. The substrate bias is either supplied externally or else generated on the *IC* chip by additional circuitry. Body biasing is an added complication that is avoided whenever possible in favor of threshold-voltage adjustment by ion implantation. The use of ion implantation for threshold-voltage adjustment is discussed briefly later in this section, and in greater detail in Section 10.6. For MOSFETs having very small dimensions, additional effects which influence the threshold voltage must be considered; these are discussed in Chapter 10.

EXAMPLE MOSFET Threshold-Voltage Considerations

A manufacturer wishes to fabricate n–channel, depletion-mode MOSFETs without using ion implantation. The process that was established produces a fixed positive–charge density $N_f = Q_f/q = 10^{11}$ cm^{-2} at the interface. The oxide thickness is 50nm and the gates are made of aluminum.

(a) If silicon wafers with doping $N_a = 10^{15}$ or 10^{17} cm^{-3} are available, which should be selected in order to be most certain of obtaining depletion–mode devices?

(b) Perform an analysis to determine whether or not the process described is capable of producing the depletion–mode MOSFETs.

Solution

In order to be assured of n–channel, depletion–mode MOSFETs, the threshold–voltage V_T must be negative so that a channel will be formed when the gate and the source are at the same potential. The surface is more easily inverted when doping is reduced; therefore, the material to choose is that doped with $N_a = 10^{15}$ cm^{-3}.

From Equations 8.3.18 and 8.4.6 with no channel or bulk bias applied, the threshold voltage V_T is

$$V_T = \Phi_{MS} - Q_f/C_{ox} + 2|\phi_p| + \frac{1}{C_{ox}'}\sqrt{2\epsilon_s q N_a (2|\phi_p|)}$$

$$C_{ox} = \epsilon_{ox}/x_{ox} = 6.9 \times 10^{-8} \text{ F cm}^{-2}$$

$$\phi_p = (kT/q)\ln(N_a/n_i) = 0.29 \text{ V}$$

$$\Phi_{MS} = .80 \text{ V}, V_T = -0.25 V$$

Hence,

$$V_T = -0.80 - 0.23 + 0.58 + 0.20 = -0.25 \text{ V}$$

Thus, the threshold–voltage is negative, and the process is capable of producing depletion–mode, n–channel devices.

Technological Evolution

The silicon wafers used for the first p–channel MOS *IC*s were (111)–oriented, and typical p–channel, aluminum-gate MOSFETs had threshold voltage of approximately -4 V. A major innovation in processing MOS *IC*s was the successful utilization of *polycrystalline silicon* (or *polysilicon*) for the gate (cf. Chapter 2). When the gate is constructed of silicon, it can be deposited prior to the source and drain diffusions, and the gate itself can serve as a mask for these diffusions. With this technology, the gate is nearly perfectly aligned over the channel. The only overlap at the source and drain is due to lateral diffusion of the dopant atoms; this *self alignment* reduces the parasitic overlap capacitances (C_{RD} and C_{RS} in Figure 9.10), and thus improves transistor performance. Additionally, fabrication with self-aligned silicon gates results in a Φ_{MS} term that aids inversion of an n–type silicon surface (as shown in Problem 8.2). This feature enabled the production of p–channel silicon-gate MOSFETs on (111) silicon with threshold voltages of roughly -2 V. A threshold voltage near this value is necessary to make MOSFET circuits that are compatible with bipolar transistor-transistor logic (T^2L) circuits—a necessity for the design of many systems. For integrated circuits, there is another important advantage to silicon-gate technology. The refractory nature of the gate material permits the complete encapsulation of the MOSFET in an SiO_2 layer. Not only does this afford excellent protection and stability to the sensitive MOSFET channel region, but it also allows the polycrystalline silicon to be used in areas other than MOSFET gates. The polysilicon can then provide an additional layer of interconnections that can be crossed by the standard metal interconnection or by yet another layer of polysilicon. Figure 9.12a shows a cross section of the glass-overcoated, p–channel silicon-gate MOSFET which was developed for the 1024-bit MOS random-access memory in 1970. The polysilicon-metal interconnection is apparent in the figure. A cross section of an n–channel, silicon-gate MOSFET made in a scanning-electron microscope is shown in Figure 9.12b. The overcoating oxide is plainly visible.

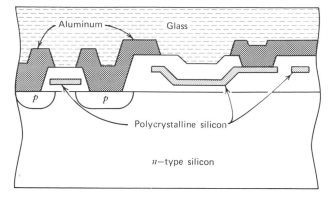

(a)

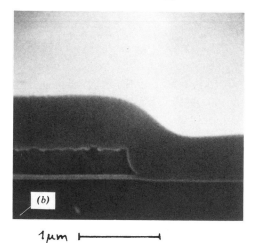

1μm ⊢━━━━━━━━━━┥

Figure 9.12 (a) Cross section showing the use of poly-crystalline silicon for a p-channel MOSFET gate (left), and for an interconnection path (right). (*Courtesy Intel Corporation*) (b) Scanning electron micrograph of a polysilicon gate overcoated with a phosphorus-glass insulating layer. (*Courtesy Siemens Corporation*)

With the advent of *very large scale integration* (*VLSI*), circuits having thousands to hundreds-of-thousands of devices are being designed. The interconnections for these circuits are typically made with cross sections as small as processing permits. As the dimensions have decreased, the conductivity achievable in polycrystalline silicon has become a limitation to its use for interconnections.

At the highest practical dopant concentrations, a 0.5 μm-thick polycrystalline-silicon film has a sheet resistance of about 20 Ω/□ (Chapter 2). The resulting resistance of interconnection paths can lead to relatively long *RC* time constants and severe dc voltage variations within a *VLSI* circuit. An approach to reducing sheet resistance of polycrystalline silicon is to anneal with a laser or in a lamp-heated furnace for a few seconds, but these methods provide limited improvement.

Another technique, which has been further developed, is to use an alternative material that retains the capabilities of polycrystalline silicon for self-aligning the MOSFET gate region and being completely oxide encapsulated while providing a lower sheet resistance. The most widely employed alternative materials are the *refractory metal silicides*, notably those of tungsten, tantalum, titanium, and molybdenum, or a refractory metal such as tungsten, itself. These materials are often placed over a thin polycrystalline–silicon layer so that the desirable properties of a silicon/silicon-dioxide interface above the gate insulator are retained. With these substitutions for a single layer of polycrystalline silicon, sheet resistances of the order of $1\ \Omega/\square$ are possible at the expense of more complicated *IC* processing.

When MOS technology was in its infancy, manufacturers carried over know-how from bipolar processing and therefore used (111)-oriented silicon wafers, for which fine control of dopant diffusion had been mastered. Later, as the influence of interface states (cf. Sections 3.5 and 8.4) on the properties of MOSFETs came to be better understood, the importance of reducing the density of these states became clear. The area density of surface atoms in (100)-oriented silicon is appreciably smaller than is that in (111)-oriented silicon, and the achievable interface-state densities for oxide-coated wafers are typically only about one third as large for (100)-oriented silicon. Because of this reduced density, (100) silicon is almost universally used at present for MOS processing. With the achievement of control over interface charge and with further refinements in processing, there was also a shift to n–channel devices, in which higher mobility leads to improve performance.

A further major innovation in the processing of MOSFETs was the introduction of ion implantation for threshold-voltage adjustment. When dopant atoms are implanted in the channel regions of a MOSFET, it is possible to adjust the depletion-layer charge Q_d (Equation 8.3.7) to a precisely determined value. By this means, the threshold voltage can be set accurately through an ion-implantation step after the gate oxide has been formed. To first order, the threshold-voltage change ΔV_T is given by $\Delta V_T = qN'/C_{ox}$, where N' is the area density of dopant atoms (*dose*) introduced *into the silicon* near its surface. The use of ion implantation has made possible the reliable production of n–channel MOSFETs. It has also permitted the use of lightly doped silicon substrates without suffering inadvertent surface inversion in the surrounding field regions. The lower is the silicon doping the smaller is the useless capacitance between the active regions of the MOSFET and the silicon bulk (C_{TS}, C_{TC}, and C_{TD} in Figure 9.11). In addition, surface mobility is higher if the dopant concentration is lower (Chapter 1). Presently, n–channel MOSFETs are typically made on (100) silicon in which $N_a \approx 10^{15}\ cm^{-3}$. Surface mobilities as high as $800\ cm^2\ V^{-1}\ s^{-1}$ are obtained.

In Chapter 10 we discuss *VLSI* MOSFETs in which channel lengths may be 1 μm or less. A major challenge in these extremely small devices is to avoid *punchthrough*, in which the depletion layers of the source and drain overlap so that a current path can exist below the channel region (Chapter 10). For short channel lengths MOS processing must be modified to avoid punchthrough. An increase in substrate doping eases problems with punchthough, but at the same time, it increases the parasitic capacitances between the substrate and the MOSFET source and drain. More desirable is a localized increase of the substrate dopant in the region where the source and drain depletion regions tend to overlap. This local doping is possible through the use of a second, higher-energy implant that can be carried out immediately after the threshold-voltage-adjusting implant. Doubly implanted MOSFETs are the basis for *high-performance* MOS (HMOS) technology.

A basic advantage of MOSFETs over bipolar transistors is their inherent self isolation; that is, (except for second-order effects) adjacent transistors do not interact unless there is a surface channel between them. This property avoids the need for the *pn*–junction-isolated wells used in conventional bipolar technology and results in a substantial area saving for MOSFET circuits compared to bipolar circuits. As a consequence, integrated circuits having the highest component density are based on MOS technologies. In practical MOSFET integrated circuits, special steps must be taken to inhibit the formation of spurious channels under inter-connection lines. This is typically accomplished by increasing the threshold voltage away from the active devices in two ways: first, by using a thick *field oxide* outside of the device regions, and second, by employing a heavier doping under this field oxide. These *channel–stopping* measures are accomplished effectively by local oxi-dation (the *LOCOS* process), as was described in Chapter 2. A schematic view highlighting features of a *LOCOS*-processed MOSFET is shown in Figure 9.13. Indicated on the figure are the use of arsenic implantation combined with phos-phorus diffusion for the source and drain electrodes and boron implantation to increase the threshold voltage in the *field*-regions (the areas on the *IC* where there are no devices). The MOSFET design shown offers excellent self-aligning of the field oxide and the channel-stop implant, and a smoothly tapered step to the field oxide that permits unbroken paths in overlying conductors. This tapered region, an advantage for the reason just cited, is now becoming a limitation for small MOSFETs because it typically cannot be reduced below about one micrometer in length.

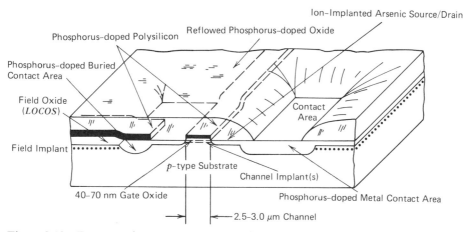

Figure 9.13   Features of a *LOCOS* processed, ion-implanted, polycrystalline silicon-gate MOSFET. (*Courtesy Siemens Corporation*)

The evolution of MOSFET technology is by no means complete. The steps described in the foregoing brief overview have accompanied the reduction in device feature sizes from about 10 to roughly 2.5 μm, with *IC* device counts expanding from the thousands to hundreds of thousands. Realistic projections for the next decade are for another 5-fold decrease in feature sizes (to about 0.5 μm) and for million-device integrated circuits.[4] Progress toward these goals demands continued introduction of technological improvements and process modifications. There is still a frontier for MOSFET processing evolution!

MOS Memory

Most of the major developments in MOSFET technology have been undertaken to advance the design of MOS memory circuits. Soon after their realization, *IC* memories displaced magnetic-core elements, which had been used for fast-access memories in digital computers, and drastically lowered the cost of this function while enhancing performance. The major portion of an *IC* memory circuit consists of a regular, repeated array of information-storage cells. The cells store *bits* or binary digits (ones or zeros) in one of several ways. For example, an elemental circuit for information storage, called a *flip-flop*, has an output that is stable at only one of two possible voltage levels. Flip-flops maintain a given state for as long as the circuit receives power, and change state according to conditions that are under the control of the circuit designer. Typically six MOSFETs are used in each flip-flop as shown in Figure 9.14*a*. These storage cells are therefore rather large in area (about 500 μm^2 for a 3-μm minimum feature size).

In the early 1970s, circuits were invented that stored bits in the form of charge packets. These circuits were built at first using only 3 MOSFETs, as shown in Figure 9.14*b*, and their introduction led to an immediate, dramatic reduction in the storage-cell size and a corresponding increase in the number of bits stored per *IC* (to 2^{10} bits equals 1*K* bit*). In contrast to flip-flop circuits, the charge-storage circuits tend to lose their information with time because of inherent leakage of charge from the electrodes on which it is stored (typically through substrate currents in MOSFET memories). To replenish the charge lost by leakage, this type of memory must be rewritten or *refreshed* at frequent intervals and is, therefore, called a *dynamic* memory; in contrast, a memory circuit that does not require a refresh cycle is called a *static* memory. Soon after the invention of the charge-storage MOSFET memory, it was discovered that only a single transistor and a capacitor are actually needed to store one bit of information as shown in Figure 9.14*c*. The one-transistor memory cell, first made with 8 μm features (minimum line widths and spaces), required only 1280 μm^2 and was used to produce a 2^{12} bit (4*K* bit) memory chip. This cell, with refinements in technology and design, is still the basis for high-density NMOS memory chips. In the next evolutionary

---

* The total bit capacity on a chip is typically a power of 2 such as 256 (2^8). For $2^{10} = 1024$ and higher densities, it is conventional to express memory capacity in units of 1024 bits and to designate the memory as *K* (capital letter) bits. For example, a 1*K* bit memory holds 2^{10} bits and a 256 *K* bit memory typically stores $2^{18} = 262,144$ bits.

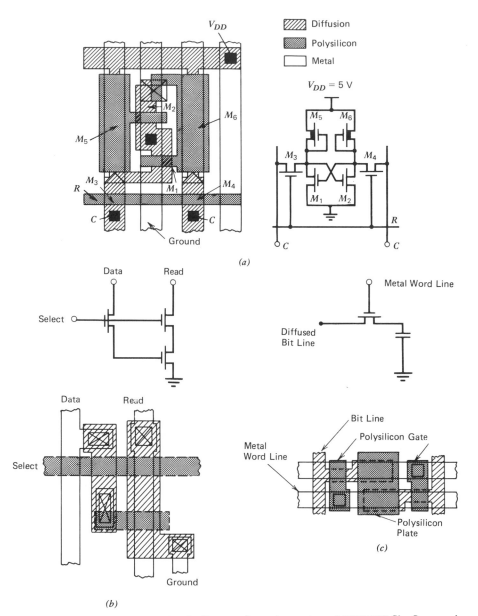

Figure 9.14 (a) Layout and circuit diagram for a 6-transistor MOSFET flip-flop used in static RAMs,[3] (b) layout and circuit for a 3-transistor MOSFET charge-storage cell for dynamic RAMs, (c) layout and circuit for a 1-transistor MOSFET charge-storage cell for dynamic RAMs.[9]

phase of one-transistor-cell memories (the 16 K bit memory), 3 μm-feature sizes were used to build a cell with an area of only 180 μm^2.

RAM, ROM, PROM, EPROM, and EAROM. If a memory is arranged so that the information can be read from any cell at random, the circuit is called a *random-access memory* (*RAM*). From the discussion above, we see that there are *static RAMs* (*SRAMs*) and *dynamic RAMs* (*DRAMs*). Frequently, the contents of a memory do not need to be changed: a circuit of this type is called a *read-only memory* (ROM). The information in some *ROMs* is permanently written on the cells according to the instructions of the user as a last step in processing or sometimes after the *IC* has already been delivered. These circuits are called *programmable read-only memories* (*PROMs*). If the memory is written by purely electrical means (as contrasted with some further *IC* fabrication steps), it is called an *electrically programmable read-only memory* (*EPROM*). Finally, the development of several new types of MOS devices has made it possible to alter the state of a *ROM* electrically. Circuits of this type are called *electrically alterable read-only memories* (*EAROMs*).

The use of MOSFETs in *IC* memories became especially intensive after the introduction of the 256-bit MOS *SRAM* in 1969. That product made use of polysilicon-gate, p–channel MOSFETs in flip/flop storage cells like that shown in Figure 9.14a. The development of silicon memory chips since that time has been extraordinarily rapid. Table 9.1 summarizes some of the important steps taken to achieve over a thousand-fold increase in memory per chip (to 256K bits in 13 years).

Shown in the table are the device types, the number of bits per chip, the circuit types, the number of MOSFETs in each memory cell, and some of the major processing innovations which led to the increased density. Most of the entries

Table 9.1 **MOS Memory Development**

Year	Device Type	Circuit Type[a]	MOSFETs/Cell	Process Changes
1962	Metal-gate PMOS	Shift Registers Logic Chips		
1969	Silicon-gate PMOS	256 Bit *SRAM*	6	*CVD* polysilicon
1970	Silicon-gate PMOS	1*K DRAM*	3	*CVD* oxide
1971	Silicon-gate NMOS	1*K SRAM*	6	
1974	Silicon-gate NMOS	4*K DRAM*	1	*CVD* nitride Ion-implant ΔV_T
1976	Silicon-gate NMOS	16*K DRAM*	1	2-layer polysilicon
1979	Silicon-gate NMOS	64*K DRAM*	1	Plasma etch Wafer stepper
1982	Silicon-gate NMOS	256*K DRAM*	1	
1982	Silicon-gate CMOS	64*K SRAM*	4	

[a] *SRAM*—static random-access memory; *DRAM*—dynamic random-access memory (*after C. N. Berglund, Intel Corp.*).

in the table refer to silicon-gate n–channel MOS (NMOS), which has dominated the development of silicon memory circuits. The last circuit in the table is an exception; it refers to *complementary MOS*, (CMOS) which is a major design interest at present and promises to be highly competitive to NMOS. Features of CMOS are discussed in Section 9.4.

The subject of MOS memory *IC*s is very much wider in scope than is appropriate for our discussion here. An up-to-date view of the field is best obtained by referring to one of the "Special Issues on Logic and Memory" that are published periodically (typically the October issue) in the *IEEE Journal of Solid-State Circuits*. Before leaving the topic of MOS memory, however, we turn from a focus on *LSI* or *large–scale integration* to a few brief remarks concerned with a specific device.

Floating-Gate Memory Element. The design of *EPROM*s and *EAROM*s has led to the invention of a number of special MOSFET structures. We discuss only one of these structures, a device that made practical a new concept for the storage of information in a *ROM*. The concept is the transfer of charge through an insulator from the silicon substrate to an insulated storage electrode, and the device is called a *floating-gate, avalanche-injection MOS transistor* (FAMOS).

The construction of the FAMOS device was made possible by the mastery of a totally encapsulated polysilicon-gate technology, such as that illustrated in Figure 9.12. The original FAMOS device consisted simply of a MOSFET with a "floating" or electrically disconnected gate, as sketched in cross section in Figure 9.15a. The only way for charge to reach the gate is by transfer through the oxide. This transfer of charge can be induced by supplying sufficient energy to the electrons to cause them to pass over the energy barrier at the silicon/silicon-dioxide interface. The energetic electrons are provided by selectively causing avalanche breakdown at the desired drain-substrate pn–junction in an array of these devices. If the FAMOS gate is charged sufficiently with electrons to cause inversion of the n–type substrate, a conducting channel forms between the source and drain, exactly as if a gate voltage were applied. Thus, binary information is stored by a FAMOS device according to the presence or absence of a conducting channel.

The buried polysilicon layer in a FAMOS device retains the charge on the gate for any practical time period unless some source of external energy is able to liberate the electrons held there. A suitable energy source to excite them is photon irradiation which can give the electrons sufficient energy to pass over the barrier at the interface between the polysilicon layer and the oxide. They can then escape to the substrate and return the FAMOS device to its off-state. Thus, complete erasure of a FAMOS memory can be accomplished by irradiating the surface with ultraviolet (*UV*) light. Memory arrays capable of being erased by *UV* light are therefore packaged with a translucent cover.

The original FAMOS design, illustrated in Figure 9.15a was suitable for p–channel MOSFETs. Later introduction of two-layer polysilicon (cf. Table 9.1) into MOS processing made possible the FAMOS structure shown in Figure 9.15b to be used with n–channel MOSFETs. The buried gates of FAMOS-related structures that have been developed more recently can be charged and discharged by

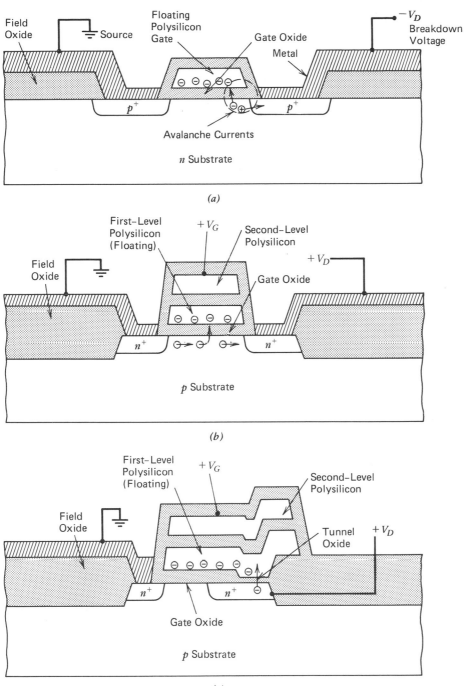

Figure 9.15 (*a*) Cross section showing the mechanism of charge injection into the gate by avalanche in a FAMOS memory element. (*b*) A FAMOS element made with two layers of polysilicon and suitable for *n*–channel MOS applications. (*c*) A floating-gate memory cell in which electrons can tunnel through a thin oxide film to alter the cell content (an *EAROM* cell).

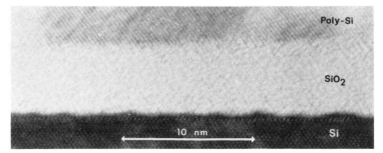

Figure 9.16   A high resolution transmission-electron-microscope (TEM) picture taken with 120 keV electrons showing a cross section of a silicon, silicon-dioxide, polysilicon structure. The point-to-point resolution is 0.33 nm. Individual dots represent pairs of atomic columns that are too close together to be spatially resolved.[10]

quantum-mechanical tunneling of electrons through very thin oxides as indicated in Figure 9.15c. In arrays of these elements, individual cells can be written and erased without changing the charge state of other cells. Such circuits are therefore suitable for *electrically alterable read-only memory* (*EAROM*) applications, where-as the basic FAMOS structure is only used in electrically progammable memories (*EPROMs*).

The design and fabrication of these memory devices relies upon an unprece-dented control over materials and processes. An example showing the precision of the analytical tools now being employed for this work is given in Figure 9.16, which is a transmission-electron micrograph (TEM) of a polysilicon-gate MOS structure having a thin oxide (6.7 nm). In the microscope picture, it is possible to see evidence of the atomic cores themselves in the partially oriented silicon gate and the monocrystalline substrate.

9.4 Devices: Complementary MOSFETs

Early in the development of MOS *IC*s, it was realized that digital circuits built with *p*– and *n*–channel MOSFETs connected in series could have low "standby" (or steady-state) power dissipation.[5] Circuits of this type are called *complementary* MOS transistor circuits or simply CMOS circuits. To understand the reason for the low power dissipation in CMOS circuits, we consider a basic building block for digital systems, the *inverter*. An inverter is a circuit whose (binary) output is the inverse of its input. By suitably interconnecting inverters, arbitrarily complex logic circuits can be built. Hence, the power consumed by an individual inverter circuit is a basic indicator of the overall power required by a digital system.

The basic CMOS inverter cell and layout are sketched in Figures 9.17a and 9.17b, along with its *voltage-transfer characteristic* (Figure 9.17c), which is a plot of the output voltage of the circuit as a function of the input voltage applied to it. In the inverter, the two MOSFETs are connected in series (*p*–channel drain

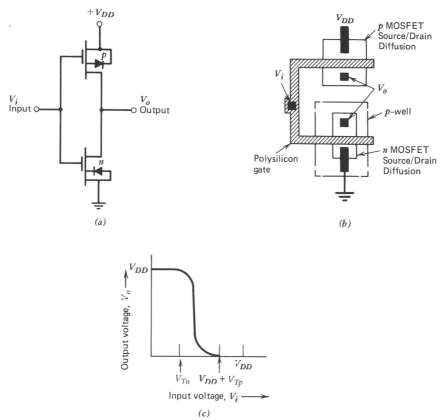

Figure 9.17   (a) CMOS inverter circuit, (b) layout of a p–well CMOS inverter, (c) voltage-transfer characteristic of a CMOS inverter.

to n–channel drain), and their gates are tied together. To understand the operation of this inverter, assume that the input voltage is lower than the n–channel threshold voltage, and sufficiently negative with respect to the bulk of the p–channel MOSFET to turn it on. Under this condition, the p–channel MOSFET provides a conducting path to the V_{DD} supply and the n–channel device is turned off. Since the output terminal is typically tied to the inputs of other inverter circuits which draw no steady-state current, the output voltage (at the drain of the p–channel MOSFET) is in its "high" state (at V_{DD}). If the input voltage is now increased, the p–channel MOSFET turns off, and as the input becomes larger than the threshold voltage of the n–channel device, its channel is turned on, pulling the output voltage toward ground. Thus, quiescently, one or the other MOSFET is always cut off, and there is no dc path to carry current from the supply except for junction leakages. For this reason, almost all of the power dissipation in CMOS circuits takes place during switching transients.

Low dc power consumption is one significant advantage of CMOS over other MOS digital *IC* technologies. Other advantages are the abrupt and well-defined voltage-transfer characteristic of CMOS inverters (Figure 9.17c), which facilitates digital design, and the noise immunity which results from the low impedance between the logic signal and either the supply voltage or ground. These desirable features of CMOS were understood long before the more demanding technological challenges of CMOS processing could be mastered.[5] Through the major technological advances of the last 15 years, CMOS has emerged as a practical and highly desirable *IC* technology, even though its fabrication is more complex and expensive.

CMOS Design Considerations. Since the basic idea in a CMOS process is to provide both *n*– and *p*–channel MOSFETs in the same *IC*, it is necessary to have both *p*– and *n*–regions on the wafer surface. If the substrate is *n*–type, and the *p*–channel devices are made directly in the substrate, *p*–type diffusions (creating what are usually called *p*–*wells*, or sometimes *p*–*tubs*) must be carried out in regions where *n*–channel MOSFETs are to be placed. CMOS can also be built on *p*–type wafers by forming *n*–wells. A cross section of an *n*–well CMOS transistor pair designed for a 64K bit *RAM* is shown in Figure 9.18.[6]

Both *n*– and *p*–well realizations have advantages and drawbacks, and neither is yet a clear choice in CMOS processing. For example, one design consideration might be that the transistor in the well is formed in compensated silicon which (because of the higher total dopant density) has a more severely degraded mobility than does the substrate device. Since nearly equal current drives in both *n*– and *p*–channel MOSFETs are desired, this consideration suggests that a *p*–well is preferable because electron mobility is higher than hole mobility. However, the width of the transistor can be adjusted to make up for the mobility difference, and other system considerations may favor the *n*–well design, particularly if much of the logic circuitry (for addressing, reading and writing) on the periphery of the CMOS *IC* is built of *n*–channel devices. It was this condition, for example, that motivated the use of *n*–wells in the CMOS design shown in Figure 9.18.

For optimal CMOS circuit performance, the threshold voltages of the two types of MOSFETs should be complements of one another (i.e., $V_{Tp} = -V_{Tn}$).

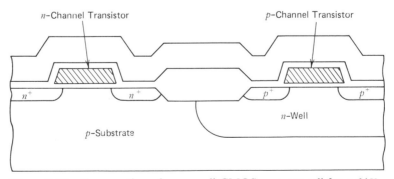

Figure 9.18   Cross section of an *n*–well CMOS memory cell for a 64K bit *RAM*.[6] The channel lengths are 1.2 μm (*n*–channel) and 1.1 μm (*p*–channel), the gate-oxide thickness is 25 nm, and the source-drain junction depths are 0.4 μm for the *p*–channel MOSFETs and 0.3 μm for the *n*–channel MOSFETs. The memory-cell area is 137 μm^2.

Ion implantation to adjust threshold voltages (described in Sections 9.3 and 10.6) has made this practical. Further comments about CMOS technology will be made after we have discussed some device and circuit considerations in CMOS design.

Since at least one of the MOSFETs in a CMOS process is built in a well, the two space-charge regions associated with the source- or drain-to well and well-to substrate junctions may reach one another, leading to a vertical punchthrough condition. For a specific example, we consider an n–well CMOS process for which (as seen from Figure 9.17a) both the source of the p–channel MOSFET and the well are connected to the positive supply voltage. The substrate is at ground potential. Thus, two depletion regions extend toward one-another in the well region. The source-to-well junction has only the built-in potential (ϕ_i) across it, while the well-to-substrate junction sustains the bias voltage V_{DD} in addition. To avoid high currents being drawn from the source, the neutral region (and thus the depth of the n–well) must be sufficient to avoid punchthrough between the source and substrate. However, the depth should not be excessive because lateral diffusion of the n–dopant during the well drive-in-diffusion step would use valuable surface area on the chip. If the n–well dopant density were increased to narrow the depletion regions and thus avoid punchthrough, it would cause the channel mobility to decrease and the capacitance of the drain to increase. This would degrade the switching performance of the circuit. Carefully balancing factors such as these is required for the optimal design of a CMOS process.

EXAMPLE   CMOS Well-Depth Design

An n–well CMOS process is designed for circuit operation at $V_{DD} = 1.5$ V. The starting wafers are p–type with $N_a = 5 \times 10^{14}$ cm^{-3}. The n–wells are to have an average dopant density $N_d = 3 \times 10^{15}$ cm^{-3}. The p–channel MOSFET sources and drains are to have junction depths $x_j = 0.8$ μm and an average dopant density $N_a = 10^{18}$ cm^{-3}. What is the minimum n–well depth that will avoid vertical punchthrough to the substrate?

Solution

Vertical punchthrough would occur in a path through the two back-to-back pn–junctions from the p–channel MOSFET source, biased at V_{DD} (1.5 V), to the grounded substrate (see Figures 9.17 and 9.18).

The source to n–well junction is essentially one-sided with a built-in voltage $\phi_i \approx 0.78$ V. From Table 4.1 or Equation 4.3.1, the depletion-layer width extends into the n–well 0.58 μm. The np–junction to the substrate has a built-in voltage $\phi_i \approx 0.58$ V, and the total depletion width at 1.5 V bias is found from Equation 4.3.1 to be 2.51 μm. Applying Equation 4.2.6, we see that one seventh of the depletion width (0.36 μm) is in the n–well. The n–well must therefore be thick enough to accommodate the depth of the drain junction (0.8 μm), as well as the total 0.94 μm (0.58 + 0.36) depleted width to avoid vertical punchthrough from the source to the substrate. Thus, the minimum well depth is 1.74 μm. Good engineering design makes it advisable to increase this dimension in order to allow a reasonable safety factor.

An additional consideration is that when the p–channel MOSFET is in the off-state, its drain is essentially at ground potential. In this condition, the depletion layer from the well to drain is wider than that from well to source because of the added V_{DD} voltage drop at the drain junction. However, even if the depletion regions between drain-well and well-substrate touch one another, no high "punchthrough" currents flow because the drain and substrate are both at ground potential. Despite this, having depletion regions touch is not good design because it may cause remote sections of the well to become pinched-off. Carrying out an analysis similar to that above shows that a well depth of 2.16 μm is needed to assure that there are charge-neutral regions throughout the well under all bias conditions.

With some added margin for safety, a reasonable design might make the well 2.5 μm deep.

CMOS Latch-up. A challenging problem in designing CMOS circuits is to avoid a condition known as *latch-up*, in which regenerative bipolar-transistor action causes a clamped, low-resistance path between the power supply and ground. Avoiding latch-up is especially challenging in small-dimension CMOS for dense *VLSI* applications.

To understand the basic latch-up phenomena, consider the p–well CMOS structure shown in Figure 9.19. Superimposed on the MOS cross sections shown in Figure 9.19 are unwanted or *parasitic npn* and *pnp* bipolar transistors. The transistors are cross-connected so that the base-collector junctions are common. From the resulting bipolar equivalent circuit shown in Figure 9.20, we see that, under active bias, the *pnp* collector delivers current to the *npn* base, and the *npn* collector delivers current to the *pnp* base. If these bipolar transistors have even

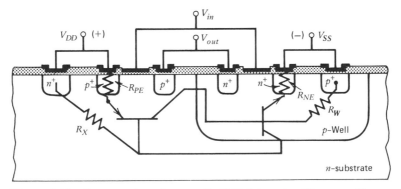

Figure 9.19   Cross section of a p–well CMOS inverter. The parasitic *pnp* and *npn* bipolar transistors are indicated along with associated substrate resistor R_X and well resistor R_W. The two resistors R_{PE} and R_{NE} represent contact and diffused-region resistance in the emitters.

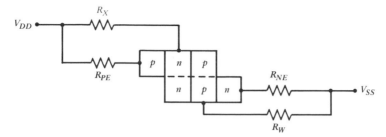

Figure 9.20   Circuit and schematic representation of the cross-coupled
parasitic *npn* and *pnp* transistors.

moderate current gains (βs), this interconnection can easily lead both devices to
saturate so that the supply voltages become connected across a low resistance in
series with two voltage drops: one, the voltage across a saturated base-collector
junction V_{CEsat}, and the other, the voltage across a saturated base-emitter junction
V_{BEsat}.*

Under normal CMOS operating conditions, the base-emitter junctions for both
bipolar transistors are reverse-biased, which makes latch-up impossible. A successful circuit design must, however, preclude latch-up under any conditions that might
be experienced by the circuit. To understand the ways in which latch-up can be
initiated, we refer to Figure 9.21, in which the cross-connected bipolar pair is
redrawn and two elements—a capacitor C_{PS} and a current source I_0—are added
in parallel across the base-collector junctions. The capacitance C_{PS} is much larger
than that of a typical base-collector junction because this capacitor represents the
large junction between the p–well and the substrate. The current source I_0 normally
models only junction leakage and is very small in magnitude. Several mechanisms,
however, can cause I_0 to increase markedly.

Among the possible sources for current through I_0 are (1) minority carriers
injected into the substrate by transient forward bias on pn junctions (typically in
input or output circuits), (2) photogeneration by ionizing radiation, and (3) impact
generation by hot carriers. The large capacitor C_{PS} can also deliver currents when
voltage transients occur, especially during the power-up phase of the circuit. Any
of these sources of current can turn on one or both of the bipolar devices. Thereafter, latch-up will take place if the gain of the cross-coupled bipolar pair is sufficient and if the V_{DD} power supply can deliver enough current.

Latch-up Models.   A simple expression that reveals conditions on device gain
which can lead to latch-up can be obtained by simplifying Figure 9.21 by omitting
I_0 and C_{PS} and considering R_{PE} and R_{NE} to be negligible (Figure 9.22). Under
this condition, the current driving the base of the pnp is equal to the base current

---

* Merged *pnp* and *npn* transistors in which the *pnp* base is driven by the *npn* collector and vice
versa are useful and important power-handling switches. These switches are frequently called *silicon
controlled rectifiers*, abbreviated *SCR*s. Since *SCR* switching has been the subject of much research,
latch-up in CMOS is often described as an *SCR* effect.

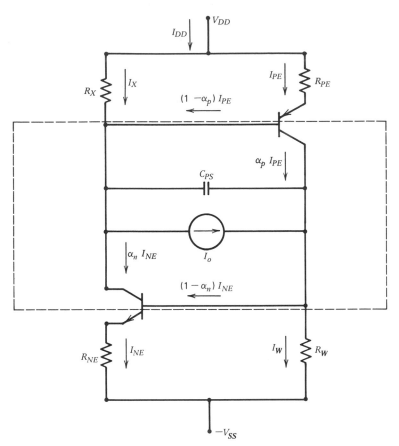

Figure 9.21   Latch-up equivalent circuit including well-to-substrate capacitor C_{PS} and parasitic current source I_0. The dashed lines surround all elements connected between the well and substrate nodes.

of the *npn* times β_n reduced by the divider action of the input resistance of the *pnp* transistor base and the substrate resistor R_X, which is in parallel with it. On a small-signal basis this is

$$\beta_n \times \frac{R_X}{r_{\pi pnp} + R_X}$$

where $r_{\pi pnp}$ is the reciprocal of δg_m, given in Equation 7.5.3. An analogous expression with the well resistor R_W in place of R_X applies for the base drive to the *npn* transistor.[11] Thus, the over-all loop gain G_L for the cross-coupled pair is

$$G_L = \beta_n \times \frac{R_X}{r_{\pi pnp} + R_X} \times \beta_p \times \frac{R_W}{r_{\pi npn} + R_W} \tag{9.4.1}$$

For a latched condition, the loop gain must equal unity; conversely, latch-up is not possible if this loop gain is less than one. This simple calculation shows that a design that avoids latch-up should reduce the bipolar transistor gains (βs) and

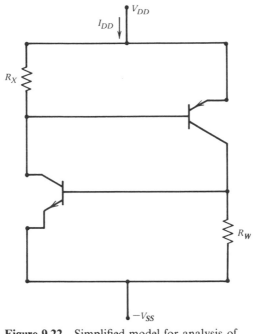

Figure 9.22 Simplified model for analysis of
gain requirements for CMOS latch-up.

also make R_X and R_W as small as possible. These basic guidelines underlie virtually
all of the ways that have been explored to design CMOS that is free of latch-up.
To prevent latch-up in very dense CMOS circuits, more complicated structures
may be needed in place of basic "bulk CMOS" design in which a single well is
diffused into the substrate, as was shown in Figure 9.18. Some techniques that
have been investigated include adding gold as a dopant or irradiating the CMOS
wafers with neutrons to reduce the lifetimes of minority carriers, thereby "spoiling"
the bipolar-transistor current gains. Closed patterns with heavy doping (*guard
rings*) are sometimes placed on the surface to collect minority carriers before they
can reach the well-to-substrate junction. Guard rings are also used to *clamp* voltages at sensitive locations. Another measure taken to avoid latch-up is to grow
epitaxial silicon on highly doped substrates so that the substrate resistance is
reduced. To make CMOS with very small dimensions, epitaxial structures and
even *twin tubs* (both $n-$ and $p-$wells formed in high-resistivity epitaxial layers) are
sometimes employed. With these advanced technologies, the tendency for latch-up
should be reduced even for submicrometer MOSFET channel lengths.

EXAMPLE Latch-up in CMOS

Use the circuit in Figure 9.21 to calculate the power-supply current I_{DD} as a
function of the current I_W in the well, the current I_X in the substrate, and the

well-substrate current source at the well junction I_0. Assume that both transistors are active, and find conditions on the transistor alpha values that would cause I_{DD} to become unbounded. Assume that any voltage changes occur slowly.

Solution

Since the voltage changes occur slowly, the capacitor C_{PS} need not be considered. By applying the Kirchhoff Current Law to the circuit of Figure 9.21, we have

$$I_{DD} = I_X + I_{PE}$$
$$I_{DD} = I_W + I_{NE}$$
$$I_W = \alpha_p I_{PE} - (1 - \alpha_n) I_{NE} + I_0$$
$$I_X = \alpha_n I_{NE} - (1 - \alpha_p) I_{PE} + I_0$$

Eliminating I_{PE} and I_{NE}, we can write

$$I_{DD} = I_X + \frac{(I_W - I_0)}{\alpha_p} + \frac{(1 - \alpha_n)}{\alpha_p} \times \left[\frac{(I_X - I_0)}{\alpha_n} + \frac{(1 - \alpha_p)}{\alpha_n} \times (I_{DD} - I_X) \right]$$

Solving this expression for I_{DD} we have

$$I_{DD} = \frac{I_0 - \alpha_p I_X - \alpha_n I_W}{I - (\alpha_n + \alpha_p)}$$

From this expression we see that I_{DD} tends toward infinity (and the circuit becomes latched) when the sum $(\alpha_n + \alpha_p)$ approaches unity. This condition on the transistor alphas can be compared to the latch-up constraint expressed through consideration of Equation 9.4.1. Equation 9.4.1 resulted from application of the small-signal equivalent circuit for the bipolar transistors, and thus expresses a condition on the circuit gain during the transient build-up to the latched condition. In this example, the steady state is considered, and the r_π resistor-divider terms are not relevant. If these terms are not considered, Equation 9.4.1 simplifies to the condition $\beta_n \times \beta_p = 1$, or

$$\left(\frac{\alpha_n}{1 - \alpha_n} \right) \left(\frac{\alpha_p}{1 - \alpha_p} \right) = 1$$

which reduces to $(\alpha_n + \alpha_p = 1)$ as concluded above.

An advanced CMOS process for very dense CMOS circuits is outlined schematically in Figure 9.23. This process utilizes both $n-$ and $p-$wells and is more complicated than that used to make the devices shown in Figure 9.18, but the complications may be justified by high performance and freedom from latch-up. Figure 9.24a shows an individual cell of a 64K bit *RAM* built using an $n-$ and p-well process on epitaxial silicon,[7] and the entire chip is shown in Figure 9.24b. Geometric, device, and circuit parameters in this *VLSI RAM* design are 2 μm

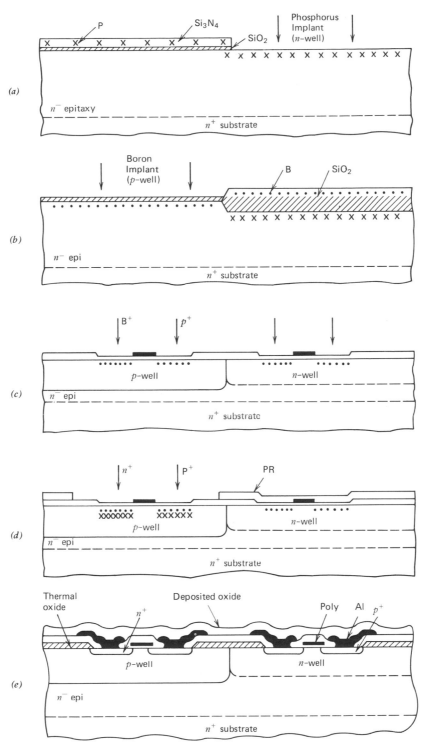

Figure 9.23 An advanced twin-well process for *VLSI* CMOS applications. The high-conductivity substrate reduces susceptibility to latch-up; the separately doped well regions provide precise control of MOSFET characteristics.[8]

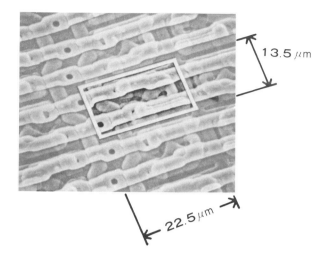

13.5 μm

22.5 μm

64K STATIC RAM

(b)

Figure 9.24 (a) Scanning electron micrograph showing a
single *RAM* cell in an advanced twin-well CMOS process.[7]
(b) Photograph of the 64K bit RAM chip with parameters
described in the text. (*Courtesy S. Masuhara, Hitachi Corporation*)

channel lengths, 2 μm-wide polycrystalline-silicon lines and spaces, 13.5 × 22.5 μm cell size, 4.8 × 7.3 mm chip size, 0.55 V threshold voltages ($V_{Tn} = -V_{Tp}$), 10 V field threshold voltages, 65 ns access time, 10 μW standby power dissipation, and 200 mW active power dissipation. The advanced CMOS characteristics just described are being explored for other *VLSI* applications in addition to high-density *RAMs*.

CMOS has become especially interesting for small-geometry devices because the removal of dissipated power in dense arrays is an increasingly difficult design challenge. Other advantages of CMOS over NMOS are (1) logical simplicity and regularity (which contributes to ease of circuit design), (2) noise immunity, and (3) steadily decreasing cost premiums of CMOS over NMOS.

Summary

In a metal-oxide-silicon field-effect transistor (MOSFET), the conductance in the *channel* between the *source–* and *drain–*electrodes can be modulated by a voltage applied to the *gate*. The MOSFET can function as an amplifier because high power in the source-drain circuit can be controlled by low power supplied to the gate-source circuit. The MOSFET can also function as a gate-controlled electronic switch that is "open" when the channel has a very low conductance, and is "closed" when the channel conductance is large. A MOSFET in which the channel can carry current between the source and the drain when the gate and source are at the same potential ($V_{GS} = 0$) is called a *depletion mode* MOSFET. If a finite value of gate voltage is necessary to induce a conducting channel, the MOSFET is called an *enhancement–mode* transistor.

The dependence of the drain current on drain voltage in a MOSFET (the output characteristic) can be divided into two regions of behavior. At low drain biases, carriers in the channel move by drift along a continuous conducting path extending from the source to the drain. At higher drain biases, the conducting path in the channel does not reach the drain so that current must flow through a high-field space-charge region near the drain. According to first-order analysis, the drain current saturates when high drain biases are applied, and this region is therefore called the *current-saturation* region. A simple derivation of the equations for MOSFET currents can be carried out by using charge-control analysis and approximating the threshold voltage in the channel region as constant. The equations derived by this method are simple to use and widely applied although they are only accurate at low drain biases. Greater accuracy in the equations for the MOSFET can be attained by using a variable depletion-charge analysis which accounts for the dependence of the depletion charge on the channel potential. Both the charge-control and the variable depletion-charge analyses predict the drain current assuming that there is a conducting path along the entire length of the channel. Hence, they are only accurate for drain voltages lower than V_{Dsat}, the voltage at which the free charge in the channel becomes zero near the drain. For $V_D > V_{Dsat}$, the current is assumed to be independent of V_D, being limited by the flow along the channel from the source to the edge of the high-field region at the drain.

Several of the parameters that describe the MOSFET are conveniently obtained by measuring drain current when the gate and drain are both tied to the same variable voltage. This technique is useful in measuring the *body effect*, which is the variation in threshold voltage that occurs when the bias between the source and the substrate is changed. The body effect becomes more influential as the substrate doping increases, but it varies inversely with the oxide capacitance. The body effect on the threshold voltage when a source-substrate bias is present is calculated using a parameter γ. Another effect of importance in practical MOSFET design is channel-length modulation which causes I_D to increase as the drain-source voltage V_{DS} is increased above V_{Dsat}. For hand calculations, channel-length modulation is usually accounted for by assuming a linear dependence on V_{DS} with a *channel-length-modulation coefficient* λ.

The transconductance g_m of a MOSFET increases linearly with drain voltage until saturation occurs, but is independent of gate voltage below saturation. Once saturation is reached, g_m becomes a linear function of gate voltage, but independent of drain voltage. The speed of response of MOSFETs, as they are presently made, is determined not by the channel transit time, but by the charging and discharging times of capacitances present in the device. A circuit model for the MOSFET consists of elements representing the dc circuit equations for the device together with associated capacitances and resistances.

Although the planar process, initially developed for bipolar-transistor circuits, is the basis of MOS circuit fabrication, a number of important modifications have been introduced to improve the performance of MOSFETs. These include the use of (100)-oriented silicon wafers to reduce oxide-charge densities, the deposition of *polysilicon* on oxide surfaces for MOSFET gates and interconnection patterns, and the use of ion implantation to adjust threshold voltages and to avoid subsurface punchthrough.

Most of the developments in MOS processing have been associated with the design of MOS memories that store *bits* of information in one of several ways. A steady and rapid growth in the bit-storage density of MOS memories has taken place through major refinements in processing, as well as through innovative circuit and device design. Information storage by charge packets in *DRAMs* makes possible the one-transistor memory cell, which is the basis of high density MOS-memory chips storing over a quarter-million bits of information. Threshold-voltage adjustment using ion implantation is a key step in the processing of *complementary* MOS (CMOS) circuits. In CMOS, both $p-$ and $n-$channel MOSFETs are fabricated on the same *IC* chip, demanding the creation of a *well* of opposite conductivity type to the substrate. Both $p-$ and $n-$well technologies exist, and there are advantages and drawbacks to each. With CMOS, digital-inverter circuits can be built that consume almost no quiescent power because only small dc leakage paths carry current from the supply voltage. This feature of CMOS is especially important at the high densities of semiconductor memory now being designed. The technological demands of CMOS production have caused its development to lag behind that of NMOS, but advances in silicon processing now make it a viable and desirable *VLSI* process. A special problem in CMOS design is to avoid *latch-up*, in which regenerative bipolar action drives the parasitic bipolar transistors (present in most CMOS processes) into saturation. With careful design, and perhaps at the expense of more complex processing, latch-up can be avoided.

The equations for MOSFET analysis and design have been collected in Table 9.2 (pages 473–474) for handy reference.

References

1. For patent enumeration, see J. T. Wallmark and H. Johnson, *Field-Effect Transistors*, Prentice-Hall, Englewood Cliffs, NJ, 1966.
2. A. S. Grove, *Physics and Technology of Semiconductor Devices*, Wiley, New York, 1967, p. 324.
3. D. A. Hodges and H. G. Jackson, *Analysis and Design of Digital Integrated Circuits*, McGraw-Hill, New York, 1983.
4. VLSI Laboratory Staff, Texas Inst. Inc., *IEEE J. Solid-State Circuits*, **SC-17**, 442 (June 1982).
5. F. M. Wanlass and C. T. Sah, *IEEE Int. Solid-State Circuits Conf.* Philadelphia, PA (Feb. 1963).
6. R. J. C. Chwang et al, *IEEE J. Solid-State Circuits*, **SC-18**, 457 (Oct. 1983).
7. O. Minato, et al, *IEEE J. Solid-State Circuits*, **SC-17**, 793 (Oct. 1982).
8. L. C. Parillo, et al. *Tech. Digest, 1980 IEEE Int. Electr. Devices Mtg.*, 752 (Dec. 1980) (figure modified).
9. C. N. Berglund, Intel Corporation.
10. A. H. Carim and A. Bhattacharyya, *Appl. Phys. Lett.*, **46**, 872 (1 May 1985)
11. K. W. Terrill, *CMOS Latch-up Modeling and Prevention*, Doctoral Thesis, Department of EECS, Univ. of California, Berkeley, Dec. 1985.

Textbooks

E. S. Yang, *Fundamentals of Semiconductor Devices*, McGraw-Hill, New York, 1978.
R. F. Pierret, *Field-Effect Devices, Volume IV of Modular Series on Solid-State Devices*, Addison-Wesley, Reading MA, 1983.

Problems

9.1 Construct a table showing the threshold voltage V_T (taking $V_S = V_B = 0$) as a function of dopant concentration for both $n-$ and $p-$channel MOSFETs. Take values of the substrate doping N_a and N_d to be 10^{15}, 10^{16}, and 10^{17} cm^{-3} and assume that there is a surface density of fixed positive charge $Q_f/q = 10^{11}$ cm^{-2} at the oxide-silicon interface in all cases. The silicon dioxide is 100 nm thick and all gates are made of aluminum so that $\Phi_M - X = 0.05$ eV. Indicate on the table whether the MOSFET is a depletion-mode or an enhancement-mode device.

9.2   (a) Fill in the steps to justify Equation 9.1.5.
      (b) Derive Equation 9.1.12.

9.3* A MOSFET for which $W/L = 5$, the gate-oxide thickness is 80 nm, and the channel mobility $\mu_n = 600$ cm^2 V^{-1} s^{-1} is to be used as a controlled resistor.
      (a) Calculate the free-electron density in the channel Q_n/q that is required for the MOSFET to present a resistance of 2.5k Ω between the source and the drain at low values of V_{DS}.
      (b) Calculate the gate voltage in excess of the threshold voltage needed to produce the desired resistance under the conditions of part a.

9.4   (a) Show that Equation 9.1.15 approaches Equation 9.1.16 as V_D approaches 0 (with $V_S = V_B = 0$).
      (b) Explain in one or two sentences why the result found in part a is expected.

9.5   (a) Verify Equation 9.1.8 under the assumptions in the text, and derive Equation 9.1.9.
      (b) Show that the maxima in the downward-facing parabolas of Figure 9.5a occur at $V_{DS} = (V_{GS} - V_T)$, and that the maxima are connected by an upward-facing parabola as shown on Figure 9.5b.

9.6   Repeat the comparison between $I_{Dsat\ 2}$ and $I_{Dsat\ 3}$ considered in the example at the end of Section 9.1 using a MOSFET for which the oxide thickness is 50 nm, the substrate doping $N_a = 2 \times 10^{15}$ cm^{-3}, the flat-band voltage $V_{FB} = -0.2$ V, and $\mu_n W/L = 5 \times 10^3$ cm^2 V^{-1} s^{-1}. Consider that $V_{GS} = 5.5, 4.5, 3.5$, and 2.5 V.

9.7† (a) Use the example comparing I_{Dsat} values carried out at the end of Section 9.1 to check whether $V_{Dsat\ 2}$ and $V_{Dsat\ 3}$ can be related by an empirical constant.
      (b) Explore further the generality of this result by considering the MOSFET described in Problem 9.6.

9.8* Consider further the n–channel MOSFET as connected in the example at the beginning of Section 9.2. Calculate the current I_D if $V_{SS} = 2, 3, 4$ and 4.5 V. (Note that the MOSFET is *not* saturated for all of these voltages.)

9.9† Repeat Problem 9.8 with $V_{SS} = 6, 8$, and 10 V. (Note that the role of source and drain becomes interchanged under these bias conditions and that $V_{GD} = 0$ V).

9.10* A series of measurements made on an n–channel MOSFET are given in the accompanying table.

V_{GS} (V)	V_{DS} (V)	V_{SB} (V)	I_D (μA)
3	4	0	120
3	6	0	130
3	4	4	76.8
4	4	0	270

Choose parameters to represent the MOSFET if it is modeled using Equations 9.2.11 and 9.2.3. Assume that $2|\phi_p| = 0.6$ V (note that ϕ_p is only a weak function of N_a).

9.11   The circuit shown in Figure P9.11 is an enhancement-load inverter. Consider that both transistors are described by Equations 9.1.17 and 9.1.18 with $k = 40 \times 10^{-6}$ AV^{-2} and $V_T = 2$ V. Take the supply voltage $V_{DD} = 8$ V and ignore body effect and channel-length modulation (that is, take γ and λ to be zero). Note that the output voltage V_o cannot exceed $(V_{DD} - V_T)$ because no current can flow in the upper (load) device unless V_o is below this value.

(a) Construct a set of output characteristics (I_D versus V_{DS}) for the lower MOSFET with gate voltages equal to 0, 2, 4 and 6 V, if the drain-voltage is varied from 0 to 8 V.

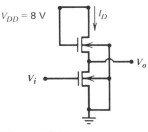

$V_{DD} = 8$ V $\quad I_D$

V_o

V_i

Figure P9.11

(b) On the characteristics of part a, plot the *load line* for the circuit; that is, draw a curve connecting the values of V_o (which is V_{DS} for the lower MOSFET) that correspond to each of the input gate voltages.

(c) Repeat part b if the upper transistor is replaced by a 20k Ω resistor.

9.12* The circuit shown in Figure P9.12 is a depletion-load inverter. Consider that the lower MOSFET (the enhancement-mode transistor for which we use subscripts E) is described by Equations 9.1.17 and 9.1.18 with $k_E = 50 \times 10^{-6}$ AV^{-2} and $V_{TE} = 1$ V. Use Equations 9.1.17 and 9.1.18 to describe the upper (depletion-mode) MOSFET, but take $k_D = 10 \times 10^{-6}$ AV^{-2} and $V_{TD} = -3$ V. In contrast to the inverter of Problem 9.11, the output voltage for this inverter can reach the supply voltage $V_{DD} = 5$ V because the load MOSFET conducts when $V_{GS} = 0$ V.

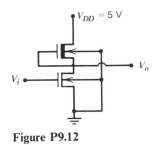

$V_{DD} = 5$ V

V_o

V_i

Figure P9.12

(a) If the input $V_i = 5$ V, calculate the output voltage V_o. (In this part, do not consider body effect or channel-length modulation.)

(b) For this part, assume that the body-effect parameter γ is 0.4 V$^{1/2}$ and $|\phi_p| = 0.3$ V Calculate the threshold voltage of the load device when V_o is at its maximum value.

(c) Use the γ given in part b and repeat the calculation called for in part a.

9.13 In the circuit shown in Figure P9.13, the MOSFET is described by Equations 9.1.17 and 9.1.18 with $k' = 25 \times 10^{-6}$ AV^{-2}, $V_T = 1$ V, and $W/L = 2$. Voltage V_i is varied from 0 to 4 V.

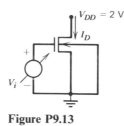

Figure P9.13

(a) Make a careful plot of $\sqrt{I_D}$ as a function of V_i showing any break points on the curve.

(b) Make a plot of the MOSFET transconductance using a solid line.

(c) On the plot of part b, use a dotted line to indicate a curve of the output conductance ($g_D = \partial I_D / \partial V_{DS}$).

9.14 Consider that a bipolar transistor and a MOSFET are being evaluated for use in a linear amplifier. The quiescent current in the device is to be 1 mA. What would the ratio between the transconductances of the two devices be if ($V_G - V_T$) for the MOSFET were 1 V? Use Equations 9.1.17 and 9.1.18 to describe the MOSFET. [The superior g_m is obtained in the bipolar transistor at any output current.]

9.15 In a typical MOSFET *IC* there are at least three different MOS structures present, corresponding to the MOSFET polysilicon gates, the interconnection polysilicon lines, and the metal interconnection lines (Figure P9.15). Consider a *p*–channel silicon-gate technology with a 5 Ω-cm substrate, $Q_f/q = 5 \times 10^{10}$ cm^{-2}, and the oxide thicknesses $x_{ox} = 80$ nm, 0.75 μm and 1.5 μm for the MOSFET, polysilicon line, and aluminum metal line, respectively.

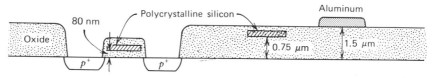

Figure P9.15

5 Ω–cm *n*–type silicon

Find the threshold voltages of the three MOS structures. Assume that the energy-band structure of the polysilicon is the same as that of similarly doped single-crystal silicon and that *p*–type polysilicon lines are heavily doped so that $E_f = E_v$.

9.16 A *p*–channel MOSFET with a heavily doped *p*–type polysilicon gate has a threshold voltage of -1.5 V with $V_{SB} = 0$ V. When a 5 V reverse bias is applied to the substrate, the threshold voltage changes to -2.3 V.

(a) What is the dopant concentration in the substrate if the oxide thickness is 100 nm?

(b) What is the threshold voltage if V_{SB} is -2.5 V?

(Hint: Note that ϕ_n changes only slowly with N_d).

9.17* An *n*–channel MOSFET used as a test structure to characterize a process has the following properties. The oxide thickness $x_{ox} = 120$ nm, $Q_f/q = 5 \times 10^{10}$ cm^{-2} and the substrate is 1 Ω-cm material. The mask gate length is 10 μm and the source and drain diffusions extend laterally 0.75 μm into the channel. The gate is heavily doped with phosphorus and $W/L = 1$.

(a) Calculate the flat-band voltage V_{FB} and the threshold voltage V_T.

(b) For the test transistor biased with $V_G = 5$ V and $V_{DS} = 0.4$ V, I_D is measured to be 18.7 μA. Calculate the channel mobility in this transistor if it has a mask gate length of 10 μm.

9.18 (a) Use Equation 9.2.7 to derive equations for $V_c(y)$ and $\mathscr{E}_y(y)$. Note that I_D is constant along y and take $V_S = V_B = 0$ V.

(b) From the expressions derived, obtain equations for $V_c(y)$ and $\mathscr{E}_y(y)$ that are valid when the MOSFET is at the edge of saturation.

9.19 (a) Using the results of Problem 9.18, sketch $\mathscr{E}_y$ as a function of y over the range $y = 0$ to $y = L$ for $V_G > V_T$ and $V_D = 0.25, 0.50, 0.75,$ and 1 times V_{Dsat}.

(b) Explain why $\mathscr{E}_y$ approaches a constant value when V_D is small.

9.20 Carry through the steps to derive Equations 9.2.8 and 9.2.10.

9.21† Obtain the expression for the channel transit time in the MOSFET (Equation 9.2.10) in a second way. Use Equation 9.1.1 to write

$$T_{tr} = -\frac{Q_N}{I_D}$$

where Q_N is the total channel calculated from the expression $Q_N = W \int_0^L Q_n(y)\,dy$.

9.22† From the expression obtained for Q_N in Problem 9.21, show that the small-signal capacitance between the gate and the source under saturation conditions is given by

$$C_{GS} \equiv \left|\frac{\partial Q_N}{\partial V_{GS}}\right| = \frac{2}{3} C_{ox} WL$$

9.23† Use the result derived in Problem 9.22 together with Equation 9.1.11 to prove that $T_{tr} = 2C_{GS}/g_{msat}$. [Note that (except for a factor 2) a similar result was found between τ_F and the diffusion capacitance in the bipolar transistor (Equation 7.5.4). The functional dependence between input capacitance and transit time is a basic charge-control result that applies generally to three terminal amplifiers.]

9.24 A MOSFET has a gate that is made of a resistive metal (nichrome) with contacts placed above both ends of the channel (Figure P9.24). The voltages at the two ends of the gate are V_{G0} and V_{G1}, respectively. Assume that the flat-band voltage is constant and take the oxide to be x_{ox} units thick. Using the depletion approximation, set up the differential equation having a solution that relates the drain current below saturation to the applied voltages. *Do NOT solve the differential equation.*

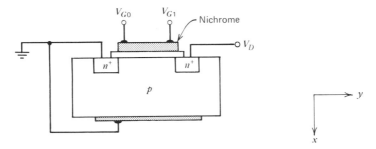

Figure P9.24

9.25[†] Consider the CMOS inverter shown in Figure 9.17a.

(a) Copy the voltage-transfer characteristic (VTC) sketched in Figure 9.17c and indicate on your copy the state of each MOSFET as V_i is changed. For example, at V_i near zero, the p–channel MOSFET is ohmic and the n–channel MOSFET is cut off. Indicate all points on the VTC where a MOSFET changes its conduction state.

(b) Calculate the voltage at all points indicated in part a if both MOSFETs are characterized by Equations 9.1.17 and 9.1.18 with the following parameters. For the n–channel MOSFET: $k_n = 40~\mu A~V^{-2}$ and $V_{Tn} = 1$ V. For the p–channel MOSFET: $k_p = 35~\mu A~V^{-2}$ and $V_{Tp} = -1$ V. For both MOSFETs, take $\lambda = \gamma = 0$. The supply voltage $V_{DD} = 5$ V.

9.26[†] Use Figure 9.22 to prove that the current (I_{DD}) that the voltage source V_{DD} must supply to keep the CMOS circuit in a latched condition is

$$I_{DD} = \frac{(V_{BE}/R_W)\beta_n(\beta_p + 1) + (V_{BE}/R_X)\beta_p(\beta_n + 1)}{(\beta_n\beta_p - 1)}$$

(This example shows that a way to keep a CMOS circuit from latch-up is to limit the current supplied to it.)

Table 9.2 **MOSFET Equations**

	n-channel	p-channel
Depletion charge density at threshold (Equation 8.3.9)	$Q_d = -qN_a x_{dmax} = -\sqrt{2\epsilon_s q N_a (2\vert\phi_p\vert + \vert V_S - V_B\vert)}$	$Q_d = +qN_{dmax} x_{dmax} = +\sqrt{2\epsilon_s q N_d (2\vert\phi_n\vert + \vert V_S - V_B\vert)}$
Flatband voltage (Equation 8.4.6)	$V_{FB} = \Phi_{MS} - \dfrac{Q_f}{C_{ox}} - \dfrac{1}{C_{ox}}\displaystyle\int_0^{x_{ox}} \dfrac{x\rho(x)\,dx}{x_{ox}}$	
Threshold voltage (Equation 8.3.18)	$V_T = V_{FB} + V_S + 2\vert\phi_p\vert + \dfrac{\vert Q_d\vert}{C_{ox}}$	$V_T = V_{FB} - V_S - 2\vert\phi_n\vert - \dfrac{\vert Q_d\vert}{C_{ox}}$
Charge-control (first level) theory Current equations below saturation (Equation 9.1.9)	$I_{D1} = \mu_n \dfrac{W}{L} C_{ox}\left[(V_G - V_T)V_{DS} - \dfrac{V_{DS}^2}{2}\right]$	$I_{D1} = -\mu_p \dfrac{W}{L} C_{ox}\left[(V_G - V_T)V_{DS} - \dfrac{V_{DS}^2}{2}\right]$
Saturation voltage (Equation 9.1.10)	$V_{Dsat\,1} = (V_G - V_T) - V_S$	
Current equations above saturation (Equation 9.1.17)	$I_D = \dfrac{k}{2}(V_G - V_T)^2$	$I_D = -\dfrac{k}{2}(V_G - V_T)^2$
Current equations below saturation (Equation 9.1.18)	$I_{D3} = k\left[(V_G - V_T)V_{DS} - \dfrac{V_{DS}^2}{2}\right]$	$I_{D3} = -k\left[(V_G - V_T)V_{DS} - \dfrac{V_{DS}^2}{2}\right]$

$[k$ proportional to $\mu C_{ox}(W/L)]$

Variable Depletion-Charge Analysis Current equations below saturation (Equation 9.1.12)

$$I_{DS} = \mu n \frac{W}{L}\left\{ C_{ox}\left[V_G - V_{FB} - 2|\phi_p| - \left(\frac{V_D}{2}\right) - \left(\frac{V_S}{2}\right)\right]V_{DS} \right.$$
$$\left. - \frac{2}{3}\sqrt{2\epsilon_s q N_a}\left([2|\phi_p| + |V_D - V_B|]^{3/2} - (2|\phi_p| + |V_S - V_B|)^{3/2}\right)\right\}$$

$$I_D = -\mu_p \frac{W}{L}\left\{ C_{ox}\left[V_G - V_{FB} + 2|\phi_n| - \left(\frac{V_D}{2}\right) - \left(\frac{V_S}{2}\right)\right]V_{DS} \right.$$
$$\left. - \frac{2}{3}\sqrt{2\epsilon_s q N_d}\left([2|\phi_n| + |V_D - V_B|]^{3/2} - (2|\phi_n| + |V_S - V_B|)^{3/2}\right)\right\}$$

Saturation Voltage (Equation 9.1.14)

$$V_{Dsat\,2} = V_G - V_{FB} - 2|\phi_p| - \frac{\epsilon_s q N_a}{C_{ox}^2}\left[\sqrt{1 + \frac{2C_{ox}^2}{\epsilon_s q N_a}(V_G - V_{FB} - V_B)} - 1\right]$$

$$V_{Dsat\,2} = V_G - V_{FB} + 2|\phi_n| + \frac{\epsilon_s q N_d}{C_{ox}^2}\left[\sqrt{1 + \frac{2C_{ox}^2|V_G - V_{FB} - V_B|}{\epsilon_s q N_d}} - 1\right]$$

Transconductance in saturation (Simple Theory) (Equation 9.2.6)

$$g_{msat} = 2k_n' \frac{W}{L}(V_G - V_T)$$

$$g_{msat} = -2k_p' \frac{W}{L}(V_G - V_T)$$

Body-effect (Equation 9.2.1)

$$\Delta V_{Tn} = \gamma\left[\sqrt{2|\phi_p| + |V_{SB}|} - \sqrt{2|\phi_p|}\right]$$

$$\Delta V_{Tp} = -\gamma\left[\sqrt{2|\phi_n| + |V_{SB}|} - \sqrt{2|\phi_n|}\right]$$

Channel-length Modulation (Equation 9.2.11)

$$I_{Dsat} = \frac{k}{2}(V_G - V_T)^2(1 + \lambda V_{DS})$$

$$I_{Dsat} = -\frac{k}{2}(V_G - V_T)^2(1 - \lambda V_{DS})$$

10

MOS FIELD-EFFECT TRANSISTORS II: LIMITATIONS AND PERSPECTIVES

In Chapter 9 the basic theory of MOSFET operation was described and equations that are frequently used in circuit design were derived. This theory contains a number of approximations that limit its accuracy. It also fails to account for several important physical effects occurring in MOSFETs. The basic theory was first derived when MOSFETs having channel lengths measured in tens of micrometers were state-of-the-art. Because MOSFET channel lengths are now approaching one micrometer, the limitations of the basic theory have become a great deal more noticeable and troublesome. In this chapter, therefore, we discuss a more detailed theory for MOSFET operation, and we also describe several influential physical phenomena that should be understood in the design of MOS VLSI systems.

One of the most useful simplifications of basic MOSFET theory is to neglect all free charge in the channel until the magnitude of the gate voltage exceeds the threshold voltage and thereafter to treat the depletion charge as constant. This simplification is useful because the free-charge densities in the channel change exponentially with the channel voltage while the fixed-charge densities change only as a fractional power of the voltage. Near the threshold voltage, however, currents are not well specified by the simple theory, as can be observed by careful measurements. The major consideration is current flowing at gate voltages smaller in magnitude than the threshold voltage, so-called *subthreshold current*. A theory that accounts for subthreshold current is not difficult; it is derived from a full accounting of both fixed and free charge near the silicon-silicon dioxide interface in Poisson's equation. The theory does not yield closed-form results, however, and we avoid presenting it both because the important behavior can be understood without it* and also because other effects are becoming more important than subthreshold current in very small channel MOSFETs.

Another important limitation to the basic theory of Chapter 9 is in the formulation used for the velocity of charge in the MOSFET channel. In deriving the differential equation for the MOSFET (Equation 9.1.5), we approximated the channel current as being carried entirely by drift with a constant mobility. Although in most cases transport along the channel is well represented by the drift process, the assumption of constant mobility is not justified unless channel fields are low. These low fields are generally found for normal bias conditions in MOSFETs having channels longer than about 15 μm. For this reason, the basic theory given in Section 9.1 is sometimes called the *long-channel theory*.

If the channel length is reduced, however, the transit time is shortened (cf. Equation 9.1.2) and several other MOSFET parameters also improve. This is one of the reasons that MOSFET channels have been reduced continually and are now approaching 1 μm for some commercial processes. As the channels have become shorter, however, the fields in the channel region have increased, and the assumption of constant mobility has become a poor approximation for MOSFET modeling.

Another assumption of basic MOSFET theory that becomes more questionable as device dimensions are reduced is the *gradual-channel approximation*. In this approximation, the quantity of charge in the channel is assumed to be controlled completely by the gate electrode, that is, by the field perpendicular to the silicon-silicon dioxide interface. In small MOSFETs, it is not possible to neglect the influence of the drain and source junctions on the quantity of charge in the channel.

It is also important to consider the effects of the high fields to which small MOSFETs are subjected. These fields can be responsible for *hot-carrier effects* that are reasonably neglected in larger devices, but important in small MOSFETs. Keeping the electric fields from becoming excessive is one of the goals of *device scaling*, for which specific rules have been developed. In this chapter we discuss MOS scaling techniques and some of their limitations.

The complexity of the small-geometry effects makes computer modeling almost imperative for modern *IC* MOSFET design. A brief discussion of numerical device modeling illustrates the insight into device operation provided by this technique.

The concluding section of this chapter describes the application of ion implantation to MOSFET technology and discusses some of the features of depletion-

---

* Fuller discussion is given in more advanced books such as S. M. Sze, *Physics of Semiconductor Devices, 2nd Edition* Wiley-Interscience, New York, 1981.

mode MOSFETs. As in Chapter 9, most of our discussion is in terms of n–channel devices. Except for special phenomena (such as hot-electron behavior), the extension of the material to p–channel MOSFETs is straightforward.

10.1 Subthreshold Current

The theories for the drain current that we have considered thus far approximate the channel conductance by assuming that the channel does not contain any free carriers unless the gate voltage exceeds the threshold voltage. These theories therefore fail to predict the *subthreshold currents* arising from the inversion charge that exists at gate voltages below the conventional threshold voltage. An approximate theory to indicate the behavior of currents in this region can be developed by considering the case of a low drain-source bias and taking ϕ_s to be close to $-\phi_p$. Under this condition, Poisson's equation can be written with terms expressing the inversion charge as well as the fixed acceptor charge. Its form is

$$\frac{d^2\phi}{dx^2} = -\frac{\rho}{\epsilon_s} \approx \frac{q}{\epsilon_s}(N_a + n) = \frac{qN_a}{\epsilon_s}\left[1 + \exp\left(\frac{q(\phi - |\phi_p|)}{kT}\right)\right] \qquad (10.1.1)$$

Multiplying Equation 10.1.1 by $d\phi$ and noting that $\mathscr{E}(d\mathscr{E}/d\phi) = d^2\phi/dx^2$, we obtain an expression for the field in the silicon near the Si–SiO$_2$ interface

$$\mathscr{E}_s = \left\{\frac{2qN_a}{\epsilon_s}\left[\phi_s + |\phi_p| + \frac{kT}{q}\left(\exp\frac{q(\phi_s - |\phi_p|)}{kT} - \exp\frac{-2q|\phi_p|}{kT}\right)\right]\right\}^{1/2} \qquad (10.1.2)$$

As in Section 8.3, Equation 10.1.2 can be used to relate the induced semiconductor charge Q_s to the surface field $\mathscr{E}_s$ through Gauss' law. The free-charge density Q_n is then obtained by subtracting the fixed depletion-layer charge $Q_d = -\sqrt{2q\epsilon_s N_a(\phi_s + |\phi_p|)}$ from Q_s. When this procedure is carried out, and approximations suitable to the range of surface potential under consideration are made, the expression for Q_n is

$$Q_n \approx -\frac{kT}{2q}\sqrt{\frac{2q\epsilon_s N_a}{\phi_s + |\phi_p|}}\exp\left[\frac{q(\phi_s - |\phi_p|)}{kT}\right] \qquad (10.1.3)$$

which indicates that just below threshold, Q_n varies nearly exponentially with the surface potential. The procedures carried out in Section 8.3 can again be employed to express Q_n as a function of gate, channel, and threshold voltages. The result is

$$Q_n \approx -\frac{kT}{2q}\sqrt{\frac{q\epsilon_s N_a}{|\phi_p|}}\exp\left[\frac{q[V_G - V_T - (V_C(y) - V_S)]}{\xi kT}\right] \qquad (10.1.4)$$

where

$$\xi = 1 + \frac{1}{2C_{ox}}\sqrt{\frac{q\epsilon_s N_a}{|\phi_p|}}$$

The subthreshold region is characterized by low free-carrier densities, so low that diffusion currents (proportional to the carrier-density gradient) are of far greater importance than drift currents (proportional to the density itself). A more complete analysis in terms of quasi-Fermi levels shows, in fact, that drift can be neglected and diffusion alone considered as being responsible for drain current in the sub-threshold regime. Since diffusion current is constant along the channel, the gradient of Q_n is also constant and, by applying Equation 10.1.4 at the source and drain, this gradient can be calculated to be

$$\frac{dQ_n}{dy} = \frac{Q_n(V_S) - Q_n(V_D)}{L} \approx \frac{Q_n(V_S)}{L}\left[1 - \exp\left(\frac{-qV_{DS}}{\xi kT}\right)\right] \quad (10.1.5)$$

The subthreshold current I_{Dst} is given by

$$I_{Dst} = \frac{\mu_{no}kTWQ_n(V_S)}{qL}\left[1 - \exp\left(\frac{-qV_{DS}}{\xi kT}\right)\right] \quad (10.1.6)$$

where we have used the Einstein relation to express the diffusion constant in terms of the mobility.

Equation 10.1.6 predicts that subthreshold current should be substantially in-dependent of V_{DS} (for $V_{DS} > \xi kT/q$) and to depend exponentially on V_G (through Q_n). The slight dependence on V_{DS} reflects the insensitivity of the density gradient to V_{DS} whenever $V_{DS} \gg \xi kT/q$.

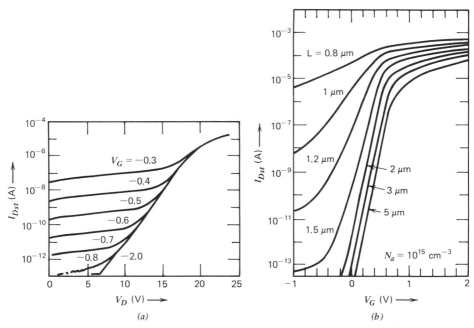

Figure 10.1 (a) Experimental low-current characteristics for a MOSFET with $L = 2.1\ \mu m$, $V_S = V_B = 0$ V. For lower values of V_D, I_{Dst} is reduced by roughly one order of magnitude for each 0.1 V reduction in V_G below the threshold voltage. At higher values of V_D, a subsurface current flows that is independent of V_G.[2] (b) Calculated behavior of I_{Dst} versus V_G for various channel lengths, $x_j = 0.33\ \mu m$, $x_{ox} = 50$ nm, $V_D = 2$ V, and $V_B = 0$ V.[20]

The behavior predicted by Equation 10.1.6 is observed in longer channel MOSFETs, as shown in Figure 10.1a.[2] As the channel length is reduced, however, variations from the subthreshold-current theory considered thus far are observed. These variations are illustrated in the calculated plots of subthreshold current as a function of gate bias for various channel lengths shown in Figure 10.1b. First, the I_{Dst} curves are displaced from each other because of a change in threshold voltage (to be discussed in Section 10.3). In addition, at channel lengths below 2 μm markedly different subthreshold characteristics are observed. Most significantly, at these short channel lengths the current becomes exponentially dependent on drain voltage instead of being independent of V_{DS}. This behavior is similar to punchthrough (discussed in Section 10.3), in which V_{DS} modulates the barrier to minority-carrier injection, except that here the barrier modulation occurs along the surface. Analyses for this region have been carried out, and the effect is sometimes called *drain-induced barrier lowering (DIBL)*.[3]

If drain-induced barrier lowering is avoided so that Equation 10.1.6 represents the subthreshold-current regime, we see that I_{Dst} decreases exponentially as V_G is reduced below V_T. Hence, the circuit designer can readily calculate the gate bias required to insure a given allowable subthreshold leakage current. Typically, designers of digital circuits specify that V_G should be about 0.5 V below V_T to minimize the subthreshold current.

EXAMPLE **Subthreshold Current**

What subthreshold current is predicted for an n–channel MOSFET in which $V_{DS} = 1$ V, $N_a = 4 \times 10^{15}$ cm^{-3}, $L = 5$ μm, $W = 50$ μm, $x_{ox} = 50$ nm, $\mu_{no} = 760$ cm^2 V^{-1} s^{-1}, and $(V_G - V_T) = -0.1$ V?

Solution

From the data given, we calculate

$$C_{ox} = 6.9 \times 10^{-8} \text{ F cm}^{-2}, \qquad \phi_p = 0.323 \text{ V}$$
$$\xi = 1.328 \text{ (Equation 10.1.4)}$$

From Equation 10.1.4

$$Q_n(V_S) = -3.16 \times 10^{-11} \text{ C cm}^{-2}$$
$$Q_n(V_D) \text{ is negligible compared to } Q_n(V_S)$$

Therefore, from Equation 10.1.6

$$I_{Dst} = 6.19 \times 10^{-9} \text{ A} \quad \text{or} \quad 6.19 \text{ nA}$$

Since I_{Dst} decreases exponentially as V_G is further reduced below V_T, we see that several hundred mV "off bias" typically reduces the subthreshold current to negligible values.

10.2 Channel Velocity Limitations

The free-charge density Q_n moves in the channel under the influence of $\mathscr{E}_y$, the field component tangential to the silicon-silicon dioxide interface. At the same time, Q_n is held near the interface by a field component $\mathscr{E}_x$ perpendicular to it. Both $\mathscr{E}_x$ and $\mathscr{E}_y$ depend on position, and both field components influence the velocity of the moving carriers, but their effects can be modeled separately.

Tangential Field. The dependence of velocity on the tangential field $\mathscr{E}_y$ can significantly affect MOSFET electronics. Experiments have shown that velocities near the surface tend to saturate at higher fields, as shown in Figure 10.2. This behavior is qualitatively similar to the velocity dependence in the bulk (Figure 1.17) and can be modeled for either n– or p–channel devices by the expression

$$v = \frac{|\mu_o \mathscr{E}_y|}{(1 + |\mathscr{E}_y/\mathscr{E}_c|^{\alpha})^{1/\alpha}} \qquad (10.2.1)$$

where v represents the magnitude of the velocity along the channel and the parameters μ_o, $\mathscr{E}_c$ and α are determined empirically. As an example, a good match to experimental data for a typical n–channel, (100)-silicon MOSFET process is ob-

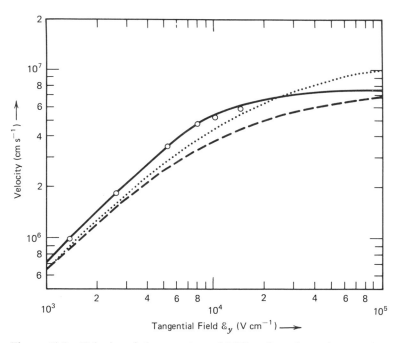

Figure 10.2 Velocity of electrons in an MOS surface channel as a function of the tangential electric field $\mathscr{E}_y$. The data (circles) are from reference 15. The solid curve is a plot of Equation 10.2.1 with $\alpha = 2$, $\mu_{no} = 710$ cm^2 V^{-1} s^{-1}, and $\mathscr{E}_c = -1.1 \times 10^4$ V cm^{-1}; the dotted curve uses the same values except for α, which is taken to be unity. For the dashed curve, $\mu_{no} = 710$ cm^2 V^{-1} s^{-1}, $\alpha = 1$, and $\mathscr{E}_c$ is -1.7×10^4 V cm$^{-1.5}$.

tained with $\mu_o = \mu_{no} = 710 \text{ cm}^2 \text{ V}^{-1} \text{ s}^{-1}$, $\alpha = 2$, and $\mathcal{E}_c = -1.1 \times 10^4 \text{ V cm}^{-1}$. (Note that for an n–channel MOSFET, both $\mathcal{E}_y$ and $\mathcal{E}_c$ are negative.)

To understand the major effects of velocity saturation,[4] we consider a simplified form for Equation 10.2.1 in which we assume an α of unity. A value of 1 for α fits the measured data of p–channel MOSFETs quite well, but is less accurate for n–channel devices. By taking $\alpha = 1$, however, we will be able to obtain closed-form solutions for the I_D–V_D characteristics of the MOSFET. We therefore make this assumption in the derivations that follow. As shown on Figure 10.2, with $\alpha = 1$ we can match the asymptotes of the experimental v-$\mathcal{E}_y$ curve for n–channel conduction at low and high $\mathcal{E}_y$ values, but we generally underestimate v in the intermediate field region (cf. dashed curve in Figure 10.2). Using Equation 10.2.1 with $\alpha = 1$ for an n–channel device, and rewriting Equation 9.1.5 (noting that $\mathcal{E}_y = -dV_C/dy$), we have

$$I_D = \frac{\mu_{no}\mathcal{E}_y W Q_n(y)}{[1 + (\mathcal{E}_y/\mathcal{E}_c)]} \tag{10.2.2}$$

where Q_n, $\mathcal{E}_y$, and $\mathcal{E}_c$ are all negative for an n–channel MOSFET. We rewrite Equation 10.2.2 as

$$I_D = \left[Q_n(y) - \frac{I_D}{\mu_{no}W\mathcal{E}_c} \right] \mu_{no}W\mathcal{E}_y \tag{10.2.3}$$

By integrating Equation 10.2.3 along the channel (Problem 10.9), we derive

$$I_{D4} = \frac{-\mu_{no}W \int_{V_S}^{V_D} Q_n \, dV_C}{L[1 + |V_{DS}/\mathcal{E}_c L|]} \tag{10.2.4}$$

where the absolute value is taken because $\mathcal{E}_c$ is negative. As before, we have added a subscript to I_D to distinguish Equation 10.2.4 from previous derivations. Comparing Equation 10.2.4 to Equation 9.1.6, we see that velocity saturation is equivalent to transport with constant mobility along a channel that is *longer than the built-in channel length L* by $|V_{DS}/\mathcal{E}_c|$.

We can use Equation 10.2.4 to consider I_{Dsat} and V_{Dsat} when velocity saturation occurs in the channel. Proceeding as in the distributed analysis, we differentiate Equation 10.2.4 to obtain the small-signal drain conductance for the MOSFET in the nonsaturated bias condition.

$$\frac{dI_D}{dV_D} = \frac{-\mu_{no}WQ_n(V_D)}{L(1 + |V_{DS}/\mathcal{E}_c L|)} + \frac{\mu_{no}W \int_{V_S}^{V_D} Q_n \, dV_C}{\mathcal{E}_c L^2 (1 + |V_{DS}/\mathcal{E}_c L|)^2} \tag{10.2.5}$$

If we set $dI_D/dV_D = 0^*$ and use Equation 10.2.4, we obtain the following relationship between $I_{Dsat\,4}$, the current at $V_{Dsat\,4}$, and $Q_n(V_{Dsat\,4})$, the free-charge density at the drain when $I_{Dsat\,4}$ is flowing.

---

* Saturation has also been defined as occurring when the velocity at the drain reaches a high value (for example, 90% of its limit), a definition that emphasizes the observed nonzero output conductance in saturation. The zero-conductance definition that we take is, however, mathematically simple, intuitive, and consistent with our earlier MOSFET analysis.

$$Q_n(V_{\text{Dsat 4}}) = \frac{I_{\text{Dsat 4}}}{W\mu_{no}\mathscr{E}_c} \tag{10.2.6}$$

If we use Equation 9.1.8 for Q_n at $V_C = V_{\text{Dsat 4}}$, we obtain

$$I_{\text{Dsat 4}} = -\mu_{no}\mathscr{E}_c W C_{ox}[V_G - V_T - (V_{\text{Dsat 4}} - V_S)] \tag{10.2.7}$$

To derive a general equation for the drain current, we use Equation 9.1.8 to specify $Q_n(V_C)$ in Equation 10.2.4 and integrate the charge along the channel. The result is

$$I_{D4} = \frac{\mu_{no}C_{ox}W}{L(1 + |V_{DS}/\mathscr{E}_c L|)}\left[(V_G - V_T)V_{DS} - \frac{V_{DS}^2}{2}\right] \tag{10.2.8}$$

which is identical to Equation 9.1.9 if the channel length L in that equation is lengthened by $|V_{DS}/\mathscr{E}_c|$.

Since Equation 10.2.8 is valid for $0 < V_D < V_{\text{Dsat}}$, an expression for $V_{\text{Dsat 4}}$ can be derived by equating Equation 10.2.7 and Equation 10.2.8 at $V_D = V_{\text{Dsat 4}}$. To simplify the algebra, we take $V_S = 0$ and find that

$$V_{\text{Dsat 4}} = |\mathscr{E}_c L|\left[\left(1 + \frac{2(V_G - V_T)}{|\mathscr{E}_c L|}\right)^{1/2} - 1\right] \tag{10.2.9}$$

Comparing Equation 10.2.9 to the charge-control result $V_{\text{Dsat 1}} = (V_G - V_T)$, we see that velocity saturation causes V_{Dsat} to vary less than linearly with $(V_G - V_T)$. Using Equation 10.2.9 in Equation 10.2.7 with $V_S = 0$, we find

$$I_{\text{Dsat 4}} = \frac{\mu_{no}C_{ox}W(V_{\text{Dsat 4}})^2}{2L} \tag{10.2.10}$$

which is similar to the form of Equation 9.1.11 obtained for the charge-control theory with constant mobility. Hence, in the presence of velocity saturation, the current in the saturation regime still varies quadratically with the saturation voltage, but, from Equation 10.2.9 the saturation voltage is no longer linearly dependent on the gate voltage.

EXAMPLE   Comparison of V_{Dsat} Values

Show that the expression for $V_{\text{Dsat 4}}$ (Equation 10.2.9) approaches that for $V_{\text{Dsat 1}}$ (Equation 9.1.10) if velocity-limitation effects are not significant.

Solution

The absence of a velocity limitation is equivalent to letting $|\mathscr{E}_c| \to \infty$ and $\alpha = 1$ in Equation 10.2.1. For these conditions, the second term in the binomial expression under the square-root sign of Equation 10.2.9 is much less than unity, and the square root can be expanded as

$$\left[1 + \frac{2(V_G - V_T)}{|\mathcal{E}_c L|}\right]^{1/2} \approx 1 + \frac{(V_G - V_T)}{|\mathcal{E}_c L|} - \cdots$$

so that $V_{Dsat\ 4}$ approaches $(V_G - V_T)$, as derived for $V_{Dsat\ 1}$ in Equation 9.1.10.

Our analysis shows that velocity saturation reduces both the saturation voltage (Equation 10.2.9) and the saturation current (Equation 10.2.10) in a MOSFET below the values predicted using constant-mobility theory. The extreme case of velocity saturation can be examined from the theory developed by allowing the critical field $|\mathcal{E}_c|$ to become small, while holding the product $|\mu_{no}\mathcal{E}_c|$ constant at a fixed value v_{sat}. From Equation 10.2.1, this condition gives rise to a constant channel velocity, independent of V_{DS}. From Equation 10.2.9, $V_{Dsat\ 4}$ then approaches $\sqrt{2(V_G - V_T)|\mathcal{E}_c L|}$, and hence, from Equation 10.2.10, $I_{Dsat\ 4}$ becomes linearly proportional to $(V_G - V_T)$ instead of varying quadratically with this "effective gate voltage". In addition, I_{Dsat} approaches $C_{ox}W(V_G - V_T)v_{sat}$, independent of the channel length L. Physically, we expect these results because the channel charge is moving at a constant velocity ($v_{sat} = |\mu_{no}\mathcal{E}_c|$). Hence, the saturation current is sensitive only to the density of charge in the channel and that, in turn, is proportional to $(V_G - V_T)$. In shorter channel MOSFETs, there is an observable trend toward a linear variation of I_{Dsat} with $(V_G - V_T)$ instead of a quadratic dependence, as is predicted without velocity saturation. This behavior is shown in Figure 10.3.

Normal Field. The velocity of carriers along the channel is also a function of $\mathcal{E}_x$, the field component normal to the channel that influences scattering at the Si–SiO$_2$

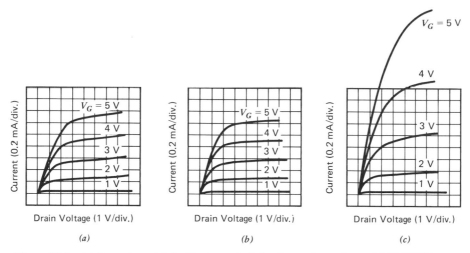

Figure 10.3 (a) Experimental I_D-V_D characteristics for a short-channel MOSFET ($L_{eff} = 2.7\ \mu$m.). (b) Characteristics calculated using velocity-saturation theory, (c) Characteristics calculated from long-channel theory.[16]

interface. The theory for the statistical and quantum-mechanical effects associated with this dependence is complicated and, as yet, unsatisfactory. However, the channel mobility can be modeled empirically as

$$\mu = \frac{\mu_o}{(1 + \Theta\mathscr{E}_s)} = \frac{\mu_o}{[1 + \Theta\epsilon_{ox}(V_G - V_C)/\epsilon_s x_{ox}]} \tag{10.2.11}$$

where μ_o is the low-field mobility (μ_{no} for n–channel MOSFETs, and μ_{po} for p–channel MOSFETs), and $\mathscr{E}_s$ is the field in the silicon at the Si–SiO$_2$ interface. The coefficient Θ depends weakly on the details of the MOS technology that is used and typically has a value between 10^{-7} and 10^{-6} cm V^{-1}.[8]

Since μ_o in Equation 10.2.11 varies with the channel voltage V_C, a varying mobility should be incorporated into a distributed expression for channel charge. In practice, adequate modeling for the effect of the normal field is possible by modifying the mobility in Equation 10.2.8. The representation is in the form of Equation 10.2.11 with the field dependence incorporated through a term proportional to V_{GS}, representing the maximum normal field along the channel.

Improved MOSFET Model[†]

To include both the normal and the tangential field effects in an expression for drain current, we modify the prefactor in Equation 10.2.8 and write

$$I_{DS} = \frac{\mu_{no}C_{ox}W}{L(1 + |V_{DS}/\mathscr{E}_c L| + \eta V_{GS})}\left[(V_G - V_T)\,V_{DS} - \frac{V_{DS}^2}{2}\right] \tag{10.2.12}$$

where η is an empirically determined parameter. Equation 10.2.12 incorporates the effects of tangential and normal electric fields on mobility by including two additional terms in the denominator. Typically, the effect of V_{DS} is greater unless a high gate-source bias is applied.

Equation 10.2.12 has been used successfully to model the behavior of MOSFETs having channel lengths as short as 1 μm. For accuracy at such short channel lengths, however, it is necessary to take special care in the extraction of the model parameters.

Parameter Extraction. The explicit parameters in Equation 10.2.12 are μ_{no}, C_{ox}, W, L, $\mathscr{E}_c$, V_T and η. Not specifically noted, but sometimes of importance to an accurate model are the series resistances at the MOSFET source and drain [($R_S + R_D$) in Figure 9.11] which are caused by contact effects and by the resistance in the silicon (which may be significant in the small cross sections of many MOS devices). Of the seven MOS parameters listed, C_{ox}, the oxide capacitance per unit area, and W, the MOSFET width, are usually straightforward to obtain or to estimate from known processing parameters. The determination of V_T, the threshold voltage, has a number of ramifications which we defer to Section 10.3.

The channel length L is of critical importance to MOS theory. Its value is usually established by the smallest practical mask dimension in the MOS process, and it also depends on the lateral spread of source and drain impurities under the gate. The channel length can generally be obtained precisely only by measurements carried out after all processing steps have been completed.

A method to determine the effective channel length L and, at the same time, to extract the series resistance R_x can be implemented if a series of MOSFETs having different channel lengths are fabricated with the same MOS process. The procedure is to measure the small-signal resistance at the drain (dV_D/dI_D) at low V_D values with $V_G \gg V_T$. In this case, Equation 9.1.4 is accurate, and the measured drain resistance R_m is

$$R_m = \frac{L_m - 2\Delta L}{\mu_n C_{ox} W(V_G - V_T)} + R_x \tag{10.2.13}$$

where L_m is the mask dimension for the channel length, ΔL is the decrease in channel length on each side of the channel resulting from the lateral spread of source and drain dopant atoms and changes in dimensions during lithography and pattern transfer, and R_x is the resistance external to the channel (R_x is independent of L_m).

We see from Equation 10.2.13 that if we take a fixed value of $(V_G - V_T)$ and plot measured values R_m versus L_m, we have a straight line. If we change $(V_G - V_T)$, we can repeat this procedure and plot a second straight line. By changing $(V_G - V_T)$ repetitively, we generate a family of straight lines that have a common intersection (at $L_m = 2\Delta L$ and $R_m = R_x$). These plots also provide a means of obtaining the mobility parameters μ_{no} and η for the MOSFET process. Figure 10.4 is an example of data plotted in this manner.

EXAMPLE   Channel Length and Series Resistance

What are the values of R_x, $2\Delta L$, μ_{no}, and η for the MOSFET from which the measurements plotted in Figure 10.4 are taken?

Solution

From Equation 10.2.13 the intercept of all curves occurs at $R_m = R_x$ and $L_m = 2\Delta L$. Hence $R_x \approx 25\ \Omega$, and $2\Delta L \approx 1\ \mu m$. To obtain the mobility value, we note from Equation 10.2.13 that

$$\mu_n = \left[WC_{ox}(V_G - V_T)\frac{dR_m}{dL_m} \right]^{-1}$$

780 cm^2 V^{-1} s^{-1} and $\eta = .013$V^{-1}. For the MOSFET of Figure 10.4, x_{ox} is 85 nm so this value of η corresponds to a value Θ in Equation 10.2.11 equal to 5.1×10^{-7} cm V^{-1}.

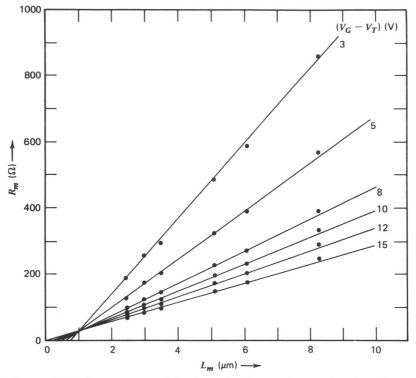

Figure 10.4 Measurements of the channel resistance R_m as a function of mask dimension L_m for the channel length. The data can be fit to Equation 10.2.13.[5,17] (W = 100 μm.)

10.3 Small MOSFET Considerations

Velocity saturation in the channel is just one effect that limits basic MOSFET theory more extensively when channel lengths are shortened. Several other effects are considered in this section. First, we examine more carefully the basic description for the threshold voltage that was presented in Chapter 8 and derive a corrected theory to account for geometric effects on the substrate depletion charge when inversion takes place. Second, we consider the effects associated with electrons that become highly energized by fields applied to the MOSFET.

Geometric Effects on Threshold Voltage

Short-Channel Effect. The threshold-voltage equations developed in Chapter 8 are based on one-dimensional theory; the space charge under the gate is assumed to be a function only of $\mathscr{E}_x$, the vertical field.* However, the pn–junctions at the source and drain in a MOSFET affect the space charge in the channel region so

_____

* This assumption is often called the *gradual-channel approximation*.

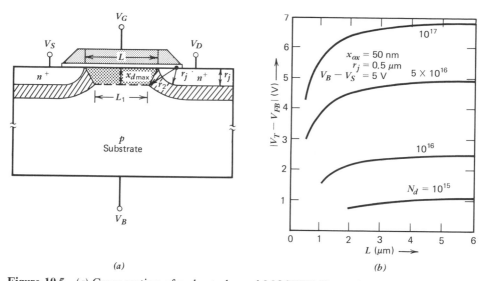

Figure 10.5 (a) Cross section of a short-channel MOSFET illustrating a geometrical model which accounts for threshold-voltage reduction. The depletion-region charge induced by the gate is approximated as that contained in the dotted region.[6] (b) Theoretical variation of threshold voltage as a function of channel length for p–channel MOSFETs having various substrate dopant concentrations. The threshold voltages have been calculated using Equation 10.3.3.[6]

that it is not solely a function of $\mathscr{E}_x$. If the channel is appreciably longer than the space-charge regions, this effect causes only a small error, but it can lead to considerable error in calculating the threshold voltage of short-channel MOSFETs. In general, the magnitude of the threshold voltage predicted using the gradual-channel approximation is too large because it assumes that all depletion-region charge is linked by field lines to the gate. In fact, some of this charge is linked to the source and drain, causing the channel region to be partially depleted without any influence of the gate voltage.

An approximate geometric technique for analyzing this effect has led to a theory that agrees moderately well with experiment.[6] For small values of V_{DS}, the technique is to consider the charge induced by V_G to be approximately contained in a volume whose cross section is the trapezoid of width x_{dmax} and length varying from L at the surface to L_1 at the substrate side of the depletion region (Figure 10.5a). The cross-sectional area is shown dotted on the figure. If this charge is called Q_{d1}, then

$$Q_{d1} = qx_{dmax}WN_a \frac{L + L_1}{2} \qquad (10.3.1)$$

where W is the width of the channel and N_a is the dopant density. The charge Q_{d1} in Equation 10.3.1 is taken to be depletion-layer charge that must be induced by the gate to bring the channel to the threshold condition. If the channel is long so that the space-charge regions at the source and drain are much smaller than L_1, then L_1 in Figure 10.5a approaches L. In that case, by Equation 10.3.1, Q_{d1} equals

$Q_d = q x_{dmax} N_a WL$, as was assumed in the first-order theory of Section 9.1. For shorter channels, L_1 becomes appreciably less than L, and Q_{d1} is therefore less than Q_d, as expected from our qualitative arguments.

For a useful theory, L_1 must be related to the geometry of the MOSFET. This can be done approximately by assuming that when $V_G = V_T$, the depletion layer is x_{dmax} units wide both in the x–direction (perpendicular to the Si–SiO$_2$ interface) and along the radius of the diffused source and drain junctions. From Figure 10.5a, $r_2 = r_j + x_{dmax}$ where r_j is the junction radius and r_2 is the radial distance to the corner of the trapezoid in this approximation for the cross section of the space-charge region. A geometric exercise (Problem 10.14) leads to

$$f \equiv \frac{Q_{d1}}{Q_d} = 1 - \frac{r_j}{L}\left(\sqrt{1 + \frac{2x_{dmax}}{r_j}} - 1\right) \tag{10.3.2}$$

The parameter f is therefore a unique function of the MOSFET geometry. The expression for the threshold voltage is written directly from Equation 8.3.18

$$V_T = V_{FB} + 2|\varphi_p| + V_S - \frac{fQ_d}{C_{ox}}$$

$$= V_{FB} + 2|\varphi_p| + V_S + \frac{f}{C_{ox}}\sqrt{2\epsilon_s q N_a(2|\varphi_p| + V_S - V_B)} \tag{10.3.3}$$

Despite the approximate nature of its derivation, Equation 10.3.3 has proven quite accurate in predicting V_T in MOSFETs with channel lengths as small as 1 μm,[6] and the equation is widely used.

Figure 10.5b shows the variation in threshold voltage predicted by Equation 10.3.3 for p–channel MOSFETs having an oxide thickness of 50 nm, and a junction depth $r_j = 0.5$ μm for several values of substrate doping. The figure shows that the threshold-voltage reduction resulting from the short-channel effect analyzed here becomes significant when the channel length is less than about 3 μm.

The analysis leading to Equation 10.3.3 has not considered the difference between the space-charge dimensions at the source and at the drain and, therefore, represents V_T for $V_{DS} = 0$. Because V_{DS} is typically much larger than the source-substrate bias V_{SB}, V_T is sensitive to V_D in short-channel MOSFETs. A more elaborate geometric analysis that includes this effect has been published.[7]

Narrow-Channel Effect. The assumption of a one-dimensional theory for V_T also becomes inaccurate if the channel is narrowed until its width is comparable to the size of the depletion layer x_{dmax} at the silicon surface. The *narrow-channel effect* causes the threshold voltage to increase in magnitude because some of the gate-induced space charge is lost in fringing fields. The narrow-channel effect is indicated together with the short-channel effect in Figure 10.6. If the fringing field is considered to be roughly cylindrical in shape,[8] then the total charge Q_{dT} in the depletion region when the threshold voltage is reached can be readily determined from geometry to be

$$Q_{dT} = q N_a WL x_{dmax}\left(1 + \frac{\pi}{2}\frac{x_{dmax}}{W}\right) \tag{10.3.4}$$

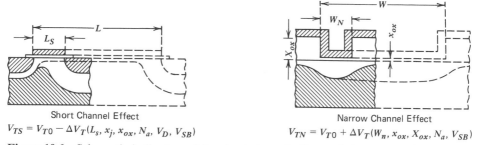

Short Channel Effect

$V_{TS} = V_{T0} - \Delta V_T(L_s, x_j, x_{ox}, N_a, V_D, V_{SB})$

Narrow Channel Effect

$V_{TN} = V_{T0} + \Delta V_T(W_n, x_{ox}, X_{ox}, N_a, V_{SB})$

Figure 10.6 Schematic indication of the short-channel effect and the narrow-channel effect.[18]

The increased depletion charge required by the fringing field is represented by the second term in the binomial in Equation 10.3.4. The increased threshold voltage ΔV_T owing to this effect is, therefore,

$$\Delta V_T = \frac{\pi q N_a (x_{d\max})^2}{2 C_{ox} W} \tag{10.3.5}$$

From Equation 10.3.4, we expect the narrow-channel effect to become significant when the second term in the binomial becomes about 0.1 or when $W \approx 16 \times x_{d\max}$. The variation in V_T shown in Figure 10.7 agrees with this estimate for the onset of a significant threshold-voltage shift because of the narrow-channel effect.

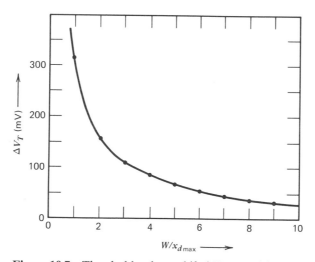

Figure 10.7 Threshold-voltage shift ΔV_T caused by the narrow-channel effect for a MOSFET with $N_a = 10^{15}$ cm^{-3}, and $x_{ox} = 50$ nm.[8] The x-axis is the ratio of the channel width W to the depletion width at inversion $x_{d\max}$.

EXAMPLE **Short- and Narrow-Channel Effects on** V_T.

Find the ratio (W/L) for a small MOSFET design such that the short-channel reduction in V_T just cancels the narrow-channel increase in V_T. For this MOSFET design let $x_{dmax} = 1\ \mu m$ and $r_j = 0.5\ \mu m$.

Solution

From Equations 10.3.2 and 10.3.3 the short-channel reduction in V_T, ΔV_{Tsc}, can be written

$$
\Delta V_{Tsc} = -\frac{r_j}{C_{ox}L}\sqrt{2\epsilon_s q N_a(2|\phi_p| + V_{SB})}\left(\sqrt{1 + \frac{2x_{dmax}}{r_j}} - 1\right)
$$

$$
= -\frac{qr_j}{C_{ox}L}N_a x_{dmax}\left(\sqrt{1 + \frac{2x_{dmax}}{r_j}} - 1\right)
$$

The narrow-channel increase in V_T, ΔV_{Tnc} is given by Equation 10.3.5. Equating the magnitudes of the two threshold-voltage shifts we find

$$
\frac{r_j}{L}\left(\sqrt{1 + \frac{2x_{dmax}}{r_j}} - 1\right) = \frac{\pi x_{dmax}}{2W}
$$

or

$$
\frac{W}{L} = \frac{\pi x_{dmax}}{2r_j\left(\sqrt{1 + \frac{2x_{dmax}}{r_j}} - 1\right)} = 2.54
$$

Note that this solution is independent of W and L. Although an exact cancellation of threshold-voltage shifts would probably not be a valid design objective, this example does emphasize that the short-channel and the narrow-channel effects tend to cancel one another in scaled-down MOSFETs.

Hot Carriers[†]

The importance of adequately describing the field in the vicinity of the drain contact was mentioned in Section 10.2 in our discussion of channel velocity limitations. The field near the corner at the silicon-silicon dioxide interface where the drain junction is directly under the gate is typically the largest in the MOSFET and has a critical influence on *hot-carrier effects*. The carriers (holes and electrons) are called "hot" because interaction with the field causes them to obtain far more kinetic energy than that corresponding to the ambient temperature (cf. Section 1.2).

A rigorous calculation of the field near the drain is only possible through computer-aided solution of the two-dimensional Poisson equation. It is, however, possible to carry out an approximate analysis by applying Gauss' law with judicious assumptions about the charge enclosed in a volume surrounding the high-field

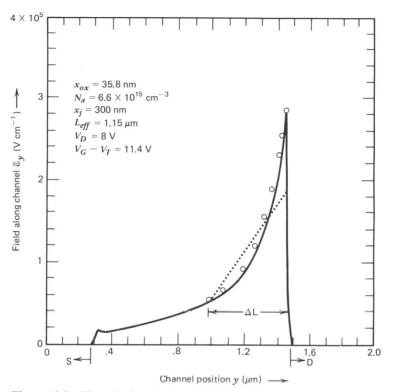

Figure 10.8 The calculated tangential channel field $\mathscr{E}(y)$ from two different models. The solid curve is calculated from a two-dimensional simulation program (*CADDET*). The dotted curve indicates the field expected for constant space charge (usually assumed in abrupt *pn*–junction analyses). The circles show points calculated from the approximate theory described in the text.[5] Electrons in the channel have saturated velocities over the region ΔL which extends from the channel pinch-off point to the drain.

region.[5] In this manner, we can derive an expression for $\mathscr{E}_y$ along the Si–SiO$_2$ interface. It is found that $\mathscr{E}_y$ has an approximate hyperbolic-cosine dependence on y between the end of the effective channel (where $V_D = V_{Dsat}$) and the metallurgical drain junction. Figure 10.8 shows that the predictions from this approximate theory compare well with those from a two-dimensional computer simulation of a short-channel MOSFET. Because of its accuracy and simplicity, the approximate analytic representation for $\mathscr{E}_y$ has proven very useful for calculation of hot-carrier effects in MOSFETs.

Injection into the Oxide. Free carriers passing through the high-field region near the drain can gain sufficient energy to cause several hot-carrier effects. Figure 10.9 shows some of these effects schematically. In the figure some hot electrons (*process* 1) are depicted as entering the oxide layer and producing a gate current I_G. The gate

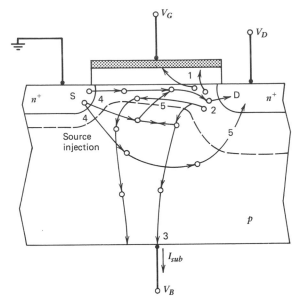

Figure 10.9 Cross section of an n-channel MOSFET showing hot-carrier processes described in the text.[9,19]

current is typically in the fA (10^{-15} A) range although for higher, but still nondestructive biases, it can grow rapidly to become several pA (10^{-12} A). If I_G does become of the order of pA, trapping of a fraction of the charge injected into the oxide can lead to a permanent change in MOSFET characteristics. Charge trapping in the oxide causes a local change in the threshold voltage, as described in Section 8.4.

EXAMPLE   Threshold-Voltage Shift due to Hot-Electron Trapping

In a short-channel MOSFET with $x_{ox} = 30$ nm, a gate current of 1 nA is momentarily caused by hot-electron injection. This gate current is estimated to flow through a section of the oxide measuring 0.2 by 10 μm near the drain end of the channel. Assume that a fraction equal to 10^{-6} of the injected electrons becomes trapped at an average distance of 0.1 x_{ox} from the Si–SiO$_2$ interface. How long must the gate current flow for the threshold voltage to change by 0.1 V in the region where the injection is taking place?

Solution

The gate current of 1 nA corresponds to a current density across the injection area of

$$10^{-9}/2 \times 10^{-8} = 5 \times 10^{-2} \text{ A cm}^{-2}.$$

Since 1 in 10^6 electrons become trapped, the rate of charge trapping in the SiO_2 is $5 \times 10^{-2} \times 10^{-6} = 5 \times 10^{-8}$ coulombs $s^{-1} cm^{-2}$. From Equation 8.4.6, trapped charge in the oxide shifts the flat-band voltage and thereby the threshold voltage (cf. Equation 8.3.18). Assuming that the charge is concentrated in a sheet, Equation 8.4.6 predicts that

$$\Delta V_{FB} = \Delta V_T = \left(\frac{1}{C_{ox}}\right)\left(\frac{0.9x_{ox}}{x_{ox}}\right)\Delta Q_{ot}$$

or

$$\Delta Q_{ot} = \frac{C_{ox}\Delta V_T}{0.9} = \frac{1.15 \times 10^{-7} \times 0.1}{0.9} = 1.28 \times 10^{-8} \ C \ cm^{-2}$$

The time required to trap this quantity of charge is

$$t = \frac{\Delta Q_{ot}}{J_{ot}} = \frac{1.28 \times 10^{-8}}{5 \times 10^{-8}} = 0.256 \ s$$

As this example shows, noticeable variations in V_T can occur rapidly if there is appreciable hot-electron injection into the oxide. A successful *IC* design must consider this possibility.

MOSFET Breakdown[†]

Avalanche Pair Production. As shown in Figure 10.9, energetic hot carriers are also responsible for *avalanche pair production (process* 2). The avalanche process produces holes that are collected by the substrate, creating a drift component of substrate current (*process* 3 in Figure 10.9). The ohmic drop associated with this current increases the voltage near the MOSFET above the substrate voltage V_B. Since the substrates used for MOS *ICs* typically have a relatively high resistivity (~ 5–30 ohm-cm) and MOSFET dimensions can be tiny, the voltage dropped in the substrate, even with small substrate currents, can have important consequences.[9]

The localized injection of holes near the drain by the avalanche effect can increase the substrate potential in the neighborhood of the nearby source junction sufficiently for the source to become forward biased. If this local forward bias approaches about 0.6 V, electrons can be injected into the substrate from the source (*process* 4 in Figure 10.9).

The source junction is much more heavily doped than the substrate and therefore has an injection efficiency for electrons (cf. Section 6.2) that is very near unity. Hence, many electrons enter the confined region under the channel as soon as the source becomes forward biased. Some of the electrons injected from the source act as in a lateral bipolar *npn* transistor (cf. Section 7.7) and are collected either through the channel region or at the drain (*process* 5 in Figure 10.9). The collected bipolar current itself results in more avalanche-generated holes, which

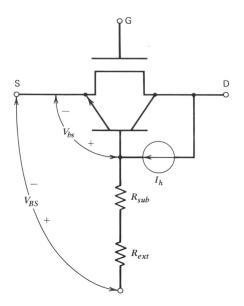

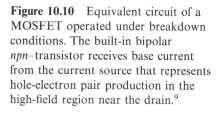

Figure 10.10 Equivalent circuit of a MOSFET operated under breakdown conditions. The built-in bipolar *npn*–transistor receives base current from the current source that represents hole-electron pair production in the high-field region near the drain.[9]

return to the substrate as additional base current for the bipolar *npn* device. An equivalent circuit for these effects is shown in Figure 10.10.

In Figure 10.10, the avalanche generation of holes is represented by the current source between the drain and the substrate. The lateral bipolar *npn* transistor with the source acting as an emitter receives base current (consisting of holes) from the avalanche-current source. The bias across the emitting substrate-source junction V_{bs} is different from the substrate-to-source voltage V_{BS} because of the resistance between the substrate electrode and the junction, as well as the (much smaller, resistance between the source electrode and the junction. Thus, an external reverse bias applied to the substrate-source junction cannot prevent the source from injecting carriers into the substrate under all conditions.

The mechanisms described may have sufficient gain to cause the holes injected into the transistor base to be themselves generated by collected current in the *npn* transistor. When this occurs, there is a rapid increase in drain current, leading to

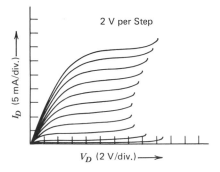

Figure 10.11 Drain current versus drain voltage characteristics in an *n*–channel MOSFET showing breakdown due to avalanche effects ($L_{eff} = 4 \, \mu m$).[9]

MOSFET breakdown as shown in Figure 10.11. Because of the complexity of the interactive mechanisms, the exact breakdown behavior depends on processing as well as on all of the MOSFET bias voltages.[9]

Punchthrough. The avalanche breakdown described above typically has only a slight dependence on channel length L (assuming that only L is varied). It is found, however, that as channel lengths are shortened to the range of $1-2\ \mu m$, leakage currents through the drain of a MOSFET occur at voltages below the avalanche-breakdown value as shown in Figure 10.12. The origin of this behavior is clearly distinct from avalanche injection, and is normally called *punchthrough*.

Punchthrough occurs when the depletion regions for the source and drain interact to reduce the barrier for electron flow between these two regions. Field mapping shows that the barrier is lowest away from the silicon-silicon dioxide interface, typically at nearly the same distance as the source– and drain–junction

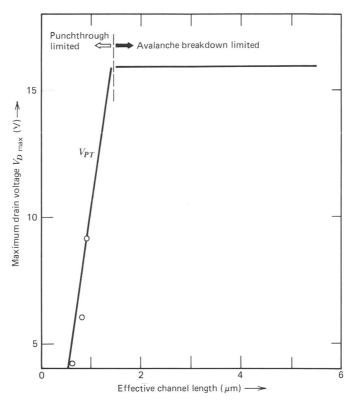

Figure 10.12 Calculated (solid line) and measured (circles) drain voltage $V_{D\,max}$ at which $I_D = 1\ \mu A$ versus effective channel length. The gate, source, and substrate are all grounded, and the channel is not inverted. The abrupt decrease in $V_{D\,max}$ as channel length is shortened occurs when punchthrough takes place.[1]

depths. The height of the barrier decreases as V_{DS} increases. Punchthrough is generally described as producing a *soft* breakdown because the leakage current does not rise as abruptly with increasing V_{DS} as in the case of avalanche breakdown. The overall current-voltage behavior in punchthrough is complex because source-drain current may be flowing at the surface (channel current) at the same time that subsurface current is flowing in the minimum barrier region. Another complication is caused by the dependence of the punchthrough barrier height on gate and substrate biases, as well as on the drain-source voltage.

Oxide Breakdown. In addition to breakdown mechanisms in the silicon, an MOS design must not exceed the breakdown strength of the insulating oxide. For thermally grown SiO_2, a field of 7×10^6 V cm^{-1} generally leads to irreversible breakdown. A voltage of 21 volts produces a field of this size across a 30 nm-thick oxide. To allow a margin for safety, oxide fields are usually limited to about 2×10^6 V cm^{-1}, which corresponds to 6 V across a 30 nm oxide.

10.4 Scaling MOSFETs to Smaller Sizes

The advantages of building *ICs* with higher and higher component densities have provided continued impetus for designing MOSFETs with ever smaller dimensions. If, however, only the surface dimensions are reduced while other circuit and device parameters remain unchanged, many MOSFET properties are degraded. As an obvious example, if V_{DS} is held constant while L is reduced, the average field along the channel becomes larger. As we have seen, higher channel fields lead to loss in gain (because of velocity saturation) and an increase in hot-carrier effects in the MOSFET. The designer can reduce the average field by simply reducing the applied bias by the same factor by which the dimensions are reduced. This guideline can be incorporated into a set of *scaling rules* to be applied to MOS design parameters when reducing the dimensions of an established MOS process. A full set of scaling rules must be sufficient to specify how each parameter is to be changed in the scaled-down MOS process. The rules are not unique; they depend upon a basic decision about the symmetries that are desired in the scaled and unscaled MOS integrated circuits.

Constant-Field Scaling. The most widely used rules are based (as in our example above) upon designing the scaled-down MOSFET so that the same (or nearly the same) fields are present as in the original unscaled device.[10] By holding the fields constant, many of the MOSFET characteristics are similar for the two devices except for a scale change. The constant-field scaling rules that are usually applied are summarized in Table 10.1.

As seen from the table, both vertical and surface dimensions are reduced by the same factor K ($K > 1$). The dopant concentrations are increased by K in order to scale the depletion-layer widths. The depletion-width scaling is only approximate, however. According to Equation 8.3.8 and Table 10.1, the scaled maximum

Table 10.1 **Scaling Rules for Constant-Field Scaling**

	Scaling Factor
Surface Dimensions	$1/K$
Vertical Dimensions, x_{ox}, x_j	$1/K$
Impurity Concentrations	K
Currents, Voltages	$1/K$
Current Density	K
Capacitance (per area)	K
Transconductance	1
Circuit Delay Time	$1/K$
Power Dissipation	$1/K^2$
Power Density	1
Power-Delay Product	$1/K^3$

depletion-layer width $x_{d\max}'$ is

$$x_{d\max}' = \sqrt{\frac{2\epsilon_s}{qKN_a}\left(2|\phi_p'| + \frac{V_{SB}}{K}\right)} \qquad (10.4.1)$$

The voltage ϕ_p' changes only logarithmically with K, so that $x_{d\max}'$ is reduced by a factor K only if $V_{SB} \gg 2|\phi_p'|$.

Using the scaling rules in the equation for a scaled threshold voltage V_T', we have

$$V_T' = \Phi_{MS}' - \frac{Q_f}{KC_{ox}} + \frac{V_S}{K} + 2|\phi_p'| + \frac{1}{KC_{ox}}\sqrt{2\epsilon_s qKN_a\left(2|\phi_p'| + \frac{V_{SB}}{K}\right)} \approx \frac{V_T}{K} \qquad (10.4.2)$$

Hence, as was the case for $x_{d\max}$, V_T only scales approximately as $1/K$ because of the contributions of Φ_{MS} and $|\phi_p|$.

Other, more subtle, physical effects can limit the validity of the variations predicted by the scaling laws. For example, insertion of the scaled dimensions and voltages into any of the drain-current expressions predicts that I_D should scale as $1/K$. This prediction, however, overlooks the fact that scaling N_a to ($K\,N_a$) causes a reduction in channel mobility so that I_D decreases by more than $1/K$.

Figures 10.13a and b illustrate the application of the constant-field scaling laws for MOSFET scaling by a factor of 5 (from an original 5 μm-channel length). These figures are based on measured device characteristics and demonstrate a successful application of constant-field scaling.

To first order, a circuit composed of devices scaled using the rules of Table 10.1 behaves similarly to an unscaled circuit because the same relation is preserved between the circuit voltages and the threshold voltage. Since reduced capacitances are driven through the same device resistance ($V'/I' = V/I$), the switching and delay times are reduced by K. The power dissipation is reduced by $K^2[P' = V'I' = (V/K)(I/K)]$, and the power-delay product is reduced by K^3; both of these results are highly favorable. Although the power per device is reduced, the number of devices per unit area can be increased by the same factor, so the power density

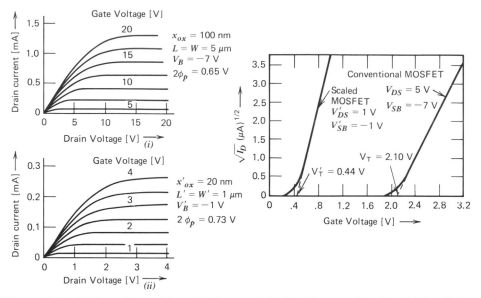

Figure 10.13 (*a*) Experimental $I_D - V_D$ characteristics for (*i*) conventional, and (*ii*) scaled-down MOSFETs having $W/L = 1$. (*b*) Experimental turn-on characteristics for conventional and scaled-down MOSFETs with $W/L = 1$.[10]

remains constant, and cooling considerations are the same as in an unscaled chip of the same overall area.

The encouraging results of scaling down a MOSFET as shown in Figure 10.13 do not imply that the scaling factor K can be arbitrarily large. The following example of a design problem encountered for a switching MOSFET illustrates a practical limitation on the scaling factor K.

EXAMPLE   Limitation to MOSFET Scaling

A circuit is designed to operate so that the current flowing in a MOSFET in the off-state (when $V_{GS} = 0$) must be 6 orders of magnitude lower than that flowing at the threshold voltage. For an unscaled MOSFET, V_T is 0.80 V, and the sub-threshold current is reduced by an order of magnitude for every 90 mV interval below V_T. What is the maximum scaling factor K that can be used while maintaining the same 10^{-6} off-current ratio at $V_{GS} = 0$ volts?

Solution

From Equations 10.1.4 and 10.1.6 we see that subthreshold current varies exponentially with $(V_G - V_T)$. Therefore, in order to maintain an off-current 10^6 times smaller than that flowing at $V_G = V_T$, the same value of $(V_G - V_T)$ is needed for both the scaled and unscaled MOSFETs. (We are overlooking here the slight variation in ξ that results from scaling.) Stated in other terms, the *rate-of-decrease* of the subthreshold current as a function of changing gate voltage does not scale with K and remains approximately one order of magnitude for every 90 mV change in $V_G - V_T$. Therefore, the specified current ratio of 10^{-6} requires that

$$|V_G - V_T'| \geq 6 \times 0.09 = 0.54 \text{ V}$$

Since $V_G = 0$ volts in the off state, the scaled value of threshold voltage V_T' must be greater than 0.54 V. From Equation 10.4.2, V_T scales approximately as K, and therefore

$$K \leq 0.80/0.54$$

$$K \leq 1.5$$

The requirement imposed by subthreshold current in the example just considered is only one of many possible limitations when scaling devices to small sizes. Some others, such as the finite thickness of the surface conducting channel (measured from the Si–SiO$_2$ interface) and its influence on the proper formulation for current in the MOSFET, the reduction in the signal-to-noise ratio which typically results from scaling down the operating voltages, or the increasing influence of fringing fields, can be treated theoretically.

Scaling may also be limited by material constraints, such as the difficulty of obtaining a thin, uniform, pinhole-free oxide. When a different set of these limitations is considered, different scaling rules may become more appropriate than the constant-field rules of Table 10.1. For feature sizes below 1 μm, for example, it has been shown that scaling rules other than those for a constant field lead to higher performance MOSFETs.[11]

10.5 Numerical Simulation[†]

The increasing complexity of the MOSFET models discussed thus far demonstrates that more accurate modeling generally leads to increased computational difficulties. To consider additional features important in MOS design (such as nonuniform dopant profiles), numerical simulation with the help of computers is almost a necessity. Even without these added complications, computer-aided simulation is especially helpful for the analysis of very small devices in which two- and three-dimensional effects can have practical significance.

In Chapter 2 the use of process-simulation programs (e.g. *SUPREM*) to obtain dopant profiles was described. Once a process simulation has been carried out, the next step in *IC* analysis is to use the calculated profiles to predict the electronic behavior of the fabricated devices. This may be accomplished by using these profiles together with the applied bias values as inputs of a transistor-simulation program. The outputs of the simulation program are the electric potential and free-carrier concentrations in the regions of the *IC* where devices are located. From these results it is generally straightforward to calculate the currents as functions of the applied voltages using the quasi-Fermi-level expressions for total currents (Equations 1.2.25 and 1.2.26).

Simulation programs typically model a transistor cross section by representing it as a grid of points. At each point, the governing equations are solved iteratively for the potential. The solutions are carried out over the entire grid until a self-consistent potential is obtained. During each iteration, the potential at one point is calculated from its previous values at adjacent points using an algebraic difference equation which is an approximation for the Poisson differential equation. Since the potential variation depends strongly on the local charge density, a large dopant-concentration gradient requires many grid points for accuracy, and the computation time can become lengthy. Thus, to obtain accuracy without overly increasing the computation time, it is usually worthwhile to employ a variable grid spacing. This allows many calculation points in regions of strongly varying dopant concentrations and correspondingly fewer points in less critical regions.

An example in which the grid spacing is critical is the simulation of a MOSFET with a shallow boron implant in the channel, as is often used for threshold-voltage adjustment. Since the boron distribution is sharply peaked near the silicon-silicon dioxide interface, the grid must be very dense in that region to obtain an accurate solution.

As with any numerical technique, the precision improves as the computation time increases. To shorten computation time, simplified models are often used to give approximate, but adequate, solutions for limited regions of operation. For example, although a simulation program may include both the Poisson equation and the free-carrier continuity equations, at low currents it is often unnecessary to solve the continuity equations repeatedly at each point in the device. Numerical models, simplified with approximations of this type, are often useful to study variations of device behavior when processing and layout are changed. The calculations may not accurately agree with device measurements, but trends indicated by the results can still serve as important aids to process development.

Once the potentials and free-carrier concentrations have been calculated at a given bias, it is possible to determine the quasi-Fermi levels from Equations 1.1.28 and 1.1.29. The hole and electron currents can then be calculated using Equations 1.2.25 and 1.2.26, and the model parameters typically employed in circuit-simulation programs can be extracted from the predicted current-voltage behavior.

EXAMPLE   Numerical Simulation of an n–channel MOSFET

Use simulation techniques to study a short, n–channel MOSFET having a gate-oxide thickness $x_{ox} = 27$ nm and an effective channel length $L' = 1.1$ μm. The oxide dimensions and dopant profiles for the MOSFET have been obtained from the process-simulation program *SUPREM*, described in Chapter 2; the dopant profile predicted by *SUPREM* along a plane near the center of the channel is shown in Figure 10.14a* There have been two boron implants: one at 70 keV with a dose

---

* Some comparison with experimental structures would be carried out to insure the validity of the process simulation.

Impurity Concentration

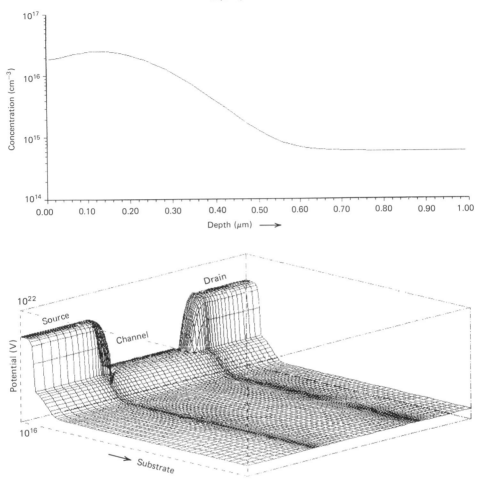

Figure 10.14 Dopant profile as predicted by process-modeling program *SUPREM* for an *n*–channel MOSFET which has been doubly implanted with boron, (*a*) through the channel region midway between the source and drain, (*b*) "bird's-eye view" of the dopant concentration throughout the MOSFET. (*Simulation courtesy of K. Cham, Hewlett-Packard*).

of 5.5×10^{11} cm^{-2} before gate oxidation, and the other at 25 keV with a dose of 5×10^{11} cm^{-2} after gate oxidation. The shallow implant adjusts the threshold voltage, and the deep implant inhibits subsurface punchthrough. Figure 10.14*b* shows a more detailed "bird's eye" view of the dopant concentration throughout the MOSFET. Although there are two implants, only one peak is seen in Figures 10.14*a* and *b* because of redistribution of the boron during subsequent heat cycles.

Investigate the potential in the device when $V_S = V_G = 0$ V, $V_B = -1.5$ V, and $V_D = 3$ V. For the same voltages applied to source, drain, and substrate, estimate the subthreshold current for $0 < V_G < 0.8$ V.

Solution

A two-dimensional transistor-analysis program[12] is used to find the potential and free-carrier distributions. This program solves the Poisson equation by considering differences between the potential at a given grid location and that at the four nearest neighbors. To limit computation time, the free-carrier continuity equations are not solved at each grid point. Consequently, the program is most useful for $V_G < V_T$, when only subthreshold current flows in the MOSFET.

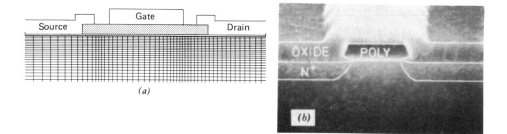

Figure 10.15 (*a*) Grid-spacing used for the device-simulation program; the grid is coarse in regions away from the channel and dense in regions where the dopant concentration varies strongly with position.[13] (*b*) Scanning-electron microscope photograph of a sectioned short-channel MOSFET. (*Courtesy Siemens Corporation*).

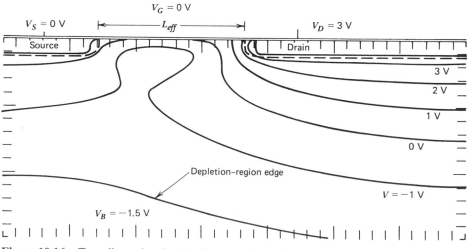

Figure 10.16 Two dimensional plot of potential as a function of position generated by the simulation program[13]; $L_{eff} = 1.1$ μm and $x_{ox} = 26.8$ nm.

Figure 10.15*a* is a sketch of the MOSFET electrodes over the variable grid to be used for the solution. The grid contains more points near the silicon-silicon dioxide interface (where the shallow implant is concentrated) and near the ends of the channel (where the heavily doped source and drain regions exist). A scanning-electron microscopic cross sectional view of a fabricated MOSFET is shown in Figure 10.15*b*.

From the computer-generated, two-dimensional plot of equipotential contours for one set of bias voltages (Figure 10.16), we see that the bias at the drain has appreciable influence over a significant fraction of the channel length.[13] A two-dimensional analysis is clearly needed to predict behavior in this short-channel MOSFET. More graphic representation potential, such as the "bird's-eye" view in Figure 10.17*a* is easily generated on the computer. In this view the potential is plotted vertically while the cross section is represented in the *xy*–plane. Readily

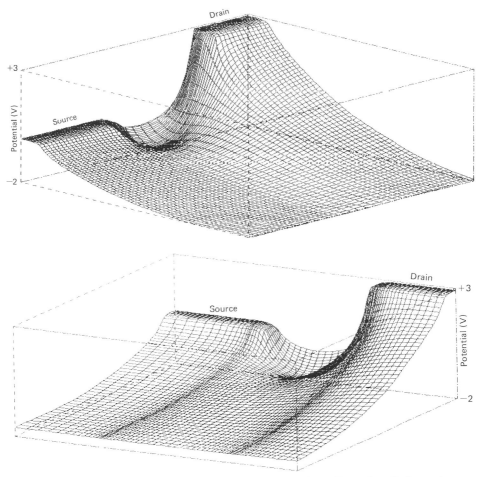

Figure 10.17 "Bird's-eye" views of the potential plotted three-dimensionally from the solutions of the simulation program. (*a*) View from the source end of the MOSFET. (*b*) View from the drain end of the MOSFET.[13]

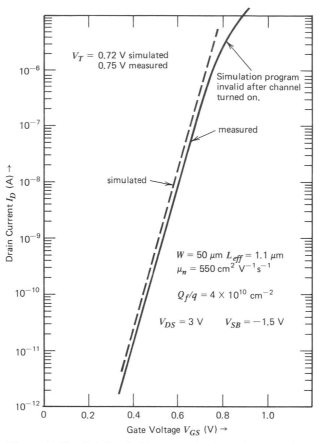

$V_T = 0.72$ V simulated
0.75 V measured

Simulation program invalid after channel turned on.

measured

simulated

$W = 50\ \mu m\quad L_{eff} = 1.1\ \mu m$
$\mu_n = 550\ cm^2\,V^{-1}s^{-1}$

$Q_f/q = 4 \times 10^{10}\ cm^{-2}$

$V_{DS} = 3$ V $\qquad V_{SB} = -1.5$ V

Drain Current I_D (A) →

Gate Voltage V_{GS} (V) →

Figure 10.18 Subthreshold current in the MOSFET of Figure 10.14 calculated by the simulation program (dashed curve) and measured (solid curve).[13]

available graphics programs allow the viewing angle to be rotated through an arbitrary angle for inspection of critical regions (cf. Figures 10.17a and b.)

The dependence of the subthreshold current (discussed in Section 10.1) on gate bias can be calculated approximately from a series of solutions for potentials and carrier densities at different values of V_G. The accuracy of these predictions is compared to experiment in Figure 10.18. From Figure 10.18 we can infer that the analysis is quite accurate until the magnitudes of currents in the device require the simultaneous solution of the continuity equations along with the Poisson equation. From Figure 10.18 we see that the theoretical value of the threshold voltage (defined as the gate voltage for which $I_D = 1\ \mu A$) is 0.72 V, compared to $V_T = 0.75$ V from measured data. The accuracy of this simulation is quite good, especially considering that an error of 1% in measuring the channel length can cause a shift in V_T of about 0.02 V, and an uncertainty in the fixed charge density Q_f of 1×10^{10} electrons cm^{-2} shifts V_T by 0.013 V.

Simulation programs that are much more complex than that employed in the example just considered are being developed to understand the complicated physical behavior of small MOS devices. These programs model such effects as avalanche pair-production, currents flowing between the channel and the gate, and currents between the channel and the substrate. By employing these programs, device optimization studies can be carried out a great deal more efficiently than would be possible without them. Simulation has already established itself as an extremely useful device– and process–design tool. It is certain that it will be more heavily exploited in the VLSI era.

10.6 Devices: Ion-Implanted MOSFETs, Depletion-Mode MOSFETs

The fundamental importance of ion implantation to the successful production of MOS *IC*s has been mentioned several times in our discussion. Because of the finely tuned control of impurity dose and depth possible with ion implantation, this process will continue to be an indispensable part of *VLSI* MOS processing as feature sizes move toward the submicrometer region.

A description of ion implantation was given in Chapter 2. The technique employs a pure beam of ionized dopant atoms that are accelerated in a strong electric field to energies of the order of 100 keV and allowed to strike the wafer surface. The ions penetrate into the silicon (or the oxide or resist film) to a depth that is usually a small fraction of a micrometer. The ions that stop in the oxide are generally electrically inert. The ions implanted in the silicon become active donors or acceptors when the wafer is annealed at a temperature that is low relative to that used for diffusion steps. Hence, the junction profiles obtained in previous diffusion steps are not significantly modified by the ion-implantation process.

Threshold-Voltage Adjustment. The implanted dopant atoms used for threshold-voltage adjustment are typically introduced through the gate-oxide layer. The implant can be used either to increase or to decrease (by compensation) the net dopant concentration at the silicon surface. For *n*–channel MOS processes, there is typically a heavy implant of boron to increase the surface dopant density in the *field region* away from the active devices. A lighter implant is also used to adjust the threshold voltage in the channel region (cf. Figure 9.13). These implants make possible the use of less heavily doped substrates, which improves MOS circuit performance. The channel-implant step is indicated in Figure 10.19.

Figure 10.20 shows the distribution of dopant atoms from a typical implant as a function of depth into the wafer. After the implant, the atoms are distributed

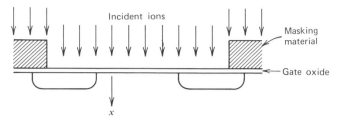

Figure 10.19 Ion implantation in a MOSFET. Dopant atoms are implanted through the gate oxide into the channel region to alter the threshold voltage and to inhibit subsurface punchthrough.

with a Gaussian profile (solid line in the figure). The implant concentration $N_i(x)$ is

$$N_i(x) = \frac{N'}{\sqrt{2\pi}\,\Delta R_p} \exp\left[-\frac{(x - R_p)^2}{2(\Delta R_p)^2}\right] \tag{10.6.1}$$

where N' is the number of implanted atoms per unit area (the *dose*), R_p is the average distance an implanted atom penetrates into the solid (the *range*), and ΔR_p is the RMS or standard deviation that describes the "width" of the distribution. Both the range and the width of the implant depend on the energy of the implanting ion beam as well as on the implanted species. Values for R_p and ΔR_p for the most common dopants in silicon are given in Figures 2.14, 2.15, and 2.16.

After the activating anneal, the implanted distribution becomes slightly broader, as shown by the dashed curve in Figure 10.20. Calculating the effect of this implant on the MOSFET threshold voltage is greatly simplified by approximating the actual distribution by the "box" distribution (dotted lines in Figure 10.20). In this approximation, the implanted dopant is assumed to have a constant density N_{ai} from the surface to a depth x_i. The total dose N' per unit area is related to N_{ai} and x_i by the equation $N' = N_{ai}x_i$. The dopant concentration in the region nearest the surface is $(N_{ai} + N_a)$ where N_a is the background acceptor concentration in the substrate. For $x > x_i$, the effective acceptor concentration is just N_a.

To calculate changes in V_T, two different cases must be considered. In the first case, the effective depth of the implant is greater than the width of the depletion region at threshold. Thus, the threshold voltage is simply given by Equation 8.3.18 with N_a replaced by $N_{ai} + N_a$. It is advisable to avoid this type of an impurity profile, however, because the increased doping near the substrate edge of the junction leads to higher substrate capacitance and lower breakdown voltage in the channel region as well as to greater sensitivity of the threshold voltage to source-substrate bias (i.e., a more troublesome *body effect*).

The second case, in which the implanted ions are all contained within the surface space-charge region, is therefore preferable. Calculation of the threshold voltage in the second case is more complicated, but it follows the same basic pattern that was used previously. The first step is to find a solution for Poisson's equation when the space charge changes abruptly, as in the approximate "box" distribution. When the solution is found, and the depletion approximation is used, an expression

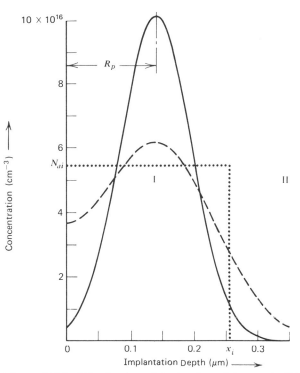

Figure 10.20 Distribution of the implanted dopant (solid curve) immediately after the implant, and after activating anneal and diffusions (dashed curve). The dotted "box" curve shows the distribution used to calculate the threshold voltage.

for the depletion-layer width x_d can be written in terms of the surface potential ϕ_s.

$$x_d = \sqrt{\frac{2\epsilon_s}{qN_a}(\phi_s + |\phi_p|) - x_i^2 \frac{N_{ai}}{N_a}} \tag{10.6.2}$$

Real solutions for Equation 10.6.2 are only possible if the second term in the square root is smaller than the first because we have assumed that the highly doped, implanted region is contained within the depletion region. Imaginary solutions occur for cases that violate this assumption. To find the threshold voltage, ϕ_s in Equation 10.6.2 must be set to the inversion value appropriate to the heavier doped surface region: $\phi_{ps} = (kT/q)\ln[(N_{ai} + N_a)/n_i]$. This leads to an expression for x_{dmax} that can be used to find the depletion-charge density at inversion Q_d.

$$Q_d = -qN_{ai}x_i - qN_a x_{dmax}$$
$$= -qN_{ai}x_i - \sqrt{2qN_a\epsilon_s(|\phi_{ps}| + |\phi_p| + V_{SB}) - q^2 x_i^2 N_a N_{ai}} \tag{10.6.3}$$

Equation 10.6.3 can easily be expressed in terms of the implantated dose by making use of the expression $N_{ai} = N'/x_i$, where N' is the density of atoms (per unit area) *penetrating into the silicon.*

As in Section 8.3, the gate voltage is then related to Q_d, and the threshold voltage can be written as

$$V_T = V_{FB} + V_S + |\phi_p| + |\phi_{ps}| + \frac{qN'}{C_{ox}}$$

$$+ \frac{1}{C_{ox}} \sqrt{2qN_a\epsilon_s(|\phi_{ps}| + |\phi_p| + V_{SB})} - q^2 x_i N_a N' \qquad (10.6.4)$$

As expressed in Equation 10.6.4, the threshold voltage has an arbitrary reference point. The zero for potential can therefore be taken at any electrode.

By comparing Equation 10.6.4 to Equation 8.3.18, we see that the implantation affects the threshold voltage V_T in three ways. (1) Instead of the voltage drop across the surface depletion region at inversion being $(2|\phi_p| + V_{SB})$, it is $(|\phi_{ps}| + |\phi_p| + V_{SB})$. (2) There is a linear dependence on dose through the term qN'/C_{ox}. (3) The square-root expression for the depletion-charge term is altered principally by the added term $-q^2 x_i N_a N'$. Of these three changes, the first depends on the dose only logarithmically and is therefore not of much consequence. The second change, which depends linearly on the dose N' is the dominant effect. It enters the expression for V_T as a direct addition to the term representing fixed surface charge Q_f, as is evident from the expression for the flat-band voltage (without distributed oxide charge) $V_{FB} = \Phi_{MS} - Q_f/C_{ox}$. The third change in the threshold-voltage expression is the only one that depends on x_i, the implantation depth. The dependence is not strong, as can be seen by inspection of Figure 10.21, which shows

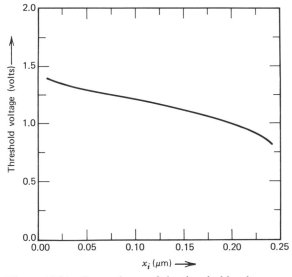

Figure 10.21 Dependence of the threshold voltage on the implantation depth x_i as predicted by Equation 10.6.4 for a MOSFET with an 87 nm-thick gate oxide, $Q_f/q = 10^{11}$ cm^{-2}, $N' = 3.5 \times 10^{11}$ cm^{-2}, and $N_a = 2 \times 10^{15}$ cm^{-3}. Both V_S and V_B are assumed to be zero.

the threshold voltage as a function of implantation depth. The values of V_T, plotted in Figure 10.21, have been calculated from Equation 10.6.4 for the process parameters given in the figure caption. Because the dependence on x_i is weak, a reasonable estimate of V_T can be obtained by disregarding the effect of the implant on the square-root term, and this is often done in practice. For the MOSFET of Figure 10.21 without implantation, the threshold voltage would nominally be -0.19 V, too close to zero for the production of reliable n–channel MOSFETs. Hence, these calculated results emphasize the importance of the ion-implantation process in fabricating a useful n–channel device.

The effectiveness both of ion implantation and of body bias in adjusting the threshold voltage of n–channel MOSFETs is illustrated in Figure 10.22. The data points on the figure represent the threshold voltages measured as the body bias V_{SB} was varied for a series of n–channel MOSFETs implanted with varying doses

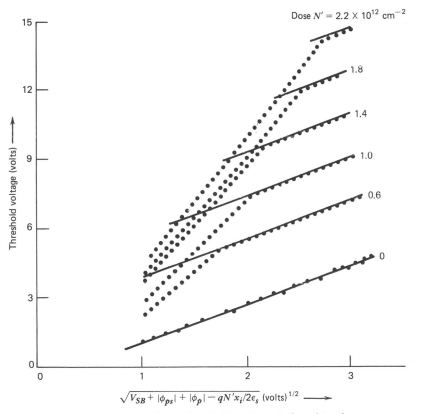

Figure 10.22 Dependence of the threshold voltage (referred to the source electrode) for ion-implanted n–channel MOSFETs. The dots represent experimental values measured as V_{SB} is varied. The solid lines are plotted from Equation 10.6.4. The properties of the MOSFETs are given in the text. (*Courtesy Hewlett-Packard*).

of boron dopant. The solid curves on the figure were calculated using Equation 10.6.4 with V_S taken as the reference. The x–axis in Figure 10.22 corresponds to the magnitude of the square root in Equation 10.6.4, and the curves are therefore linear. The substrate material was (100)-oriented silicon doped with 1.2×10^{16} boron atoms cm^{-3}. The MOSFETs had an oxide thickness of 100 nm, an oxide-charge density $Q_f/q = 8 \times 10^{10}$ cm^{-2}, and $\Phi_{MS} = -0.92$ V.

In order for Equation 10.6.4 to predict the experimental data, increasing values of body bias V_{SB} are seen from the figure to be necessary as the dose is increased. The higher body bias is needed to cause the depletion layer to penetrate completely through the implanted silicon, a condition required to validate the analysis presented. In the region for which the analysis is valid, the correspondence between theory and experiment is good, and Equation 10.6.4 is thus demonstrated to be useful for predicting the threshold voltage in an ion-implanted MOSFET.

Depletion-Mode MOSFETs.[†] In depletion-mode MOSFETs there is a conducting channel between the source and the drain when the gate and source are at the same potential ($V_{GS} = 0$). This zero-bias channel is especially useful if a MOSFET serves as a load element in a digital inverter stage because it permits fast switching, a maximum voltage swing for a given supply voltage, and well-defined waveforms to represent binary ones and zeroes. Since the switching transistor in an inverter circuit should be an enhancement device, the use of a depletion-mode load requires fabricating MOSFETs with two different threshold voltages in the same *IC*. Hence, the first MOS logic circuits had only enhancement-mode MOSFET loads. Selective ion implantation has made possible enhancement-switch, depletion-load (*E/D*) MOSFET circuits, now the most frequently used NMOS *IC* logic technology.

Looked at most simply, depletion-mode MOSFETs result from shifting a threshold voltage sufficiently to change its sign. For example, the positive threshold voltage of an n–channel enhancement-mode MOSFET is reduced in magnitude and ultimately changed to a negative value when the donor-implant dose is continuously increased. Considering only the threshold-voltage change, we model drain current in the depletion MOSFET by merely changing the sign of V_T in the equations derived in Chapter 9. This model accurately represents the transistor provided that the donors implanted in the silicon are located in a very thin sheet near the Si–SiO$_2$ interface. Under these conditions, the implant has the same effect as modifying the fixed interface-charge density Q_f. For n–channel MOSFETs with implanted donors, the change in V_{Tn} is negative:

$$\Delta V_{Tn} = -qN'/C_{ox} \tag{10.6.5}$$

For p–channel MOSFETs, implanted acceptors cause V_{Tp} to increase from its initial negative value:

$$\Delta V_{Tp} = +qN'/C_{ox} \tag{10.6.6}$$

Although Equations 10.6.5 and 10.6.6 are approximate, modeling depletion-mode MOSFETs in terms of only a threshold-voltage shift is often adequate for circuit design.

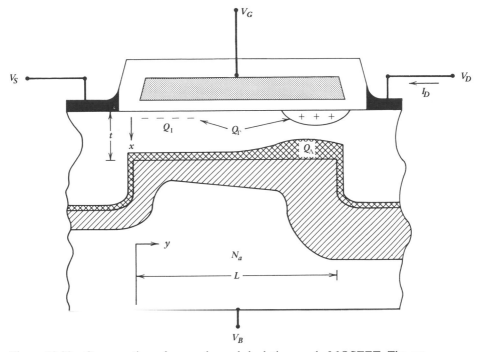

Figure 10.23 Cross section of an n–channel depletion-mode MOSFET. The net channel implant charge Q_I forms a pn junction t units below the Si–SiO$_2$ interface; Q_Γ is the charge modulated by the gate, Q_Λ is the charge depleted at the substrate-channel junction.[14]

In practical enhancement/depletion processes, even though a low-energy (shallow) implant is used, subsequent heat treatments tend to drive dopant atoms initially near the silicon-silicon dioxide interface deeper into the substrate. In this case, the depletion-mode MOSFET contains a thin n–type region in the channel itself, as shown for an n–channel MOSFET in Figure 10.23. The conductance through this "built-in" channel can be reduced (as in a JFET) by an applied negative value of V_{GS} or it can be enhanced by an applied positive bias. In enhancement, the channel surface region becomes accumulated and there are two parallel conducting paths, one through the built-in channel, and one through the accumulated surface. As the drain voltage is increased, a partial enhancement condition may exist in which the region near the source is accumulated while depletion occurs near the drain. The sketch in Figure 10.23 shows this condition.

A theory for drain current in the depletion-mode MOSFET has been formulated in terms of three charge densities (charge per unit area).[14] For an n–channel MOSFET, the charge densities consist of (1) the net donor charge implanted in the channel Q_I (a positive number), (2) the charge density depleted or enhanced by the gate-to-channel voltage $Q_\Gamma(y)$ (can be positive or negative), and (3) the depleted portion of the implanted charge density at the channel-to-substrate junction $Q_\Lambda(y)$ (a positive number). In terms of these three charge densities, the drain current is expressed by

$$I_D = W[\mu_1 Q_1 - \mu_\Gamma Q_\Gamma(y) - \mu_\Lambda Q_\Lambda(y)] \frac{dV}{dy} \qquad (10.6.7)$$

where the mobilities denoted by subscripted μ-symbols in Equation 10.6.7 can be characteristic either of bulk or of surface conduction, depending on the location of the corresponding charge.

A model for the depletion-mode MOSFET is built up by expressing the charge densities in Equation 10.6.7 for all conditions of bias. The method is straightforward, but the equations derived are algebraically complicated. The model is derived in reference 14, where it is shown to agree well with measurements.

Summary

The basic MOSFET theory of Chapter 9 contains many approximations that are useful in providing a tractable set of equations for designers. An important simplification comes from the assumption that there are no free carriers in the channel if the magnitude of the gate voltage is less than the magnitude of the threshold voltage. The free-carrier densities do not, however, change abruptly at $|V_G| = |V_T|$, instead they are a continuous exponential function of the surface potential. The channel free charge that is present when $|V_G| < |V_T|$ is responsible for *subthreshold current*. In contrast to the drain current flowing when $|V_G| > |V_T|$ (which is a drift current), subthreshold current is transported by diffusion.

A number of the approximations in the basic MOSFET theory of Chapter 9 become increasingly in error as device dimensions are reduced to the order of micrometers. One particularly limiting assumption is that free charge moves along the channel by drift with a constant mobility. Although drift is the predominant transport mechanism when $|V_G| > |V_T|$, mobility is not constant in the channel. Instead, the mobility depends on both the local field $\mathscr{E}_y$ along the channel, and the field $\mathscr{E}_x$ normal to the Si–SiO$_2$ interface. Even in the basic or *long-channel theory*, $\mathscr{E}_y$ becomes very large near the drain, and therefore an assumption that mobility remains constant over the entire channel length is only approximate.

For short channels, saturation of drift velocity (leading to a decreasing mobility) becomes significant at low voltages, and the velocity is nearly constant over most of the channel length. Saturation of drift velocity in the channel reduces the current below that predicted for a constant mobility. Because of velocity saturation, I_{Dsat} varies more slowly than $(V_G - V_T)^2$ and V_{Dsat} is smaller than $(V_G - V_T)$. An adequate model for these short-channel effects can be obtained by incorporating two new parameters into the MOSFET drain current-drain voltage equations: $\mathscr{E}_c$ a critical field introduced from the velocity-saturation model, and η a parameter representing the velocity reduction due to the normal field in the channel region.

Several additional effects become significant in MOSFETs having small dimensions. Of particular importance are threshold-voltage variations, hot-carrier effects, and new modes of device breakdown. The threshold voltage is a function both of the channel length and width in MOSFETs having small dimensions. As the channel length decreases, the magnitude of the threshold voltage tends to become smaller because of the influence of the source and drain space-charge regions on the charge in the channel region. If the channel is made narrow, however, the threshold voltage increases in magnitude because of the fringing field from the gate electrode to the substrate.

The high-field region at the end of the channel nearest the drain strongly influences the production of *hot carriers* having very high energies. Some of these carriers can surmount the barrier at the silicon-silicon dioxide interface and cause gate currents. A fraction of the charge corresponding to this gate current can be trapped in the gate oxide, where it modifies the effective value of the oxide fixed charge Q_f and alters the threshold voltage. Hot carriers can also generate extra substrate current through the junction avalanche process. Substrate current can give rise to a positive feedback process in which avalanche generation in the drain region stimulates minority-carrier injection from the emitter of the lateral bipolar *npn* transistor formed by the source, substrate, and drain regions. These coupled mechanisms combine to produce a characteristic "$C-$"shaped $I_D - V_D$ breakdown characteristic in which the drain breakdown voltage first decreases with increasing gate voltage at low gate voltages and then begins to increase with further increase of the gate voltage. For short-channel MOSFETs, breakdown can also be caused by *punchthrough* in which the depletion regions at the source-substrate and drain-substrate junctions reach one another as the drain voltage is increased. Under this condition the barrier to majority carrier flow between source and drain can be reduced directly by an applied drain-source bias, giving rise to a rapidly increasing leakage current as V_{DS} is increased.

Scaling rules aid in the design of MOSFETs as device dimensions are reduced. One set of frequently used scaling rules has been derived in an attempt to keep fields in the MOSFET constant as the dimensions are reduced. Because of the range of variations of device parameters on device dimensions, however, it is impossible to scale exactly according to this or any other set of scaling rules. Approximate scaling is nonetheless often used advantageously for the design of reduced-size MOSFETs. Numerical simulation is useful for studying the fields and carrier densities in MOSFETs. Typical simulations solve the governing physical equations (usually Poisson's equation and the continuity equations for both carrier types) in a device structure. Simulation can speed the design of an MOS process and reduce (but not eliminate) test runs and experimental trials.

Threshold-voltage adjustment by ion implantation makes possible the production of *enhancement-depletion* (*E/D*) integrated circuits, presently the highest volume MOSFET digital *IC* family. The current-voltage relationships in a depletion-mode MOSFET are complicated because channel current in this device is carried by several charge components that are affected differently by gate voltage. Most *E/D* digital circuits are, however, designed adequately with the first-order MOSFET model (Equations 9.1.17 and 9.1.18) using a positive threshold voltage for a *p*-channel MOSFET, and a negative threshold voltage for an *n*-channel MOSFET.

References

1. J. J. Barnes, K. Shimohigashi, and R. W. Dutton, *IEEE J. Solid-State Circuits*, **SC-14**, 368 (April 1979).
2. R. R. Troutman, *IEEE J. Solid-State Circuits*, **SC-9**, 55 (April 1974).
3. R. R. Troutman, *IEEE J. Solid-State Circuits*, **SC-14**, 383 (April 1979).
4. B. T. Murphy, *IEEE J. Solid-State Circuits*, **SC-15**, 325 (June 1980).
5. P. K. Ko, *Hot-Electron Effects in MOSFETs*, Doctoral Thesis, Department of EECS. University of California, Berkeley, June, 1982.
6. L. D. Yau, *Solid-State Electronics*, **17**, 1059 (1974).
7. L. M. Dang, *IEEE J. Solid-State Circuits*, **SC-14**, 358 (1975).
8. G. Merckel, *Process and Device Modeling for IC Design*, (F. Van de Wiele, W. L. Engl, and P. G. Jespers, Editors), Noordhoff, Leyden (1977), p. 705.
9. F. C. Hsu, (a) *Breakdown Mechanisms and Related Effects in MOSFETs,* Doctoral Thesis, Department of EECS, University of California, Berkeley, June, 1983, (b) *IEEE Trans. Electr. Devices*, **ED-29**, 1735 (Nov. 1982).
10. R. H. Dennard et al., *IEEE J. Solid-State Circuits*, **SC-9**, 256 (Oct. 1974).
11. VLSI Laboratory Staff, Texas Inst. Inc., *IEEE J. Solid-State Circuits*, **SC-17**, 442 (June, 1982).
12. J. A. Greenfield and R. W. Dutton, *IEEE Trans. Electr. Devices*, **ED-27**, 1520 (Aug. 1980).
13. Simulation courtesy of K. Cham, Hewlett-Packard.
14. J. S. T. Huang and G. W. Taylor, *IEEE Trans. Electr. Devices*, **ED-22**, 995 (Nov. 1975).
15. R. Coen and R. S. Muller, *Solid-State Electronics*, **23**, 35 (Jan. 1980).
16. K. Yamaguchi, *IEEE Trans. Electr. Devices*, **ED-26**, 1068 (July 1979).
17. J. G. J. Chern, P. Chang, R. F. Motta, and N. Godinho, *IEEE Electron Devices Letters*, **EDL-1**, 170, (Sept. 1980).
18. Courtesy F. Gaensslen, IBM Corporation.
19. E. Sun, J. Moll, J. Berger, and B. Alders, *IEEE Int. Electron Devices Mtg.*, Washington, DC (Dec. 1978) p 478.
20. N. Kotani and S. Kawazu, *Solid-State Electronics*, **22**, 63 (Jan. 1979).

Textbooks

S. M. Sze, *Physics of Semiconductor Devices, 2nd Edition*, Wiley-Interscience, New York, 1981.
E. H. Nicollian and J. R. Brews, *MOS Physics and Technology*, Wiley-Interscience, New York, 1982.

Problems

10.1 Draw an energy-band diagram along the x-direction (perpendicular to the silicon-silicon dioxide interface) to represent the channel region in an n-channel enhancement-mode MOSFET under the condition $\phi_s \approx -\phi_p$ but $|\phi_s|$ slightly less than $|\phi_p|$. Use the diagram to develop arguments that justify the form of Equation 10.1.1.

10.2* Recalculate I_{Dst} as described in the example on subthreshold current in the text if N_a is changed to 2×10^{15} cm^{-3}, x_{ox} is reduced to 40 nm, and L is made 2 μm. The other parameters are unchanged from the example.

10.3 Drain-induced barrier lowering (*DIBL*) is described in the text. Illustrate *DIBL* by sketching an energy-band diagram along the channel (*y*–direction) at the surface for two cases: (1) a biased long-channel MOSFET in which *DIBL* does not take place, and (2) a short-channel MOSFET in which *DIBL* leads to increased currents as V_{DS} is increased.

10.4[†] Complete the analysis of subthreshold currents in the text by deriving Equations 10.1.3 and 10.1.4.

10.5[†] (a) Consider Poisson's equation (Equation 4.1.10) in the channel region of a MOS-FET. Using the condition that there is charge neutrality in the substrate, show that the impurity doping terms $N_d - N_a$ can be written $n_i[\exp u_B - \exp(-u_B)] = 2n_i \sinh u_B$, where u represents the potential normalized to (kT/q) and u_B is the substrate potential $(u_B = -q\Phi_p/kT)$ in a *p*-type substrate.

(b) Using the form derived in part a, show that the Poisson equation can be written in terms of u as

$$\frac{d^2u}{dx^2} = \frac{1}{L_{Di}^2}\left[\sinh(u) - \sinh(u_B)\right]$$

where L_{Di} is the intrinsic Debye length

$$L_{Di} = \left(\frac{\epsilon_s kT}{2q^2 n_i}\right)^{1/2}$$

(c) Calculate L_{Di} at 300 K.

10.6[†] (a) Solve the form of the Poisson equation derived in Problem 10.5 using an integrating factor equal to $2du/dx$ which multiplies both sides of the equation so that the left-hand side becomes a perfect differential for

$$\left[\frac{du}{dx}\right]^2 = \mathscr{E}^2$$

(b) Show that the field in the silicon at the silicon-silicon dioxide interface $\mathscr{E}_s$ where $u = u_s$ is

$$\mathscr{E}_s = \pm\frac{kT}{qL_{Di}}F_s(u_s, u_B)$$

where the positive sign is used if $(u_B - u_s) > 0$ and the negative sign is used if $(u_B - u_s) < 0$. The function F_s is defined as the positive root in the following expression

$$F_s \equiv \sqrt{2}[(u_B - u_s)\sinh u_B - (\cosh u_B - \cosh u_s)]^{1/2}$$

(c) Use a band-diagram sketch to show that $(u_B - u_s) > 0$ corresponds to accumulation of the surface and $(u_B - u_s) < 0$ occurs in depletion and inversion for a *p*-type substrate.

10.7[†] The charge Q_s at the silicon surface is related to $\mathscr{E}_s$ by $Q_s = \epsilon_s \mathscr{E}_s$, hence

$$Q_s = \pm\frac{\epsilon_s kT}{qL_{Di}}F_s$$

where the sign choice is made as in Problem 10.6. Consider a substrate doped with $N_a = 10^{15}$ cm^{-3} at 300 K. Sketch a semilogarithmic plot of $|Q_s|$ as the surface potential u_s is changed from accumulation ($u_B - u_s$ positive) through flat-band ($u_s = u_B$)

to inversion ($u_s > -u_B$). Make the range of u_s be from -20 to $+20$. Identify the flat-band voltage and conventional threshold voltage on your plot. [Problems 10.5 to 10.7 provide the basis for an exact treatment of subthreshold currents since the full specification of surface space charge is made in the Poisson equation.]

10.8   Show that Equation 10.1.3 is a special case of Q_s as derived in Problem 10.7. (Note that Equation 10.1.3 applies when $\Phi_s \approx -\Phi_p$).

10.9   Derive Equations 10.2.4, 10.2.6, and 10.2.7.

10.10   Derive Equation 10.2.8.

10.11   Derive Equation 10.2.10.

10.12   (a) Use Equation 10.2.9 to plot $[V_{Dsat\,4}/\mathscr{E}_c L]$ as a function of $[(V_G - V_T)/\mathscr{E}_c L]$.
   (b) Use Equation 10.2.10 to calculate $I_{Dsat\,4}$ as a function of $(V_G - V_T)$ for $\mu_{no} C_{ox} W/L = 40\ \mu\text{A V}^{-2}$, $\mathscr{E}_c = 2 \times 10^4\ \text{V cm}^{-1}$, $L = 1.5\ \mu\text{m}$, and $(V_G - V_T) = 2$, 4, and 6 V.
   (c) Plot $(I_{Dsat\,4}/I_{Dsat\,1})$ as a function of $(V_G - V_T)$ for the conditions of part b.

10.13*   According to Equation (10.2.12), both normal and tangential fields can be interpreted in terms of an effective lengthening of the MOSFET channel. Calculate the effective channel lengthening (as a percentage of the actual length) in a MOSFET for which $|\mathscr{E}_c L| = 4$ V, $V_T = 0.2$ V, and $\eta = 0.02$ if $V_{Dsat\,4} = 1.51$ V (Equation 10.2.9) and $V_{DS} = 2$ V.

10.14   Derive Equation 10.3.2 for the short-channel effect on threshold voltage.

10.15   Derive Equation 10.3.4 for the narrow-channel effect on threshold voltage.

10.16†   One theory for oxide "wearout" is that hot electrons become trapped in the oxide increasing the field until ultimately the breakdown value of $7 \times 10^6\ \text{V cm}^{-1}$ is reached. Consider the example in Section 10.3 concerned with threshold-voltage shifting by hot electrons and assume that the hot-electron injection described there occurs continuously. The oxide is initially free of charge and the voltage across it is maintained at 20 V.
   (a) How long does it take for breakdown to occur in the oxide?
   (b) Sketch a plot of the field and voltage versus distance through the oxide showing the initial condition and the condition at breakdown.
   (c) Repeat parts a and b for a new oxide in which all parameters are the same except that the trapping sites are concentrated 3 nm away from the gate electrode.

10.17*   (a) Show that the number of electrons on a MOSFET gate that can cause an oxide to break down is a function only of the gate area and the oxide permittivity.
   (b) If a MOSFET having $L = 6\ \mu\text{m}$ and $W/L = 5$ is charged by static electricity, how many electrons could be transferred to the gate before the oxide field reaches the breakdown value?
   (c) How long would it take to deliver these electrons to the gate if the average current is 1 pA?

10.18   Check the scaling rules for a constant field in Table 10.1 for each of the variables.

10.19†   Derive Equation 10.6.2 for the depletion-layer width for the "approximate" dopant profile from the ion-implantation process discussed in Section 10.6.

10.20   Carry through the steps indicated in the text to obtain Equations 10.6.3 and 10.6.4 for the depletion-charge density at inversion and the threshold voltage, respectively, in an ion-implanted MOSFET.

ANSWERS TO
SELECTED PROBLEMS

The following are numerical solutions to those problems marked with an asterisk in the text. A solutions manual showing calculations is available (to teaching faculty only) by direct request from the publisher.

Chapter 1

1.2 (a) $E_i - E_f = 0.35$ eV, $n = 2.1 \times 10^4$ cm^{-3}, (b) $E_f - E_i = 0.29$ eV,
$p = 2.1 \times 10^5$ cm^{-3}.

1.4 (a) 2.3×10^5 cm^{-3}, (b) 4×10^9 cm^{-3}, (c) 9.5×10^{16} cm^{-3}.

1.6 160 cm^2 V^{-1} s^{-1}.

1.8 (a) 207 collisions, (b) 162 mV.

1.10 -825.6 Acm^{-2}.

1.14 (a) 1.85 μm (*IR*), (b) 1.10 μm (*IR*), (c) 0.87 μm (*near IR*), (d) 0.14 μm (*UV*).

Chapter 2

2.1 (a) 1.36×10^{15} cm^{-3}, (b) 2.21×10^{15} cm^{-3}.

2.4 1.1 hour.

2.7 1 μm.

2.9 (a) 1.79 μm, (b) 1.88 hours.

2.18 34.5 ns.

2.20 (a) 147 Ω per square, (b) an added *acceptor* density of approximately
6.5×10^{16} cm^{-3} (implant *dose* of 2.6×10^{13} cm^{-2}).

Chapter 3

3.1 (a) 4.094 eV, 4.261 eV, (b) 0.167 V.

3.5 (a) $C = 0.282$ pF, (b) $V_R = -0.813$ V.

3.7 (a) $N_{dmax} = 2.9 \times 10^{17}$ cm^{-3}, (b) $\rho > 0.04$ Ω cm.

3.16 (a) for *n*-type doping, a Schottky barrier is obtained if $N_d > 7.45 \times 10^{11}$ cm^{-3};
this is smaller than any practical *IC* doping level, (b) since $\Phi_M - X_S \approx E_g/2q$,
any level of *p*-doping will result in a blocking contact.

Chapter 4

4.1 (a) $\phi_i = 0.72$ V, $x_d = 0.97$ μm, $\mathscr{E}_{max} = 1.48 \times 10^4$ V cm^{-1} (at 0 V),
(b) $x_d = 3.73$ μm, $\mathscr{E}_{max} = 5.75 \times 10^4$ V cm^{-1} (at -10 V).

4.3 $\mathscr{E}_{max} = 648$ V cm^{-1}; for abrupt junction $\mathscr{E}_{max} = 1.53 \times 10^4$ V cm^{-1}.

4.8 (a) $N_{do} = 6.1 \times 10^{17}$ cm^{-3}, (b) $a = 1.83 \times 10^{21}$ cm^{-4},
(c) $\mathscr{E}_{max} = 3.4 \times 10^4$ V cm^{-1}.

4.10 (a i) 0.33 eV, (a ii) 0.36 eV, (b i) 0.62 V, (c) total depletion possible at
$V_R = 23.5$ V.

Chapter 5

5.1 (a) 7.25×10^{16} cm^{-3} s^{-1}, (b) 2.6×10^{11} cm^{-3} s^{-1}.

5.2 (a) $n = 10^{16}$ cm^{-3}, $p = 10^{13}$ cm^{-3}, (b) 10 ns.

5.9 $\gamma = 1.55 \times 10^{-4}$.

5.11 (a) 0.738 V, (b) $n_p(-x_p) = 1.37 \times 10^{14}$ cm^{-3}, $p_n(x_n) = 6.83 \times 10^{13}$ cm^{-3},
(c) $J_T = 0.472$ A cm^{-2}, (d) currents are equal at 1.35 μm from the physical
junction (in p-type region).

5.21

	V_a (V)	r_d (Ω)	Capacitance (pF)
(a)	0.1	2.95×10^{10}	3.36
	0.5	6.14×10^3	6.68
	0.7	2.8	3580
(b)	0	1.38×10^{12}	3.15
	-5	5×10^{95} *	1.18
	-20	10^{346} *	0.63

* other mechanisms actually limit the resistance

Chapter 6

6.1 9.9×10^{14} cm^{-3}.

6.5 (a) 1.346 μC cm^{-2}, (b) 0.84 μm.

6.8 $\phi_i = 0.91$ V, $K_A = 0.163$ pC V$^{-1/2}$, $Q_{VE}(-50$ V$) = -1$ pC,
$Q_{VE}(0.3$ V$) = +0.028$ pC.

6.12 $V_{CEsat} = 0.048$ V.

6.17 (a) $\gamma = 0.99722$, (b) $\alpha_T = 0.99993$, (c) $\beta_F = 360$, error is 2.57%.

Chapter 7

7.3 $\dfrac{V_A \text{ (constant doping)}}{V_A \text{ (exponential doping)}} = 0.74$.

7.7 $\beta_F = 1.333$.

7.9 $\beta_F = \beta_o[0.75 + 0.25\sqrt{1 + 4I_F/I_K}]^{-1}$ where I_F is the forward-injected current and $I_K = qD_nA_EN_a/x_B$.

7.23 (a) $Q_F = 24$ pC, (b) $Q_F = 81.1$ pC, $Q_R = 156$ pC, (c) charge ratio $= 9.88$.

7.30

	constant doping	exponential doping
g_m	77.5 mS	77.5 mS
C_D	1.94 pF	1.61 pF
δ	2.5×10^{-4}	2.08×10^{-4}
η	1.5×10^{-4}	2.2×10^{-4}

Chapter 8

8.1 (a) $V_{FB} = -0.17$ V, (b) $V_{FB} = -0.90$ V.

8.12 (a) 1.18 V, (b) -2.02 V.

8.15 $C_{MOS}/C_{junction} = 22.8$.

8.16 $N_{st} = 2 \times 10^{10}$ cm^{-2}.

Chapter 9

9.3 (a) 8.3×10^{11} cm^{-2}, (b) 3.1 V.

9.8 (a) 18.7 μA, (b) 15.5 μA, (c) 12.2 μA, (d) 7.56 μA.

9.10 $V_{T0} = 1$V, $k = 50$ μAV^{-2}, $\gamma = 0.3$ V$^{1/2}$, $\lambda = 0.05$ V^{-1}.

9.12 (a) $V_o = 0.232$ V, (b) $V_{TD} = -2.36$ V, (c) 0.223 V.

9.15 (a) MOSFET: $V_T = -0.225$ V, (b) polysilicon line: $V_T = -4.3$ V, (c) aluminum line: $V_T = -10.1$ V.

9.17 (a) $V_{FB} = -1.188$ V, $V_T = 1.61$ V, (b) 434 cm^2 V^{-1} s^{-1}.

Chapter 10

10.2 8.34 nA.

10.13 54%

10.17 (b) 2.72×10^7 electrons, (c) 4.35 s.

INDEX

Selected List of Symbols

Symbol	Definition	Section
$A_{C,E}$	BJT collector, emitter area	6.3
B	magnetic field	1.3
B	parabolic rate coefficient	2.3
B	characteristic field in tunneling	4.4
B/A	linear rate coefficient	2.3
BV	breakdown voltage	4.4
C	small-signal capacitance	3.2
C_D	diffusion capacitance in hybrid-pi model	7.5
C_d	diode diffusion capacitance	5.4
C_g	impurity concentration in gas stream	2.5
C_j	diode junction capacitance	5.4
$C_{jc,je}$	small-signal capacitance at BC, BE junctions	7.1
C_o	concentration of oxidant in gas	2.3
C_{ox}	oxide capacitance	8.2
C_p	peak dopant concentration	2.5
C_s	impurity concentration at solid surface	2.6
C^*	equilibrium gas-phase oxidant concentration	2.3
D_{eff}	effective diffusivity	2.5
D_{it}	density of interface traps (per area and energy)	8.4
$D_{n,p}$	diffusion constant for electrons, holes	1.2
D_s	surface-state density	3.5
E_1	minimum energy for ionizing collision	4.4
E_a	acceptor energy	1.1
E_a	activation energy	2.6
E_f	Fermi-level energy	1.1
$E_{fn,fp}$	quasi-Fermi levels for electrons, holes	1.1
E_g	band-gap energy	1.1
E_i	intrinsic Fermi-level energy	1.1
E_o	energy of free electron	3.2

Symbol	Definition	Section
E_{st}	energy of surface states	5.2
G_A	Auger generation rate	5.2
$G_{n,p}$	generation rate for electrons, holes	5.1
G_{sp}	spontaneous generation rate	5.2
GN	Gummel number	6.2
I_{0E}	emitter-base saturation current in BJT	6.4
$I_{CO,CBO}$	collector current (emitter open-circuited)	6.4
I_{CEO}	collector current (base open-circuited)	6.4
I_D	drain current	9.1
I_{Dst}	subthreshold current	10.1
$I_{KF,KR}$	"knee" currents in Gummel-Poon model	7.6
I_{pE}	emitter hole current in a BJT	6.2
I_{rB}	base-region recombination current	6.2
I_{Deat}	FET current in the saturation region	4.5
J	current density	1.1
J_0	diode saturation-current density	5.3
J_1	characteristic current for Kirk effect	7.1
J_S	linking-current density (BJT)	6.1
J_g	generation density current	5.3
$J_{n,p}$	electron, hole current density	1.2
J_r	recombination current	5.3
J_t	total current in a diode	5.3
K	scaling factor for MOSFET dimensions	10.4
L	pattern length	2.9
L_D	Debye length	3.4
L_{Di}	intrinsic Debye length	4.2
$L_{p,n}$	diffusion length for holes, electrons	5.3
L'	undepleted length of FET channel	4.5

Symbol	Definition	Section
M	avalanche multiplication factor	4.4
N_{ai}	"box" approximation to implant density	10.6
N_c	effective density of states (conduction band)	1.1
$N_{d,a}$	donor, acceptor atomic density	1.1
N_{st}	area density of surface states	5.2
N_v	effective density of states in the valence band	1.1
N'	area density of dopant (implant dose)	2.5
Q_B	base majority charge	6.1
$Q_{F,R}$	forward, reverse charge-control variables	7.3
Q_N	total free charge in channel	9.1
$Q_{VE,VC}$	emitter, collector stored junction charge	6.3
Q_d	depletion-charge density	8.1
Q_f	fixed interface charge density	8.4
Q_{it}	interface trapped-charge density	8.4
Q_m	mobile charge density (oxide)	8.4
Q_n	channel free-electron charge	8.1
Q_{nB}	minority charge storage in base (npn BJT)	7.3
Q_{ot}	oxide trapped-charge density	8.4
Q_{ox}	charge in an oxide layer	8.4
Q_{pE}	minority charge storage in emitter (npn BJT)	7.3
$Q_{p,n}$	stored minority charge (diode)	5.4
Q_s	semiconductor space charge	3.2
R_A	Auger recombination rate	5.2
R_B	base spreading resistance	7.2
R_D	drain-region resistance (MOSFET)	9.2
R_H	Hall coefficient	1.3
R_S	source-region resistance (MOSFET)	9.2
R_W	well resistance (CMOS)	9.4
R_X	substrate resistance (CMOS)	9.4
R_d	deposition rate	2.6
R_p	implant range	2.5
R_{sp}	spontaneous recombination rate	5.2
T	temperature	1.1
T_{tr}	transit time	9.1
U	net rate of recombination	5.2

Symbol	Definition	Sectio
U_A	net rate of Auger recombination	5.2
U_s	surface recombination rate	5.2
$V_{A,B}$	Early voltage (forward, reverse bias)	7.1
V_C	voltage in MOS channel	8.3
V_{FB}	flat-band voltage	8.1
V_H	Hall-effect voltage	1.3
V_T	MOSFET threshold voltage	9.2
V_{Tn}	threshold voltage (n-channel MOSFET)	9.3
V_{Tp}	threshold voltage (p-channel MOSFET)	9.3
V_a	applied voltage	3.2
V_t	thermal voltage (kT/q)	7.5
V_{CEsat}	collector-emitter voltage when BJT is saturated	6.4
V_{Dsat}	drain voltage at edge of saturation	4.5
W	pattern width	2.9
$e_{n,p}$	emission-rate constant	5.1
f_D	Fermi-Dirac distribution function	1.1
f_T	frequency at which β_F equals unity	7.5
g	conductance of a square resistor pattern	2.9
g_m	transconductance	4.5
g_{msat}	transconductance in saturation	4.5
$g(E)$	density of available electron energy states	1.1
h	Planck's constant	1.1
k	Boltzmann's constant	1.1
k	MOSFET current multiplying parameter	9.1
k_s	surface reaction-rate coefficient	2.6
k'	MOSFET current parameter	9.1
l	mean-free path	4.4
m	mass	1.1
n	electron density	1.1
n_i	intrinsic free-carrier density	1.1
n_{ie}	effective intrinsic-carrier density	1.1
n_p'	excess electron density (p region)	5.3
n_{po}	equilibrium electron density (p-region)	5.3
n_s, p_s	surface electron, hole density	3.3